LIST OF ERRATA FOR RIGGS' "CONTROL THEORY AND PHYSIOLOGICAL FEEDBACK MECHANISMS"

N.B. In the following list, the more important changes are preceded by an asterisk.

PRELIMINARY MATERIAL

End paper table: In the lower right corner of the box for "INDUCTORS" in the "ROTATIONAL" column, "τ" should be "τ^2".

Page xv, 5th line: "noctural" should be "nocturnal".

CHAPTER 1

* Page 23, first line of Example 1.11.1: "$H_1 = \frac{d^2[\ ]}{dt^2} + 5$" should be

"$H_1 = \frac{d^2[\ ]}{dt^2} + 5[\ ]$"

Page 29, 8th line from bottom: "Galetea" should be "Galatea".

Page 30, equation 1.15.1: "Adt" after the second integral sign should be "$Ad\tau$".

Page 32, equation in part D: "$(\rho_s - \rho_m$" should be "$[\rho_s - \rho_m]$"

CHAPTER 2

* Page 54: Add the following paragraph between the mid-page table and "B. Solution using loop equations" etc.:

"Actually it is unnecessary (and hence wrong from the standpoint of efficient computation) to include the first row and the first column in the above matrix because the three unknown voltages can be calculated from the three equations represented by the last three rows of the matrix. The current, i_s, can then be calculated easily from one of the voltage drops across one of the known resistors."

Page 58, next-to-last line of paragraph headed "I. The REAL SYSTEM and its PHYSICAL MODEL".: "of $\dot{Q}_{1n}(t)$ mg/hr" should be "of S at a constant rate of $\dot{Q}_{1n}(t)$ mg/hr".

Page 58, 2nd and 3rd lines of paragraph headed "II. The CIRCUIT DIAGRAM".: "how many different values the across variable can assume" should be "how many different values of the across variable we have to define".

Page 60: A line should be ruled across the page below the two-line legend of of Fig. 2.15.2 and the running text which begins with "The root of the characteristic equation", etc.

* Page 61, last term of equation 2.15.9: " $\frac{\dot{Q}_{in}(t)}{[\dot{V}cl]_{tot}}$ "

should be " $\frac{\dot{Q}_{in}(t)}{[Vdist]}$ "

Page 78, next-to-last line: Delete "description of the automobile suspension".

CHAPTER 3

* Page 84: The top of the page should be

"$\lim_{t \to \infty} \frac{\epsilon^{t^2}}{\epsilon^{bt}} = \lim_{t \to \infty} \epsilon^{t[t-b]}$

But for any finite b, whether positive, zero, or negative, t[t-b] will $\to \infty$ as $t \to \infty$. Hence $\lim_{t \to \infty} \epsilon^{t[t-b]} = \infty$. So ϵ^{t^2} is not of exponential order and has no Laplace transform."

* Page 90: In equation 3.6.5, the exponents should be "-kt" not "kt", and the third line below equation 3.6.5 should be

"2k = 3.74 k = 1.87"

not

"-2k = 3.74 k = -1.87"

Page 102, sixth line under SECTION 3.9: Insert comma after "a given period".

* Page 109: Equation 3.11.13 should be

$$C(t) = \frac{0.0125}{0.084}\left\{\left[1 - \epsilon^{-0.084t}\right] u(t) - \left[1 - \epsilon^{-0.084[t-48]}\right] u(t-48)\right\}$$

* Page 120, equation 3.13.9: The numerator of the term on the right should be "1", not "R_a".

* Page 130, just above "Example 3.16.2" insert the following:

"As a matter of fact, steps 1 and 2 can usually be reduced to a single step, since if we are given any sinusoidal function,

$f(t) = a\cos(\omega t) + b\sin(\omega t)$,

we can write the corresponding phasor by inspection as

$\mathfrak{F}(j\omega) = a - jb$"

ELECTRICAL	HYDRODYNAMIC	THERMAL	CONCENTRATIONAL
$e(t)$ = voltage	$P(t)$ = pressure	$T(t)$ = temperature	$C(t)$ = concentration of solute
$i(t)$ = current	$\dot{V}(t)$ = flow	$\dot{q}(t)$ = heat flow	$\dot{Q}(t)$ = solute flow
$p(t) = e(t)i(t)$	$p(t) = P(t)\dot{V}(t)$	N.B.! NOT "AT" $p(t) = \dot{q}(t)$	
t) = magnetic flux (t) $= \int e_L(t) \cdot dt$	$\Gamma(t)$ = pressure momentum $\Gamma_I(t) = \int P(t) \cdot dt$		
$q(t)$ = charge $q_C(t) = \int i_C(t) \cdot dt$	$V(t)$ = volume $V_C(t) = \int \dot{V}_C \cdot dt$	$H(t)$ = quantity of heat $H(t) = \int \dot{q}(t) \cdot dt$	$Q(t)$ = quantity of solute $Q(t) = \int \dot{Q}(t) \cdot dt$
electrical energy of capacitor magnetic energy of inductor	$\mathcal{E}_k$ = kinetic energy of fluid flow $\mathcal{E}_p$ = potential energy of pressure head	$\mathcal{E}_p$ = thermal potential energy of stored heat	
$_D$ = R = resistance $e_R = e_1 - e_2$  e_2 i_R R	$1/k_D$ = R = resistance to flow $P_R = P_1 - P_2$ 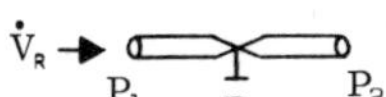 $\dot{V}_R$ P_1 R P_2	$1/k_d$ = R = resistance to flow of heat  T_1 T_2 $\dot{q}_R$ R	$k_d = [\dot{V}cl]$ = clearance (solute conductance) or $k_d = \dot{Q}$ per unit C diff.  C_1 C_2 $\dot{Q}$ $\frac{1}{[\dot{V}cl]}$
= C = capacitance $e_C = e_1 - e_2$ 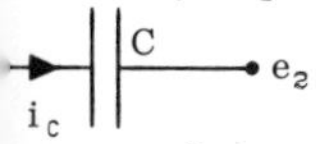 C e_2 i_C $\mathcal{E}_e = Ce^2/2$	k_A = C = fluid capacitance $P_C = P_1 - P_2$ 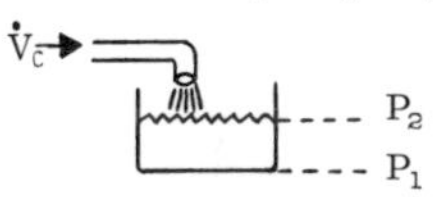 $\dot{V}_C$ P_2 P_1 $\mathcal{E}_p = CP^2/2$	k_A = C = heat capacitance $T_C = T_1 - T_2$ 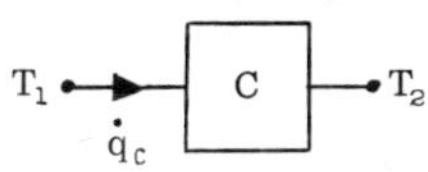 T_1 C T_2 $\dot{q}_C$ $\mathcal{E}_p = CT_C$	k_A = [Vdist]=solute capacitance (volume of distribution) $\Delta C = C_1 - C_2$ 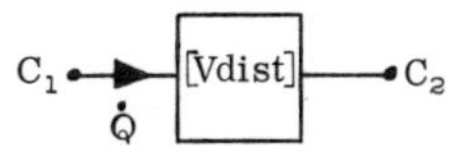 C_1 [Vdist] C_2 $\dot{Q}$
= L = inductance $e_L = e_1 - e_2$  e_2 i_L L $= \frac{1}{2} \cdot \frac{1}{L} \Psi^2 = \frac{1}{2} Li^2$	k_I = I = fluid inertance $P_I = P_1 - P_2$  P_1 P_2 $\dot{V}_I$ I $\mathcal{E}_k = \frac{1}{2} I\dot{V}^2 = \frac{1}{2} \cdot \frac{1}{I} \Gamma^2$		
y cell with "zero" ernal resistance	Large reservoir supplying constant pressure head	Constant-temperature bath of large volume	Large volume of dialyzing solution
toelectric cell with inite" internal istance.	Constant-volume pump (e.g., mammalian heart under certain conditions)	Heat production by metabolism	Intravenous infusion of solute

* Page 138, just below equation 3.17.2 insert the following:

"In actual practice, instead of finding the overall impedance or admittance of a system (as was illustrated by Example 3.16.2), it is often simpler to find the transfer function, G(s), in the usual way, substitute $j\omega$ for s, and use equation 3.17.2 to find the phasor output. The numerical value of ω is, of course, the radian frequency of the sinusoidal input."

CHAPTER 4

*Page 152: Delete the last line, i.e. equation 4.3.6 on this page.

* Page 169, last two lines: The numerical values of s should all be negative.

* Page 171, lines 5 and 6 beneath the NOTE: "S + 0.01 = 0" should be "s + 0.01 = 0", and "$S = 100 \cdot 10^{4}$" should be "$s = -100 \cdot 10^{-4}$".

Page 177, third line beneath equation 4.10.3: "$\tan^{-1}(\beta\omega)/\alpha(\omega)$" should be "$\tan^{-1}(\beta(\omega)/\alpha(\omega))$

* Page 183, equation 4.12.1: The right side of the equation should be

$$\left\{\frac{0.08a}{a^2+b^2}\right\} - j\left\{\frac{0.08b}{a^2+b^2}\right\}$$

Page 189: In example 4.14.1, the equations should be numbered 4.14.10 and 4.14.11.

Page 199, Exercise 1, line 6: Between "the right." and "The voltages at" insert "(There must, of course, be a pathway for current to flow from e_{out} to e_{in}, but for simplicity this pathway has been omitted.)"

Page 199, Exercise 1: The first sentence of Part A should be "Find an expression for e_b as a function of the input and the parameters."

Page 199, Exercise 1: Part C should be "Write a set of working equations for this system, and draw the corresponding block diagram to show that it is indeed a feedback system."

CHAPTER 5

* Page 218, Fig. 5.4.1: The direction of the i(t) arrowhead at the extreme left should be reversed.

Page 224, equation 5.4.43: In the fraction immediately after the equals sign, the last term in the denominator should be "0.003" not "0.0003".

Page 236: The end of the line just below equation 5.6.27 should be "equation 5.6.27", not "equation 5.6.26".

* Page 237: Equation 5.6.34 should be

$$[OLG](j0) = \left|\frac{k_1 k_2 k_3}{k_I k_2 k_H}\right| = 0.2667 \quad \text{not} \quad [OLG](j0) = \left|\frac{k_1 k_2 k_3}{k_1 k_2 k_H}\right| = 0.2667$$

CHAPTER 6

* Page 247, equation 6.2.16: The numerator of the fraction in the exponent should be "1", not "t".

Page 248: Just above equation 6.2.19, delete "equation 6.2.17.a gives".

Page 249, third line of last full paragraph: After "proportional control" insert "where $\overline{T}_{bath}$ is taken to be the steady-state bath temperature, $T(\infty)$." Throughout the rest of the paragraph, "T_{bath}" should be "$\overline{T}_{bath}$".

Page 260: Last word in line above equations 6.9.3 should be "become", not "becomes".

Page 261: In equations 6.9.7, the first number in square brackets should be "0.220", not "0.221", and the arithmetical results should be revised accordingly.

Page 268: In equation 6.11.9, the denominator on the right should be "x(t)", not "x".

Page 271: The four equations for k_1, k_2, k_3, and k_4 near the bottom of the page should be labelled at the right as equations 6.12.9.

*Page 281: The scale of ordinates at the left of Fig. 6.13.2.b should run from 1.0 to 1.8, not 10 to 18, and it should be headed "$\dot{V}(L)$", not "$\dot{V}(LX)$".

Page 286: Delete "$= x_0$", "$= \dot{x}_0$", and "$= \ddot{x}_0$" from equations 6.15.2.

Page 288: The subscript for each of the terms just to the right of the equals sign in equations 6.15.20 should be "n - 1".

Page 291: In the last line, delete "for testing the stability of an equilibrium point".

Page 292: First and second lines should be "will be discussed briefly. These are the perturbation method (Sections 6.17 and 6.18) and the second method of Liapunov (Section 6.19)."

Page 292: At the extreme left of the last equation of the 6.17.2 set, "$h_n(t)$" should be "$\dot{h}_n(t)$"

* Page 296, fourth and third lines above equations 6.18.5: "from whatever inputs are already present" should be "from the corresponding constant inputs, $U_1, U_2, \cdots, U_n$, already present".

* Page 296: in equations 6.18.5 $\Delta U_1(t)$, $\Delta U_2(t)$, $\Delta U_n(t)$, should be replaced by

$$\left.\frac{\partial f_1}{\partial U_1}\right|_{\underline{X}_{eq}} \Delta U_1(t),\quad \left.\frac{\partial f_2}{\partial U_2}\right|_{\underline{X}_{eq}} \Delta U_2(t),\ \text{and}\ \left.\frac{\partial f_n}{\partial U_n}\right|_{\underline{X}_{eq}} \Delta U_n(t)$$

respectively.

Page 307, next-to-last line: Delete "or voltage".

Page 309, exercise 10, part A: Delete "principal".

CHAPTER 7

* Page 312, equation 7.3.1: Place vertical bars around the product of derivatives to show that the absolute value is to be taken.

Page 319, equation 7.4.11: Delete the circled dot above the first subscript "op".

* Pages 321 and 322: Delete the last sentence (which begins "From equation 7.4.20 ---"), on page 321, and the first part of page 322 up to, but not including equation 7.4.23. For the deleted material, substitute the following:

"Now for very small changes,

$$\frac{\Delta A}{\Delta U_A} = \frac{dA}{dU_A} \qquad 7.4.21$$

Solving equation 7.4.20 for this ratio, and rearranging the result slightly,

$$\left.\frac{dA}{dU_A}\right|_{op,\ closed} = \frac{\beta_1}{1 - \dfrac{\beta_2\beta_5 + \beta_3\beta_8 + \beta_2\beta_6\beta_8 + \beta_3\beta_5\beta_9}{1 - \beta_8\beta_9}} \qquad 7.4.22$$

Now suppose that we want to compare the closed-loop behavior with the behavior which will occur when all of the loops influencing the response of ΔA to ΔU_A have been opened, for example by letting $\beta_5 = \beta_8 = 0^*$. Then equation 7.4.22 becomes"

* Page 322, equation 7.4.25: Insert a minus sign just before the ratio and just after the first vertical bar.

Page 327, next-to-last line: "variable" should be "variables".

Page 342: In equation 7.8.19, change "P" to "P(∞)".

Page 361: In the block diagram at the lower right, the line from $\Delta\dot{G}$ to the $1/V\delta$ block should have an arrowhead.

Page 366, exercise 9, B: Change "antogonistic" to "antagonistic".

PART II

* Page 387. Figure 2 should be replaced by the following:

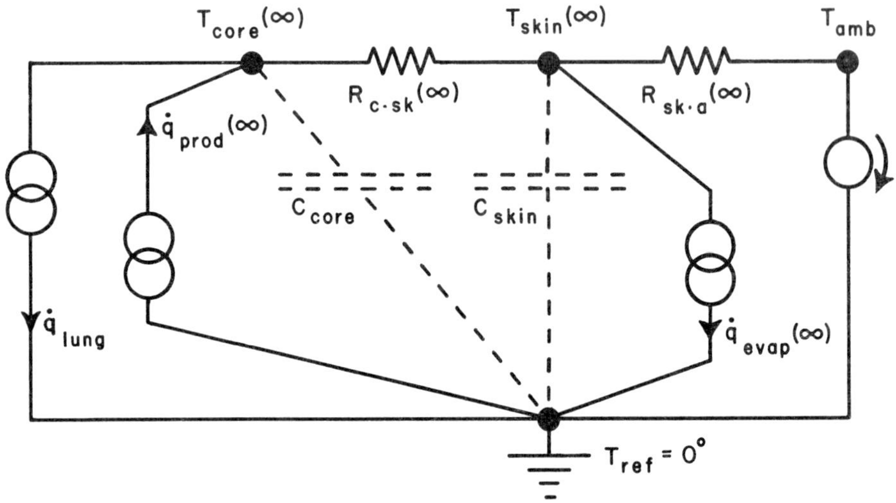

Page 392, fourth line under Fig. 5: "T_{skin}" should be "$T_{skin}(\infty)$"

Page 406: Equation "11.A" should be "11.T".

Page 424: In the expression on the right of equation 4.S, "$V(\infty)$" should be "$V_{ven}(\infty)$".

* Page 428: At the end of the eighth line, "$\dot{V}_{bl}$" should be "V_{bl}".

Page 440: The example for $\dot{Q}_{max}$ should be "1.5 ml (STPD)/min", not "1.5 ml (STPD)/ml".

Page 443: In Fig. 3, the lower line of Loop I should have an arrowhead pointing towards "$\dot{Q}_{cons}(\infty)$".

* Page 501: The heavy arrow pointing from "$V_{edema\ lungs}$" to "Diffusing capacity" should be broken, not solid.

APPENDIX

Page 530, first line below the equation near the top of the page: "k = 3 n = 1" should be "k = 3, n = 1".

* Page 534: A series of steps is diagrammed in the middle of the page. At the top left of that series,

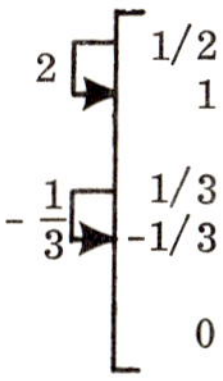

(etc.) should be

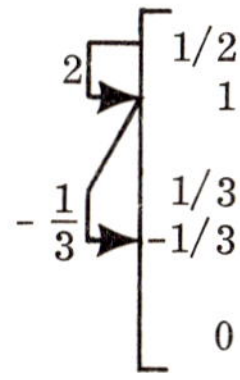

(etc.)

Page 542: The first equation on the page should be

$$P(s) = s^n + a_{n-1}s^{n-1} + a_{n-2}s^{n-2} + \cdots + a_1 n + a_0$$

Page 568: The line just above "DAMPING RATIO" should be "and λ_2 are negative or to systems where the real parts of λ_1 and λ_2 are zero".

			GENERAL	TRANSLATIONAL	ROTATIONAL
VARIABLES	PRIMARY	**A** = Across variable	$v_E(t)$ = velocity, voltage (etc.) difference across element E	$v(t)$ = velocity	$\Omega(t)$ = angular velocit
		T = Through variable	$f_E(t)$ = force, flow, (etc.) through element E	$f(t)$ = force	$\tau(t)$ = torque
		AT = Product	$p_E(t)$ = power supplied to element E $p_E(t) = f_E(t)v_E(t)$	$p(t) = f(t)v(t)$	$p(t) = \tau(t)\,\Omega(t)$
	INTEGRATED	$\int A \cdot dt$	$x_E(t) \int v_E(t) \cdot dt$	$x(t)$ = displacement $x(t) = \int v(t) \cdot dt$	$\theta(t)$ = angular displacen $\theta(t) = \int \Omega(t) \cdot dt$
		$\int T \cdot dt$	$h_E(t)$ = momentum or quantity of thing flowing $h_E(t) \int f_E(t) \cdot dt$	$h(t)$ = momentum $h_m(t) = \int f_m(t) \cdot dt = mv_m$	$h(t)$ = angular momentu $h_J(t) \int \tau_J(t) \cdot dt = J\Omega(t$
		$\int AT \cdot dt$	$\mathcal{E}(t)$ = energy = work done on system $\mathcal{E}(t) = \int p(t) \cdot dt$	$\mathcal{E}_k$ = kinetic energy of moving mass $\mathcal{E}_p$ = potential energy of compressed spring	$\mathcal{E}_k$ = kinetic energy of rotating mass $\mathcal{E}_p$ = potential energy o twisted spring
ELEMENTS	PASSIVE	RESISTORS (D-elements) Dissipate energy as heat	$f(t) = k_D v(t)$ $v(t) = \frac{1}{k_D} f(t)$	$k_D = D$ = damping constant $v_D = v_1 - v_2$ 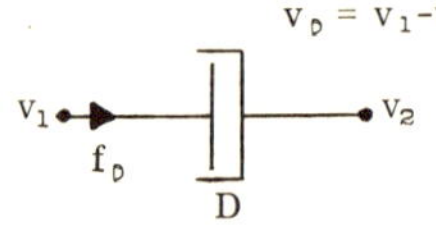	$k_D = D$ = damping cons $\Omega_D = \Omega_1 -$ 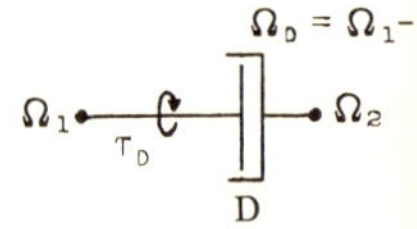
		CAPACITORS (A-elements) Store energy by virtue of their A-variable	$f(t) = k_A \frac{dv(t)}{dt}$ $v(t) = \frac{1}{k_A} \int f(t) \cdot dt$	$k_A = m$ = mass $v_m = v_1 - v_{Ref}$ 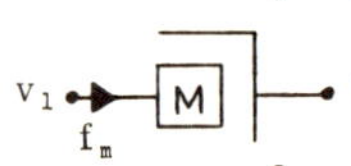 $\mathcal{E}_k = mv^2/2$	$k_A = J$ = Inertia $\Omega_J = \Omega_1 - \Omega_{Ref}$ 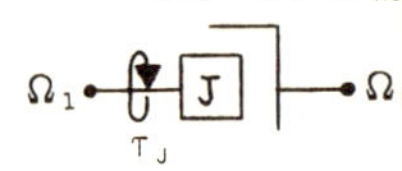 $\mathcal{E}_k = J\Omega^2/2$
		INDUCTORS (T-elements) Store energy by virtue of their T-variable	$f(t) = \frac{1}{k_T} \int v(t) \cdot dt$ $v(t) = k_T \frac{df(t)}{dt}$	$1/k_T = k$ = spring stiffness $v_k = v_1 - v_2$  $\mathcal{E}_p = \frac{1}{2} kx^2 = \frac{1}{2} \cdot \frac{1}{k} f^2$	$1/k_T = k$ = torsional stiffnes $\Omega_k = \Omega_1 - \Omega$ 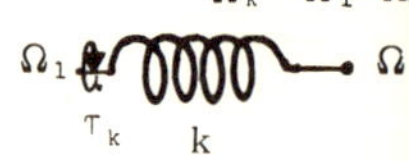 $\mathcal{E}_p = \frac{1}{2} k\theta^2 = \frac{1}{2} \cdot \frac{1}{k}\ \tau$
	SOURCE	Constant input A-sources (examples)	$v_{in}(t)$ is determined by the source; $f_{in}(t)$ by the passive elements	Hydraulic piston moved by constant flow pump 	Large flywheel with constant angular veloc
		Constant input T-sources (examples)	$f_{in}(t)$ is determined by the source; $v_{in}(t)$ by the passive elements	Weight of a given mass in constant gravitational field	Shaft turned by weight: 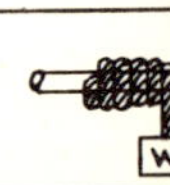

CONTROL THEORY
and
PHYSIOLOGICAL FEEDBACK MECHANISMS

CONTROL THEORY and PHYSIOLOGICAL FEEDBACK MECHANISMS

Douglas S. Riggs, M.D.

Professor of Pharmacology,
School of Medicine,
State University of New York
at Buffalo

The Williams and Wilkins Co.
Baltimore

Made in the United States of America

Library of Congress Catalog Card Number 72-131305
SBN 683-07244-7

Printed at the
Waverly Press, Inc.
Mt. Royal and Guilford Aves.
Baltimore, Md. 21202, U.S.A.

To

Joyce

and

Ginger

who made this book

The road to wisdom?–Well, it's plain
and simple to express:
Err
and err
and err again
but less
and less
and less.

PREFACE

My interest in biological mechanisms which control and regulate was first aroused while I was preparing a short chapter on feedback relationships for my previous book, "The Mathematical Approach to Physiological Problems," Williams and Wilkins, Baltimore, 1963. (Paperback edition, M.I.T. Press, 1970.) I was aware that systems engineers had developed a variety of powerful mathematical methods for analyzing the behavior of feedback control systems, and that these methods were being used more and more frequently in the life sciences. But I soon realized that my own competence in mathematics was far too limited to let me understand even the simplest aspects of control theory. I was thus faced with two alternatives: either I could abandon my interest in physiological feedback mechanisms, or I could go back to school to get some training in mathematics and engineering. I went back to school.

This book is an attempt to share with other life scientists, and particularly with graduate students in the biological disciplines, what I learned during twelve months of intensive formal course work. In twelve months I obviously could not become an expert in systems analysis, or in control theory, or in anything else. But I did develop a working knowledge of some aspects of control theory, both linear and nonlinear, and I did learn how to use certain concepts and techniques of control theory in analyzing biological feedback mechanisms. These are the things I write about in this book.

Part I is an introduction to control theory. It is designed primarily as a textbook for students who have had the usual two-year sequence of college mathematics, including differential equations. However, even students who lack formal training in differential equations should not be too severely handicapped. Solving differential equations by the method of undetermined coefficients is outlined in the Appendix, and solving differential equations by using Laplace transforms is covered in the text. I have found that the material of Part I (with the exception of the matrix methods described in Chapter 5) can be fitted into a one-semester course of three credit-hours. But I have also found that ample time must be allowed for the students to do homework exercises, and for the thorough discussion of homework as soon as the instructor has had a chance to correct it. Many of the exercises in this book are difficult--as they should be for graduate students. Hence, in addition to working some of the problems at the end of each chapter, the reader should himself devise, solve, and check as many easy warm-up exercises as he needs for drill. The fewer, the better.

Nothing is more deadly than trivial, run-of-the-mill, two-for-the-penny exercises, except as they prepare one to attack more complicated, more realistic, more challenging problems.

Part II describes the steady-state characteristics of a number of physiological feedback systems. I originally intended Part I to be a mere introduction; Part II was to be the main show. I now see Part I as by far the more important, -- and by far the less troublesome. Part II already clamors for revision, deletion, and extension. I have asked Williams and Wilkins to print it on paper which will automatically disintegrate in three years, but I am not sure they will honor my request. In any event, I shall welcome suggestions about revising Part II.

The Appendix describes a number of techniques which supplement the text and which make the entire book reasonably self-contained. Only the section on fitting empirical equations includes material not easily found elsewhere.

Like any textbook, Control Theory and Physiological Feedback Mechanisms has its particular strengths and weaknesses. It is a weakness that I say almost nothing about computers, particularly digital computers. In defense, I only remark that before he entrusts his data to the brainless machinations of a computer, the biologist should thoroughly understand the technique of computation to be used, and must know how and when he can legitimately use it. As in my previous book, so also in this it is a weakness that I pay too little attention to, and give too little credit for work previously published by others. The reason (which is no excuse) is that I get my kicks from working things out by myself, not from reading other authors. This is bad because it makes for poor balance. I believe it a strength that I insist upon the careful definition of such terms as "feedback," "transfer function," and "gain," terms all too frequently used with little understanding of their true meaning. I believe it a strength that I discuss in detail the steady-state behavior of multiple-loop, nonlinear, feedback systems. I am afraid it is the only kind of behavior most of us biologists will get to study for a long time to come.

I have been helped by so many people that I scarcely know where to begin. I am grateful to The State University of New York at Buffalo, and more specifically to Dr. Douglas M. Surgenor, then Dean of the Medical School, for letting me resign from the Chairmanship of the Pharmacology Department and for granting me two years' leave of absence. I am indebted to Drs. Paul L. Munson, James F. Dickson, III, and James W. Dow for helping me to make suitable arrangements for my period of retraining, and to Drs. Edward A. Carr, Jr., Alfred E. Farah, Avram Goldstein and Fred M. Snell for writing letters in support of my fellowship application. I am grateful to the National Institute of General Medical Sciences for awarding me a generous two-year Special Fellowship. Professor Robert Thrall, his colleagues, and his graduate students at the University of Michigan introduced me to modern mathematics during a special twelve-week program (supported by NIH) for life scientists in the summer of 1966. This gave me, to use an already obsolete term, a jet-assisted takeoff. I am much indebted to Professor Lawrence Stark for accepting me as his most ancient and least visible graduate student at the Chicago Circle

Campus of the University of Illinois during the academic year 1966-67. I am also indebted to his staff, especially Professor Gyan Agarwal, whose year-long course in systems analysis was the focal point of my entire period of training. I am most grateful to Professor W. D. M. Paton, Chairman of the Pharmacology Department, Oxford University, England, for giving me an academic home while I wrote the first draft of this book. I also wish to thank Mr. B. B. Lloyd of the Physiology Department, Oxford, for carefully reviewing the section on the chemical control of breathing in Part II, and Mr. David Witt of the Engineering Department, Oxford, for reviewing an early draft of Chapter 7. I wish to thank Carolyn Shrauger for drawing most of the illustrations in Part I. I am heavily indebted to Mr. Gary Leininger of the School of Engineering at Buffalo for reviewing all of Part I thoughtfully and critically, and for making many suggestions for revision. I also thank my student-colleagues Wes Baumgardner, Bill Hayton, Mike Hlastala, Rick Moran, Vijay Shahi, Lynn Spees, Nathan Strahl, Dr. William Werkheiser, and Barry Winkler for suffering through my obscurities and helping me clarify them when I first taught a course based upon the material of this book.

But by far my deepest gratitude goes to Joyce Opala and to Ginger Moronski. With imperturbable calm, Ginger typed, and often retyped, and sometimes even re-retyped until the whole horrible manuscript was finally done. Joyce made many of the figures, all of the block diagrams and all of the symbol-and-arrow diagrams. She cut into the manuscript thousands of corrections whose invisibility is a tribute to her endless patience. She computed my numerical examples, corrected by spelling, struggled with my erratic punctuation, proofread the entire manuscript, and teased me out of my all-too-frequent fits of despondency. Who but Ginger would say "I love to type the equations"? Who but Joyce would say, "I'll really be sorry when it's finished"? This is as much their book as mine.

After careful consultation with Mr. Dick M. Hoover, Editor-in-Chief of the Williams and Wilkins Company, I agreed to prepare "camera copy" from which the book could be printed directly in an inexpensive paperback edition which I hope graduate students can afford. But now that I am finally about to deliver the manuscript, I am assailed by the kind of apprehension which a parturient mother must feel when she has to deliver her infant without benefit of obstetrician. I am told that such infants are usually normal and healthy, showing no ill effects from their unsupervised entry into the world. But to an author who thinks that even one error is too many, this book, born without redaction, seems dreadfully blemished. δ's masquerade as d's. Overdots disappear. Little m's become big M's. [Vdist] in one chapter turns to V_{dist} in the next. The discrepancies in punctuation boggle the imagination. While it may be true, as Emerson said, that "a foolish consistency is the hobgoblin of little minds," it does not follow that disregard of consistency is the hallmark of great ones. Besides, in mathematics, consistency is never foolish, and I have fought as valiantly as I could against my lack of it. But there comes a time when yet another review of the manuscript cannot be faced, and the book must go to press. I would like to believe that the number of errors per page has been declining exponentially towards some tolerable asymptote, but I am now beyond caring. On the seventh (or is it the seventeenth?) re-reading, I find the text boring, the humor contrived, the examples trivial, the illustrations childish. I can only hope that the reader, coming new to the book, will discover in it some of the freshness and excitement which I myself felt on first writing it.

Douglas S. Riggs, M.D.
Professor of Pharmacology
The School of Medicine
State University of New York
at Buffalo

Eden, New York

Preparation of this book was aided in part by the General Research Support Grant awarded to the School of Medicine, State University of New York at Buffalo, by the National Institutes of Health.

CONTENTS

PART I. CONTROL THEORY

PART II. PHYSIOLOGICAL FEEDBACK MECHANISMS

APPENDIX

PART I

CONTROL THEORY

Time is but the stream I go a-fishing in.

Thoreau, WALDEN

1

INTRODUCTION

SECTION 1.1. Control Theory in Engineering and in Biology

The complex and versatile artifacts, --automobiles, television sets, atomic power plants, orbiting satellites, hydrogen bombs, --which bless and curse modern society are the numerous progeny of an extraordinarily fertile marriage between mathematics and physics. If mathematics be the queen of the sciences, physics must surely be king. Indeed, at the germinal level of theoretical physics, the union between these royal partners is so complete that it becomes difficult to distinguish one from the other. At a more prosaic level, when an engineer designs the guidance system for a missile, or sketches part of the circuit diagram for a new digital computer, he relies upon a variety of powerful mathematical techniques which enable him to predict how some complicated array of individual components will behave in response to an external stimulus. This body of mathematical techniques has come to be called, with more brevity than clarity, systems analysis. Alternatively, it is often referred to as control theory because it deals largely with how one component controls the activity of the next in an interlinked series.

In the living systems which they study, biologists are also confronted by complicated arrays of interlinked components, and they too would like to discover how such systems, and their component parts, behave in response to external stimuli. It is only natural, therefore, that biologists should want to exploit the mathematical methods which have been used to such good effect in engineering. Yet in spite of a few notable successes, progress in the application of control theory to biological problems has been hindered by two major difficulties. First, until recently most students in the life sciences could not understand even elementary control theory because they lacked the necessary background in mathematics. This hindrance is gradually disappearing. Second, the most fully-developed and most easily comprehended methods of control theory deal with linear systems, (defined in Section 1.13), while most biological systems are non-linear. This hindrance is intrinsic, and it is no good trying to sneak around it by applying inappropriate linear methods to grossly non-linear systems, simply because the linear methods are easy to use.

The present book is intended as a text for students in the life sciences who have had enough calculus so that they can solve simple differential equations, i.e. who have had the minimum basic training in college mathematics now considered essential for any kind of scientific career. My aim is to help such students develop a real working knowledge of some of the techniques of linear and non-linear control theory so that they can actually use these techniques in studying the behavior of living systems. While biological systems, particularly biological feedback systems, will be emphasized, much of the early discussion will center upon familiar physical (especially electrical) systems, because they are simple and easy to understand.

SECTION 1.2. Real Systems and Model Systems

The primary definition of "system" given in Webster's New Collegiate Dictionary (1953 edition) can hardly be bettered: "An assemblage of objects united by some form of regular interaction or interdependence." This is what we mean when we speak of any real system such as the solar system, or the cardiovascular system or even those incredibly complex astronautical systems which guide our space ships to the moon and beyond. It is the behavior of real systems which we would like to investigate. Unfortunately the component parts of any such real system are so numerous and the interactions between the component parts so complex that we can never hope to analyze them in full. So we necessarily begin by introducing various simplifying assumptions which enable us to represent the real system by an abstract, but still three-dimensional model system which is simple enough to describe in detail, but which retains certain essential features of the real system.* With respect to those features, a good physical model must behave sufficiently like the real system so that conclusions based upon a study of the model can be applied with tolerable accuracy, and within specified limits, to the real system. An ideal gas is a familiar example of a good model. In this book, the unqualified term "system" will usually refer to such a model system.

[*"A model, like a map, cannot show everything. If it did, it would not be a model but a duplicate. Thus, the classic definition of art as 'the purgation of superfluities' also applies to models, and the model-maker's problem is to distinguish between the superfluous and the essential." (1.)]

A three-dimensional model system can be fully characterized by defining a few variables and by stating how each of these changes with time as it is influenced by other variables within the system or by external disturbances. It is natural, therefore, to speak of the model system as "a system of variables." And since the changes occurring in these variables are described by a set of equations which constitute the mathematical model of the system, a final bit of metonymy leads to the term "system of equations." But as we shall soon see (Section 1.4), the set of equations does not necessarily tell us all we need to know about the behavior of the model system.

SECTION 1.3. Constants, Parameters, and Variables

The quantities with which we deal in systems analysis are either absolute constants, parameters, space-time coordinates, other independent variables.

or dependent variables. (Figure 1.3.1.) The distinctions between these five categories are fundamental and may best be discussed in conjunction with a concrete example.

Example 1.3.1. Consider a simple electrical system consisting of an uncharged capacitor whose capacitance is C farads, in series with a resistor whose resistance is R ohms. (Figure 1.3.2.a) At time zero, a switch is closed so that the system is connected to a constant voltage source supplying $e_s(t)$ volts. We want to find an equation for $e_c(t)$, the voltage across the capacitor, as an explicit function of time, t.

The steps leading to a solution of this kind of problem will be discussed in detail in the next chapter. At present, we are interested only in the solution itself:

$$e_c(t) = e_s(t)\left\{1 - \epsilon^{-[1/RC]t}\right\}; \qquad t \geq 0 \qquad 1.3.1.$$

The behavior of $e_c(t)$ described by equation 1.3.1. is illustrated in Figure 1.3.2.b for a specific example where $e_s(t)$ is 10 volts, R is 10^5 ohms, C is $2 \cdot 10^{-6}$ farads, and t is in seconds.

All five kinds of quantities are illustrated by equation 1.3.1 :

1) An absolute constant always retains the same numerical value whatever happens. (With dimensional constants such as the ideal gas law constant we must add, "for any particular choice of units of measurement.") In equation 1.3.1., the pure numbers 1 and ϵ (the base of natural logarithms) are absolute constants. Like the telephone directory, absolute constants are important but frightfully dull.

2) A parameter (sometimes called an arbitrary constant) is a quantity whose value, though not invariant in an absolute sense, is a fixed characteristic of any particular system. It must therefore be kept constant while we calculate how the variables of that system change with the passage of time. In equation 1.3.1 , R and C are parameters. Like many parameters, R and C can be regarded as constants of proportionality (or their inverses). When given specific numerical values they characterize some particular system of the kind we are currently examining. For example, the behavior of a capacitor can be fully described by saying that the voltage difference across the capacitor at any time, t, is directly proportional to the integral of the current $i_c(t)$ which has been flowing through it:

$$e_c(t) \propto \int_{-\infty}^{t} i_c(\tau)\,d\tau \qquad 1.3.2.$$

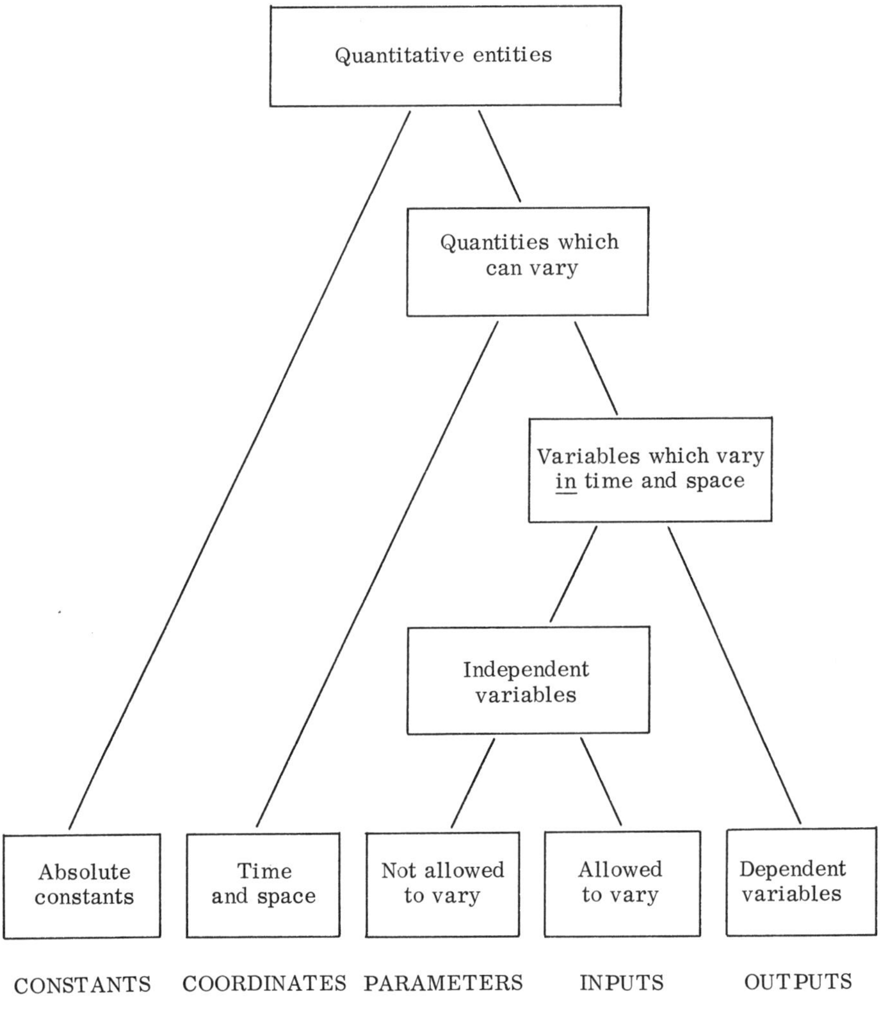

Figure 1.3.1. A classification of quantitative entities. (See Section 1.3.)

A succession of binary partitions divides quantitative entities into the five categories shown here. Note, in particular, the meaning of the much-abused term, "parameter."

The above classification suffices for deterministic systems (See Section 1.9.) which are the only ones discussed in this book. If we were to consider "noisy" stochastic systems, we would have to deal with certain additional quantities such as probabilities, expected values, correlations, and the like, which are not included in the figure.

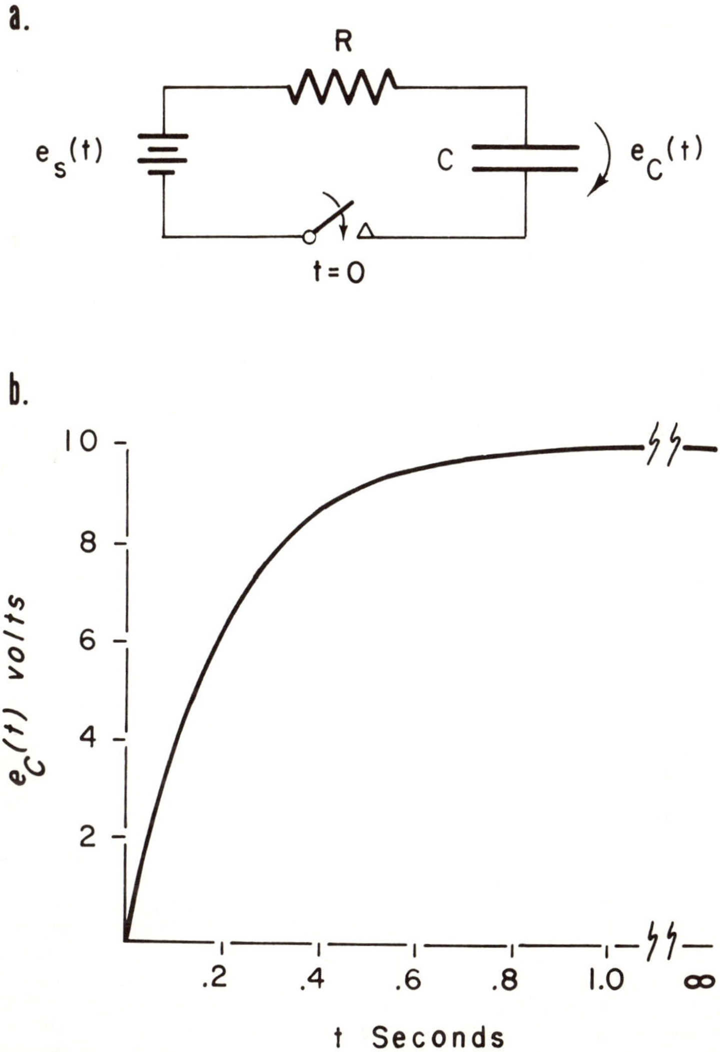

Figure 1.3.2. a, Circuit diagram, and b, voltage across the capacitor as a function of time, for the electrical system of Example 1.3.1.

At time zero the switch is closed so that current from the battery can begin to flow through the resistor into the previously uncharged capacitor. The curve in part b is plotted according to equation 1.3.1 with $e_s(t) = 10$ volts, $R = 10^5$ ohms, and $C = 2 \cdot 10^{-6}$ farads.

or

$$e_c(t) = \frac{1}{C} \int_{-\infty}^{t} i_c(\tau)\, d\tau \qquad 1.3.3.$$

where $1/C$ is the constant of proportionality, and τ is a "dummy" variable of integration. Now equation 1.3.3. as written above, is a general expression for the voltage difference across any linear capacitor, and we are free to assign whatever value we wish to the capacitance, C. But whenever we actually want to use equation 1.3.3. in calculating $e_c(t)$ from $i_c(t)$, we must treat C as a constant throughout the calculation whether we assign it the specific numerical value, say 2 microfarads, which characterizes some particular capacitor, or continue, for the sake of generality, to call it C.

Unfortunately, quantities such as resistances and capacitances are sometimes still referred to as parameters even when they are allowed to vary with time. This leads to the self-contradictory term, "time-variable parameter." Worse still is the slovenly habit of talking about "feedback through parameters," for, as will soon become clear, every participant in a feedback loop must itself be dependent upon some other member of the loop. Therefore if a "parameter" is a member of a feedback loop, it cannot remain constant so long as the other variables are changing, and it should not be called a parameter. Both of these confusing uses of the term "parameter" will be avoided in this book. But we shall not hesitate to change the value of a parameter between calculations so that we can see what influence the parameter has upon the result of the calculation.

3. Time and space. The remaining quantities in equation 1.3.1. are variables whose changing values are the true objects of our study. Consider such commonplace biological variables as concentration of glucose in blood, or blood pressure, or body length, or interval between two heart beats. Like the voltages $e_c(t)$ and $e_s(t)$ in equation 1.3.1. such quantities may change from time to time and from place to place. Indeed, in the real world the very idea of change is inseparably bound up with the concepts of time and space. Real changes can occur only as we move from one instant to the next, and real objects can exist only as they extend from one point to another. Time and space are thus the framework within which things exist and events occur. This confers upon them, and especially upon time, a unique position in our catalog of variables; for if we wish to study the transient behavior of a system, that is, its behavior while it is actually changing we must treat time as an independent variable because the changes we want to investigate are inevitably changes with respect to time. This is why the starting point for our analysis will so often be one or more differential equations expressing the rate of change of some variable, say x(t), with time. Note that x(t) may itself be a time (e.g. interval between heart beats) or a distance (e.g. body length). But

such quantities are finite intervals, not the infinitely-extending space-time coordinates we are here discussing.

Now although time and space constitute the essential milieu within which changes occur, these coordinates are in no sense responsible for the initiation or perpetuation of the changes themselves. Nor can the changes taking place in the other variables of a system alter the space-time coordinates in any way. Consequently, when we come to examine the causal relationships among the variables of a system, we shall not need to consider space-time at all. Time and space are indeed independent variables; independent, imperturbable, unperturbing!

4. Other independent variables, (= Inputs), and 5. Dependent variables (= Outputs). These are the quantities which change with time, and which we shall accordingly symbolize as functions of time, e.g. $e_c(t)$, $e_s(t)$. Granting that all such variables are "dependent upon" time in the peculiar sense described above, it is often convenient to distinguish between dependent variables, for example $e_c(t)$, which we view as belonging to the system under study, and independent variables, for example $e_s(t)$, which we regard as exerting an effect upon the system from outside, and hence as being independent of whatever happens in the system itself. An independent variable of this sort corresponds to the experimental variable in a laboratory investigation; it is independent of the system under study, but it is not independent of us. We manipulate it to suit our convenience.

We must often deal simultaneously with time and with external variables such as $e_s(t)$. Both are "independent variables," but in quite different senses. Therefore, to avoid confusion, it is now customary to use the term input, or input forcing function, or forcing function for any such independent variable as $e_s(t)$, and to use the term output for any such dependent variable as $e_c(t)$. These terms are precisely synonymous* with "variable influencing the system from outside" (=input) and "variable belonging to the system itself" (=output). They do not by any means necessarily designate something which is actually being "put into," or is actually "coming out of" the system. The input to our capacitor-resistor system is the external source voltage, $e_s(t)$, not an actual flow of electrons into the system. If we want to study the system in which antidiuretic hormone helps to control water exchange we shall undoubtedly regard the output as the osmolality of the plasma, and not as the volume of water being "put out" by the kidneys.**

[*at least for stationary systems. (See Section 1.9.)]

[**It is worth noting that, in the steady state, the total rate at which any substance is being "put out" of the body by metabolism and excretion must exactly equal the total rate at which it is being "put in" from exogenous and endogenous sources. Hence, any drug which initially alters the rate of metabolism or excretion of a substance without altering its rate of entry is bound to "lose its effectiveness" on long-continued administration if we judge "effectiveness" by the quantity of the substance metabolized and excreted per unit of time. For example, "refractoriness" is said to develop to diuretic agents because, after prolonged administration, they no longer produce a negative sodium balance. But this does not necessarily mean that they have lost their effect upon the kidney; it may simply mean that, in spite of their continued inhibition of sodium reabsorption, a balance between sodium intake and sodium excretion has been re-established. Indeed, if a diuretic were so "effective" that "refractoriness" never occurred, no such balance could be re-established, and chronic administration of the drug would eventually kill the patient by producing severe sodium depletion.]

SECTION 1.4. Mathematical vs. Working Relationships. The Method of Concomitant Variations

It is essential to understand that, by themselves, the equations which describe a system tell us nothing about how the system actually works in the sense of what causes what. Mathematical analysis can never show whether $x_1(t)$ depends on $x_2(t)$, (i.e., whether a change in $x_2(t)$ causes a change in $x_1(t)$) or whether $x_2(t)$ depends on $x_1(t)$, or whether both directions of dependence coexist. Such questions can only be settled by observing how the system behaves when different variables are altered experimentally. Yet the observed behavior is by no means irrelevant to the mathematical analysis. For it is this working behavior alone which determines, for example, whether one or more feedback loops are present in the system, and hence whether the mathematical techniques appropriate to the analysis of feedback systems should be used.

Two simple examples will illustrate the difference between mathematical relationships and working relationships, and will show how the method of concomitant variations is used to identify the working relationships.

Example 1.4.1. Consider a spring which is fixed at one end and which has a weight hanging on the other end (Figure 1.4.1). Let F be the force exerted by the weight. Let x_0 be the unstretched length of the spring when the weight is absent, and let $x(\infty)$ be the steady-state length of the spring when the weight is present. Then the change in the length of the spring is $\Delta x = x(\infty) - x_0$. Let k be the stiffness of the spring. If the spring obeys Hooke's law, the mathematical relationship among the variables defined above may be expressed, with equal validity, as:

$$\Delta x = \frac{1}{k} F, \tag{1.4.1.}$$

or as:

$$F = k\Delta x, \tag{1.4.1.a}$$

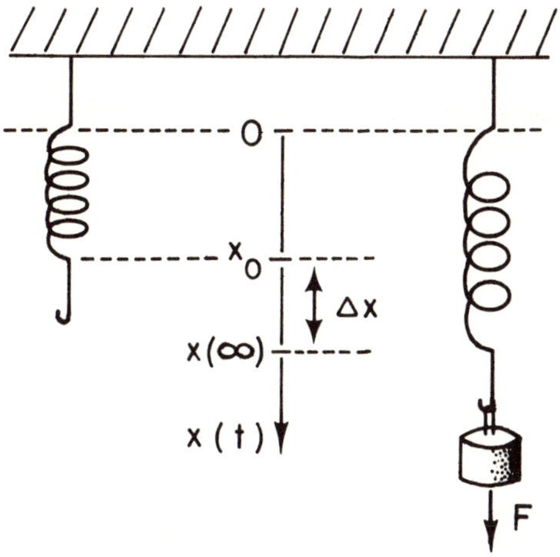

Figure 1.4.1. If a simple spring of initial length x_0 obeys Hooke's law, the steady-state extension, Δx, caused by hanging a weight on the spring is proportional to the force exerted by the weight.

or as:

$$k = \frac{F}{\Delta x}, \qquad 1.4.1.b$$

and we are free to choose whichever one of these variants we wish, according to whether we want to calculate the change in length from the spring stiffness and the force, or the force from the stiffness and the change in length, or the stiffness from the force and the change in length.

But closer investigation of how the system actually behaves will show that the working relationship between the three variables is also characterized by a directionality which is not revealed by the mere mathematical formulae given above. For this closer investigation, we use the method of concomitant variations. To apply the method of concomitant variations, we take each variable in turn, and make a primary change in it and it alone. We then note the concomitant (or subsequent) secondary changes which occur in the other variables of the system. In the present example, if we increase F we observe no change in k but we observe an increase in Δx. If we increase k (for example by using a stiffer spring) we observe no change in F but we observe a decrease in Δx. But now if we try to make a primary change in Δx, we find that it is impossible. Δx can be altered only via F or k! In this simple system, k and F can be treated as independent variables; Δx cannot. It is this observed behavior, and not the mathematical relationship, which justifies saying that (with k held constant as a parameter) a change in F causes a change in Δx, and that this is a strictly unidirectional relationship, i.e., that a change in Δx does not cause a change in F. Note, by the way, that the causes and effects which are identified by the method of concomitant variations are, in general, changes in the variables concerned, not the variables themselves. It would be absurd to say "the length of the spring is caused by the weight!"

Example 1.4.2. As a second example, let us examine the relationship between e_R, the voltage difference across a resistor whose resistance is R, and i_R, the flow of current through the resistor. Provided the resistor obeys Ohm's law,

$$i_R = e_R/R \qquad 1.4.2.$$

As in the previous example, we could solve equation 1.4.2. for e_R, or for R, rather than for i_R without destroying its mathematical validity. But equation 1.4.2. does not by itself tell us anything about the cause-and-effect relationships among the three quantities involved. To find out how the system actually works let us apply the method of concomitant variations.

Consider first the situation when the resistor is connected to a source of constant voltage, say a dry cell, whose internal resistance we shall disregard. If we increase R, we observe no change in e_R, but i_R decreases. If we increase e_R (for example by adding a second dry cell in series with the first) we observe no change in R, but i_R increases. However, when we try to make a primary change in i_R, we find that it is impossible, for i_R can be altered only by changing e_R or R. So again the method of concomitant variations has revealed a directionality of behavior such that i_R is unequivocally dependent on e_R and R.

Consider next the situation when the resistor is connected to a source of constant current, say a photoelectric cell under uniform illumination, whose internal resistance we shall disregard. If we increase R, we observe no change in i_R but e_R increases. If we increase i_R (for example, by increasing the illumination) we observe no change in R, but e_R increases. However, when we try to make a primary change in e_R we find it is impossible, for now e_R can be altered only by changing i_R or R. Hence in this situation e_R is the dependent variable, i_R and R the independent variables.

From the preceding example, it is clear that the observed direction of dependence is not inherent in Ohm's law itself, but is contingent upon the circumstances under which Ohm's law is applied. So if we accept the view (2) that the method of concomitant variations identifies causal relationships we shall have to say that when the resistor is connected to a voltage source, a change in e_R causes a change in i_R, but that when the resistor is connected to a current source, a change in current flow causes a change in e_R. I would defend both of these statements. But I would also defend the assertion that, even with a current source, the flow of current through the resistor is caused by the voltage difference across its terminals and not vice versa. For if the voltage difference between the ends of the resistor were caused by the current flowing through it, the voltage where the current leaves the resistor would surely have to be higher than the voltage where the current enters. No self-respecting simple resistor ever behaves in such a perverse way. In fact, such a "resistor" would not be a resistor at all but would be a source of current! Resolution of this paradox is good clean intellectual fun, (see exercise 2.5), but not essential to our main argument. For in the systems we examine we shall always be interested only in the observed working behavior as revealed by the method of concomitant variations. Let us therefore avoid controversy about ultimate causes by using the simple descriptive term "working relationship" for any relationship which has been identified by the method of concomitant variations. However, in talking about such relationships I shall not hesitate to use such straightforward statements as "a change in A(t) causes a change in B(t)" whenever it seems natural to do so.

SECTION 1.5. Working Equations and Symbol-and-Arrow Diagrams

We have just seen that even when the mathematical relationships among the variables of a system can be precisely specified by an equation, the equation itself does not tell us what the working relationships among the variables are. This deficiency could be remedied if we agreed to solve every equation for its dependent variable(s). I shall term such an equation a working equation, because it clearly suggests how the system actually works, i.e. how the independent variables on the right of the equal-sign influence the dependent variables on the left. For example, equation 1.4.1. is a working equation because Δx (on the left) really does depend on k and F (on the right) in the manner specified by the equation. But equation 1.4.1.a is not a working equation because if we increase k, F does not increase; and equation 1.4.1.b is not a working equation because if we increase F, k does not increase. However it is not always convenient to keep our equations in the form of working equations as we subject them to various mathematical manipulations. Furthermore, if a system has many variables, it is not easy to visualize their complex interrelationships if we simply look at the working equations which describe the system. It is therefore desirable to have some other way of summarizing the working relationships which are identified by the method of concomitant variations. A convenient device for this is the symbol-and-arrow diagram.

Consider a particular pair of variables, A(t) and B(t). If a primary increase in A(t) causes a secondary increase in B(t), we shall symbolize this relation by a solid arrow pointing from A(t), the independent variable, to B(t) the dependent variable:

A(t) ——➤ B(t)

But if a primary increase in A(t) causes a secondary decrease in B(t) we shall use a broken arrow:

A(t) - - -➤ B(t)

Notice that the direction of the arrow clearly indicates the direction of influence of the one variable upon the other as revealed by the method of concomitant variations. For example, the working relationship summarized by equation 1.4.1. would be diagrammed:

$$F \longrightarrow \Delta x \leftarrow - - k$$

which shows at a glance not only the direction of dependence but also the direction of the change which will occur in the dependent variable when either the independent variable F, or the parameter, k, is changed.

SECTION 1.6. The Four Fundamental Types of Working Relationships Which Can Exist Between Two Variables

Using symbol-and-arrow notation, it is easy to define the four fundamentally different types of *immediate* working relationships which can exist between any two variables, A(t) and B(t). (3) (An immediate working relationship is one which does not involve some intermediate third variable).

1. *No immediate relationship*: A primary change in A(t) causes no change in B(t); a primary change in B(t) causes no change in A(t). Symbol-and-arrow diagram:

$$A(t) \qquad B(t)$$

(Note that even when A(t) and B(t) are not directly related to each other they may be *correlated* with each other if they both depend, either immediately or remotely, upon some third variable, C(t). For example:

$$A(t) \longleftarrow C(t) - - \rightarrow B(t)$$

would represent an inverse correlation between A(t) and B(t).)

2. *A unidirectional relationship*: A primary change in the first variable causes a change in the second, but a primary change in the second (if it be possible at all) causes no change in the first. Symbol-and-arrow diagrams:

$$A(t) \longrightarrow B(t) \qquad A(t) \longleftarrow B(t)$$

$$A(t) - - - \rightarrow B(t) \qquad A(t) \leftarrow - - - B(t)$$

3. *An equilibrium*: A primary change in A(t) causes a change in B(t), *and* a primary change in B(t) causes a change in A(t), both relationships being described by the *same* equation. Symbol-and-arrow diagrams:

$$A(t) \longleftrightarrow B(t) \qquad A(t) \leftarrow - \rightarrow B(t)$$

Using a single arrow between the symbols shows that only a single relationship, described by a single equation, is involved, while using two heads on the arrow shows the reversibility of the equilibrium. In an equilibrium, either variable may serve as the independent variable. For example, if we assume that CO_2 reaches diffusion equilibrium across the alveolar-capillary wall and if we neglect arterio-venous shunts across the lung, the working relationship between $C_{CO_2,alv}(t)$, the concentration of carbon dioxide in alveolar air, and $C_{CO_2,art}(t)$, the concentration of carbon dioxide in arterial plasma may be described by the working equation,

$$C_{CO_2,art}(t) / C_{CO_2,alv}(t) = S_{CO_2} \qquad 1.6.1.$$

where S_{CO_2} is the solubility of CO_2 in arterial plasma. The corresponding symbol-and-arrow diagram:

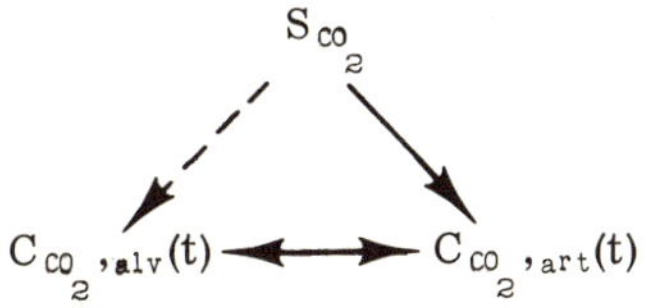

shows that a primary increase in $C_{CO_2,alv}(t)$ (such as would be produced by increasing the concentration of CO_2 in the inspired air) will cause an increase in $C_{CO_2,art}(t)$, while a primary increase in $C_{CO_2,art}(t)$ (such as would be produced by an increase in metabolic CO_2 production) will cause an increase in $C_{CO_2,alv}(t)$, *both changes being described by one and the same equation,* namely equation 1.6.1.

4. *A feedback relationship*: A primary change in A(t) causes a change in B(t), *and* a primary change in B(t) causes a change in A(t). But these are two *different* relationships which must be described by two different equations. Hence in the corresponding symbol-and-arrow diagram there must be two different arrows, one pointing from A(t) to B(t), the other pointing from B(t) to A(t). With an arrangement of this kind, the influence exerted by one variable on the second variable via some particular mechanism is, so to speak, "fed back" from the second to the first via a different mechanism; hence the term "feedback." There are two general types of feedback: negative and positive.

In a symbol-and-arrow diagram for negative feedback between A(t) and B(t), one arrow is solid, the other broken. For example:

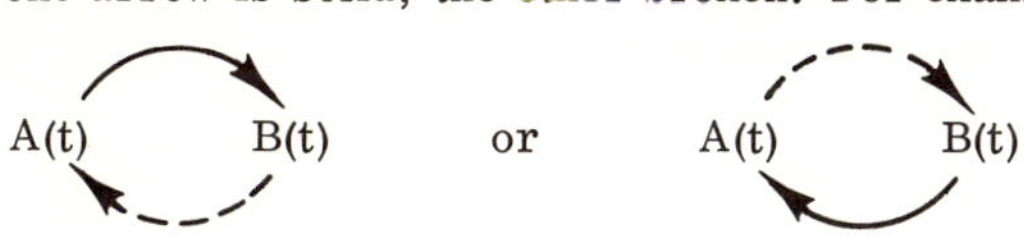

To see how a negative feedback loop operates, consider the diagram on the left. Suppose that some external variable (= input) causes a primary increase in A(t). The increase in A(t) will cause an increase in B(t) as indicated by the solid arrow. But this increase in B(t) will, in turn, cause a decrease in A(t) as indicated by the broken arrow, thus tending to oppose or minimize the primary change. Such loops are called negative because, with the feedback loop in operation the change produced in A(t) or B(t) by any external disturbance is less than it would be without feedback, i.e. with one or the other of the two arrows missing, so that the influence of the first variable upon the second can no longer be "fed back" to the first variable again. In the jargon of control theory, the change in A(t) or B(t) caused by a change in the input is smaller in the closed-loop system than in the open-loop system.

In a symbol-and-arrow diagram for positive feedback between A(t) and B(t), both arrows are solid or both are broken. For example:

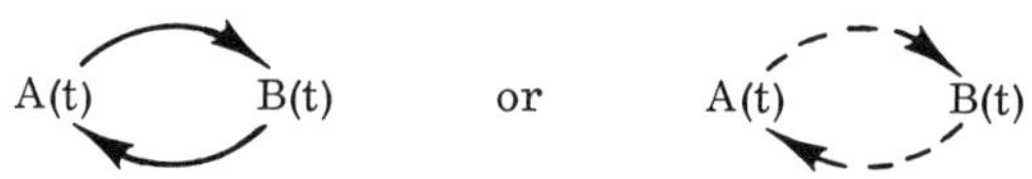

The same reasoning as before will show that a positive feedback loop tends to enhance or magnify the change produced in A(t) or B(t) by a change in the input to the system.

SECTION 1.7. Working Relationships in Multivariable Systems

In extending symbol-and-arrow notation to systems of many variables, the only new feature which requires comment is that both equilibria and feedback loops may comprise more than two variables. Indeed, feedback loops consisting of a dozen or more variables are by no means uncommon. Moreover, particularly in biological systems, a single variable often belongs to several different loops. In dealing with multivariable feedback loops, two simple rules are useful. First, any closed cycle of variables in a symbol-and-arrow diagram represents a feedback loop, if, and only if, the arrows can be followed all the way around the cycle in but one direction. Notice that this rule allows any of the arrows in the loop to be double-headed provided at least one arrow is unidirectional. For if all of the arrows were two-headed we would have a reversible equilibrium, not a feedback loop. Second, a feedback loop is positive if the number of broken arrows is even, negative if the number of broken arrows is odd. The reader should find it easy to verify this rule.

It is worth re-emphasizing that feedback is a particular kind of working relationship, not a particular kind of mathematical relationship. No set of equations, however precise, can tell us whether or not a system contains any feedback loops. As I have pointed out elsewhere (3), failure to recognize this fact has caused a great deal of unnecessary confusion.

The set of equations and the symbol-and-arrow diagram for a given system, though distinct, are by no means unrelated. They are, in fact, complementary ways of describing the system. As the next section will indicate, this complementarity may be put to good use when we are investigating a newly-discovered or poorly-understood biological system. It is therefore important to discuss the relation between equation and diagram for the three kinds of equations . . . differential, time-dependent, and steady-state --with which we shall commonly deal.

As a specific example, consider the factors which control C(t), the concentration of a drug, D, in the body. For simplicity, we assume that D is instantaneously and uniformly distributed through [Vdist], a constant volume of distribution, and that [$\dot{V}$cl], its clearance from [Vdist], is also constant. We shall investigate the behavior of C(t) as it declines from C(0), the initial concentration at time zero. Again we are not now concerned with the detailed steps in the derivation but only with the resulting equations.

The <u>differential equation</u> we begin with is

$$[\text{Vdist}]\,\frac{dC(t)}{dt} + [\dot{V}\text{cl}]\,C(t) = 0 \qquad 1.7.1.$$

which specifies the quantitative relationship between the two variables, $\frac{dC(t)}{dt}$ and C(t). To facilitate applying the method of concomitant variations, let us solve equation 1.7.1. for each variable. Solving first for C(t),

$$C(t) = -\,\frac{[\text{Vdist}]}{[\dot{V}\text{cl}]}\left\{\frac{dC(t)}{dt}\right\} \qquad 1.7.1.a$$

Viewed as a working equation, this makes no sense, for it would have us believe that a primary increase in [$\dot{V}$cl] would produce an increase in C(t) while a primary increase in dC(t)/dt (assuming it were possible) would decrease C(t). But

$$-\,\frac{dC(t)}{dt} = \frac{[\dot{V}\text{cl}]}{[\text{Vdist}]}\,C(t) \qquad 1.7.1.b$$

is a perfectly valid working equation because it correctly shows that a primary increase in [$\dot{V}$cl] or in C(t) will hasten the rate at which C(t) declines, while a primary increase in [Vdist] will have the opposite effect. (In a first-order differential equation the dependent variable is usually the derivative, as it is here.) The corresponding symbol-and-arrow diagram,

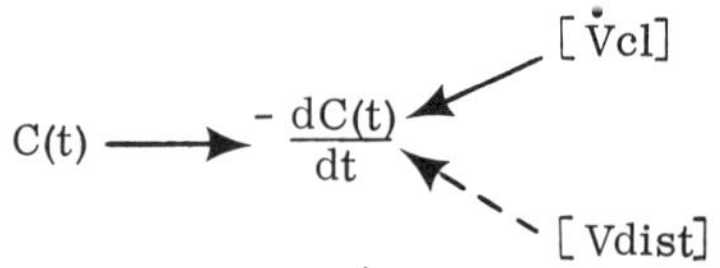

shows how a change made at any time in any of the other quantities, whether parameters or variables, would influence - dC(t)/dt, the rate of decrease of drug concentration. Notice that we can let the parameters (which cannot change with time) assume different values because time does not enter equation 1.7.1. as an explicit variable. But neither equation 1.7.1.b nor the corresponding diagram describes the system completely because they do not show what factors determine C(t) which is certainly a dependent variable.

To obtain C(t) as an explicit function of time, we solve equation 1.7.1.

$$C(t) = C(0)\,\epsilon^{-\frac{[\dot{V}cl]}{[Vdist]}t} \qquad 1.7.2.$$

where ϵ is the base of the natural logarithms. Except for time itself, C(t) is the only variable. Hence, equation 1.7.2. as it stands ought to be in the form of a working equation. The symbol-and-arrow diagram corresponding to equation 1.7.2. would therefore appear to be

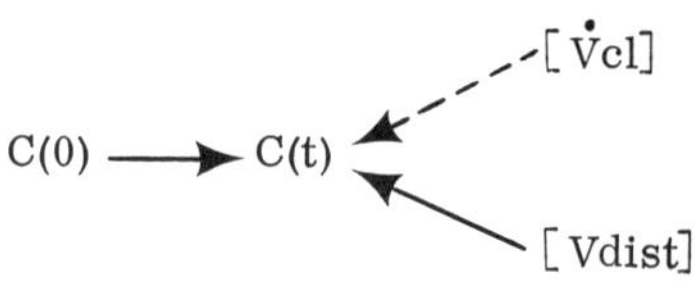

This diagram seems to suggest that if we make a primary increase in [Vdist] alone, C(t) will increase, which is nonsense! The difficulty is that equation 1.7.2. was obtained by integrating equation 1.7.1. with respect to time under the assumption that [Vdist] and [V̇cl] remain constant as parameters while C(t) is changing. If we change these parameters "in midstream," we cannot use equation 1.7.2. to predict what will happen to C(t), for equation 1.7.2. is no longer valid. So we have no right to ask, "What effect will altering [V̇cl] or [Vdist] have on C(t)?" while C(t) is changing. Nor can we change C(0) in midstream, for it is an initial condition only. So if we insist on drawing a symbol-and-arrow diagram corresponding to the equation for a transient state where time is an explicit variable, the arrows no longer reliably indicate immediate effects. The arrow pointing from [Vdist] to C(t) does not mean "an increase in [Vdist] will produce an increase in C(t)" but rather "If the parameter [Vdist] had been larger throughout the entire time C was changing, the value reached by C at time t would have been greater." I find such diagrams too confusing to be useful.

Finally, if an input to the system is changed at time zero but is kept constant thereafter (= a "step" input), the dependent variables may themselves approach new constant values as time approaches infinity. For such constant, time-infinity values, we can write a steady-state equation in which time is no longer a variable. The corresponding symbol-and-arrow diagram will then show how the steady-state values of the variables are influenced by changes in the input or in the parameters. We cannot use the present example as an illustration, for equation 1.7.2. shows that C(t) approaches 0 as t approaches ∞, i.e., that $C(\infty) = 0$. Let us therefore consider the behavior of C(t) during the infusion of drug D at a constant rate, $\dot{Q}_{in}$. If there is no drug D in the body at time zero when the infusion is started, the value of C(t) will be given by

$$C(t) = \frac{\dot{Q}_{in}}{[\dot{V}cl]} \left\{ 1 - \varepsilon^{-\frac{[\dot{V}cl]}{[Vdist]} t} \right\} \qquad 1.7.4.$$

Now if we let t approach infinity we have

$$C(\infty) = \frac{\dot{Q}_{in}}{[\dot{V}cl]} \qquad 1.7.5.$$

The corresponding symbol-and-arrow diagram is

$$\dot{Q}_{in} \longrightarrow C(\infty) \longleftarrow - - [\dot{V}cl]$$

which shows that if we increase the value of Q_{in} (or decrease the value of $[\dot{V}cl]$) to a new value and keep it there, C(t) will approach a new and higher steady-state value, $C(\infty)$, which is independent of [Vdist]. (Notice that again we are free to change the parameters because equation 1.7.5., like equation 1.7.1., does not depend upon time as an explicit variable.) However, neither the steady-state equation nor the diagram tell us anything about the transient response of C(t) as it moves from one steady state to another. For although the arrows in a diagram for the steady state usually also indicate correctly the direction in which the variables will start to change immediately after imposition of a step input at time zero, in some systems, the final steady-state change is actually in the direction opposite to that of the initial change. For example, if the environmental temperature is suddenly changed from about 4° C to 0° C, the already very low body temperature of a golden hamster in deep hibernation will at first decrease even further. But this initial decrease will usually interrupt hibernation so that body temperature will rise to a steady-state value of about 37° C even though the ambient temperature remains at 0° C (4). Hence it is safest to regard the arrows in a symbol-and-arrow diagram for the steady state as showing only the ultimate effects caused by changing an input or a parameter. Many familiar "laws,"--e.g., Ohm's law, the ideal gas law, Poiseuille's law,--are really examples of steady-state equations, the corresponding transient changes occurring so rapidly that they can be regarded as instantaneous. Symbol-and-arrow diagrams are useful chiefly for depicting such steady-state relationships.

SECTION 1.8. The Practical Utility of Symbol and Arrow Diagrams

Since the working behavior of a system (as summarized by a symbol-and arrow diagram) and the mathematical description of the same system (as specified by a set of equations) are two quite different things, it is often possible to construct at least a crude symbol-and-arrow diagram for a system long before we know enough about it to derive a set of equations describing the quantitative relationships among the variables. Besides showing how the system works, such a diagram often suggests what equations will eventually be needed, since for each variable towards which one or more arrows point there should be at least one working equation showing how that variable is influenced by others. When investigating a new system, therefore, construction of a symbol-and-arrow diagram and derivation of equations should be regarded as complementary procedures, a crude initial diagram suggesting what equations to seek; the equations suggesting what additional variables should be added to the diagram. Ideally, the final result will be a set of equations and a symbol-and-arrow diagram which, taken together, completely describe the system. Neither alone suffices.

Yet it would be wrong to imply that construction of a symbol-and-arrow diagram is always a necessary or even a desirable step in our analysis. In linear systems the same purpose is better served by a block diagram (See Sec. 3.13). Furthermore, even when the behavior of a system is well understood, difficulties may arise in trying to depict that behavior in a symbol-and-arrow diagram. For example: (1) A single dependent variable may be subject to several different constraints, corresponding to two or more different equations. (2) Dependent behavior which is not monotonic throughout the range of interest cannot be represented by either a solid arrow (which indicates a monotonic increasing relationship) or a broken arrow (which indicates a monotonic decreasing relationship.)* (3) The working relationships among the variables may depend partly on what external variables (= inputs) are acting on the system. (An illustration of this was discussed previously. See Example 1.4.2.) Because of these and other difficulties, symbol-and-arrow diagrams are not always worth troubling to construct. However, they will often prove to be extremely useful when we analyze the complicated non-linear interrelations among the many variables of a biological system in the steady state. We shall therefore use them extensively in Part II of the present book.

[*When a dependent variable, say B(t), seems to be influenced nonmonotonically by some other variable, say A(t), it often turns out that A(t) has both a direct effect, tending to alter B(t) in one direction, and an indirect effect, via some neglected third variable, say C(t), which tends to alter B(t) in the opposite direction. By revising our diagram to include C(t), we avoid the difficulty while making the diagram itself more complete and specific.]

SECTION 1.9. Classification of Systems

Different kinds of systems are usually distinguished from each other by differences in the equations which are needed to describe them, and by corresponding differences in the mathematical techniques used to analyze their behavior. In conformity with this practice, we shall classify any system according to five independent attributes, each of which leads to a binary division.

1. Distributed-parameter vs. lumped-parameter systems. In a distributed-parameter system, the elements of the system (e.g. resistors, capacitors, etc.) are so far extended ("distributed") in space that the variables they influence, (e.g. currents and voltages) must be treated as differing not only from time to time but also from one point in a resistor, or in a capacitor, etc, to another. Consequently, the values of such variables depend upon both time and space coordinates, and the system must be analyzed by solving partial differential equations. For example, any reasonably sophisticated analysis of the cardiovascular system would require treating it as a distributed-parameter system. After all, the resistance offered by the blood vessels to the flow of blood is distributed over arteries, arterioles, capillaries, venules, and veins, and we would like to be able to calculate how pressure and flow vary from one place to another under a variety of circumstances. A second example is the diffusion of a solute through a liquid where the concentration gradient (and hence the rate of diffusion) varies from point to point.

In a lumped-parameter system, the capacitors, resistors, etc, are small enough (or the objectives of our analysis are limited enough) so that we can treat each such element as if it were concentrated ("lumped") at one particular point in the system. This greatly simplifies our analysis because ordinary differential equations suffice when the variables change only with time. In this book, we shall confine the discussion strictly to lumped-parameter systems.

2. Stationary vs. nonstationary systems. In a stationary (= fixed) system, the quantities such as resistances, capacitances etc. which characterize the elements of the system remain constant as parameters. In a nonstationary (= time-varying) system, at least one such quantity is a variable function of time, (but not of any other variables in the system). It is thus the system itself which is "nonstationary." Consider, for example, the disappearance of penicillin from [Vdist], its volume of distribution in the body. Normally, the kidneys remove the penicillin by two "supply-limited" (5) processes: glomerular filtration and tubular secretion. Consequently, [Vcl], the renal plasma clearance of penicillin, remains constant as a parameter. If we assume that [Vdist] also remains constant, C(t), the concentration of penicillin, would decrease with time according to equation 1.7.1:*

[*To be honest, penicillin is excreted so rapidly that this equation is quite inaccurate. To improve the accuracy, we would have to treat the system as a distributed-parameter system.]

$$[\text{Vdist}]\frac{dC(t)}{dt} + [\dot{V}\text{cl}]\,C(t) = 0 \qquad 1.9.1.$$

which describes a stationary system. But suppose that a small and gradually-decreasing concentration of probenecid, a drug which interferes with the tubular secretion of penicillin, were also present in the plasma. Then $[\dot{V}\text{cl}]$ would no longer be constant but would increase with time, and equation 1.9.1. would have to be written:

$$[\text{Vdist}]\frac{dC(t)}{dt} + [\dot{V}\text{cl}]\,(t)C(t) = 0 \qquad 1.9.2.$$

In differential equations describing stationary systems, the unknown function, here C(t), and its derivatives, have <u>constant coefficients</u>, e.g. $[\dot{V}\text{cl}]$ and [Vdist] in equation 1.9.1. But in a similar equation for a nonstationary system there is at least one <u>variable coefficient</u>,* e.g. $[\dot{V}\text{cl}](t)$ in equation 1.9.2. We shall deal only with stationary systems.

[* Sometimes colloquially, but erroneously, called a "time-variable parameter."]

3. <u>Discrete vs. continuous systems</u>. In a <u>discrete</u> (= discrete-time) <u>system</u>, changes in one or more of the variables occur only at <u>specified instants</u> of time between which the variable is either undefined or remains constant. For example, money left at interest in a savings bank which compounds interest quarterly increases "instantaneously" every three months but remains constant between these instants of compounding. Sudden changes of this sort, strictly limited to discrete instants of time, are rarely encountered in nature. Certainly the vast majority of biological systems are <u>continuous</u> (= continuous time) <u>systems</u> in which changes are not limited to specific instants even when they occur intermittently. For example, the ejection of blood from the heart, though periodic, is not instantaneous but takes place over an interval of time. If we want to study it carefully, therefore, we must treat it as a function of continuous time.

Notice that it is not the time-dependent variable but <u>time itself</u> which is treated as continuous or discrete. If the time-dependent variable can assume only discrete values, but can change from one discrete value to another at any time, the system is <u>quantized</u> but continuous. For example the number of atoms of a radioactive isotope remaining in a small sample decreases with time by discrete and instantaneous steps of unity. But these steps may occur at <u>any</u> instant, not just at certain specified instants, so the number of atoms remaining must be treated as a function of continuous time.

Though truly discrete <u>systems</u> are rare in nature, discrete <u>data</u> are commonplace. It is often inconvenient, or downright impossible, to collect and record data continuously. Furthermore, digital computers, for all their speed and accuracy, can only deal with a finite number of discrete data. Therefore observations of continuous systems are often made only at discrete times the frequency of such data sampling depending upon how rapid are the continuous changes we want to study.

Continuous-time systems are described by differential equations. Discrete-time data, whether from a truly discrete system or from a "sampled data system" are described by difference equations. In this book, consideration will be given only to continuous systems. However the numerical methods for solving non-linear differential equations which are discussed in chapter 6 really depend upon treating a continuous system as if it were discrete.

4. Deterministic vs. stochastic systems. A purely deterministic system is one whose response to a specified input applied under specified conditions, is completely predictable from relatively simple physical laws, and is therefore precisely reproducible time after time. Strictly deterministic systems exist only as abstract models, for the variables in any real system will always be more-or-less subject to those unpredictable and random variations collectively known as "noise." A purely stochastic system is one which behaves in an entirely random noisy fashion. The output of such a system can be described, but only by statistical laws and in terms of probabilities and expected values. A microscopic particle exhibiting Brownian movements might be viewed as a purely stochastic system. Fortunately, stochastic behavior in the small tends to become deterministic behavior in the large, so that simple physical laws can often be relied upon in macroscopic systems. Nevertheless, noise is an unavoidable fact of life, and a variety of methods have been devised to take account of, or even to make use of random variations.

In this book, we shall be deaf to noise, and we shall consider deterministic systems only.

5. Linear vs. nonlinear systems. In this book, we shall deal with both linear and nonlinear systems. The distinction between these can best be made after discussing operators in the next section.

In summary, all naturally-occurring systems are, in the final analysis, distributed-parameter, non-stationary, quantized, stochastic, nonlinear systems. That we can sometimes obtain useful information by treating these wayward creatures as if they were lumped-parameter, stationary, continuous, deterministic, linear systems is little short of miraculous.

SECTION 1.10. Operators

I am ashamed to say that for a long time I was confused about the meaning of the term "operator," failing to grasp either its great simplicity or its great generality. The present section is included for those few readers who may also be suffering from the entirely false impression that a mathematical operator is some kind of remote and mysterious abstraction. In fact it is no more remote and mysterious than the recipe for a cake. A cake recipe is a set of directions which tells you exactly how to manipulate the ingredients, the oven temperature and the time of baking to obtain a cake. An operator is a set of directions, i.e. a rule of operation, which tells you exactly how to manipulate one function to get another.* Any such rule is an operator; any operator is such a rule.

[*"In its widest sense the word operator is synonymous with function, transformation, etc., denoting a mapping of one set of elements (the domain of the operator) onto another (its range). But when the word operator is used, it is generally understood that both the domain and the range consist of functions (i.e., functions from number to number)." (6).]

Let H represent any rule of operation. Then

$$Hu(t) = v(t) \qquad 1.10.1.$$

simply means that if we treat the function u(t) exactly as the operator H tells us to, (i.e. if we let H "operate" on u(t)), we shall obtain the function v(t). For example, if H_1 tells us that to get v(t) we must differentiate u(t) twice with respect to time and add five times the value of u(t) we would have,

$$H_1u(t) = \left\{D^2 + 5\right\} u(t) = \frac{d^2u(t)}{dt^2} + 5u(t) = v(t) \qquad 1.10.2.$$

Where D is defined as the differential operator, $\frac{d[\ ]}{dt}$, and D^2 is defined as $\frac{d^2[\ ]}{dt^2}$. If H_2 tells us that to obtain v(t) we must subtract 4 from u(t) we would have:

$$H_2u(t) = u(t) - 4 = v(t) \qquad 1.10.3.$$

If H_3 tells us that to obtain zero at any time t we must integrate u(t) with respect to time between limits of $-\infty$ and t, and then add Ru(t) to the result of the integration we would have:

$$H_3u(t) = \left\{R + \int_{-\infty}^{t} [\ ]d\tau\right\} u(t) = Ru(t) + \int_{-\infty}^{t} u(\tau)d\tau = 0 \qquad 1.10.4.$$

In formulations of this sort, the empty brackets, [ ], merely show where u(t), the operand, fits into the operator.

SECTION 1.11. Linear vs. Nonlinear Operators

An operator L is a linear operator if, and only if,

$$L\{au_1(t) + bu_2(t)\} = aLu_1(t) + bLu_2(t) \qquad 1.11.1.$$

where $u_1(t)$ and $u_2(t)$ are any two functions which lie in the domain of L, (i.e. which can legitimately be operated upon by L), and a and b are any two scalar constants. Equation 1.11.1. states that with a linear operator, L, 1.) adding two functions together and then letting L operate on their sum yields the same result as letting L operate on each separately and then adding; and 2.) multiplying a single function by a constant and then letting L operate on the product yields the same result as letting L operate on the function and then multiplying by the constant.

To test any operator for linearity, we actually perform the operations which are called for in equation 1.11.1. If the result of operating as shown on the left side of equation 1.11.1 is identical to the result of operating as shown on the right side, the operator is linear. If the two results differ, the operator is non-linear.

Example 1.11.1. Is the operator $H_1 = \frac{d^2[\ ]}{dt^2} + 5$ of equation 1.10.2 a linear operator?

Applying this operator to $\{au_1(t) + bu_2(t)\}$ as on the left of equation 1.11.1,

$$\frac{d^2}{dt^2}\left\{au_1(t) + bu_2(t)\right\} + 5\left\{au_1(t) + bu_2(t)\right\} =$$

$$au_1''(t) + bu_2''(t) + 5au_1(t) + 5bu_2(t) \qquad 1.11.2.$$

Applying the same operator in the manner prescribed on the right of equation 1.11.1,

$$a\left\{\frac{d^2}{dt^2}u_1(t) + 5u_1(t)\right\} + b\left\{\frac{d^2}{dt^2}u_2(t) + 5u_2(t)\right\} =$$

$$au_1''(t) + 5au_1(t) + bu_2''(t) + 5bu_2(t) \qquad 1.11.3.$$

Since this is the same function we obtained in 1.11.2., the operator is linear.

Example 1.11.2. Is the operator $\frac{d[\ ]^2}{dt}$, which tells us to square the operand and then take its first derivative, linear?

$$\frac{d}{dt}\left\{au_1(t) + bu_2(t)\right\}^2 = \frac{d}{dt}\left\{a^2u_1^2(t) + 2abu_1(t)u_2(t) + b^2u_2^2(t)\right\}$$

$$= 2a^2u_1(t)u_1'(t) + 2ab\left[u_1(t)u_2'(t) + u_1'(t)u_2(t)\right] + 2b^2u_2(t)u_2'(t) \qquad 1.11.4.$$

But:

$$a\frac{d}{dt}u_1^2(t) + b\frac{d}{dt}u_2^2(t) = 2au_1(t)u_1'(t) + 2bu_2(t)u_2'(t) \qquad 1.11.5.$$

which is not identical with the right side of 1.11.4. at all! The operator is therefore not linear.

Further examples will be found in Exercise 3.

It is easy to prove (see Exercise 2) that the sum of any number of linear operators is itself a linear operator. Note, by the way, that an operator may have no numerical value and no dimensions (e.g. the operator ln, meaning "take the natural logarithm of"), a numerical value but no dimensions (e.g. the numeral 5 in such an expression as 5y(t) meaning "multiply y(t) by 5"), dimensions but no numerical value (e.g. the differential operator, d^n/dt^n, which has the dimension of $[\text{time}]^{-n}$), or both dimensions and a numerical value, (e.g. the operator 5 ohms in Ohm's law applied to a particular resistor: 5i(t) = e(t)). Operators, no less than constants and variables, must obey dimensional rules (5,7). Hence all of the individual operators in a sum of operators must have the same dimensions.

SECTION 1.12. Mathematical vs. Physical Operations

It is natural to regard any real physical system as operating on its inputs to produce its outputs. For example, we may view a capacitor as a device which operates on an inflowing current to produce a voltage across the capacitor. We may view the cook-book recipe as a device for operating on the ingredients, the oven temperature, and the time of baking to produce a cake. We may view the β cells of the islets of Langerhans as a device for operating on the blood glucose concentration to produce insulin secretion. Physically, then, we are commonly faced with the following question: Given a particular system, whose working behavior is known, and a particular input, which is a known function of time, how can we calculate the unknown output?

To answer this question, it would seem most logical to describe the physical operation of the system by a mathematical operator which would operate on the known input to generate the unknown output. We could then obtain the output just by letting the operator operate! Unfortunately, in practice we shall not often be able to adopt such a straightforward approach. It is true that the mathematical operators we shall use do indeed specify the relationship between the known input and the unknown output. But usually that relationship must be stated in the form of a differential equation in which the differential operator operates mathematically on the unknown output to "produce" the input, exactly the reverse of what actually happens in the physical system.* Hence, although we may know the operator, we do not know its operand, so we cannot "just let the operator operate." Instead, we must answer the much more difficult question: Upon what unknown output must our differential operator operate so that the output will correspond to the known input? The vast majority of our mathematical work will be devoted to ways of answering this question.

[*We are thus reminded again that mathematical relationships, being purely quantitative, take no account of the directionality of working behavior which is so important a characteristic of the real world.]

SECTION 1.13. Linear Systems and the Principle of Superposition

A linear system is a system whose input-output relationships can be fully described by one or more linear equations; i.e., equations whose operators are all linear operators. The most important attribute of a linear system is that it obeys the law or principle of superposition. The principle of superposition states that in any linear system the output produced by applying several inputs simultaneously is equal to the sum of the outputs produced by applying each input separately. This simple additive behavior, such that when two or more inputs are "superposed" their outputs are also superposed, is what makes linear systems so tractable mathematically; failure of nonlinear systems to obey the law of superposition accounts for the difficulties which beset nonlinear analysis.

Proof that linear systems do indeed obey the law of superposition is elementary: Any linear system may be described by an equation which is either of the general form:

$$Lx(t) = y(t) \tag{1.13.1.}$$

or, more frequently, of the general form:

$$Ly(t) = x(t) \tag{1.13.2.}$$

where x(t) is the input, y(t) is the output, and L is a linear operator. (To include systems with inputs at several different points and with several outputs in this formulation, we must recognize that, as will be discussed in Chapter 5, x(t) and y(t) may be vectors, L then specifying certain vector-matrix operations.) Let $y_1(t)$ be the output produced by a particular input, $x_1(t)$, and let $y_2(t)$ be the output produced by a second input, $x_2(t)$. Applying equation 1.13.2. to each of these particular cases,

$$Ly_1(t) = x_1(t) \qquad 1.13.3.$$

$$Ly_2(t) = x_2(t) \qquad 1.13.4.$$

Adding equations 1.13.3. and 1.13.4. together,

$$Ly_1(t) + Ly_2(t) = x_1(t) + x_2(t) \qquad 1.13.5.$$

But since L is a linear operator, equation 1.11.1. shows that

$$Ly_1(t) + Ly_2(t) = L\left\{y_1(t) + y_2(t)\right\} \qquad 1.13.6$$

Combining equations 1.13.5. and 1.13.6.,

$$L\left\{y_1(t) + y_2(t)\right\} = x_1(t) + x_2(t) \qquad 1.13.7.$$

which states that $y_1(t) + y_2(t)$, the output when the two inputs are applied simultaneously is just the sum of the outputs obtained when each input is applied singly. But this is the law of superposition, applied to two inputs. Generalization to any number of inputs merely requires iteration of these steps, for the combined input $[x_1(t) + x_2(t)]$ may now itself be treated as a single input which can be combined with any other.

Proof of the law of superposition for equation 1.13.1. follows the same lines; indeed it is practically self-evident from the definition of a linear operator (Equation 1.11.1.).

SECTION 1.14. Analog Computers in the Analysis of Biological Systems

I shall conclude this brief discussion of real systems and their physical and mathematical models by a few remarks about the role of yet a third kind of model--the analog computer--in the analysis of biological systems.

Systems analysis pays far more attention to how a system behaves then to what it is. As a consequence, the mathematical techniques of systems analysis can be applied to an astonishing variety of different systems whose actual components may be as unlike as the mass of an automobile, the capacitor of an electric circuit and the volume of distribution of deuterium oxide in the body. In the next chapter, we shall see how easy it is to describe the behavior of such disparate objects (or, more precisely, their physical models) by a single mathematical formulation, which, except for particularization of symbols, is the same for all. The resulting mathematical model may then be solved so as to find the unknown output of the system produced by any specified input. If the mathematical solution can be obtained analytically, i.e. without recourse to tedious numerical methods such as are described in Chapter 6, so much the better, for an analytical solution is both exact and general. But the equations describing complicated, obstinately nonlinear biological systems cannot be dealt with analytically, and for solving such equations we shall often want to turn to an electrical analog computer. In an analog computer, the physical operations of the electrical components can be made to simulate the mathematical operations dictated by the equations so closely that the voltage inputs and outputs of the computer are, to a very good approximation, directly proportional to the mathematical inputs and outputs of the equations. In this sense, the electrical model "solves" the equations, and we should be duly grateful.

But there's a bit of the Pygmalion in all of us. We draw our computer diagram, we patch up an impressive tangle of leads, we correct our inevitable mistakes, and we finally see a wonderfully smooth curve traced out by the X-Y plotter. How easily we can collect "data" from this array of capacitors and transistors! How easily we can slip in another integrator or add yet another input so that our beautiful curves will better match those scattered and error-prone biological observations which we originally set out to explain! In fact, why bother any more with a physical model of the real system? Why not let this elegant electronic creature be our model? Why not study the computer instead of the biological system?

If the reader thinks I exaggerate the risk of succumbing to the charms of an analog computer program, let him ponder the following quotations drawn from a single field but originating from two different laboratories on opposite sides of the Atlantic: In the first, the authors have just introduced into their mathematical model an equation describing the rate of formation and destruction of a substance "postulated to control glucose production in the liver":

> "Whether this regulatory substance is a biological fact or a mathematical fiction is irrelevant to an appraisal of the over-all success of this mathematical model: it suffices to produce the type of oscillations so frequently observed in glucose infusion studies." (8)

The second quotation is even more revealing:

> "The [computer] model presented here is the latest one in a series of prototypes. Each of many steps in the modification of the model has increased its capacity to simulate the glucose-insulin curves. This necessarily implies that the modifications have made the model more "physiological," a fact that makes it possible in the future to investigate each specific step of the glucose-insulin interrelationship with the aid of deductions drawn from the model." (9)

O my beloved Galatea! How beautiful you are as the currents surge through your veins!

Let me be very specific about where the danger lies. Data obtained from biological experiments are usually quite limited, both in number and accuracy.

Such data can always be fitted "satisfactorily" by some small set of equations containing a few adjustable parameters. It is therefore always possible to interconnect the elements in an analog computer so that the output of the computer "satisfactorily" matches the biological observations. Unfortunately, however, there are many different sets of equations, and correspondingly many different computer programs, each of which will describe the data equally well. (See Section 6.14. in Reference 5) Among these many "successful" models, there is little to choose mathematically. It is only the biological plausibility of a particular model which justifies preferring it to all others.

In Figure 1.14.1. I have tried to illustrate on the left the proper role of a mathematical model and its analog computer representation in the analysis of a biological system. On the right I have diagrammed the sequence of steps used by investigators suffering from the Galatea Delusion. I hope the distinction between the two sides is clear.

SECTION 1.15. Remarks on Symbology

In a previous book (5) I discussed a few general principles which I felt should guide the choice and use of symbols. In the present book I have tried to follow these principles, but in one respect I am hoist by my own petard: "It would be folly," I solemnly declared, "to call. . . the base of natural logarithms anything but e." Now, having been brainwashed by electrical engineers, I still call it "e," but write it ϵ. I shall also follow their convention in using j, not i, to represent $\sqrt{-1}$.

In mathematical expressions, square brackets will be used to indicate multiplication (for example, $a[b + c] = ab + ac$) or, occasionally, to bind several letters together into a single symbol (for example, $[\dot{V}cl]$ = clearance). Parentheses will be reserved for the arguments of functions (for example, $e(t)$ = the voltage, e, as a function of time, t, or the value of e at time t; $\cos(\varphi + \omega t)$; etc.).

In general expressions, the indefinite integral will often be used. For example, $e_c(t) = \frac{1}{C}\int i_c(t)dt$. This integral equation really tells us no more than the differential equation, $de_c(t)/dt = [1/C]i_c(t)$, from which it was derived, namely that the rate of change of voltage, $e_c(t)$, across a capacitor is directly proportional to $i_c(t)$, the current flowing through the capacitor, and inversely proportional to the capacitance, C. But when we wish to find an explicit expression for $e_c(t)$ as a function of time by actually performing the integration, we shall want to let the independent variable, t, be the upper limit of integration. Then we shall use a "dummy" variable, for example τ, to indicate that the value of the definite integral is independent of the variable of integration, and that t affects the result only as a variable upper limit of integration. For example, if $i_c(t)$ is a constant current of A amperes, starting to flow into an uncharged capacitor at time zero:

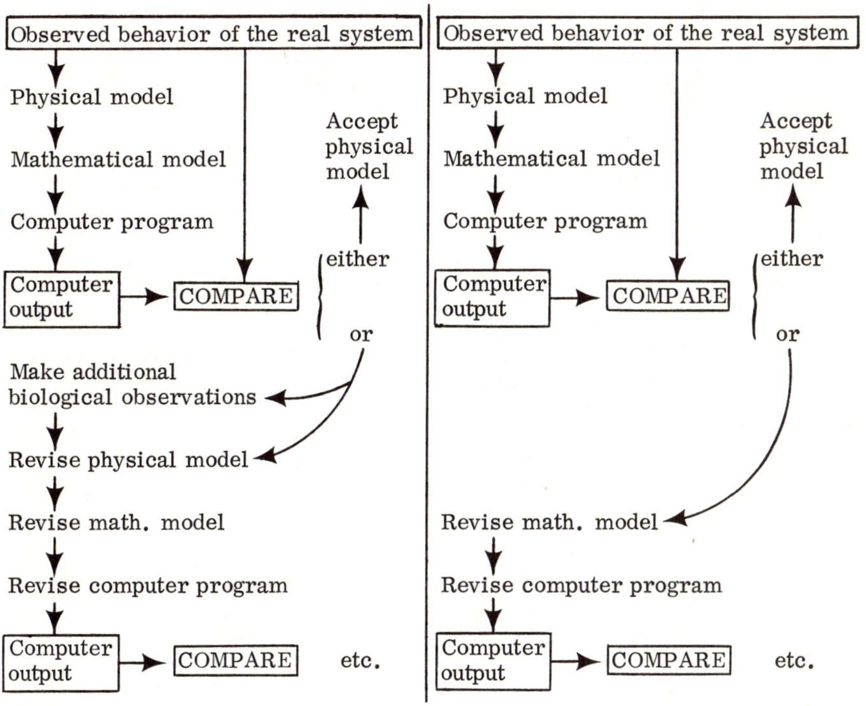

Figure 1.14.1. Correct and incorrect ways of using an analog computer in analyzing a biological system

LEFT: An analog computer may properly be used to solve a set of equations, (the mathematical model,) describing how a simple physical model of the real system would work. If the output of the computer and the observed behavior of the real biological system agree, the physical model can be tentatively accepted. If they disagree, the physical model must be revised, often with the help of data from new experiments which have been suggested by failure of the previous model.

RIGHT: Investigators who suffer from the Galetea Delusion become so infatuated with the mathematical model and its electrical representation which they have created that they forget all about the real system and its physical model. They repeatedly revise their mathematical model and the computer program until the output of the computer agrees with the available biological observations. Then they claim that the biological system must actually contain components which correspond to various parts of the mathematical model, because "Now the computer behaves just like the real biological system!"

$$e_c(t) = \frac{1}{C}\int_0^t i(\tau)d\tau = \frac{1}{C}\int_0^t A dt = \frac{A}{C}\tau \Big|_0^t = \frac{A}{C}t\,;\ t \geq 0 \qquad 1.15.1.$$

(Note, in passing, that even when it remains constant throughout the period of interest, i(t) is still a function of time, because to every value of the independent variable, t, there corresponds one and only one value of the dependent variable, i(t). The fact that all of these dependent values happen to be A for $t > 0$, so that i(t) does not vary with time, in no way prevents us from regarding i(t) as a function of time.)

There are many ways of symbolizing derivatives. They are used interchangeably, depending upon convenience. Consider, for example, the third derivative of y(x) with respect to x:

$$\frac{d^3y(x)}{dx^3} \equiv D^3y(x) \equiv y'''(x) \equiv y^{(3)}(x) \qquad 1.15.2.$$

Of these variants, I much prefer the first. It horrifies me that a whole generation of students is being brought up on the colorless "y-prime" notation. y' utterly fails to convey the fundamental idea that the first derivative of y(x) with respect to x is the rate at which y(x) changes as x changes, evaluated at some particular value of x. y' may be ever-so-pure mathematically, but physically it is a notational fiasco which does not even display its proper dimensions! Less objectionable is the "y-overdot" notation which, by convention, is restricted to derivatives with respect to time. For example,

$$\frac{d^2y(t)}{dt^2} \equiv \ddot{y}(t) \qquad 1.15.3.$$

I shall use the overdot notation chiefly, but not exclusively, in symbols for physical entities such as flow (= $\dot{V}(t)$) when they are to be treated as variables in their own right rather than as first derivatives of some other variable with respect to time. But the student has doubtless already suffered from--and survived--enough symbolic redundancies, contradictions and arbitrary conventions to have developed considerable tolerance for such notational peculiarities.

When the various symbols of equation 1.15.2. are modified so as to refer to the third derivative of y with respect to x evaluated at some specific value of x, say x = 5, we have:

$$\frac{d^3y(x)}{dx^3}\Big|_5 \equiv D^3y(x)\Big|_5 \equiv \dddot{y}(5) \equiv y^{(3)}(5) \qquad 1.15.4.$$

The two somewhat clumsy notations on the left are needed because $\frac{d^3[\ ]}{dx^3}$ and $D^3[\ ]$ tell us to take the third derivative of their operand. If the operand is y(5), meaning "the value of y(x) when x = 5," then $\frac{d^3y(5)}{dx^3} = D^3y(5) = 0$ because y(5) is a constant. In contrast, $y'''(5)$ and $y^{(3)}(5)$ are defined to mean "first differentiate y(x) three times, and then evaluate the result at x = 5." I must concede the economy of that notation.

SECTION 1.16. A Preview of Subsequent Chapters

The remainder of Part I is devoted to a step-by-step introduction to some of the principal topics and techniques of systems analysis. In Chapter 2, five simple "building-block" elements are described. The similarities in the behavior of analogous simple elements from different systems is discussed, and we see that these similarities make it possible to develop much of linear systems analysis from just three equations in two variables. In Chapter 3, Laplace transforms, block diagrams and transfer functions, and phasor analysis in the frequency domain are introduced as related ways of dealing expeditiously with more complex linear systems. This paves the way for a discussion of linear feedback systems and stability criteria in Chapter 4. Chapter 5 reviews and consolidates the previous material by introducing matrix techniques for solving sets of first-order differential equations and for dealing with multiple inputs and outputs. In Chapter 6, various ways of studying non-linear systems are introduced. Numerical methods of solving non-linear differential equations, and linearization about equilibrium points are stressed. In Chapter 7, the advantages and disadvantages of confining attention to steady states are discussed with particular emphasis on systems with multiple feedback loops. This serves as an introduction to the specific physiological feedback systems described in Part II. Finally, certain manipulative techniques, designed to make the book reasonably self-contained, are described in the Appendix.

I have striven throughout for clarity, sometimes at the expense of brevity. But the student will get precious little out of any book of this kind, however clear it may purport to be, unless he carries out the various procedures himself. This is why exercises are given at the end of each chapter. In addition, whenever he encounters some portion of the text which he does not understand, the student should try to work out by himself a concrete numerical example which illustrates the point at issue. This may be difficult; it may be dull; it may be time-consuming. But it is marvelously instructive. A single such example, mastered in detail, is worth a hundred re-readings of the text material. You will get most from such a book as this when you have shoved it aside so that you can follow up your own train of thought with pencil and paper.

Exercises

1. In each of the following examples, (i) decide what kind of quantity each symbol represents (Cf. Section 1.3.) and (ii) construct a symbol-and-arrow diagram showing the working relationships among the parameters and variables. (Most of the equations have deliberately been cast in a form which is completely ambiguous as far as working relationships and independent variables are concerned.)

A. The ideal gas law:

$$RCT/P = 1$$

$$n/CV = 1$$

n = number of moles of the gas, C = molar concentration of the gas, V = volume of the gas, T = temperature of the gas, degrees Kelvin, P = pressure exerted by the gas, R = ideal gas law constant.

B. The law of gravitation:

$$Fx^2/m_1m_2k = 1$$

F = force of attraction between two bodies whose masses are m_1 and m_2, separated by a distance, x. k = gravitational constant.

C. Newton's second law:

$$F/ma = 1$$

F = force, m = mass, a = acceleration

D. Stoke's law arranged for calculating the radius of small spheres falling at constant velocity through a fluid medium:

$$r = \left\{\frac{9\eta\,v}{2g(\rho_s - \rho_m}\right\}^{1/2}$$

r = radius of sphere of density ρ_s falling at velocity v through a medium of density ρ_m and viscosity η, under the influence of gravity. (g = acceleration due to gravity).

E. The behavior of a simple pendulum for small angular displacements:

$$4\pi^2x/t_p^2\,g = 1$$

$$x_o[1 + \alpha T]/x = 1$$

t_p = time of a complete oscillation (= period), x = length of the pendulum at temperature T, x_0 = length of the pendulum at standard temperature, α = coefficient of thermal expansion, g = acceleration due to gravity.

F. Dissociation of a buffer in aqueous solution:

$$K_A[\text{Acid}]/[H^+][\text{Base}] = 1$$

$[H^+]$, [Acid], [Base] = molar concentration of hydrogen ion, Brønsted acid (= proton donor), and conjugate base (= proton acceptor), K_A = acid dissociation constant.

G. Poiseuille's law for streamline flow through small tubes:

$$\pi[\Delta P]r^4/8x\eta\dot{V} = 1$$

ΔP = difference in pressure between two points along the tube separated by a distance x, r = radius of the tube, η = viscosity of flowing fluid, $\dot{V}$ = flow.

a.) consider the input to be a constant ΔP

b.) consider the input to be a constant $\dot{V}$

H. Metabolic clearance of thyroid hormone:

$$\dot{Q}/[\dot{V}cl]C = 1$$

$$kC/[\dot{V}cl] = 1$$

$\dot{Q}$ = rate of secretion of thyroid hormone, $[\dot{V}cl]$ = clearance of thyroid hormone, C = concentration of thyroid hormone in plasma, k = a parameter.

I. Distension of a thin-walled elastic tube of fixed length:

$$r\left\{\frac{1}{r_0} - \frac{P}{Yw}\right\} = 1$$

$$r_0 w_0/rw = 1$$

r, w = radius and wall thickness of distended tube, r_0, w_0 = radius and wall thickness of undistended tube (i.e. for P = 0), P = transmural distending pressure, Y = Young's modulus of elasticity.

2. Given that the sum $\{L_1 + L_2\}$ of two <u>linear</u> operators is defined by the equation:

$$\{L_1 + L_2\}\ y(t) = L_1 y(t) + L_2 y(t)$$

where y(t) is an operand in the domain of both L_1 and L_2, prove that the sum of any number of linear operators is itself a linear operator.

3. In each of the following equations, one function u(t) is transformed into another function v(t) by some operator. Is the operator linear? (a, b, c, are constants.)

i) $v(t) = au(t)$

ii) $v(t) = a + bu(t)$

iii) $v(t) = au^2(t)$

iv) $v(t) = \sqrt{u(t)}$

v) $v(t) = \sin u(t)$

vi) $v(t) = au(t) + b\frac{du(t)}{dt} + c\int u(t)dt$

vii) $u(t)\frac{du(t)}{dt} = a + [v(t) - b]\,u(t)$

viii) $v(t)u(t) / [a - v(t)t] = du(t)/dt$

ix) $v(t) = \ddot{u}(t) + g(t)\dot{u}(t) + h(t)u(t)$

x) $v(t) = t^2 \frac{d^2u(t)}{dt^2} + t\frac{du(t)}{dt} + [1 - \frac{a^2}{t^2}]u(t)$

4. Is the recipe for making a cake a linear operator?

References

1. Editorial. Working models in medicine. J. Am. med. Ass., 174, 407-408, 1960.

2. Wolf, A.: Causality or causation. In The Encyclopedia Brittannica, Vol. 5, p. 61. Encyclopedia Brittannica, Inc., Chicago, 1955.

3. Riggs, D.S.: Feedback: fundamental relationship or frame of mind? Adv. Enzyme Reguln, 5, 357-382, 1967.

4. Lyman, C.P.: The oxygen consumption and temperature regulation of hibernating hamsters. J. exp. Zool., 109, 55-78, 1948.

5. Riggs, D.S.: The Mathematical Approach to Physiological Problems. Williams and Wilkins Co., Baltimore, 1963.

6. The International Dictionary of Applied Mathematics. p. 653. D. van Nostrand Co., Inc. Princeton, N.J., 1960.

7. Abramson, A.B.: Dimensional Analysis for Students of Medicine. Josiah Macy, Jr. Foundation, New York, 1950.

8. Seed, J.C., Acton, F.S., and Stunkard, A.J.: A model for the appraisal of glucose metabolism. Clin. Pharmac. Ther., 3, 191-215, 1962.

9. Cerasi, E.: An analogue computer model for the insulin response to glucose infusion. Acta endocr., 55, 163-183, 1967.

Order and simplification are the first steps toward the mastery of a subject.

Thomas Mann, THE MAGIC MOUNTAIN

2

THE RUDIMENTS OF LINEAR SYSTEMS ANALYSIS

SECTION 2.1. The general kinds of physical systems

The behavior of any physical system is determined by how some form of energy is generated, transmitted, stored, released, and transformed by the component parts or elements of the system. For each different form of energy--mechanical (translational or rotational), electrical, hydraulic, thermal or concentrational--there is a different set of elements, and a correspondingly different set of variables is used to describe the elements and their behavior. If each of these different sets of elements and variables also required a unique mathematical approach, the task of the systems engineer would indeed be formidable. Fortunately, however, if the variables used to describe different systems are properly chosen, the seemingly disparate elements exhibit features so similar that (with a few exceptions) a single general description suffices for all. The analogies between the elements and variables of the different types of systems have been summarized in the table on the endpapers of this book. The table contains definitions of the symbols used, and should be consulted freely during the following discussion. Table 2.1.1 is a condensed version of the end-paper table.

SECTION 2.2. Primary variables

Only three primary variables are needed for the general description of the lumped-parameter linear systems to which the discussion in this chapter will be confined; time, t; an "across" variable, v(t); and a "through" variable, f(t).

Time, viewed as the continuum within which events occur and changes take place, is always a completely independent variable. (See Section 1.3.)

The across variable (= A-variable) is so called because it is measured across an interval between two points within the system itself, or between some point in the system and a reference point outside the system. For example, an electrical potential difference is measured by connecting a voltmeter across two terminals. The velocity of a train (e.g., "70 miles per hour") is measured across the interval between the train and some point of reference on the surface of the earth, but the velocity of a passenger (e.g., "2 miles per hour") who is walking towards the dining car is measured with respect to some point on the train. Even when we do not say so explicitly, the across variable commonly

Table 2.1.1. Analogous variables and passive elements in different kinds of systems

GENERAL	TRANSLATIONAL	ROTATIONAL	ELECTRICAL	HYDRODYNAMIC	THERMAL	CONCENTRATIONAL
ACROSS variables velocities, voltages, etc. $v(t)$	velocity, $v(t)$	angular velocity, $\Omega(t)$	voltage, $e(t)$	pressure, $P(t)$	temperature, $T(t)$	concentration, $C(t)$
THROUGH variables forces, flows, etc. $f(t)$	force, $f(t)$	torque, $\tau(t)$	current, $i(t)$	flow, $\dot{V}(t)$	heat flow, $\dot{q}(t)$	solute flow, $\dot{Q}(t)$
D-ELEMENTS $v(t) = \frac{1}{k_D} f(t)$	k_D = damping constant, D	k_D = damping constant, D	$\frac{1}{k_D}$ = resistance, R	$\frac{1}{k_D}$ = fluid resistance R	$\frac{1}{k_D}$ = thermal resistance, R	k_D = solute conductance, 1/R or clearance, [$\dot{V}$cl]
A-ELEMENTS $v(t) = \frac{1}{k_A} \int f(t) \cdot dt$	k_A = mass, m	k_A = inertia J	k_A = capacitance, C	k_A = fluid capacitance, C	k_A = thermal capacitance, C	k_A = volume of distribution [Vdist]
T-ELEMENTS $v(t) = k_T \frac{df(t)}{dt}$	$\frac{1}{k_T}$ = spring stiffness, k	$\frac{1}{k_T}$ = torsional stiffness, k	k_T = inductance, L	k_T = fluid inertance, I		

represents the difference between two velocities, or voltages, or pressures, etc. If there is any chance of misunderstanding, the two points between which the variable is measured should be specified. (See Section 2.4.)

The symbol v(t) for the generalized across variable has been chosen to suggest velocities and voltages, though for this mnemonic device to be fully successful the word "voltage" must be taken to represent not only the electrical potential differences which cause the flow of electric current, but also the analogous differences in pressure (causing fluid flow), temperature (causing heat flow) and concentration (causing solute flow).

The through variable (T-variable) is so called because it can be measured only by passing all of it through the measuring instrument. For example, an electric current is measured by passing it through an ammeter. To measure your weight, the whole downward force exerted by the earth's gravitational field upon your body must act through the bathroom scales. The entire flow of blood in an artery must pass through an electromagnetic flowmeter.

The symbol f(t) has been chosen for the generalized through variable to suggest forces and flows.

SECTION 2.3. Derived variables

Except in thermal and concentrational systems, the product of the A-variable and the T-variable is the power supplied to the system or to an element thereof. Three other important sets of derived variables are obtained, as indicated in the tabular summary on the endpapers, by integrating the across variable, or the through variable, or the power supplied, with respect to time. In particular, the integral of power with respect to time is the energy supplied to (= the work done upon) the system or an element thereof.

The reader will recognize that convenience alone justifies selecting v(t) and f(t) as "the" primary variables from which all of the other variables are to be derived. There is no fundamental reason why some different pair of variables, such as v(t) and energy, or f(t) and power, could not have been chosen as primary. The lack of any basic significance in the choice of v(t) and f(t) is further emphasized by considering the working relationships (Section 1.6.) between these "primary" variables in the different types of systems, In a mechanical system, a change in the force acting through a damping element causes a change in velocity ($f(t) \rightarrow v(t)$). But in an electrical system, it is the other way around; a change in the voltage across a resistor causes a change in current flow ($v(t) \rightarrow f(t)$). See also footnote in Section 2.5.

SECTION 2.4. Subscript notation for variables

When only a single element is being considered, it is obvious enough that the across variable is being measured across the element, the through variable, through (or at least into) the element under discussion. But when several elements have been interconnected to form a more complicated system, we shall often need to specify precisely where the variables are being measured. This is done by using subscripts in one of two ways: most commonly a particular element is designated (e.g., $i_{R2}(t)$ = the current flowing through resistor number 2); alternatively, the two points between which the variable is measured, or between which it is flowing, may be designated, the origin first, the destination second. (E.g., $P_{1,2}$ = the pressure drop from point 1 to point 2; $q_{A,B}$ = the heat flow from A to B.)

SECTION 2.5. Passive elements

A D-element (= Dissipative element) is an element which merely dissipates energy in the form of heat. In a dissipative element, the through variable, $f_D(t)$, is directly proportional to the across variable, $v_D(t)$:

$$f_D(t) = k_D v_D(t) \qquad 2.5.1.$$

where the constant of proportionality, k_D, depends upon the characteristics of the particular element under consideration. For example, in an electrical system, equation 2.5.1. becomes Ohm's law:

$$i(t) = [1/R]e(t) \qquad 2.5.2.$$

where i(t) is the current (amperes) flowing through the resistor, e(t) is the voltage difference (volts) across the resistor, and [1/R], the constant of proportionality, is the conductance (mhos) of the resistor. (R, the resistance (ohms) is the reciprocal of the conductance.) In a translational system,

$$f(t) = Dv(t) \qquad 2.5.3.$$

where f(t) is the force (dynes) acting through a damping element, v(t) is the velocity (cm/sec) caused by f(t), and D is the damping constant (dyne-sec/cm) which is a measure of the viscous frictional resistance of the element.*

[*In equation 2.5.3., the damping constant, D, measures the viscous resistance to motion. Yet the analogous constant of proportionality in Ohm's law (equation 2.5.2.) is the reciprocal of the resistance to current flow. This paradox arises because equation 2.5.2., but not equation 2.5.3., is written with the dependent variable on the left. Unlike equation 2.5.3., equation 2.5.2. therefore expresses a true working relationship as well as a true mathematical relationship. (See Section 1.4.). If we solve equation 2.5.3. for its dependent variable, v(t);

$$v(t) = [1/D]f(t) \qquad 2.5.3.a$$

it becomes obvious that the larger the damping constant, the smaller is the velocity produced by a given force. Unfortunately, it is impossible to write the analogous equations for all types of systems in the form $f_D(t) = k_D v_D(t)$ and at the same time always have the dependent variable on the left, because for the translational and rotational systems, f(t) is not the dependent variable. It seems best to retain the same form of equation for all systems and to accept 1/D as mathematically analogous to R, even though both R and D measure resistance in their particular systems and thus measure the ability of those systems to dissipate energy in the form of heat.]

An expression for $v_D(t)$ in terms of $f_D(t)$ may be obtained by solving equation 2.5.1. for $v_D(t)$:

$$v_D(t) = [1/k_D]f_D(t) \qquad 2.5.1.a$$

An A-element is an element which stores energy by virtue of the A-variable. When the energy stored has reached a maximum, the across variable is maximum and the through variable is zero. In an A-element, the through variable is directly proportional to the rate at which the across variable changes with time:

$$f_A(t) = k_A[dv_A(t)/dt] \qquad 2.5.4.$$

where the constant of proportionality, k_A, depends upon the characteristics of the particular A-element under consideration. For example, in a translational system k_A is the mass, m, of some object, and equation 2.5.4. becomes:

$$f(t) = m[dv(t)/dt] \qquad 2.5.5.$$

or, since dv(t)/dt is the acceleration, a(t), of the mass caused by the force,:

$$f(t) = ma(t) \qquad 2.5.6.$$

which is Newton's second law of motion. In a concentrational system, k_A is the volume, [Vdist], throughout which the solute in question is distributed (volume of distribution), and equation 2.5.4. becomes:

$$\dot{Q}(t) = [Vdist]\,[dC(t)/dt] \qquad 2.5.7.$$

where $\dot{Q}(t)$ is the rate of addition of solute to [Vdist], and dC(t)/dt is the rate of change of the solute concentration.

An expression for $v_A(t)$ in terms of $f_A(t)$ may be obtained by separating the variables in equation 2.5.4. and solving for $v_A(t)$:

$$v_A(t) = (1/k_A)\int f_A(t)dt \qquad 2.5.8.$$

A T-element is an element which stores energy by virtue of the through variable. When the energy stored has reached a maximum, the T-variable is maximum and the A-variable is zero. In a T-element, the A-variable is directly proportional to the rate at which the T-variable changes with time:

$$v_T(t) = k_T[df_T(t)/dt] \qquad 2.5.9.$$

where the constant of proportionality, k_T, is determined by the characteristics of the particular T-element under consideration. For example, in an electrical system, k_T is the inductance, L, of an inductor, and equation 2.5.9. becomes:

$$e(t) = L[di(t)/dt] \qquad 2.5.10.$$

In a rotational system k_T is (1/k), where k is the torsional spring constant (= torsional stiffness), and equation 2.5.9. becomes:

$$\Omega(t) = [1/k]\ [d\tau(t)/dt] \qquad 2.5.11.$$

An expression for $f_T(t)$ in terms of $v_T(t)$ may be obtained by separating the variables in equation 2.5.9. and solving for $f_T(t)$:

$$f_T(t) = [1/k_T] \int v_T(t)dt \qquad 2.5.12.$$

There are no T-elements in thermal and concentrational systems. Indeed, the endpaper table makes it clear that the mechanical, electrical, and hydrodynamic systems form one subgroup, the thermal and concentrational systems form a second subgroup. The similarities among the members of the same subgroup are more striking than are the similarities between the two groups.

The two types of energy-storing elements (A-elements and T-elements) are passive in the sense that they are incapable of generating energy, or of releasing energy continuously for long periods of time. But the limited amount of energy which they can store while it is being supplied by some active source (Section 2.6.) is stored reversibly, and may, under suitable conditions, be released to other elements in the system. Storage elements can thus behave transiently as if they were source elements. This is a most important characteristic, and accounts for the ability of a passive system which contains stored energy to exhibit a natural (= unforced) response as the energy is gradually released. Natural responses are described by homogeneous equations, and the quantity and location of the stored energy is indicated, directly or indirectly, by the initial conditions of the system.

SECTION 2.6. Source elements

A system composed entirely of passive elements can be kept in an active state for indefinitely long periods of time only if some source element feeds energy, continuously or intermittently, into the system so as to replenish the energy being dissipated as heat by the D-elements which are invariably present in a passive system.* A system with one or more source elements is thus an active system, and it eventually exhibits only a pure forced response (sometimes called a steady-state response**) to the energy supplied by the source. Forced responses are described by nonhomogeneous equations. The nonhomogeneous part of the equation is the forcing function (= input) which specifies the A-variable or T-variable supplied by the source as a function of time.

[*It is possible to imagine a passive system, composed entirely of storage elements, which would dissipate no energy as heat but would conserve without loss all of the energy initially present. By exchanging this energy between its various elements, such a conservative system could remain active indefinitely. However, with the exception of superconducting systems at temperatures close to absolute zero, perfectly conservative systems do not exist in real life, and we shall always have one or more D-elements in the systems which we study.]

[**An unfortunate synonym, since the "steady"-state response forced upon a system by a time-varying source will also vary with time. To an engineer, a box of dice being shaken violently is, after the first second or two, in a steady state! I find this confusing. In this book, I shall use the term "steady state" only when the variables of the system are no longer changing with time but are, in fact, genuinely steady.]

There are two kinds of source elements: A-sources and T-sources. An A-source supplies energy in such a way that the magnitude of the A-variable delivered by the source, be it constant or time-varying, is determined entirely by the source itself and is independent of the system which it supplies. In contrast, the magnitude of the T-variable is not controlled by the A-source, but depends upon the characteristics of the system being supplied. A familiar electrical example of an A-source (voltage source) is a dry cell. Neglecting internal resistance (See Section 2.8.), the voltage across the terminals of a dry cell is kept at a constant value which is determined only by the nature of the chemical reaction within the cell, and which is therefore independent of an external resistance placed across the terminals. But the current is inversely proportional to the resistance.

A T-source supplies energy in such a way that the magnitude of the T-variable delivered by the source, be it constant or time-varying, is determined entirely by the source itself and is independent of the system which it supplies. In contrast, the magnitude of the A-variable is not controlled by the T-source, but is determined by the characteristics of the system being supplied. For example, in a hydrodynamic system, a pump forcing water from a well into a storage tank at a constant rate represents a flow source, but the pressure difference across the pump between tank and well at any instant of time will depend upon the size of the tank, the size of the air cushion above the water in the tank, and the rate at which water is being withdrawn from the tank. As a more physiological example, the evaporation of 100 ml of sweat per hour will supply the body with minus 58 Kcal of heat per hour, but the temperature difference between body and air caused by this negative heat source will depend upon a large number of other factors.

The reader may find it useful to supplement the discussion in the preceding sections by considering additional examples of various elements which may be suggested by the endpaper table. In particular, he should note in that table the conventional diagrams which are used to represent the elements.

SECTION 2.7. Transformers and transducers

In addition to the three passive and two active elements already described, there are two other important kinds of passive elements which should be mentioned, even though we shall not make much explicit use of them.

A transformer is a device which changes the ratio of the A-variable to the T-variable without altering either their product (= power in most types of systems) or the form of energy being dealt with. For example, a step-down electrical transformer accepts a small input current of high voltage, and puts out a large current at low voltage. A simple mechanical lever such as a crowbar placed against a fulcrum and used to elevate a heavy stone trades off a comparatively small torque applied with comparatively large angular velocity at the long end for a smaller angular velocity but a proportionately larger torque delivered at the small end.

A transducer is a device which changes one form of energy into another. Transducers are commonly encountered in feedback control systems, whether physiological or man-made, where several different forms of energy may be involved in a single feedback loop. For example, if a room whose temperature is regulated by a wall thermostat becomes too warm, the thermal energy of a coiled bimetallic strip is transduced into rotational energy which rotates a capsule in which a globule of mercury is translated by the force of gravity away from the contacts of an electrical relay which cuts off the supply of current to the mechanical pump which has been producing hydrodynamic flow of oil to the furnace for conversion of its chemical bond energy into thermal energy used to heat the room, and, with it, the thermostat. From this single example, it is obvious that transducers are too various and complex to analyze in full detail. We shall usually have to content ourselves with an empirical description of the (we hope linear!) relationship between the input to, and the output from a transducer.

SECTION 2.8. Real elements, ideal elements

In the preceding sections, the elements which have been described are not the real off-the-shelf resistors, capacitors, dry cells and the like which one can purchase at any electrical supply shop. They are instead ideal or model elements. An ideal element is a convenient abstraction whose behavior is completely specified by a single simple equation. Real elements are not that simple. An ideal fluid capacitor presented with a constant inflow of fluid will continue to accumulate a greater and greater volume at higher and higher pressure without limit. A real fluid reservoir would eventually overflow or burst. An ideal mechanical damper is a massless device which does nothing but turn kinetic energy into heat. The real shock absorbers on my car have not only mass but some degree of springiness. An ideal electrical inductor does nothing but store energy reversibly in a magnetic field. A real electrical inductor dissipates part of the energy as heat, thus exhibiting resistive, as well as inductive characteristics. But the major point to be emphasized is not that the real elements fail to behave precisely as expected from their simple-minded ideal counterparts; it is rather that the analysis of a system composed of imaginary ideal elements usually succeeds in predicting the behavior of the corresponding real system. The ideal elements are abstract models to be sure, but, used intelligently, they are highly successful models.

Sometimes, however, the simple description of the active source elements given in Section 2.6. is grossly inadequate. Consider first an A-source, specifically a dry cell. Suppose that it really did behave as an ideal source of constant voltage. If we connect its terminals by a short circuit of practically zero resistance, Ohm's law tells us that the current through the short circuit and the output of power from the dry cell would be fantastically large. Of course this does not happen, because the internal resistance of the battery itself opposes the flow of current. A practical A-source must therefore be represented by an ideal A-source plus a D-element in series with it. (See Figure 2.8.1.) This representation should be used whenever the external (= load) value of k_D (resistance, or inverse of damping) is less than, or of the same order of magnitude as the internal k_D of the source itself.

Consider next a T-source, specifically a photoelectric cell under constant illumination. Suppose that it really did behave as an ideal source of constant current. Suppose we leave an "open circuit" (i.e., an air gap of very high resistance) between its terminals. Ohm's law tells us that to maintain a constant current through a very high resistance, the voltage difference across the resistance must be very high, requiring a fantastic output of power by the photoelectric cell. Of course this does not happen because there are always internal resistive pathways for current flow within the source itself. A practical T-source must therefore be represented by an ideal T-source plus a D-element in parallel with it. (See Figure 2.8.1.) This representation should be used whenever the external k_D (resistance, or inverse of damping) is of the same order of magnitude as, or greater than the internal k_D.

SECTION 2.9. Circuit diagrams

Having characterized the several linear elements, we are now ready to analyze the behavior of simple networks* formed by interconnecting two or more elementary components. This will lay the foundation for the analysis of the considerably more complicated systems to be discussed in the next chapter.

[*Properly speaking, the term "circuit" means a single closed path. But a network may contain any number of circuits interconnected in any way.]

Everyone knows that an electric current can flow only through a closed path (= circuit). Because most of us in this electronic age are already somewhat familiar with electrical circuits, and because electrical elements tend to behave a bit more ideally than do the elements of other systems, I shall illustrate most of the following discussion with electrical examples. But the ideas developed are, with suitable modifications, equally applicable to other systems. In particular, circuit diagrams, showing the routes followed from element to element by the through variable, are useful in non-electrical as well as in electrical systems. However, as we shall see, the rule that the system can operate only when at least one closed path is available requires reinterpretation when applied to other systems. (See Section 2.11. and especially Figure 2.11.2).

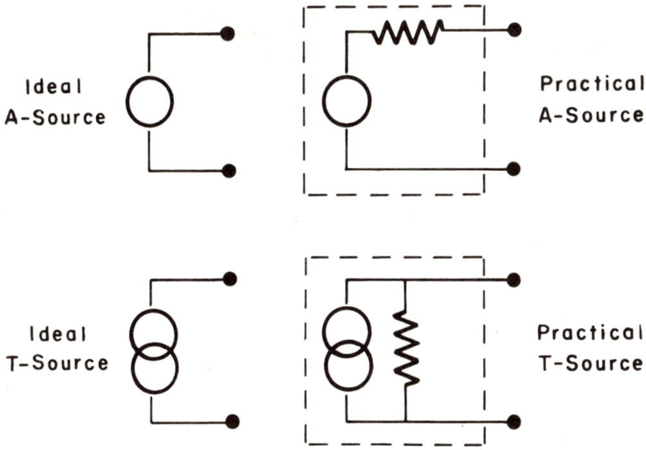

Figure 2.8.1. Ideal versus practical sources.

In an ideal source element, the internal resistance in series with the source is zero, the internal resistance in parallel with the source is infinite. But real (= practical) sources have finite internal resistance, as shown within the broken lines. This internal resistance must be taken into account whenever the external (= load) resistance for an A-source is not much larger than, or for a T-source not much smaller than the internal resistance.

Consider the generalized circuit diagram in Figure 2.9.1.a. consisting of an A-source and five arbitrary passive elements, 1,2,3,4, and 5. The lines connecting the several elements are leads. Leads, however long or crooked, represent inert pathways for the T-variable, neither storing nor dissipating energy. The solid dots in Figure 2.9.1.a. represent junctions or terminals, you could think of them as blobs of solder. Like the leads, they neither store nor dissipate energy. As they are placed in Figure 2.9.1.a., they have no particular significance because there is no reason to distinguish C from F or D from G. Solid dots should not be placed on a circuit diagram arbitrarily as they have been in Figure 2.9.1.a., but should be used only to represent genuine nodes. At a genuine node, the value of the A-variable is (or at least may be) unique. C and F cannot represent two different nodes because no element lies between them. They are connected by a mere lead, and, far from being unique at each, the value of the A-variable must always be exactly the same at C as at F. Nodes are not always explicitly marked on circuit diagrams, but when they are shown they should be so placed that every node is directly connected only with elements; every element is directly connected only with nodes. Accordingly, the circuit diagram of Figure 2.9.1.a. should be revised as in Figure 2.9.1.b. or Figure 2.9.1.c. or in any other equivalent way which clearly shows the five genuine nodes of the network.

The two nodes between which an element lies are the terminals of that element.

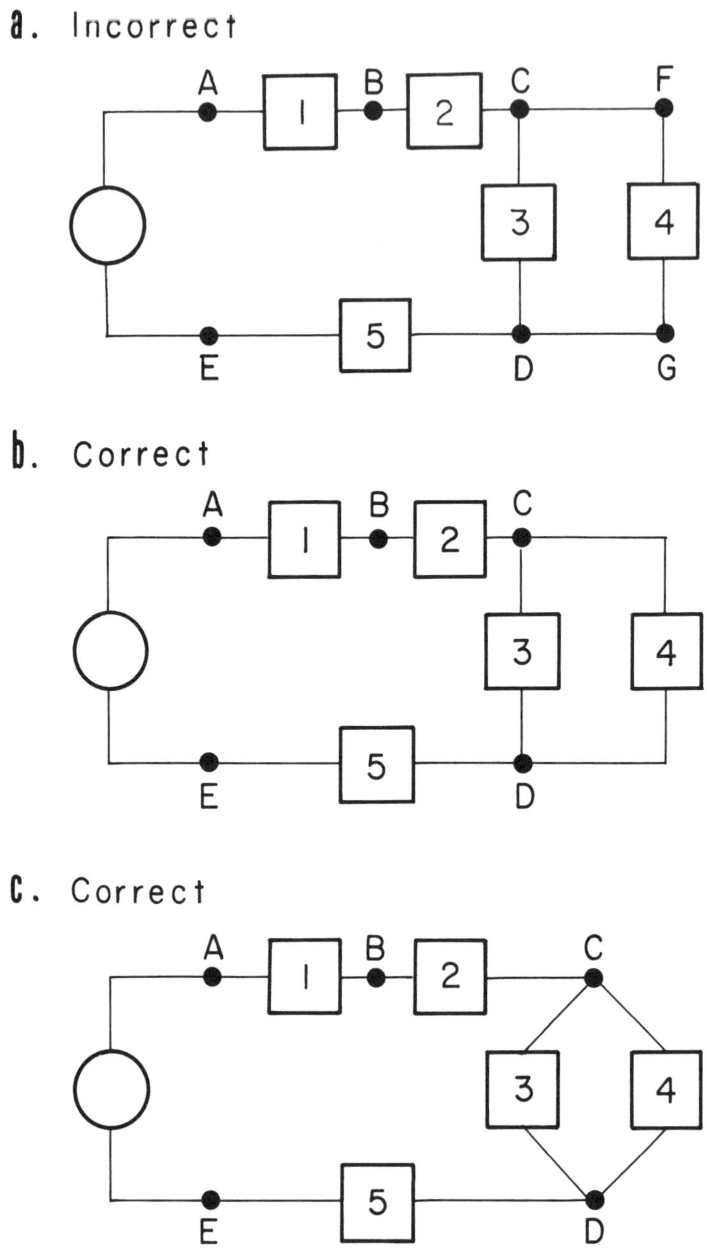

Figure 2.9.1. Relation of nodes to elements in a circuit diagram.

SECTION 2.10. Sign conventions

a) Polarity of the charge flowing in electrical circuits. Benjamin Franklin guessed wrong, and that's that. We are therefore obliged to pretend that positive charges are what flow through an electric circuit. Accordingly, in response to a voltage gradient, current flows from the more positive to the more negative region. Fortunately, no similar decision about the polarity of force or flow is necessary in other systems.

b) Sign convention for the direction of the T-variable. Mark each different path in the circuit diagram with an arrow-head pointing in the actual (or assumed) direction of the T-variable through that path. If the actual direction of the T-variable is the same as the direction marked, the value of the T-variable is positive. If the actual direction of the T-variable is opposite to the direction marked, the value of the T-variable in the marked direction is negative.

c) Sign convention for the direction of change of the A-variable across an element. Let us think of ourselves as travelling around a circuit in some arbitrary direction which does not have to be the same as the direction of the T-variable. The A-variable difference between any two points on our path is obtained by subtracting the value of the A-variable at the second point we encounter from its value at the first point we encounter, so that a decrease in the A-variable in the direction we are travelling is defined as positive. For example, suppose that the voltage, e_1, at terminal 1 of an element (measured with respect to some zero reference voltage such as ground voltage) is +20 volts, while the voltage, e_2, at terminal 2 is -10 volts. If we traverse the element from 1 to 2, the voltage change is $e_1 - e_2 = (+20) - (-10) =$ plus 30 volts. But if we traverse the element from 2 to 1, the voltage change is $e_2 - e_1 = (-10) - (+20) =$ minus 30 volts. Notice that when our direction of passage through the element is the same as that of the T-variable, the difference in the A-variable is positive when the element is storing or dissipating energy, negative when the element is a source or is releasing stored energy. This agrees with the convention previously mentioned (Section 2.3.) which specifies that the power supplied to, and the work done upon an element or a system are positive.

We shall often place an arrow next to an element to indicate the direction of decrease of the A-variable.* Such an arrow may be labelled with the symbol for, or the value of the A-variable decrease which it represents. This value will be positive if we traverse the element in the direction of the voltage drop arrow.

[*Though the orientation of the A-variable arrow described here seems to be the most logical, it is by no means universal. Some prefer to point the arrow towards the lead with the higher value of the A-variable. Others prefer to point the arrow towards the lead through which the T-variable enters the element. But whatever rule for orienting the arrow is adopted, the corresponding rules for its interpretation must ensure that the sign of the A-variable difference across a given element in a given circuit at a given time will be as here described.]

The above conventions, applied specifically to electrical systems, may be summarized by saying that if we mark the direction of flow of positive charges through any element, a current in that direction is positive, and a voltage drop in that direction is positive.

The direction of the T-variable will usually be clear from the types of elements, and the arrangement of the elements in a network. But when the actual direction for one or more pathways is not clear, simply assign an arbitrary direction to each such pathway and proceed with the analysis. If the value of the T-variable obtained from the analysis turns out to be positive, the assumed direction was correct. But if the calculated value of the T-variable turns out to be negative, the direction arbitrarily assigned was wrong. (See exercise 2.2).

SECTION 2.11. Kirchhoff's laws. Loop equations. Node equations

The laws governing the A-variable and the T-variable in single elements have already been discussed. For the analysis of circuits and networks containing several elements, we need two additional laws which place fundamental constraints upon the values which these variables can assume when several elements are interconnected. The laws are almost self-evident:

> Kirchhoff's voltage (A-variable) law. (Figure 2.11.1.a.) At any instant of time, the sum of the voltage (or, more generally, A-variable) differences around any closed path in a network is zero.

We shall make continual use of Kirchhoff's A-variable law by writing a loop equation for the sum of the A-variable differences around each closed path of interest in a given network.

> Kirchhoff's current (T-variable) law. (Figure 2.11.1.b.) At any instant of time, the sum of the currents (or, more generally, T-variables) entering any node is equal to the sum of the currents (T-variables) leaving that node.

This law is a direct consequence of applying the principle of the conservation of mass or of energy at a point (i.e., a node) where neither mass nor energy can be created, stored, released, or dissipated. We shall make continual use of Kirchhoff's current law by writing a node equation, stating the equality between T-variables entering and T-variables leaving, for each node of interest in a given system.

The statement of Kirchhoff's voltage law given above implies that the law holds for A-variables in general; but we have yet to examine the question of what constitutes a closed path in nonelectrical systems, and the even more fundamental question of whether the existence of a closed path for the T-variable in other systems is a sine qua non as it is in electrical systems. At first, it seems tempting to stress the obvious similarities between blood, pumped by the heart, flowing in response to pressure differences through closed-circuit blood vessels back to the heart again, and electric current,

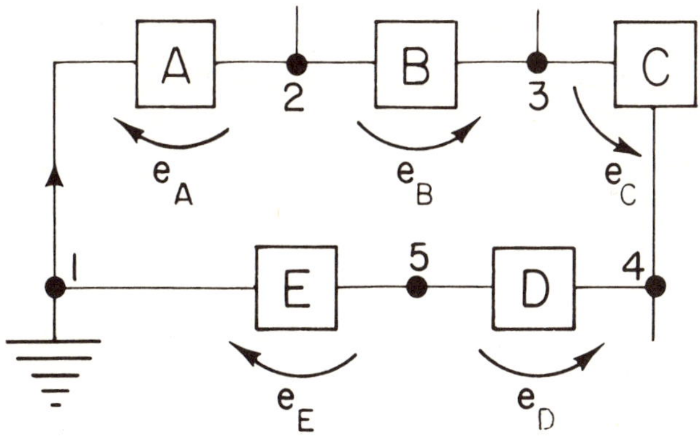

Figure 2.11.1.a. Kirchhoff's voltage law. The sum of the voltage drops across the elements of any closed circuit is zero. For example,

Node #	voltage	Element	voltage drop across element
1.	0 (ground)	A	$e_A = e_1 - e_2 = -6$ volts
2.	6 volts	B	$e_B = e_2 - e_3 = +1$ volt
3.	5 volts	C	$e_C = e_3 - e_4 = +2$ volts
4.	3 volts	D	$e_D = e_4 - e_5 = -4$ volts
5.	7 volts	E	$e_E = e_5 - e_1 = +7$ volts

$$[e_A + e_B + e_C + e_D + e_E] = \sum_{i=1}^{5} e_i - \sum_{i=1}^{5} e_i = 0 \text{ volts}$$

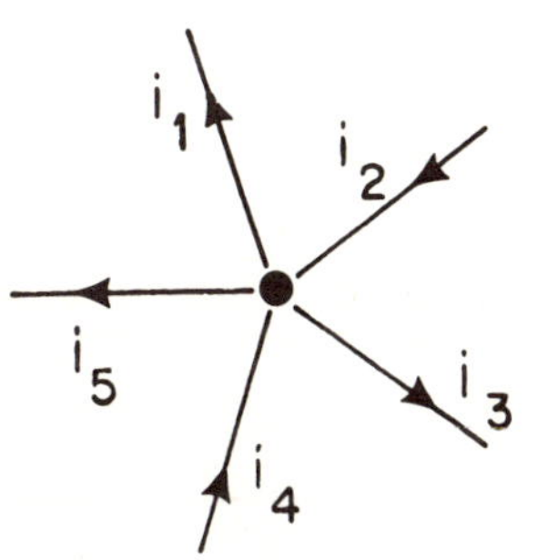

$$i_2 + i_4 = i_1 + i_3 + i_5$$

Figure 2.11.1.b. Kirchhoff's current law. The current entering any node equals the current leaving that node.

"pumped" by a battery, flowing in response to voltage differences through the closed-circuit pathways of a network back to the battery again. But if we sever the two circuits at some point, we observe a profound difference: the electric current stops flowing at once; the blood continues to flow, gushing out at an alarming rate. In similar fashion, heat can "leak out" of the cut end of an insulated thermal conductor, and certain solutes can diffuse out of the renal tubules. Furthermore, the mechanical force produced by the engine of my car, and transmitted as torque to the wheels through a system of pistons, shafts, and gears, does not in any intelligible sense "return" to its source in the engine. These examples make it clear that of all the T-variables in different systems, electric current alone requires a truly closed path. Figure 2.11.2. provides an explanation of this unique behavior. Fortunately, however, we can still use Kirchhoff's A-variable law in the analysis of non-electrical systems. For we now note that it is Kirchhoff's law for nodes which has to do with the T-variable, and that the Kirchhoff law for loops has to do only with adding up the successive A-variable differences around the loop. Hence, even when no physically-closed path exists for the T-variable, we may always imagine (and draw a circuit diagram for) a closed sequence of A-variable differences. And this is precisely what we shall do, so that we can continue to exploit the powerful analogies which exist between the different types of systems. In circuit diagrams for non-electrical systems, the imaginary leads will be shown as broken lines.

SECTION 2.12. The analysis of dissipative networks

Now at last we have the tools needed for a complete analysis of simple systems. A complete analysis is one which lets us calculate the value of every variable in the system at any instant of time. (See Section 2.18.)

Let us begin by considering the simplest sort of active system, namely one which has nothing but D-elements (e.g., electrical resistors) and a single source element (e.g., a 6-volt battery). Such a network can be analyzed without using differential or integral equations. We shall arrange the resistors in a pattern sufficiently intricate to illustrate several important features of circuit analysis. (Figure 2.12.1.a.)

Example 2.12.1. Analysis of a resistive network. Given the configuration of elements shown in the circuit diagram of Figure 2.12.1.a., the value of the six resistances, and the voltage of the ideal source, find the thirteen unknowns of the following table where the voltage drops are in the direction of current flow.

element	Source	$R_{int} = 2\Omega$	$R_1 = 7\Omega$	$R_2 = 3\Omega$	$R_3 = 20\Omega$	$R_4 = 30\Omega$	$R_5 = 12\Omega$
voltage drop (volts)	$e_s = -6$	e_{int}	e_1	e_2	e_3	e_4	e_5
current (amps)	i_s	i_{int}	i_1	i_2	i_3	i_4	i_5

In a metallic electrical conductor, the thing actually flowing (electrons) is a physical part of the conductor itself. If the conductor is cut, the electrons cannot "leak out" because they are bound to the atoms of the conducting metal by strong electrostatic forces. Electric current can flow only when the electrons can be pushed along from one atom to the next all the way around a closed circuit, with no gross accumulation or deficit of electrons occurring anywhere. In a conductor for fluid flow, (i.e., a tube), the conductor merely contains the thing actually flowing. If the conductor is cut at some point, flow continues and the fluid leaks out. Hence,

1. Severing an electrical conductor is equivalent to clamping a blood vessel.

2. A capacitor can be charged only if electrons can both enter and leave simultaneously. But a fluid reservoir can be filled even when there is no route by which the fluid can flow out.

3. A capacitor contains no more electrons when charged than when uncharged. The electrons are simply redistributed. But a fluid reservoir contains more fluid when full than when empty.

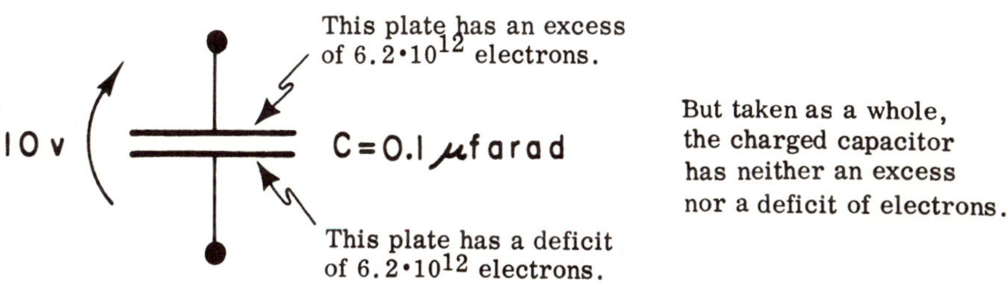

Figure 2.11.2. The peculiarities of electrical flow through a metallic conductor

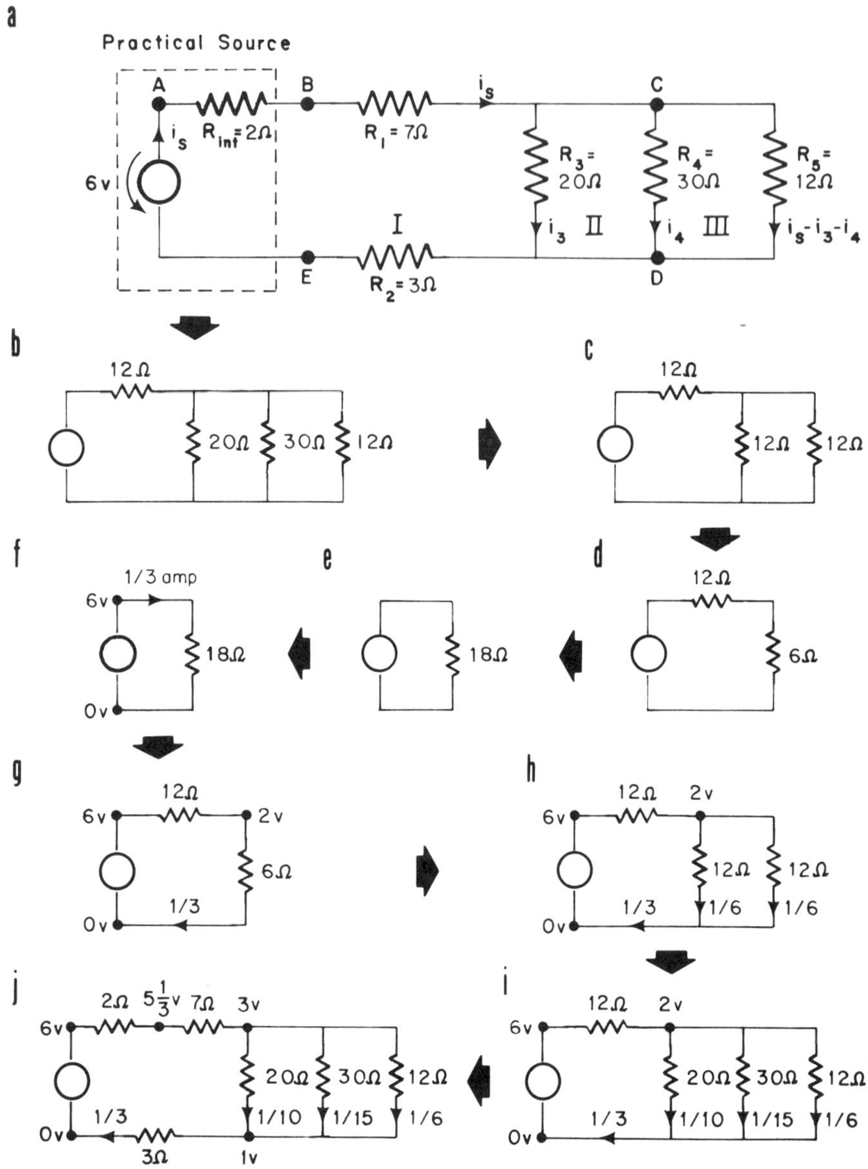

Figure 2.12.1. The resistive network of example 2.12.1.a.

By repeatedly applying the rules for combining resistances in series or in parallel we can reduce the network of part a to the single equivalent resistance of part e. The further steps needed to find the unknown currents and voltages are shown in parts f through j.

(Notice that the drop in voltage across the source in the direction of current flow is minus 6 volts, which is equivalent to a voltage rise of plus 6 volts.)

A substantial reduction in the number of unknowns may be achieved at once by using Kirchhoff's current law:

element	Source	$R_{int}=2\Omega$	$R_1=7\Omega$	$R_2=3\Omega$	$R_3=20\Omega$	$R_4=30\Omega$	$R_5=12\Omega$
voltage drop (volts)	- 6	e_{int}	e_1	e_2	e_3	e_4	e_5
current (amps)	i_s	i_s	i_s	i_s	i_3	i_4	$i_s-i_3-i_4$

Further reduction may be achieved by arbitrarily selecting some node, say node E, as a reference node, assigning it a voltage of zero, and expressing the voltage at every other node with respect to this zero-voltage reference node:

element	Source	$R_{int}=2\Omega$	$R_1=7\Omega$	$R_2=3\Omega$	$R_3=20\Omega$	$R_4=30\Omega$	$R_5=12\Omega$
voltage drop (volts)	-6	$6 - e_B$	$e_B - e_C$	$e_D - 0$	$e_C - e_D$	$e_C - e_D$	$e_C - e_D$
current (amps)	i_s	i_s	i_s	i_s	i_3	i_4	$i_s-i_3-i_4$

We now have only six unknowns: i_s, i_3, i_4, e_B, e_C, and e_D. Furthermore it is obvious that if we can find the voltages, we can apply Ohm's Law to each resistor to obtain the currents; if we can find the currents, Ohm's Law will give us the voltages.

Three methods of solution will be exemplified.

A. Solution using node equations (from Kirchhoff's current law) written in terms of the unknown voltages and i_s.

Node	Current entering (amps)	= Current leaving (amps)
A	i_s	$= [6 - e_B]/2$
B	$[6 - e_B]/2$	$= [e_B - e_C]/7$
C	$[e_B - e_C]/7$	$= \dfrac{e_C - e_D}{20} + \dfrac{e_C - e_D}{30} + \dfrac{e_C - e_D}{12}$
D	$\dfrac{e_C - e_D}{20} + \dfrac{e_C - e_D}{30} + \dfrac{e_C - e_D}{12}$	$= e_D/3$
E	$e_D/3$	$= i_s$

This is a set of 5 simultaneous equations in 4 unknowns. Hence one equation must be dependent in the sense that it can be derived directly from the others as some linear combination of them. Such a dependent equation is redundant, for it contributes no additional information about the system. Once we have chosen a particular node (here E) as the zero reference node in an n-node network, we need only solve the set of n - 1 equations written for the non-reference nodes. Accordingly, in the present system, we shall deal only with the equations for nodes A, B, C, and D. By putting the corresponding augmented matrix:

	i_s	e_B	e_C	e_D	Non-homog.
A	1	1/2	0	0	3
B	0	-9/14	1/7	0	-3
C	0	1/7	-13/42	1/6	0
D	0	0	1/6	-1/2	0

into row-echelon form (see Appendix) we obtain: i_s= 1/3amp, e_B= 16/3 volts, e_C= 3 volts, e_D= 1 volt. From these values, the remaining unknowns are easily calculated:

element	Source	$R_{int}= 2\Omega$	$R_1= 7\Omega$	$R_2= 3\Omega$	$R_3= 20\Omega$	$R_4= 30\Omega$	$R_5= 12\Omega$
voltage drop (volts)	- 6	2/3	7/3	1	2	2	2
currents (amps)	1/3	1/3	1/3	1/3	1/10	1/15	1/16

B. Solution using loop equations (from Kirchhoff's voltage law) written in terms of the unknown currents.

There are six different closed loops in the network. But we need only 3 equations to solve for the unknown currents. It would seem reasonable to choose the three loops, labelled I, II, and III, which are the meshes of the network. Proceeding either clockwise or counter-clockwise, we sum the voltage drops around each loop and set the sum equal to zero. Notice that what is a voltage drop when we proceed with the current is a voltage rise (and therefore negative) when we proceed against the current.

Arbitrarily-chosen Direction	Loop	Equation
Clockwise	I	$-6 + 2i_s + 7i_s + 20i_3 + 3i_s = 0$ volts
Clockwise	II	$-20i_3 + 30i_4 = 0$ volts
Clockwise	III	$-30i_4 + 12i_s - 12i_3 - 12i_4 = 0$ volts

This is a set of 3 simultaneous equations in 3 unknowns. By putting the corresponding augmented matrix:

	i_s	i_3	i_4	Non-homog.
I	12	20	0	6
II	0	-20	30	0
III	12	-12	-42	0

into row-echelon form (see Appendix) we obtain $i_s = 1/3$, $i_3 = 1/10$, $i_4 = 1/15$. From these values, the remaining unknowns are easily calculated.

C. Solution by calculating equivalent resistances.

To apply this method we need two rules (see Exercise 2.3) for combining several resistances into a single equivalent resistance:

When two or more resistors are connected <u>in series</u> (i.e., so that the same current always passes through each), they are equivalent to a single resistor with a <u>resistance</u> equal to the sum of the individual resistances.

When two or more resistors are connected <u>in parallel</u> (i.e., between the same nodes, so that the same voltage drop occurs across each) they are equivalent to a single resistor with a <u>conductance</u> (= 1/R) equal to the sum of the individual conductances.

By applying these rules repeatedly (Figure 2.12.1.b through e) it is often possible to reduce a complicated network to a much simpler array of one or more equivalent resistances. (Figure 2.12.1.e.) Analysis of the simplified network (which can sometimes be done by inspection) serves as the starting point for a series of easy calculations as one retraces the steps used in the simplification. (Figure 2.12.1.f through j).

Engineers have developed a variety of useful shortcuts and specialized topological techniques to deal with dissipative networks far more complicated than the one just discussed. These need not detain us. The physiologist will rarely be able to obtain quantitative information about dissipative biological systems complex enough to require analytical tools more sophisticated than the ones described above. However, he should not dismiss the solution of dissipative networks as mere algebraic drudgery, unrelated to his main interests. For as we shall see later (Section 3.16.), <u>phasor analysis</u>, one of the most versatile techniques for dealing with all sorts of linear systems, draws its strength from reducing A-elements and T-elements to quasi-dissipative elements, so that simple algebraic methods such as those of example 2.12.1. can still be applied.

SECTION 2.13. The analysis of systems containing storage elements

Purely dissipative systems are creatures of the moment, whose state at any instant of time is completely determined by the activity of the source at that particular instant. In the jargon of the systems engineer, they have no memory, because their present state is independent of their previous history. This is obviously not true of systems which are capable of storing energy over a period of time, and which are therefore said to have a memory. Because of their memory, such systems exhibit behavior which is much more varied and interesting than is the behavior of systems which merely dissipate energy.

Systems containing one or more storage elements can manifest four general types of behavior:

Inactivity is described by a homogeneous differential equation with zero initial conditions, whose solution specifies the state of the system at any instant of time equal to or greater than zero. An inactive system does nothing; it has only the when-and-if interest of a sleeping beauty waiting for an energetic kiss.

A natural response to energy stored in the system itself is described by a homogeneous differential equation with at least one non-zero initial condition, whose solution specifies the state of the system at any instant of time equal to or greater than time zero. (The term "transient response" is sometimes used as a synonym for "natural response," but it is also used to denote the initial part of the total response. To avoid confusion, I shall use it only in the latter sense.)

A pure forced response (= non-transient response) is the pattern of behavior to which the system eventually settles down when kept in continuous activity for a long time by one or more source elements. The input from the source is described by the nonhomogeneous part of a differential equation, with any initial conditions, whose solution, evaluated in the limit as time approaches infinity, specifies the state of the system at any instant of time sufficiently far beyond time zero.*

[*"Sufficiently far beyond time zero" is commonly interpreted to mean $t > (5/k_{min})$ where t is time measured from time zero, and k_{min} is the exponential decay constant (= negative coefficient of time in the exponent) with the smallest absolute value.]

A total response is the response of the system at any time to one or more source elements and also to any energy which may be stored initially in the system. It is described by a nonhomogeneous differential equation with any initial conditions, whose solution specifies the state of the system at any instant of time equal to or greater than time zero.

As the reader will recall, the general solution of a nonhomogeneous differential equation (such as the one used to describe a total response) consists of the sum of two parts: the general solution of the corresponding complementary (homogeneous) equation, and the particular solution (= particular integral) required by the specified forcing function. Since the complementary equation describes the natural response and the particular solution depends upon the forced response, it is often stated that "the total response is the sum of the natural response and the forced response." But this statement is true only of the general solution, because the initial conditions appropriate for a specific system can only be applied after the general complementary solution and the particular solution have been found and added together. We cannot get the correct total response by first calculating the natural response for specific initial conditions, next calculating the pure forced response, and finally adding these two responses together.

The features which characterize these four types of response are summarized in Table 2.13.1. Because of its generality, the total response is the one we shall most often study. The other types of response really represent special cases of the total response.

Table 2.13.1.

Type of behavior	Type of equation	Initial conditions	Valid at times
Inactivity	Homogeneous	$= 0$	> 0
Natural response	Homogeneous	$\neq 0$	> 0
Pure forced response	Nonhomogeneous	any	$\to \infty$
Total response	Nonhomogeneous	any	> 0

SECTION 2.14. Initial conditions and "instantaneous" changes at time zero

We shall often be interested in the response to an input which is imposed upon a system "instantaneously" at time zero. Throwing a switch, dropping a weight, opening a valve, injecting a priming dose of a drug are all examples of instantaneous (or, more often, practically instantaneous) changes which can initiate a response. Depending upon what kind of input is imposed at time zero such an instantaneous change may elicit either a natural response or a total response. But in either case, we shall find it convenient to assume that the single point in time defined as "time zero" can be split into "time zero-minus," symbolized 0^-, just before the instantaneous change, and "time zero-plus," symbolized 0^+ just after the instantaneous change. With this convention, the initial conditions used for solving the differential equation will always be the conditions at $t = 0^+$, just after the instantaneous change has occurred.

Splitting a single instant of time in two is as illegitimate mathematically as it is impossible practically, for it implies that certain functions of time can have two different values at one and the same instant. Strictly speaking, such functions have no right to be called functions at all, for a mathematically legitimate function can have but a single value at any given point in the domain of its independent variable. Yet this momentary violation of mathematical rules is essential if we are to admit the convenient notion of an "instantaneous" change. "Zero-minus" and "zero-plus" are such innocent little fictions that we should have no qualms about including them in our bag of tricks.

WARNING: "Zero-minus" and "zero-plus" are really exactly the same instant. Hence, no energy can be stored or dissipated between them; for to store or dissipate a finite amount of energy during a zero time interval would require the expenditure of an infinite amount of power.

SECTION 2.15. Stepwise analysis of simple linear systems

In this section we shall consider in minute detail two examples of linear systems analysis so as to illustrate the reasoning which guides each individual step as one proceeds in a logical sequence from the real system to its physical model; from the physical model to the circuit diagram; from the circuit diagram to the initial conditions, and to a differential equation for one of the unknowns; from the equation to its solution and the checking of the solution; from the solution for one unknown to the calculation of the other unknowns. When one is thoroughly familiar with the system, and with the analytical procedure, some of these individual steps may be lumped together or performed almost subconsciously, with a considerable saving of time. The individual steps are there, nevertheless, and the beginner does well to follow them faithfully until he is sure where the safe shortcuts lie.

Example 2.15.1. Clearance of a solute from its volume of distribution in the body.

I. The REAL SYSTEM and its PHYSICAL MODEL. Consider a solute, S, which, after intravenous injection into an anesthetized dog, is distributed fairly rapidly throughout a single volume of distribution (i.e. a single "compartment") of [Vdist] liters, part of which is blood plasma. S is removed from [Vdist] both by renal excretion and by metabolic degradation. The renal plasma clearance of S is $[\dot{V}cl]_{ren}$ liters per hour, and the plasma clearance due to metabolism of S is $[\dot{V}cl]_{met}$ liters per hour. We want to calculate the total response of C(t), the plasma concentration of S, when a priming dose of Q_p mg is given at time zero, and a continuous intravenous infusion of $\dot{Q}_{in}(t)$ mg/hr is also begun at time zero. We shall assume that C(t) = 0 when $t = 0^-$.

To obtain a sufficiently simple physical model, we shall assume that the priming dose is injected instantaneously, that the solute is distributed instantaneously and uniformly throughout its volume of distribution, and that [Vdist], $[\dot{V}cl]_{ren}$ and $[\dot{V}cl]_{met}$ all remain constant. With these assumptions, the physical model can be depicted as a single well-stirred compartment (Figure 2.15.1.) with S entering and leaving by the routes described above.

II. The CIRCUIT DIAGRAM. In constructing a circuit diagram corresponding to the physical model, the first step is to decide how many different values the across variable can assume, for this will determine the number of nodes in the diagram. In the present system, the across variable is concentration, and the only concentration involved would seem to be C(t), the concentration of S in [Vdist]. But we must remember that the across variable for any element is always the difference between the values which it has at the two nodes which are the terminals of the element. In this system, C(t) is measured with respect to zero concentration, so we must include a zero-concentration reference

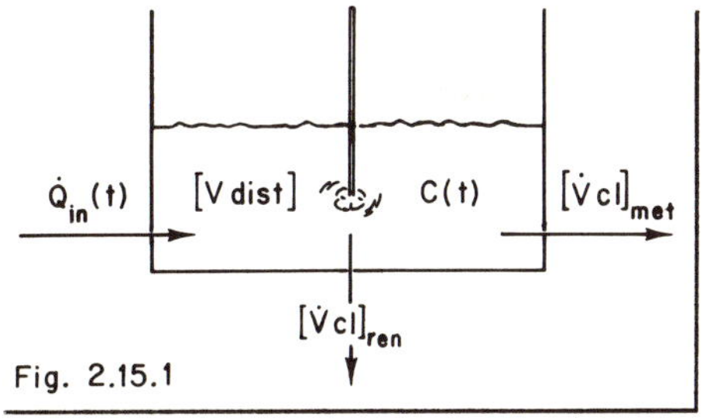

Fig. 2.15.1

node in the diagram. The <u>second step</u> is to identify the elements of the system. From the physical model and from the column for concentrational systems in the end-paper table, it is clear that we have a T-source, supplying $\dot{Q}_{in}(t)$ from time zero onwards; we have a capacitative element with a solute capacitance of [Vdist]; and we have two dissipative elements whose solute conductances are $[\dot{V}cl]_{ren}$ and $[\dot{V}cl]_{met}$ (corresponding to solute resistances of $1/[\dot{V}cl]_{ren}$ and $1/[\dot{V}cl]_{met}$). The <u>last step</u> is to place each element between the proper nodes. In the present system, it is impossible to go wrong! For with only two nodes, <u>all</u> elements must be connected in parallel between them.

The resulting circuit diagram is shown in Figure 2.15.2. The imaginary "lead" through which no solute actually passes but which is included merely to "close the circuit" (See Section 2.11) is depicted as a broken line.

III. The <u>INITIAL CONDITION.</u> According to our assumptions about the physical model, Q_p, the priming dose of S, is injected at t = 0 so as to raise C(t) instantaneously from 0 at $t = 0^-$ to $Q_p/[\text{Vdist}]$, the concentration which will result from distributing Q_p instantaneously throughout [Vdist]. Our initial condition is therefore $C(0^+) = Q_p/[\text{Vdist}]$.

IV. The <u>DIFFERENTIAL EQUATION</u> is just the node equation for the C(t) node. By Kirchhoff's T-variable law,

$$[\text{Vdist}]\,\frac{dC(t)}{dt} + [\dot{V}cl]_{ren}\,C(t) + [\dot{V}cl]_{met}\,C(t) = \dot{Q}_{in}(t) \qquad 2.15.1.$$

Defining the total clearance (= total solute conductance) as

$$[\dot{V}cl]_{tot} = [\dot{V}cl]_{ren} + [\dot{V}cl]_{met}, \qquad 2.15.2.$$

and dividing both sides of equation 2.15.1. by [Vdist] we obtain

$$\frac{dC(t)}{dt} + \frac{[\dot{V}cl]_{tot}}{[\text{Vdist}]}\,C(t) = \frac{\dot{Q}_{in}(t)}{[\text{Vdist}]} \qquad 2.15.3.$$

subject to the initial condition

$$C(0) = Q_p/[\text{Vdist}] \qquad 2.15.4.$$

V. <u>SOLUTION</u> of the <u>DIFFERENTIAL EQUATION.</u> The characteristic equation (See Appendix) corresponding to equation 2.15.3. is

$$\lambda + \frac{[\dot{V}cl]_{tot}}{[\text{Vdist}]} = 0 \qquad 2.15.5.$$

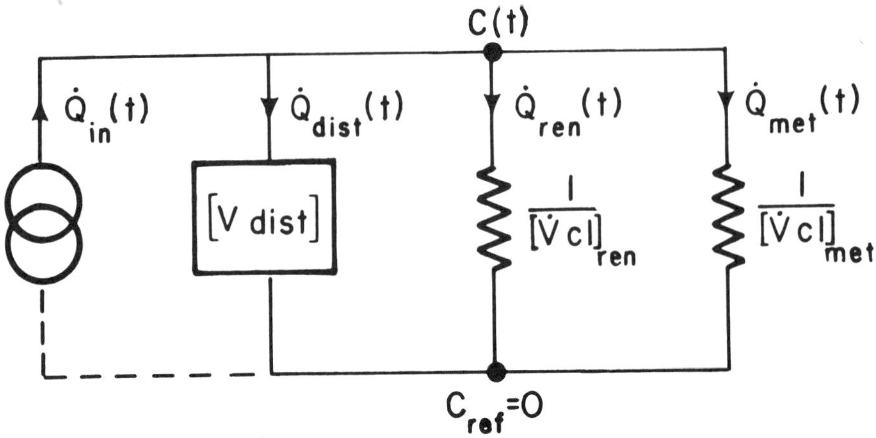

Figure 2.15.2. A circuit diagram for the single-compartment model of Example 2.15.1 and Figure 2.15.1.

The root of the characteristic equation (= eigenvalue; see Appendix) is

$$\lambda = -\frac{[\dot{V}cl]_{tot}}{[Vdist]} \qquad 2.15.6.$$

Using the method of undetermined coefficients (See Appendix) the solution will be of the general form

$$C(t) = A\epsilon^{-\frac{[\dot{V}cl]_{tot}}{[Vdist]}t} + \frac{\dot{Q}_{in}(t)}{[\dot{V}cl]_{tot}} \qquad 2.15.7.$$

where $\dot{Q}_{in}(t)/[\dot{V}cl]_{tot}$ is the particular solution (= particular integral) corresponding to $\dot{Q}_{in}(t)/[Vdist]$, the constant forcing function which appears as the non-homogeneous term in equation 2.15.3. Finally, by requiring equation 2.15.7. to satisfy the initial condition of equation 2.15.4. at t = 0, we find $A = \frac{Q_p}{[Vdist]} - \frac{\dot{Q}_{in}(t)}{[\dot{V}cl]_{tot}}$, so the complete solution is

$$C(t) = \left\{\frac{Q_p}{[Vdist]} - \frac{\dot{Q}_{in}(t)}{[\dot{V}cl]_{tot}}\right\}\epsilon^{-\frac{[\dot{V}cl]_{tot}}{[Vdist]}t} + \frac{\dot{Q}_{in}(t)}{[\dot{V}cl]_{tot}} \qquad 2.15.8.$$

VI. CHECKING the SOLUTION. Differentiating equation 2.15.8. with respect to time:

$$\frac{dC(t)}{dt} = -\frac{[\dot{V}cl]_{tot}}{[Vdist]}\left(\frac{Q_p}{[Vdist]} - \frac{\dot{Q}_{in}(t)}{[\dot{V}cl]_{tot}}\right)\epsilon^{-\frac{[\dot{V}cl]_{tot}}{[Vdist]}t}$$

$$= -\frac{[\dot{V}cl]_{tot}}{[Vdist]}C(t) + \frac{\dot{Q}_{in}(t)}{[\dot{V}cl]_{tot}} \qquad 2.15.9.$$

or, rearranging terms,

$$\frac{dC(t)}{dt} + \frac{[\dot{V}cl]_{tot}}{[Vdist]}C(t) = \frac{\dot{Q}_{in}(t)}{[Vdist]} \qquad 2.15.3.$$

So the solution satisfies the original differential equation. Furthermore, evaluating 2.15.8. at $t = 0$,

$$C(0) = \frac{Q_p}{[Vdist]} \qquad 2.15.4.$$

So the solution also satisfies the initial conditions.

VII. CALCULATION of the OTHER UNKNOWNS. Knowing C(t), it is now a simple matter to calculate the quantity of solute following each of the three pathways in Figure 2.15.2. These are just the terms on the left of our original node equation, 2.15.1. $\dot{Q}_{ren}(t)$, the rate of renal excretion of S, and $\dot{Q}_{met}(t)$, the rate of metabolism of S, are given by

$$\dot{Q}_{ren}(t) = [\dot{V}cl]_{ren}\,C(t) \qquad 2.15.10.$$

and

$$\dot{Q}_{met}(t) = [\dot{V}cl]_{met}\,C(t) \qquad 2.15.11.$$

Where C(t) is to be calculated from equation 2.15.8. $\dot{Q}_{dist}(t)$, the rate at which S is being stored in [Vdist], can now be obtained by combining equations 2.15.1., 2.15.10., and 2.15.11.

$$\dot{Q}_{dist}(t) = [Vdist]\,dC(t)/dt = \dot{Q}_{in}(t) - \dot{Q}_{ren}(t) - \dot{Q}_{met}(t) \qquad 2.15.12.$$

Example 2.15.2. A second-order mechanical system.

In the previous example, symbols were used throughout. For variety, specific numerical values will be used in the present example. Ordinarily, symbols are to be preferred since the solution can then be applied to any specific set of values.

I. The REAL SYSTEM and its PHYSICAL MODEL. An automobile's real suspension system consists of various springs and shock absorbers interposed between the main mass of the automobile and its wheels and axles. To test the ability of the rear shock absorbers to damp out the vertical oscillations caused by a rough road, we shall lift the rear bumper a few centimeters and then suddenly let go, observing how much bouncing occurs before the body comes to rest in its original position. Since the force of gravity is exerted in the downward direction, we shall define downward displacements, velocities, and accelerations as positive. To obtain a sufficiently simple physical model, we shall neglect the wheels and axles, and we shall treat the entire system as if it consisted of a single mass, $m = 8 \times 10^5$ grams, equal to half the mass of the automobile, connected to the ground by a single spring of stiffness $k = 10^7$ dynes per cm., and a single damping element with a damping constant of $D = 2.99 \times 10^6$ dyne-sec/cm. (Figure 2.15.3a.)

II. The CIRCUIT DIAGRAM. In constructing a circuit diagram for this model system, notice first that (as in the preceding example) there is only one A-variable difference to consider, namely, v(t), the velocity of the mass with respect to $v_{ref} = 0$, the velocity of the ground. This same velocity, v(t), must also be the velocity across the spring and across the shock absorber because their upper terminals are rigidly attached to the mass, while their lower terminals are attached to the ground. Next, notice that in addition to the three passive elements, mass, damper, and spring, we have two T-variable sources: the earth's gravitational field, exerting a force of F_g dynes downward, and the person lifting the bumper, exerting an opposing force of F_{lift} dynes upward. Finally, since there are only two nodes, all five elements must be connected in parallel between them. The resulting circuit diagram is shown in Figure 2.15.3.b.

III. INITIAL CONDITIONS. Consider first the situation at $t = 0^-$, immediately before letting go of the bumper at time zero. At $t = 0^-$, both $v(0^-)$, the velocity downward, and $dv(t)/dt\,\big|_{0^-}$, the acceleration downward, are zero because the car bumper is being held still. Hence there is no difference in velocity across the elements, and $F_m(0^-) = m\,[dv(t)/dt]_{0^-}$ and $F_D(0^-) = Dv(0^-)$ are both zero. Thus the entire net downward force of $7.25 \cdot 10^8$ dynes (= F_g minus F_{lift}) must equal $F_k(0^-) = k\int_{-\infty}^{0^-} v(t)dt$ through the spring, as is evident from applying Kirchhoff's T-variable law to the v(t) node when $t = 0^-$ (Figure 2.15.3.c).

$$7.85 \cdot 10^8 = 6 \cdot 10^7 + k\int_{-\infty}^{0^-} v(t)dt \quad \text{dynes} \qquad 2.15.13$$

or*,

$$F_k(0^-) = k\int_{-\infty}^{0^-} v(t)dt = 7.25 \cdot 10^8 \quad \text{dynes} \qquad 2.15.14$$

[*Note that although the velocity at $t = 0^-$ is zero, the $k\int_{-\infty}^{0^-} v(t)dt$ term for the force through the spring is not zero, because the integral of all of the previous velocities (which integral is merely the downward displacement of

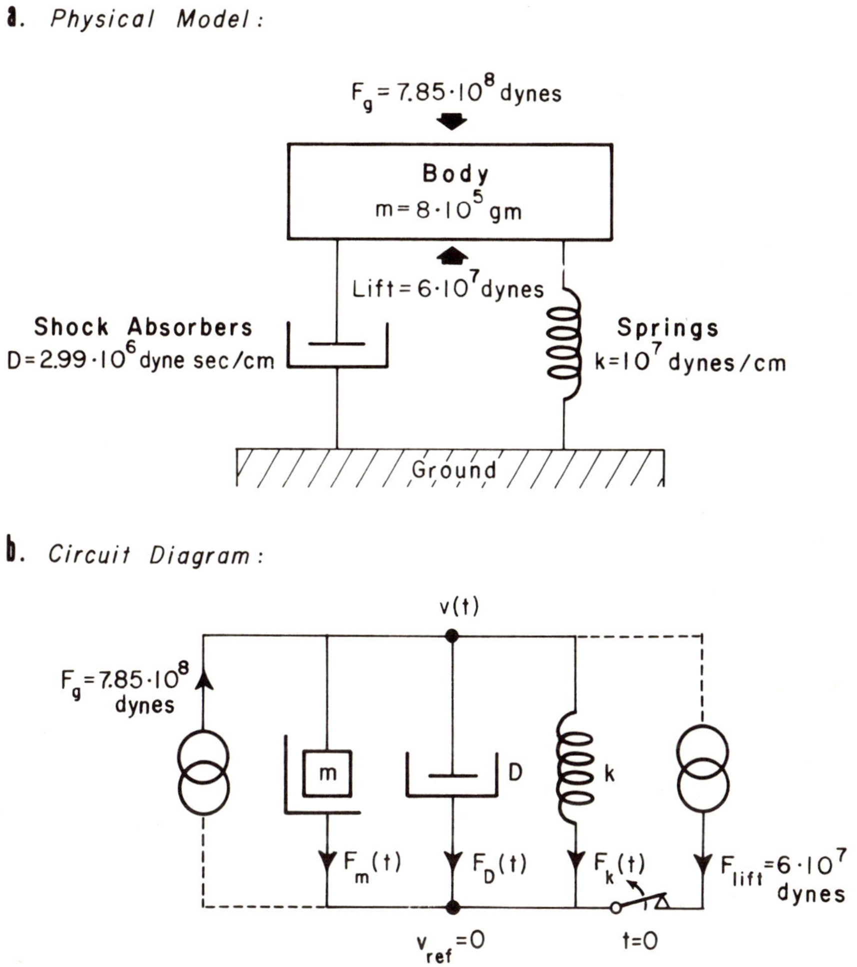

Figure 2.15.3.a. The physical model, and b the corresponding circuit diagram for the automobile suspension problem of Example 2.15.2.

The physical model shows that the velocity measured across all three passive elements with respect to ground (=zero) velocity must be the same. In the circuit diagram, that velocity (defined as positive downwards) is represented by the v(t) node. F_g, the gravitational force, is depicted as a T-source oriented so as to increase the downward forces through the passive elements. F_{lift}, the lifting force, is depicted as a second T-source oriented so as to decrease the downward forces through the passive elements. Cessation of the lifting force at time zero is represented by opening the switch.

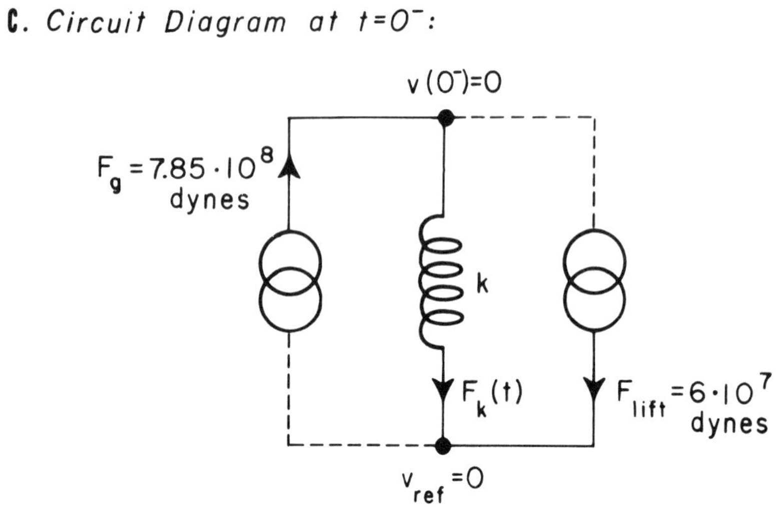

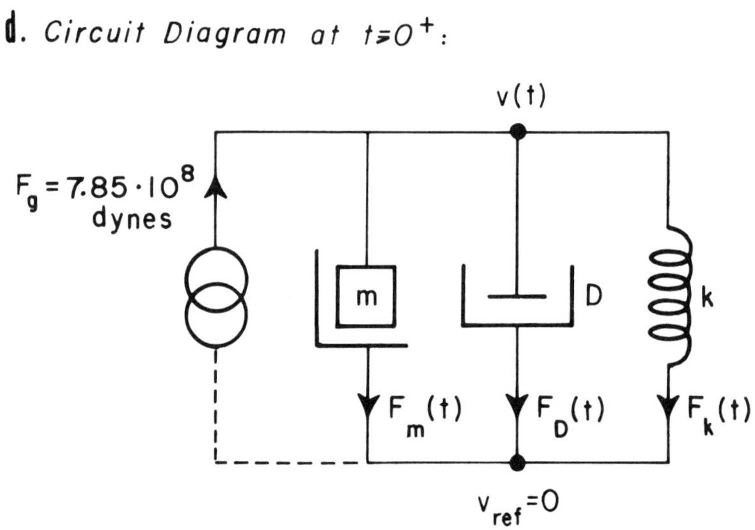

Figure 2.15.3.c and d. Circuit diagrams showing conditions in the automobile suspension problem just before (panel c) and at any time after (panel d) instantaneous cessation of the lifting force at time zero.

At $t = 0^-$, both the downward velocity and the downward acceleration are zero. Hence the passive system can be represented by just the spring through which all of the force from the T-sources is exerted. But for $t \geq 0^+$, (i.e. at any time after the lifting force has ceased) we must consider all three of the passive elements.

the spring from its uncompressed position) is positive. As a result of these "remembered" past velocities, some energy is already stored in the partially compressed spring at $t = 0^-$.]

Next, consider the situation at $t = 0^+$, immediately after letting go of the bumper at time zero. In the general circuit diagram (Figure 2.15.3.b.) this abrupt cessation of the lifting force is symbolized by opening the switch, leaving us at $t \geq 0^+$ with the network shown in Figure 2.15.3.d. But we must now recall that neither the velocity of a mass, nor the length of a spring can be changed instantaneously between $t = 0^-$ and $t = 0^+$, because, as was pointed out in Section 2.14, no energy can be stored in such storage elements during the zero interval of time between 0^- and 0^+. Hence, $v(0^+) = v(0^-) = 0$, $F_D(0^+) = Dv(0^+) = Dv(0^-) = 0$ dynes and $F_k(0^+) = k\int_{-\infty}^{0^+} v(t)dt = k\int_{-\infty}^{0^-} v(t)dt = 7.25 \cdot 10^8$ dynes, so that the forces through the damping element and through the spring are the same at $t = 0^+$ as at $t = 0^-$. Therefore, at $t = 0^+$, the increased downward force due to cessation of the lifting force must be exerted entirely through the mass, producing an initial acceleration which can be calculated by evaluating the node equation for the network of Figure 2.15.3.d

$$m \frac{dv(t)}{dt} + Dv(t) + k\int v(t)dt = F_g \qquad 2.15.15,$$

at $t = 0^+$:

$$[8.10^5] \left.\frac{dv(t)}{dt}\right|_{0^+} + 0 + 7.25 \cdot 10^8 = 7.85 \cdot 10^8 \text{ dynes} \qquad 2.15.16.$$

Solving for $dv(t)/dt \Big|_{0^+}$,

$$\left.\frac{dv(t)}{dt}\right|_{0^+} = 75 \text{ cm per sec}^2 \qquad 2.15.16.a.$$

Our initial conditions are therefore $v(0^+) = 0$; $\left.\frac{dv(t)}{dt}\right|_{0^+} = 75 \text{ cm/sec}^2$

IV. The DIFFERENTIAL EQUATION. To analyze the behavior of this two-node system, we need only solve the single node equation, equation 2.15.15., subject to the initial conditions which we have just found. As it stands, equation 2.15.15. is not a differential equation, but is an integrodifferential equation. Let us convert it to a purely differential equation, using either of two simple methods. In the first, we change the dependent variable from a velocity, $v(t)$, to a distance, $x(t)$, by using the following definition of velocity:

$$v(t) = dx(t)/dt \qquad 2.15.17.$$

where $x(t)$ must be taken as positive in the same direction as velocity, i.e., downwards in this example. Substituting $dx(t)/dt$ for $v(t)$ in equation 2.15.15.,

$$m \frac{d^2x(t)}{dt^2} + D \frac{dx(t)}{dt} + k\int dx(t) = F_g = 7.85 \cdot 10^8 \text{ dynes} \qquad 2.15.18.$$

But

$$k\int dx(t) = kx(t) + C \qquad 2.15.19.$$

Where C, the constant of integration, may be found by evaluating equation 2.15.19 at time zero when $x(t) = x(0^+)$. From equation 2.15.17., it is clear that $k\int_{-\infty}^{0^+} dx(t) = k\int_{-\infty}^{0^+} v(t)dt$, which, as we already know from equation 2.15.14., is $7.25 \cdot 10^8$ dynes. Substituting this value into equation 2.15.19., and solving for C,

$$C = 7.25 \cdot 10^8 - 10^7\, x(0^+) \text{ dynes} \qquad 2.15.20.$$

Combining equations 2.15.19. and 2.15.20. by eliminating C,

$$k\int dx(t) = kx(t) + 7.25 \cdot 10^8 - 10^7 x(0^+) \text{ dynes} \qquad 2.15.21.$$

Substituting this value for $k\int dx(t)$ into equation 2.15.18.,

$$m \frac{d^2x(t)}{dt^2} + D \frac{dx(t)}{dt} + kx(t) = 6 \cdot 10^7 + 10^7 x(0^+) \text{ dynes} \qquad 2.15.22.$$

In the second method, we differentiate equation 2.15.15. with respect to time, thus obtaining the homogeneous differential equation,

$$m \frac{d^2v(t)}{dt^2} + D\frac{dv(t)}{dt} + kv(t) = 0 \qquad 2.15.23.$$

V. <u>SOLUTION</u> of the <u>DIFFERENTIAL EQUATION</u>. The solution of equation 2.15.22. is assigned as exercise 9. We shall here solve equation 2.15.23. Substituting the values of the parameters for our system into equation 2.15.23., we get

$$8 \cdot 10^5 \frac{d^2v(t)}{dt^2} + 2.99 \cdot 10^6 \frac{dv(t)}{dt} + 10^7 v(t) = 0 \quad \frac{\text{dynes}}{\text{sec}} \qquad 2.15.24.$$

whose characteristic equation (see Appendix) is

$$8\lambda^2 + 29.9\lambda + 100 = 0 \quad \text{dynes/cm} \qquad 2.15.25.$$

with a damping ratio (see Appendix) of

$$\rho = \sqrt{\frac{[29.9]^2}{4[8]100}} = 0.529$$

Since $\rho<1$, the system is underdamped.

The roots of the characteristic equation (= eigenvalues) are:

$$\lambda = \frac{-29.9 \pm \sqrt{[29.9]^2 - 3200}}{16} = -1.87 \pm j3 \text{ sec.}^{-1} \qquad 2.15.26.$$

So the solution will be of the form,

$$v(t) = A\epsilon^{-1.87t}\cos 3t + B\epsilon^{-1.87t}\sin 3t \text{ cm/sec} \qquad 2.15.27.$$

Applying the first initial condition, $v(0^+) = 0$, to equation 2.15.27.

$$v(0^+) = A = 0 \text{ cm/sec} \qquad 2.15.28.$$

so that equation 2.15.27 becomes,

$$v(t) = B\epsilon^{-1.87t}\sin 3t \text{ cm/sec} \qquad 2.15.29.$$

Differentiating equation 2.15.29. with respect to time,

$$\frac{dv(t)}{dt} = -1.87B\epsilon^{-1.87t}\sin 3t + 3B\epsilon^{-1.87t}\cos 3t \text{ cm/sec}^2 \qquad 2.15.30.$$

Applying the second initial condition, $\left.\frac{dv(t)}{dt}\right|_{0^+} = 75 \text{ cm/sec}^2$

$$\left.\frac{dv(t)}{dt}\right|_{0^+} = 3B = 75 \text{ cm/sec}^2; \; B = 25 \text{ cm/sec} \qquad 2.15.31.$$

So the velocity downwards at any $t \geq 0^+$ is,

$$v(t) = 25\epsilon^{-1.87t}\sin 3t \text{ cm/sec} \qquad 2.15.32.$$

VI. CHECKING the SOLUTION. Differentiating equation 2.15.32 twice,

$$\frac{dv(t)}{dt} = -1.87[25]\epsilon^{-1.87t}\sin 3t + 3[25]\epsilon^{-1.87t}\cos 3t \text{ cm/sec}^2$$

2.15.33.

$$\frac{d^2v(t)}{dt^2} = [1.87]^2\, 25\epsilon^{-1.87t}\sin 3t - 3[1.87]25\epsilon^{-1.87t}\cos 3t$$

$$- 3[25]\,1.87\epsilon^{-1.87t}\cos 3t - 3[3]25\epsilon^{-1.87t}\sin 3t$$

$$= \left\{[1.87]^2 - [3]^2\right\} 25\epsilon^{-1.87t}\sin 3t - 6[1.87]25\epsilon^{-1.87t}\cos 3t$$

$$\text{cm/sec}^3$$

Substituting these values into the original differential equation, (equation 2.15.23),

$$8\cdot 10^5 \left\{[1.87]^2 - [3]^2\right\} 25\epsilon^{-1.87t}\sin 3t - [8\cdot 10^5]\,6[1.87]\,25\epsilon^{-1.87}\cos 3t$$

$$+ 2.99\cdot 10^6[-1.87]\; 25\epsilon^{-1.87t}\sin 3t \quad + \quad 2.99\cdot 10^6\,[3]\; 25\epsilon^{-1.87}\cos 3t$$

$$+ 10^7\cdot\; 25\epsilon^{-1.87t}\sin 3t \qquad = 0$$

So our solution satisfies the differential equation.

Furthermore, at time zero plus, equation 2.15.32. gives $v(0^+) = 0$ cm/sec., and equation 2.15.33 gives $\left.\frac{dv(t)}{dt}\right|_{0+} = 75$ cm/sec^2, so our initial conditions are also fulfilled.

VII. CALCULATION of the OTHER UNKNOWNS. From equation 2.15.32. for v(t), and the relation $F_D(t) = Dv(t)$ we obtain,

$$F_D(t) = 2.99\cdot 10^6\; [25]\epsilon^{-1.87t}\sin 3t = 7.48\cdot\; 10^7\epsilon^{-1.87}\sin 3t \quad \text{dynes} \qquad 2.15.34.$$

By differentiating equation 2.15.32. with respect to time, we find the acceleration,

$$\frac{dv(t)}{dt} = [75\cos 3t - 46.75\sin 3t]\epsilon^{-1.87t} \qquad \text{cm/sec}^2 \qquad 2.15.33.a.$$

From equation 2.15.33.a. and the relation $F_m(t) = m\frac{dv(t)}{dt}$ we get,

$$F_m(t) = [6\cdot\; 10^7\cos 3t - 3.74\cdot\; 10^7\sin 3t]\epsilon^{-1.87t} \qquad \text{dynes} \quad 2.15.35.$$

From equations 2.15.34. and 2.15.35. and the relation $F_k(t) = F_g(t) - F_m(t) - F_D(t)$, we obtain,

$$F_k(t) = 78.5\cdot\; 10^7 - [6\cdot\; 10^7\cos 3t + 3.74\cdot\; 10^7\sin 3t]\epsilon^{-1.87} \quad \text{dynes} \qquad 2.15.36.$$

In addition to calculating these variables, we might also be interested in $x(t)$, the downward displacement. From the relation $F_k(t) = k\int v(t)dt = k\int dx(t)$, and equation 2.15.21. we obtain,

$$x(t) = \frac{1}{10^7}\left[F_k(t) - 7.25 \cdot 10^8 + 10^7\, x(0^+)\right] \quad \text{cm} \qquad 2.15.37.$$

or, combining equations 2.15.36. and 2.15.37,

$$x(t) = x(0^+) + 6 - [6 \cos 3t + 3.74 \sin 3t]\epsilon^{-1.87t} \quad \text{cm} \qquad 2.15.38.$$

where $x(0^+)$, the initial position, may be defined as zero, or may be measured from any convenient zero reference point such as the ground or the final position, $x(\infty)$. For example, if we choose to let $x(\infty) = 0$, we have, from equation 2.15.38., $x(\infty) = x(0^+) + 6 = 0$, whence $x(0^+) = -6$ cm measured downwards (= +6 cm measured upwards) from $x(\infty)$.

In Figure 2.15.4.b., velocity (equation 2.15.32.), acceleration (equation 2.15.33.a.), and displacement (equation 2.15.38., with $x(0^+)$ defined as - 6 cm) have been plotted against time. The slight overshoot is characteristic of a slightly underdamped system. For comparison, similar plots are shown for a markedly underdamped suspension, $\rho = 0.20$, (Figure 2.15.4.a) a critically damped suspension, $\rho = 1.00$, (Figure 2.15.4.c.) and a markedly overdamped suspension, $\rho = 3.00$, (Figure 2.15.4.d.) of the same mass, using the same spring constant, k, but different damping factors.

SECTION 2.16. The total response of a simple linear system to a sinusoidal input

In both of the examples of the previous section, an input forcing function was imposed at time zero, but was kept constant thereafter. As an example of how a linear system responds to a time-varying input beginning at $t = 0$, we shall analyze the behavior of a simple electrical circuit activated by an alternating current.

Example 2.16.1. An electrical RC (= Resistor and Capacitor) circuit with an a-c (= alternating current) voltage source.

Consider the electrical network in Figure 2.16.1. Before time zero, the switch has been in the position shown for a very long (i.e., "infinite") time. At time zero it is flipped into the broken-line position. Find $i(t)$ and $e_c(t)$ for all $t \geq 0^+$ while the system is being activated by a voltage source varying sinusoidally about a constant baseline of +10 volts.

By inspection of Figure 2.16.1., at $t = 0^-$, $i(0^-) = 0$, $e_c(0^-) = 2$ volts. At $t \geq 0^+$, the behavior of the circuit is described by Kirchhoff's voltage law,

$$e_R(t) + e_c(t) = e_{in}(t) \qquad 2.16.1.$$

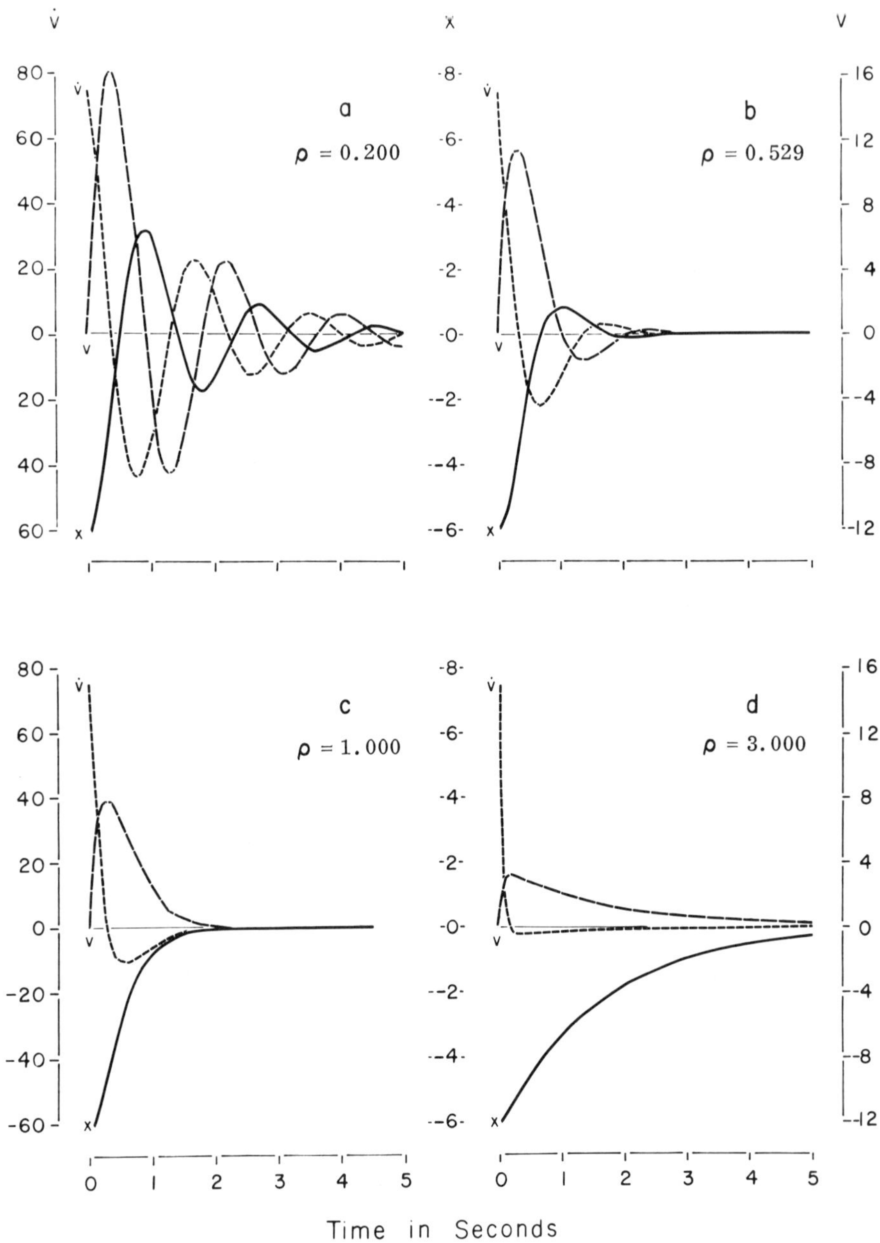

Figure 2.15.4. Response of the automobile suspension system for various values of ρ, the damping ratio.

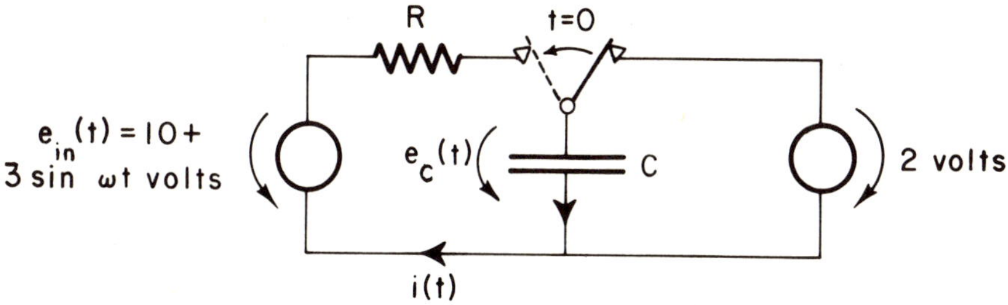

Figure 2.16.1. Circuit diagram for the system of Example 2.16.1.

or, more specifically,

$$Ri(t) + \frac{1}{C}\int i(t)dt = 10 + 3 \sin \omega t \quad \text{volts} \qquad 2.16.2.$$

when $t = 0^+$, equation 2.16.2. becomes

$$Ri(0^+) + \frac{1}{C}\int_{-\infty}^{0^+} i(t)dt = 10 \text{ volts} \qquad 2.16.3.$$

But since a capacitor cannot accumulate additional charge in zero time,

$$\frac{1}{C}\int_{-\infty}^{0^+} i(t)dt = \frac{1}{C}\int_{-\infty}^{0^-} i(t)dt + \frac{1}{C}\int_{0^-}^{0^+} i(t)dt = 2 + 0 = 2 \text{ volts}$$

Substituting this value into equation 2.16.3. and solving for $i(0^+)$, we get the initial condition,

$$i(0^+) = 8/R \quad \text{amp} \qquad 2.16.4.$$

Differentiating equation 2.16.2. with respect to time, and dividing through by R,

$$\frac{di(t)}{dt} + \frac{1}{RC}\, i(t) = \frac{3\omega}{R} \cos \omega t \quad \text{amp/sec} \qquad 2.16.5.$$

Whose general solution by classical methods (See Appendix) is

$$i(t) = \alpha \epsilon^{-\frac{1}{RC}t} + \frac{3\omega C}{1+[\omega RC]^2}[\cos \omega t] + \frac{3R[\omega C]^2}{1+[\omega RC]^2}[\sin \omega t] \quad \text{amp}$$

2.16.6.

At $t = 0^+$, equation 2.16.6., together with our initial condition (equation 2.16.4.) give

$$i(0^+) = \frac{8}{R} = \alpha + \frac{3\omega C}{1+[\omega RC]^2} \quad \text{amp} \qquad 2.16.7.$$

Whence

$$\alpha = \frac{8}{R} - \frac{3\omega C}{1+[\omega RC]^2} \quad \text{amp} \qquad 2.16.7.a.$$

Substituting this into equation 2.16.6., we obtain as our final result the equation for the total response of i(t),

$$i(t) = \left\{\frac{8}{R} - \frac{3\omega C}{1+[\omega RC]^2}\right\} \epsilon^{-\frac{1}{RC}t} + \frac{3\omega C}{1+[\omega RC]^2}[\cos \omega t + \omega RC \sin\omega t]$$

$$\text{amp} \qquad 2.16.8.$$

Knowing i(t) and $e_{in}(t)$, we can easily calculate the total response of $e_c(t)$ from equation 2.16.2.,

$$e_c(t) = \frac{1}{C}\int i(t)dt = 10 + 3 \sin \omega t - i(t)R \quad \text{volts} \qquad 2.16.9.$$

In Figure 2.16.2., $e_c(t)$ has been plotted against time, using the specific values: $R = 3 \cdot 10^4$ ohms, $C = 10^{-4}$ farads, $\omega = 1$ radian per second. Several important features of the figure, as indicated in its legend, deserve careful study. In particular, notice the rapid approach of $e_c(t)$ to the pure forced response described by equation 2.16.9. as $t \to \infty$,

$$\underset{t\to\infty}{e_c(t)} = 10 + 3 \sin \omega t - \frac{3\omega RC}{1+[\omega RC]^2}[\cos\omega t + \omega RC \sin \omega t]$$

$$= 10 - \frac{3}{1+[\omega RC]^2}[\omega RC \cos\omega t - \sin\omega t] \quad \text{volts} \qquad 2.16.10.$$

SECTION 2.17. Sign conventions when storage elements are releasing energy

The beginner is often (and understandably!) confused about signs when he is confronted by a passive network exhibiting a natural response to stored energy. A simple example will illustrate the confusion and show how to avoid it. Consider the circuit of Figure 2.17.1. Note that the voltage arrow indicates a voltage rise across the capacitor in the direction of current flow, so the capacitor must be discharging.

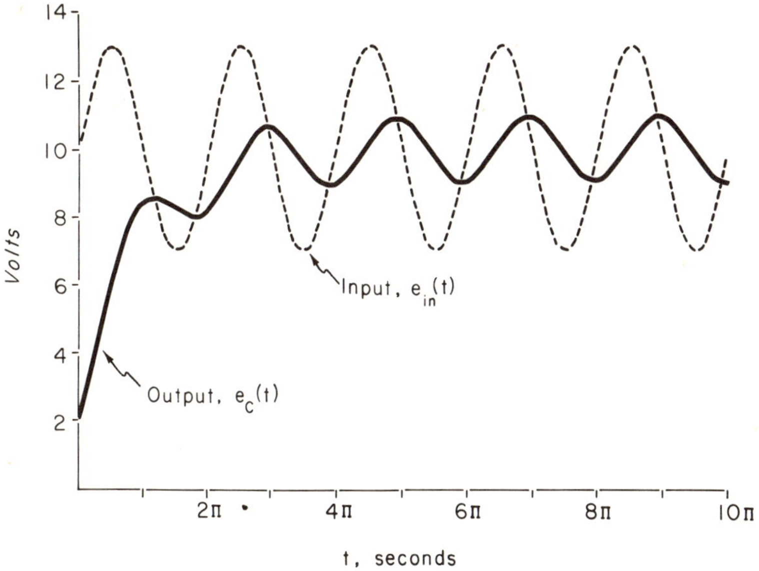

Figure 2.16.2. The response of $e_c(t)$ in the system of Figure 2.16.1. during the first half-minute after the switch has been thrown. Parameters: $R = 3 \cdot 10^4$ ohms; $C = 10^{-4}$ farads; $\omega = 1$ radian per second. Notice that:

1. $e_c(t)$ (solid line) rises rapidly from its initial value of 2 volts to a mean value of 10 volts imposed by the forcing input voltage, $e_{in}(t)$ (broken line). The rate of this exponential approach to the pure forced response is determined by the exponential decay constant, $1/RC = 1/3$ per second. The corresponding time constant, RC, is 3 seconds. In accordance with the rule-of-thumb mentioned in the footnote in Section 2.13., the transient part of the total response is practically complete when $t = 5 \cdot 3 = 15$ secs.

2. Whenever $e_c(t) = e_{in}(t)$, the two curves cross. These crossings always occur where $de_c(t)/dt = 0$. Why?

3. Consequently, the peak-to-peak interval for $e_c(t)$ during the transient part of the response is not a reliable indicator of the forcing frequency, ω.

4. When $e_c(t)$ has settled into the pattern of the pure forced response:

   a.) a sinusoidal input ($e_{in}(t)$) yields a sinusoidal output ($e_c(t)$).

   b.) the frequency of the periodic oscillations of the output is exactly the same as the frequency of the input, namely, $\omega = 1$ radian/sec.

   c.) the magnitude of the oscillations is different for output and input. In this example the oscillations of the output are damped (= decreased).

   d.) in any given cycle, the output reaches its maximum value somewhat later than the input does. The output is said to lag (= to lag behind) the input by a constant fraction of the full cycle. So the output is out of phase with the input by that fraction of 2π radians.

These matters will be studied in detail when we discuss phasor analysis and the frequency domain in the next chapter.

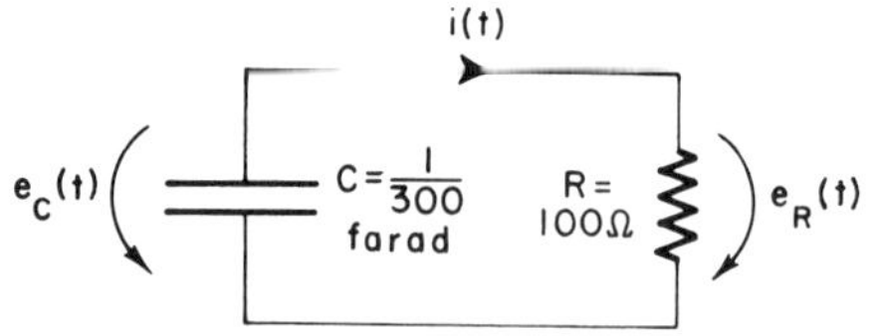

Figure 2.17.1. Discharging RC circuit.

Let $e_C(0^+) = 10$ volts. Find i(t) for all $t \geq 0^+$.

The wrong procedure will be presented first. By Kirchhoff's voltage law, the sum of the voltage drops around the loop is zero:

$$e_R(t) - e_C(t) = 0 \qquad 2.17.1.$$

Furthermore, from the laws governing the individual elements,

$$e_R(t) = Ri(t) = 100\, i(t) \text{ volts} \qquad 2.17.2.$$

$$e_C(t) = \frac{1}{C}\int i(t)dt = 300\int i(t)dt \quad \text{volts} \qquad 2.17.3.$$

Combining equations 2.17.1., 2.17.2., and 2.17.3.,

$$100\, i(t) - 300\int i(t)dt = 0 \quad \text{volts} \qquad 2.17.4.$$

Differentiating equation 2.17.4. with respect to time, and dividing through by 100,

$$\frac{di(t)}{dt} - 3i(t) = 0 \quad \text{volts/sec} \qquad 2.17.5.$$

Solving equation 2.17.5. subject to the initial condition,

$$i(0^+) = \frac{e(0^+)}{R} = \frac{10}{100} = 0.1 \quad \text{amp} \qquad 2.17.6.$$

we obtain

$$i(t) = 0.1\epsilon^{3t} \quad \text{amp} \qquad 2.17.7.$$

which indicates, quite wrongly, that current flow will increase without bound as time increases.

The mistake lies in equation 2.17.3., which incorrectly states that during discharge of the capacitor the current which flows in the direction marked on the circuit diagram will increase the voltage across the capacitor. But in fact, current flowing in the direction marked is carrying positive charge away from the capacitor so as to decrease $e_C(t)$. The correct version of equation 2.17.3. for this example is therefore

$$e_C(t) = -\frac{1}{C}\int i(t)dt = -300\int i(t)dt \quad \text{volts} \qquad 2.17.8.$$

which says correctly that $e_c(t)$ decreases with time as current flows in the direction indicated. If we combine this correct equation with equations 2.17.1. and 2.17.2., we obtain

$$100i(t) - [-300\int i(t)dt] = 100i(t) + 300\int i(t)dt = 0 \qquad 2.17.9.$$

which leads to the correct solution,

$$i(t) = 0.1\epsilon^{-3t} \qquad 2.17.10.$$

Similar care must be taken with T-elements such as inductors. A passive RL circuit in which an inductor is using the energy stored in its electromagnetic field to cause current flow may be correctly represented by the circuit diagram of Figure 2.17.2., where the numerical value of the voltage drop, $-L\frac{di(t)}{dt}$, will turn out to be positive (as it must for a correctly-oriented voltage drop arrow) because the current is decreasing and therefore $di(t)/dt$ itself is negative.

In practice, difficulties of this sort are easily avoided. Treat all storage elements as though they were storing energy, and mark the direction of the T-variable accordingly. Equations written for such networks will be valid whatever the actual direction of the T-variable. For example, consider a capacitor of capacitance 0.3 farads which at some particular time is being charged by a current such that $de_c(t)/dt = 2$ volts/sec. Since $i(t) = C\,[de_c(t)/dt]$, the charging current will be +0.6 amp. Suppose that the same capacitor at some other time is discharging at such a rate that $de_c(t)/dt = -2$ volts/sec. Then the charging current will be -0.6 amp, i.e., there will be a discharge current of +0.6 amp. So expressions such as

$$i(t) = C\,\frac{de_c(t)}{dt}$$

correctly represent the T-variable defined in the energy-storing direction whether the element is, in fact, storing energy or is, in fact, releasing energy.

Plusses and minuses are nasty, troublesome things. It is therefore useful to know that in a correct differential or integrodifferential equation for a stable linear system, when the terms which are homogeneous in the unknown variable have been collected on one side of the equation, they must all be of the same sign (commonly positive). This was not true of the incorrect equation 2.17.4. It was true of the correct equation 2.17.9. Mixed signs should always arouse the darkest suspicion of error in the derivation because they forecast one or more positive exponential terms in the solution (Cf. equation 2.17.7.) which therefore increases without limit in time, and contradicts the assumption of stability (See Section 4.7.). Finally, it is comforting to realize that, as pointed out above, if an equation is correct for a force or a flow in one direction, it will be equally correct for a force or a flow in the opposite

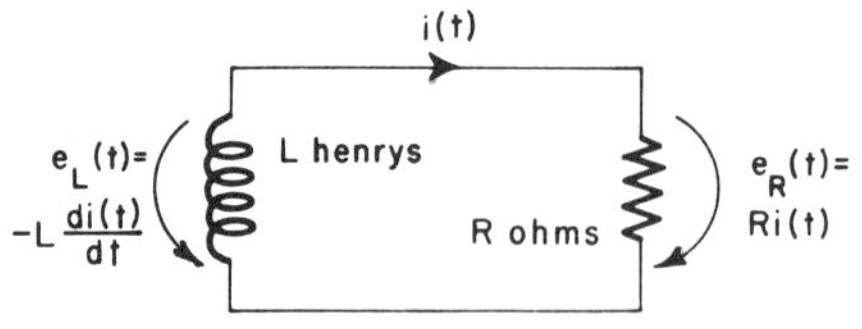

Figure 2.17.2. Discharging RL circuit.

direction. We are thus spared the mind-boggling prospect of switching equations every 1/120 sec. as we try to keep up with our domestic a-c.

SECTION 2.18. The information needed for a complete analysis of a system

The preceding examples show that, armed with enough information about a system, we may indeed be able to calculate the value of every dependent variable at any instant of time. It is important to understand precisely what kinds of information we must have if we are to achieve such a complete analysis We must know:

1. What individual elements compose the system, and what laws govern their behavior.

2. The parameters which characterize the elements.

3. The pattern of interconnections among the individual elements.

These three items of information allow us to write one or more homogeneous equations, usually differential or integro-differential, which specify how the unknown variables change with time when the system is activated. To specify how the system is activated, we must also know:

4. The actual value of each unknown variable at some particular instant of time, usually time zero. More specifically, for an nth- order system we need n such initial conditions, i.e., one for each storage element.

5. The A-variable or T-variable input to the passive elements of the system from one or more sources, as specified by a forcing function which gives the value of the input variable at every instant of time $\geq 0^+$.

To a mathematician, all of this information, neatly packaged in the form of a clean-cut, well-defined, nonhomogeneous equation together with an appropriate set of initial conditions, is regarded as "given." But to a physiologist, the information is far from given; it must all be extracted, by means of carefully controlled experiments, from a living system. In planning his experiments, therefore, the wise physiologist will keep constantly in mind the types of information he is going to need when he finally switches off the polygraph, disconnects the respirator, washes up the instruments and trudges home with his notebook and slide-rule under his arm. What are the elements of my system, and how are they interrelated? Will the parameters remain constant? What are the most suitable initial conditions, and how can I establish them? What kind of experimental change (= input forcing function) shall I use to activate the system?

Can I continue observations long enough to observe the pure forced or steady-state response? Above all, the physiologist should choose the simplest forcing function suitable to his purpose. Otherwise, by piling the Pelion of a complicated input upon the Ossa of a complex and non-linear biological system he will raise such an insurmountable barrier to mathematical analysis that his results will be suitable for publication only in the already overcrowded pages of The International Archives of Uninterpretable Observations.

Exercises

1. If a 1.5 volt dry cell really behaved as an ideal voltage source, how many watts of power would be consumed while the current from the dry cell was being passed through a copper wire, 2 mm in diameter and 3.14 cm long, maintained at a temperature of 20° C where its resistivity is 1.72 microhm•cm?

2. For the following circuit, find an expression in terms of x for the current flowing through the resistor between nodes A and B. Check your answer at $x = 0\ \Omega$, $x = 13.5\Omega$, and $x = \infty\ \Omega$.

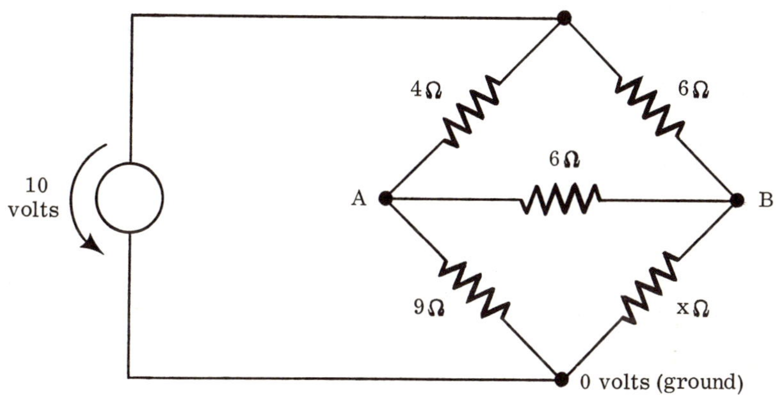

3. Use Ohm's law and Kirchhoff's laws to prove that when n resistors whose individual resistances are R_1, R_2, . . . R_n are connected in series, R_{ser}, the resistance of a single equivalent resistor is

$$R_{ser} = R_1 + R_2 + \cdots + R_n$$

but when connected in parallel, the equivalent resistance is

$$R_{par} = \frac{1}{\frac{1}{R_1} + \frac{1}{R_2} + \cdots + \frac{1}{R_n}}$$

[Hint. Analyze a circuit which includes the resistors and an appropriate source element.]

Derive the corresponding expressions for mechanical damping elements connected in series and in parallel.

4. In the following problems, the ideal source element is to be connected instantaneously at time zero so as to activate the designated passive elements, none of which have any stored energy at $t = 0^-$. When possible, derive an equation for the unknown function of time listed in the last column. If it is not possible, explain why.

Problem	Ideal source element supplying	Ideal passive elements*	Unknown function
A.	constant current	RC in parallel	$e_C(t)$
B.	constant current	RC in series	$i_C(t)$
C.	constant voltage	RC in parallel	$i_R(t)$
D.	constant voltage	RC in series	$e_R(t)$
E.	constant current	RL in parallel	$e_R(t)$
F.	constant current	RL in series	$e_L(t)$
G.	constant voltage	RL in parallel	$i_R(t)$
H.	constant voltage	RL in series	$i_L(t)$

*R = resistor, C = capacitor, L = inductor.

5. In example 1.4.2 of the previous chapter, the method of concomitant variations showed that when a resistor is activated by a current source, the steady-state voltage drop across the resistor was dependent upon the current flow. Yet I asserted that even with a current source, the flow of current through the resistor was caused by the difference in voltage across its terminals. Can you resolve this paradox? (Hint. Be a realist, not an idealist! Any real resistor will certainly behave as an inductor as well as a resistor. Any real (= practical) current source will certainly--what? Now connect the real current source with the real resistor at time zero and see what happens!)

6. A snail, whose best speed was 10 cm per min, challenged an electron, moving in a one-ampere direct current through a cylindrical copper wire 2 mm in diameter, to run a race over a 20 cm course. Which contestant should you put your money on, assuming that only one electron per atom participates in current flow? How long will it take the average electron to cover the distance?

7. Use the sequence of steps:

Real system → physical model → circuit diagram → {differential equation, initial conditions} → solution for unknown → checking solution

as exemplified in the text description of the automobile suspension to analyze the following system.

(a.) At time zero, 6 grams of inulin are distributed through a volume of 12 liters of extracellular fluid. Assuming that the volume of distribution is well-stirred at all times, find C(t) the concentration of inulin in the extracellular fluid at any time ≥ 0 when the renal plasma clearance of inulin is 125 ml of plasma per min.

(b.) Consider the same system as in part (a.), with the same parameters and initial conditions. At time zero, a continuous intravenous infusion of inulin is begun at a rate of 0.1 gram per min. Find C(t) at any time ≥ 0.

8. In the system of example 2.15.1, let $[\dot{V}dist] = 3.0$ liters; $[\dot{V}cl]_{ren} = 100$ ml/hr, $[\dot{V}cl]_{met} = 400$ ml/hr; $Q_p = 15$ mg; $\dot{Q}_{in}(t) = 0$ for $t < 0^+$, $\dot{Q}_{in}(t) = 2$ mg/hr for $t \geq 0^+$. Plot C(t) as a function of time in the interval [-4 hr, +24 hr].

9. Solve equation 2.15.22 with $x(0^+)$ defined as zero cm displacement. Compare your solution with Figure 2.15.4.b.

10. Consider an inductor where L = 2 henrys in parallel with a resistor where R = 6 ohms and an ideal current source of zero amperes for $t \leq 0^-$, and 8 cos 4t amperes for $t \geq 0^+$.

a.) Find e(t) for all $t > -2$ seconds.

b.) What does e(t) approach as t approaches infinity?

11. Read and criticize Section 6-3, pages 80-81, in Milhorn's text (1). You should be able to state exactly why the whole thrust of the argument in this section is wrong. (NOTE: You will have to decide what the variable y represents, because the author never defines it for us.)

References

1. Milhorn, H.T.: The Application of Control Theory to Physiological Systems. W.B. Saunders Co., Philadelphia, 1966.

General Reference:

Hayt, W.H. Jr. and Kemmerly, J.E.: Engineering Circuit Analysis. McGraw-Hill, N.Y., 1962.

This is a particularly lucid text, containing many drill exercises and problems. Although the discussion is confined to electrical systems, the methods of analysis presented are, of course, equally applicable to other linear systems.

"Come then, my sailors:
Loose your dripping hawsers, from their deep-sunk graves
Haul up your anchors, raise your brave new sails."

Antipater of Sidon, PRIAPOS OF THE HARBOR (Translated by Dudley Fitts)

3

MORE POWERFUL METHODS OF ANALYSIS OF LINEAR SYSTEMS

SECTION 3.1. On crossing the Atlantic in a sailing dinghy

In principle, the student who has mastered the methods described in the previous chapter should be able to analyze any stationary linear system, however complex, just as, in principle, the sailor who has mastered the essentials of navigation should be able to cross any body of water, however wide, in a sailing dinghy. But in actual practice, crossing the Atlantic in a sailing dinghy is a hazardous venture, costly of time, uncertain of destination. The student who tries to analyze a fourth-order system using classical methods alone is, like the sailor, prone to lose his bearings, or to be overwhelmed by tedium. In this chapter we shall begin to study certain more powerful methods which will greatly facilitate the analysis of a wide variety of linear systems. They do not guarantee a safe voyage, but they increase its probability. They will be our "brave new sails."

SECTION 3.2. The complex frequency, s

The investigation of complicated systems is often easier if we analyze their behavior not in the time domain where time, t, is the independent variable, but in the s-domain (= complex frequency domain) where the complex frequency, s, is the independent variable. s may be defined by the equation,

$$s = \sigma + j\,\omega \qquad 3.2.1.$$

where σ is the neper frequency (e.g. nepers per second) at which an exponential change is occurring

ω is the radian frequency (e.g. radians per second) at which a sinusoidal change is occurring

j is $\sqrt{-1}$

A neper is the $\log_\epsilon$ unit ($\epsilon = 2.71828...$) which is used to measure the extent of an exponential change. The reader undoubtedly knows that the statement, "x has increased by three $\log_{10}$ units" means exactly the same thing as "x has increased by a factor of 10^3" or "x has increased 1000 fold." Similarly, "x has increased by three nepers," "x has increased by three $\log_\epsilon$ units," "x has increased by a factor of $(2.71828\ldots)^3$," and "x has increased a bit over 20-fold" are equivalent statements. In mathematical analysis (as distinguished from mere calculation, or the presentation of data to people with ten fingers on their hands) the base used is always ϵ, never 10. However, nepers are rarely talked about; they are usually called "$\log_\epsilon$ units" (= ln units).

We have already encountered examples of the neper frequency, σ, for the neper frequency is the same thing as the familiar exponential rate constant. In example 2.15.2, σ was -1.87 ln units per sec; in example 2.16.1, σ was -1/RC ln units per unit time. In your savings-bank account (assuming 5 per cent interest compounded continuously) σ is + 0.05 per year. σ is thus a proportion per unit time, and it has the dimension of reciprocal time. It commonly appears in the solution of a differential equation as a coefficient of time in the exponent of ϵ.

A radian is the unit which is used to measure angles. An angle of one radian with its vertex at the center of a circle subtends an arc which is $\frac{1}{2\pi} = 0.15915\ldots$ of the circumference. Hence there are 2π radians in a full circle (or in a single sinusoidal cycle). In mathematical analysis (as distinguished from engineering, or the presentation of data to people with $90°$ in their right angles) the angular measure is always radians, never degrees.

We have already encountered examples of radian frequency, ω, for it is how the rate of any sinusoidal (or cosinusoidal) change is expressed. In example 2.15.2., ω was 3 radians per second. In a 60-cycle current there are sixty cycles of sinusoidal current fluctuation per second, so $\omega = 60[2\pi] = 377$ radians/sec. ω is thus an angular change per unit time, and, like σ, it has the dimensions of reciprocal time. In the solution of a differential equation, it may appear, preceded by j, as a coefficient of time in the exponent of ϵ, or (adroitly stripped of j, see Appendix) as the coefficient of time in the argument of a trigonometric function.

We have, in fact, already encountered an example of the complex frequency, s, in toto, though at the time we called it λ, not s. For the roots, or eigenvalues, $-1.87 \pm j3$, of the characteristic equation 2.15.25. are indeed frequencies of the form $\sigma + j\omega$ and $\sigma - j\omega$.

I have deliberately tried from the very beginning to endow the real and imaginary parts of s with the down-to-earth meanings which they will almost always have in this particular book. I have also tried to suggest that, again within the context of this book, "complex frequency, s" and "eigenvalue, λ" are closely related terms. I do not wish to leave the impression that they are identical, and the subsequent discussion should serve to distinguish them. (See Section 3.19.). But I am convinced that for the beginner to mistakenly regard s and λ as synonymous would be far less disastrous than for him to regard the complex variable, $s = \sigma + j\omega$ and the differential operator,

$D = d[\ ]/dt$ as synonymous. Yet the latter pseudosynonymy has actually been sanctioned (1), or implied (2) in print! Let it be stated once and for all that the complex variable s is NOT equivalent to the differential operator $d[\ ]/dt$, that s and $d[\ ]/dt$ cannot, in general, be interchanged in equations, and that nothing but confusion can arise from treating them as closely-related entities. If, later on, the informed student notices certain restricted and superficial similarities between $d[\ ]/dt$ and s (See Theorem vi in Section 3.5.), he may want to remember them as a purely mnemonic device with no fundamental significance.

SECTION 3.3. Laplace transforms

The actual changes which occur in any real physical system take place in the time domain and are therefore described by one or more equations in which time is the independent variable. If we want to exploit the advantages of s-domain analysis, we need a rule (= an operator) which will transform each different function of time into a unique function of s, and an inverse operator which will transform each different function of s into a unique function of time. While no such universal operator is available, the Laplace operator (a linear operator) is capable of transforming any member of a large class of time-domain functions---including most of the ones important in linear systems analysis---into a corresponding function in the s-domain. And the inverse Laplace operator (also a linear operator) can perform the inverse transformation from the s-domain to the t-domain.

Let F(t) be a suitable function of time. Let f(s) be the Laplace transform of F(t), i.e. the function of s which results from letting the Laplace operator, $\mathcal{L}$, operate on F(t). Then the equation

$$\mathcal{L}\{F(t)\} = \int_0^\infty \varepsilon^{-st} F(t)dt = f(s) \qquad 3.3.1.$$

defines the transforming operation. Notice that the integration extends only over the semi-infinite interval beginning at $t = 0$. (This is the usual restriction, though there does exist a "two-sided" Laplace transform over the entire range from $-\infty$ to $+\infty$).

How do we know whether a particular f(t) is "a suitable function," i.e., one for which the Laplace transform exists? Let t_1 be any finite time ≥ 0. Then if F(t) is 1.) at least piecewise continuous on the interval $0 \leq t \leq t_1$ for all values of t_1, and 2.) of exponential order as $t \to \infty$, it has a Laplace transform, f(s). "Piecewise continuous" is practically self-explanatory: a graph of F(t) plotted against time over the interval $0 \leq t \leq t_1$ may have any finite number of discontinuities but must be continuous between them. "Of exponential order" means that as $t \to \infty$, the ratio $F(t)/\varepsilon^{bt}$ does not approach $+\infty$ or $-\infty$, but approaches zero or a finite value for some positive constant, b. In other words, as $t \to \infty$, the absolute value of F, $|F(t)|$, must not increase more rapidly than an exponential of the form ε^{bt} where b is some constant, $0 < b < \infty$. While these two attributes are sufficient to guarantee that F(t) has a Laplace transform, they are not strictly necessary, for some functions such as $F(t) = t^{-1/2}$ which fail to satisfy the first criterion just stated nevertheless have a Laplace transform. To include such functions, the first criterion may be replaced by the broader (but less easily tested) criterion that the integral of $\epsilon^{-st}F(t)$ with respect to time between the limits $t = 0$ and $t = t_1$ must exist for all t_1 between zero and plus infinity.

Example 3.3.1. Does the function t^2 meet the criteria for the existence of a Laplace transform, and, if so, what is $\mathcal{L}\{t^2\}$?

(i) t^2 is obviously continuous on the interval $0 \le t \le t_1$ for all finite $t_1 \ge 0$

(ii) Is t^2 of exponential order? Use L'Hospital's Rule (See Appendix) to evaluate

$$\lim_{t\to\infty} \frac{t^2}{\epsilon^{bt}} = \lim_{t\to\infty} \frac{2t}{b\epsilon^{bt}} = \lim_{t\to\infty} \frac{2}{b^2\epsilon^{bt}} = 0 \text{ for all } b > 0$$

Therefore t^2 is of exponential order and $\mathcal{L}\{t^2\} = f(s)$ exists.

By equation 3.3.1,

$$\mathcal{L}\{t^2\} = \int_0^\infty \epsilon^{-st} t^2 dt = f(s)$$

Integrating by parts,

$$\int_0^\infty \epsilon^{-st} t^2 dt = -\frac{t^2}{s} \epsilon^{-st}\Big|_0^\infty + \frac{2}{s}\int_0^\infty \epsilon^{-st} t\, dt$$

$$= 0 + \frac{2}{s}\left\{-\frac{t}{s}\epsilon^{-st}\Big|_0^\infty + \frac{1}{s}\int_0^\infty \epsilon^{-st} dt\right\} = 0 + \frac{2}{s}\left\{0 - \frac{1}{s^2}\epsilon^{-st}\Big|_0^\infty\right\} = \frac{2}{s^3}$$

Hence,

$$\mathcal{L}\{t^2\} = 2/s^3$$

Example 3.3.2. Does the function ϵ^{t^2} meet the criteria for the existence of a Laplace transform, and, if so, what is $\mathcal{L}\{\epsilon^{t^2}\}$?

(i) ϵ^{t^2} is obviously continuous on the interval $0 \le t \le t_1$ for all $t_1 \ge 0$

(ii) Is ϵ^{t^2} of exponential order? We wish to evaluate

$$\lim_{t \to \infty} \frac{\epsilon^{t^2}}{\epsilon^{bt}} = \lim_{t \to \infty} \frac{\epsilon^t}{\epsilon^b}$$

But for any finite b, whether positive, zero, or negative, ϵ^b is a constant > 0, while ϵ^t increases without limit as time increases. Hence $\lim_{t \to \infty} \frac{\epsilon^t}{\epsilon^b} = \infty$. So ϵ^{t^2} is not of exponential order and has no Laplace transform.

SECTION 3.4. Inverse Laplace transforms and tables of Laplace transforms

The operator which transforms a function of s, f(s), back to the corresponding function of t, F(t), is the inverse Laplace operator, $\mathcal{L}^{-1}$, whose defining equation will not be given here since it involves complex integration and is difficult to use. It is far easier to look up inverse Laplace transforms in a table. As a matter of fact, even equation 3.3.1. is not used very often, because again it is usually easier to find the transform in a table. For example, the Handbook of Chemistry and Physics (3) lists many functions of t and the corresponding functions of s in two parallel columns so that one can go with equal facility from the t-domain to the s-domain and from the s-domain to the t-domain. Table D-6 in Korn and Korn's Handbook (4) is even more extensive.

SECTION 3.5. Theorems about Laplace transformations

We cannot use Laplace transforms until we know something about their properties. The following list of theorems, though far from exhaustive, includes those we shall need most frequently. Others may be found in the Handbook of Chemistry and Physics (3) under the heading "Laplace Operations." In the equations below, a and b are constants, and n is a non-negative integer.

(i) Definition of, and symbols for the Laplace transform.

$$\mathcal{L}\{F(t)\} \equiv f(s) \equiv F_{\mathcal{L}}(s) \equiv \int_0^\infty \epsilon^{-st} F(t)dt \qquad 3.5.1.$$

(ii) $\mathcal{L}$ is a linear operator.

$$\mathcal{L}\{aF(t) + bG(t)\} = a\mathcal{L}\{F(t)\} + b\mathcal{L}\{G(t)\} \qquad 3.5.2.$$

NOTE: The inverse Laplace operator, $\mathcal{L}^{-1}$, is also a linear operator.

(iii) Laplace transforms of constants.

$$\mathcal{L}\{a\} = a/s \qquad 3.5.3.$$

EXAMPLE: $\mathcal{L}\{3\} = 3/s$

(iv) Laplace transforms of exponential functions.

$$\mathcal{L}\left\{\epsilon^{at}F(t)\right\} = f(s-a) \qquad 3.5.4.$$

EXAMPLE: Find $\mathcal{L}\left\{8\epsilon^{-2t}\right\}$.

$F(t) = 8;\ f(s) = 8/s;\ a = -2.$

Therefore, $\mathcal{L}\left\{8\epsilon^{-2t}\right\} = 8/[s+2]$

NOTE: Remember that f (s-a) means "the function f for arguments of s-a instead of s." So we simply replace s in the function $f(s) = 8/s$ by s-a. Since $a = -2$, this gives us $8/[s+2]$.

(v) Laplace transforms of terms with an integral, non-negative power of t as a factor.

$$\mathcal{L}\left\{t^nF(t)\right\} = [-1]^n \frac{d^nf(s)}{ds^n} \qquad 3.5.5.$$

EXAMPLE: Find $\mathcal{L}\left\{t^2\epsilon^{-2.7t}\right\}$.

$F(t) = \epsilon^{-2.7t};\ f(s) = 1/[s+2.7];\ n = 2;\ [-1]^n = 1$

$$\frac{df(s)}{ds} = \frac{-1}{[s+2.7]^2} = \frac{-1}{s^2+5.4s+7.29}$$

$$\frac{d^2f(s)}{ds^2} = \frac{2s+5.4}{[s+2.7]^4} = \frac{2[s+2.7]}{[s+2.7]^4} = \frac{2}{[s+2.7]^3}$$

Therefore, $\mathcal{L}\left\{t^2\epsilon^{-2.7t}\right\} = 2/[s+2.7]^3$

NOTE the special case where $F(t) = 1$.

$$\mathcal{L}\left\{t^n\right\} = [-1]^n \frac{d^ns^{-1}}{ds^n} = \frac{n!}{s^{n+1}} \qquad 3.5.6.$$

NOTE: For non-integral but positive powers of t, say t^k, $k > 0$, there is a similar transform in which n! is replaced by $\Gamma(k+1)$, the Gamma function of $k+1$.

(vi) Laplace transforms of derivatives.

$$\mathcal{L}\left\{dF(t)/dt\right\} = sf(s) - F(0^+) \qquad 3.5.7.$$

or, more generally,

$$\mathcal{L}\left\{d^nF(t)/dt^n\right\} = s^nf(s) - s^{n-1}F(0^+) - s^{n-2}F'(0^+) - s^{n-3}F''(0^+) - \cdot\cdot\cdot\cdot - s^0F^{(n-1)}(0^+) \qquad 3.5.8.$$

EXAMPLE: Find $\mathcal{L}\left\{d^2x(t)/dt^2\right\}$

$$\mathcal{L}\left\{\frac{d^2x(t)}{dt^2}\right\} = s^2x_{\mathcal{L}}(s) - sx(0^+) - \left.\frac{dx(t)}{dt}\right|_{0^+}$$

NOTE that each term beyond the first in equation 3.5.8. embodies one of the initial conditions.

(vii) Laplace transform of an integral.

$$\mathcal{L}\left\{\int_{-\infty}^{t} F(\tau)d\tau\right\} = \frac{1}{s}f(s) + \frac{1}{s}\int_{-\infty}^{0} F(t)dt \qquad 3.5.9.$$

EXAMPLE: Find the Laplace transform of the integral which represents the voltage across a capacitor of capacitance C, when the current flow through the capacitor is i(t).

$$\mathcal{L}\left\{\frac{1}{C}\int_{-\infty}^{t} i(\tau)d\tau\right\} = \frac{1}{C}\mathcal{L}\left\{\int_{0}^{t} i(\tau)d\tau\right\} + \frac{1}{C}\mathcal{L}\left\{\int_{-\infty}^{0} i(t)dt\right\}$$

$$= \frac{1}{sC}\, i_{\mathcal{L}}(s) + \frac{1}{sC}\int_{-\infty}^{0} i(t)dt$$

NOTE that the second term is the Laplace transform of a constant, namely the initial condition $\frac{1}{C}\int_{-\infty}^{0} i(t)dt$, which is the voltage across the capacitor at $t = 0$.

(viii) The shifting theorem.

$$\mathcal{L}\left\{F(t - \alpha)\, u(t - \alpha)\right\} = f(s)\varepsilon^{-\alpha s} \qquad 3.5.10.$$

where α = a particular time ≥ 0

$$u(t - \alpha) = \begin{cases} 0 & \text{for } t < \alpha \\ 1 & \text{for } t \geq \alpha \end{cases}$$

EXAMPLE: (See Figure 3.5.1.) Find the Laplace transform of $5\epsilon^{-0.3[t-2]}u(t-2)$, where time is in seconds.

$$F(t) = 5\epsilon^{-0.3t}; \ \alpha = 2 \text{ sec}; \ f(s) = 5/[s + 0.3] \text{ sec};$$

$$\mathcal{L}\left\{5\epsilon^{-0.3[t-2]}u(t-2)\right\} = \frac{5}{s+0.3}\,\epsilon^{-2s} \text{ sec}$$

NOTE: u(t), called the unit step function, is defined as zero for all negative values of its argument, unity for all non-negative values of its argument. Figure 3.5.1. gives a graphical explanation of the effect of u(t) upon the exponential function of the present example. See also Section 3.8.

(ix) The convolution theorem. Let

$$F_1(t) * F_2(t) \equiv \int_0^t F_1(t-\tau)\,F_2(\tau)\,d\tau \qquad 3.5.11.$$

be the convolution of $F_1(t)$ with $F_2(t)$. Then

$$\mathcal{L}\left\{F_1(t) * F_2(t)\right\} = f_1(s)\,f_2(s) \qquad 3.5.12.$$

EXAMPLE: A detailed example illustrating the use of the convolution theorem is discussed in Section 3.12.

In the preceding theorems, a capital letter has been used for the symbol of the time-domain function, the same letter, in lower case, for the corresponding s-domain function. This accords with the usage in the Handbook of Chemistry and Physics and in many textbooks, but electrical engineers reverse the convention, preferring f(t) and F(s). To preserve freedom of choice of either capital or lower-case letters as primary symbols for variables, in the remainder of this book I shall append a subscript "$\mathcal{L}$" to the time-domain symbol for a function to denote its Laplace transform. This convention has already been defined in equation 3.5.1. and used in the examples illustrating Theorems (vi) and (vii).

SECTION 3.6. Solving equations by using Laplace transforms

We are now prepared to consider the steps used in the solution of integral, differential, or integro-differential equations by the Laplace transform technique, and to illustrate these steps with an example. Suppose we have an equation where the unknown function of time (perhaps representing an A-variable or a T-variable) is x(t).

1. Define $x_{\mathcal{L}}(s) \equiv \mathcal{L}\{x(t)\}$
2. Take Laplace transforms of both sides of the equation, term by term, using a table of Laplace transforms freely. The result will be an algebraic (not differential) equation in $x_{\mathcal{L}}(s)$.

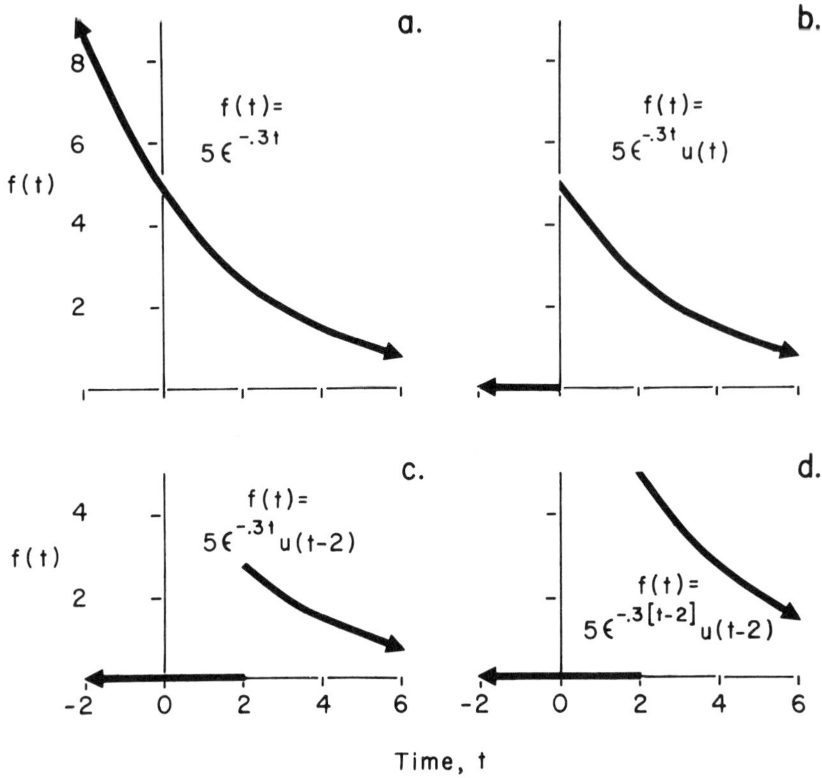

Figure 3.5.1. The effect of multiplying an ordinary function of time by the unit step function.

(a.) The graph for the unrestricted exponential function of time depicted here would extend an infinite distance on either side of $t = 0$. (b.) The unit step function, u(t), suppresses the function prior to time zero. (c.) The unit step function u(t-2) suppresses the function prior to time 2. (d.) Replacing t by t-2 both in the exponent and in the argument of u(t) shifts the entire curve of panel b two time units to the right along the time scale without otherwise altering it.

3. Solve the algebraic equation for $x_{\mathcal{L}}(s)$. You will usually obtain an expression of the form

$$x_{\mathcal{L}}(s) = \frac{N(s)}{D(s)}$$

where the numerator N(s), and the denominator, D(s) are polynomial functions of s with the order of the numerator equal to or less than the order of the denominator.

4. Factor the denominator, (See Appendix) and break N(s)/D(s) into the corresponding partial fractions (See Appendix) so as to express $x_{\mathcal{L}}(s)$ as the sum of a series of partial fractions.

5. Making free use of a table of Laplace transforms, take the inverse Laplace transform of the sum, term by term. This will give the desired unknown function, x(t).

Notice that initial conditions have not been mentioned explicitly in any of these steps. The reason is that, by virtue of theorems (vi) and (vii) the initial conditions are automatically included when we take Laplace transforms in step 2.

Example 3.6.1. Solve equation 2.15.15. with the parameters of example 2.15.2. using Laplace transforms. We have,

$$m\frac{dv(t)}{dt} + Dv(t) + k\int v(t)dt = F_g \qquad 3.6.1.$$

Putting in the numerical values of the parameters,

$$8\cdot 10^5 \frac{dv(t)}{dt} + 2.99\cdot 10^6\, v(t) + 10^7\int v(t)dt = 7.85\cdot 10^8 \text{ dynes} \qquad 3.6.2.$$

with the initial conditions,

$$v(0) = 0 \text{ cm/sec} \ ; \ F_k(0) = k\int_{-\infty}^{0} v(t)dt = 7.25\cdot 10^8 \text{ dynes}$$

1. Define $v_{\mathcal{L}}(s) \equiv \mathcal{L}\{v(t)\}$

2. Take Laplace transforms of both sides of equation 3.6.2.,

$$8\cdot 10^5 s v_{\mathcal{L}}(s) - 0 + 2.99\cdot 10^6 v_{\mathcal{L}}(s) + \frac{1}{s}10^7 v_{\mathcal{L}}(s) + \frac{1}{s}\,7.25\cdot 10^8 = \frac{1}{s}\,7.85\cdot 10^8$$

$$\text{dyne}\cdot\text{sec} \qquad 3.6.3.$$

3. Solve equation 3.6.3. for $v_{\mathcal{L}}(s)$,

$$v_{\mathcal{L}}(s) = \frac{75}{s^2 + 3.74s + 12.5} \text{ cm} \qquad 3.6.4.$$

4. Factor the denominator in equation 3.6.4. and break the right hand side of the equation into partial fractions. (See Appendix).

A little mental checking shows that the denominator cannot be factored further except into complex factors. Hence,

$$v_{\mathcal{L}}(s) = \frac{75}{s^2 + 3.74s + 12.5} = \frac{[\alpha]s + [\alpha k + \beta\omega]}{s^2 + [2k]s + [k^2 + \omega^2]} =$$

$$\mathcal{L}\left\{\alpha\epsilon^{kt}\cos\omega t + \beta\epsilon^{kt}\sin\omega t\right\} \text{ cm} \qquad 3.6.5.$$

From which

$$\left.\begin{aligned} \alpha &= 0 \\ \beta\omega &= 75 \\ -2k &= 3.74 \\ k^2 + \omega^2 &= 12.5 \end{aligned}\right\} \Longrightarrow \left\{\begin{aligned} \beta &= 25 \\ k &= -1.87 \\ \omega &= 3 \end{aligned}\right.$$

Substituting these values into equation 3.6.5.,

$$v_{\mathcal{L}}(s) = \mathcal{L}\left\{25\epsilon^{-1.87t}\sin 3t\right\} \text{ cm} \qquad 3.6.6.$$

5. Take inverse transforms on both sides of equation 3.6.6.

$$v(t) = 25\epsilon^{-1.87t}\sin 3t \text{ cm/sec} \qquad 3.6.7.$$

which, of course, is precisely the solution found by the "classical" approach. (Cf. equation 2.15.32.).

Example 3.6.2. Solve equation 2.16.2.,

$$Ri(t) + \frac{1}{C}\int i(t)dt = 10 + 3\sin\omega t \quad \text{volts} \qquad 3.6.8.$$

subject to the initial conditions,

$$e_c(0^+) = \frac{1}{C}\int_{-\infty}^{0} i(t)dt = 2 \text{ volts}$$

1. Let $i_{\mathcal{L}}(s) \equiv \mathcal{L}\{i(t)\}$

2. Take Laplace transforms of both sides of equation 3.6.8.,

$$Ri_{\mathcal{L}}(s) + \frac{1}{Cs}i_{\mathcal{L}}(s) + 2\frac{1}{s} = 10\frac{1}{s} + 3\frac{\omega}{s^2 + \omega^2} \text{ volt}\cdot\text{sec} \qquad 3.6.9.$$

3. Solve equation 3.6.9. for $i_{\mathcal{L}}(s)$,

$$i_{\mathcal{L}}(s) = \frac{8Cs^2 + 3\omega Cs + 8\omega^2 C}{RCs^3 + s^2 + RC\omega^2 s + \omega^2} \quad \text{amp}\cdot\text{sec} \qquad 3.6.10.$$

4. Factor the denominator in equation 3.6.10. and break the right hand side of the equation into partial fractions.

Using Lin's method (See Appendix), we find that

$$RCs^3 + s^2 + RC\omega^2 s + \omega^2 = [RCs + 1][s^2 + \omega^2] \qquad 3.6.11.$$

Hence,

$$i_{\mathcal{L}}(s) = \frac{8Cs^2 + 3\omega\ Cs + 8\omega^2 C}{[RCs + 1][s^2 + \omega^2]} = \frac{\frac{8}{R}s^2 + \frac{3\omega}{R}s + \frac{8\omega^2}{R}}{[s + \frac{1}{RC}][s^2 + \omega^2]} =$$

$$\frac{\left\{\frac{8}{R} - \frac{3\omega C}{1 + [\omega RC]^2}\right\}}{s + \frac{1}{RC}} + \frac{\left\{\frac{3\omega C}{1 + [\omega RC]^2}\right\} s + \left\{\frac{3[\omega RC]^2}{R\{1 + [\omega RC]^2\}}\right\}\omega}{s^2 + \omega^2}$$

$$\text{amp}\cdot\text{sec} \qquad 3.6.12.$$

5. Take inverse transforms on both sides of equation 3.6.12.

$$i(t) = \left\{\frac{8}{R} - \frac{3\omega C}{1 + [\omega RC]^2}\right\} \epsilon^{-\frac{1}{RC}t}$$

$$+ \left\{\frac{3\omega C}{1 + [\omega RC]^2}\right\} [\cos \omega t + \omega RC \sin \omega t] \quad \text{amp} \qquad 3.6.13.$$

as was found before (equation 2.16.8.).

The reader should not be deceived into thinking that the Laplace transform technique of solving equations is always shorter and easier than the classical techniques. Indeed, if he will actually compare the time needed to perform all of the necessary steps in example 2.16.1. (classical method) with the time needed to perform all of the necessary steps in example 3.6.2. (Laplace transform method) he will probably find that the classical method for this example is much the shorter. The true power of the Laplace transform method will not really become evident until we apply it to more complicated problems.

SECTION 3.7. Transfer functions

Up to this point, our attention has been focused upon single elements and how they can be connected together to form simple systems. We shall now begin to investigate methods for analyzing the behavior of more general systems whose component parts may be either simple elements of the sort already described or subsystems consisting of several elements. In such a general system, the form of energy may well change from component to component so that we can no longer think in narrow terms of voltages and currents only, or forces and velocities only, or pressures and flows only. In a generalized system what is passed along or transferred from component to component is simply called a signal, in recognition of its role in carrying information from one component to another.

Each separate component in a general system accepts an input signal (= input) (which may be the output of some other component of the system or may come from outside the system) and operates physically upon the input to produce an output signal (= output) which may, in turn, become the input for one or more other components of the system. This is always a unidirectional process; for a given component, the input is always an independent variable, described by a forcing function, and the output is a dependent variable. Unidirectionality of signal flow is, in fact, the essential feature which allows us to treat each individual component as a separate entity which can be influenced by other parts of the system, or by external forcing, only via its input.

In Section 2.14 it was pointed out that the type of activity which a linear system of passive elements can potentially exhibit is completely specified by a homogeneous equation with zero initial conditions. This equation is derived entirely from information about the elements of the system and how they are interlinked; it is not influenced by any previous activating inputs which may have left their mark in the form of stored energy, (initial conditions), or by any present activating input, (forcing function). In a more general system, whose components are linear subsystems, the potential behavior of each individual component can likewise be specified by its homogeneous equation, with zero initial conditions. Knowing that equation, we can certainly calculate the particular output of the component which corresponds to any particular input. But we would like to do better than this. We would like to find a general expression for the output, in terms of the input and the equation describing the characteristics of the component. We could then use this general expression as the input forcing function for the next component in the system. By repeating this process, we would hope to find a general equation for the system as a whole which would let us calculate from the external inputs alone some final output of interest, without having to bother with the intermediate outputs and inputs linking the individual components within the system itself. To find out whether we can actually do this, let us consider, in the simplest and most general way, the equation which characterizes a single component.

In general, the homogeneous equation which characterizes the nth component in a linear system will be of the form,

$$L_n y_n(t) = 0 \qquad 3.7.1.$$

where L_n is the linear operator which completely specifies the nature of the component, $y_n(t)$ is the output, and all initial conditions are zero. Let $x_n(t)$ be any input to the general component described by equation 3.7.1. Then the general nonhomogeneous equation for the nth component will be,

$$L_n y_n(t) = x_n(t) \qquad 3.7.2.$$

But if we naively "solve" this equation for $y_n(t)$, the general output we are after, we obtain,

$$y_n(t) = \frac{x_n(t)}{L_n}, \qquad 3.7.3.$$

an "equation" which is sheer nonsense since it asks us to "divide" the input function of time by an operator whose proper domain is the output!* On the other hand, we cannot use classical methods for handling differential equations to find a general expression for $y_n(t)$ as a straightforward function of $x_n(t)$ because we cannot solve equation 3.7.2. except for some particular kind of input. Since each different kind of input will give a different kind of output, we are right back where we started.

[* See also the discussion in Section 1.12.]

Fortunately, Laplace transforms provide a simple and practical way out of our dilemma. By taking Laplace transforms on both sides of equation 3.7.2., we obtain an algebraic equation which can be solved for $y_{n,\mathcal{L}}(s)$, the Laplace transform of the generalized output,

$$y_{n,\mathcal{L}}(s) = G_n(s)\, x_{n,\mathcal{L}}(s) \qquad 3.7.4.$$

where $x_{n,\mathcal{L}}(s) \equiv \mathcal{L}\{x_n(t)\}$ is the Laplace transform of the generalized input, and the function $G_n(s)$ is the transfer function of the nth component.

Notice how very general were the assumptions used in deriving equation 3.7.4. We began with the homogeneous equation 3.7.1., carefully stipulating zero initial conditions. Zero initial conditions are essential for generality, because if energy were already stored in the system from previous inputs, the response to any new input would not be determined solely by the inherent nature of the component itself but would be determined in part by its particular past history. When we did allow an input (equation 3.7.2.) we defined it as any forcing function of time. Finally, the Laplace transforms which we defined are just as general as the corresponding functions of time. Consequently, the transfer function $G_n(s)$ is determined entirely by the linear operator, L_n, and thus reflects only the nature of the nth component itself. Equation 3.7.4. tells us that we can get the Laplace transform of the output for any input simply by multiplying the Laplace transform of the input, $x_{n,\mathcal{L}}(s)$, by the transfer function, $G_n(s)$. We may then use the Laplace transform of the output so obtained as the transformed input for the next component in line, thus accomplishing, with Laplace transforms, the task we originally set ourselves, namely, to find a general expression for the time-dependent behavior of the system as a whole. For it is obvious that at any stage we can take inverse transforms, thereby translating our result back to time domain.

By solving equation 3.7.4. for G(s) we obtain a formal definition of the term "transfer function,"

$$\text{Transfer function} = G(s) = \frac{\text{Laplace transform of output}}{\text{Laplace transform of input}} \qquad 3.7.4.a.$$

The transfer function for any linear component or any linear system is the Laplace transform of the output divided by the Laplace transform of the input, with zero initial conditions assumed. No other definition is valid.*

[*"Transfer function" has sometimes been defined as the output divided by the input (1,5,6). This is no more correct than saying that the product of two numbers is the sum of their logarithms.]

In Section 3.10 below, we shall see that we lose little by insisting upon zero initial conditions, for we can define a special kind of input, the impulse input, which plays exactly the same role as does an initial condition. In transfer function analysis, therefore, impulse inputs can be used to establish initial conditions.*

[*In Chapter 2 it was stressed that in solving differential equations, the initial conditions to be used are always the conditions at $t = 0^+$, i.e., immediately after the instantaneous imposition of an input at time zero. But in transfer function analysis, the system must initially be inactive, and the equivalent of initial conditions may have to be established by imposing a suitable impulse input instantaneously at time zero as described in Section 10. Hence, in transfer function analysis, the statement that all initial conditions must be zero refers to the conditions at $t = 0^-$, just before the input has been applied. At $t = 0^+$, immediately after the input has been applied, the input will already have established precisely the conditions which, in Chapter 2, we called initial conditions.]

The actual derivation of transfer functions will now be illustrated by two examples,

Example 3.7.1. What is the transfer function for the system of example 2.16.1., (a) if the input is $e_{in}(t)$ and the output is $i(t)$? From equation 2.16.2., we have,

$$Ri(t) + \frac{1}{C}\int i(t)dt = e_{in}(t) \qquad 3.7.5.$$

whose Laplace transform (assuming zero initial conditions) is,

$$Ri_{\mathcal{L}}(s) + \frac{1}{Cs}\, i_{\mathcal{L}}(s) = e_{in,\mathcal{L}}(s) \qquad 3.7.6.$$

Solving equation 3.7.6. for $i_{\mathcal{L}}(s)/e_{in,\mathcal{L}}(s)$,

$$\text{transfer function} = \frac{i_{\mathcal{L}}(s)}{e_{in,\mathcal{L}}(s)} = \frac{Cs}{RCs+1} = \frac{[1/R]s}{s+[1/RC]} \qquad 3.7.6.a.$$

(b) if the input is $e_{in}(t)$ and the output is $e_c(t)$? Since

$$e_c(t) = \frac{1}{C}\int i(t)dt \qquad 3.7.7.$$

equation 3.7.5. can be written as

$$e_c(t) = e_{in}(t) - Ri(t) \qquad 3.7.8.$$

whose Laplace transform is

$$e_{c,\mathcal{L}}(s) = e_{in,\mathcal{L}}(s) - Ri_{\mathcal{L}}(s) \qquad 3.7.9.$$

combining equations 3.7.6.a. and 3.7.9. to eliminate $i_{\mathcal{L}}(s)$,

$$\frac{e_{in,\mathcal{L}}(s)Cs}{RCs+1} = \frac{e_{in,\mathcal{L}}(s)}{R} - \frac{e_{c,\mathcal{L}}(s)}{R} \qquad 3.7.10.$$

Solving equation 3.7.10. for $e_{c,\mathcal{L}}(s)/e_{in,\mathcal{L}}(s)$,

$$\text{transfer function} = \frac{e_{c,\mathcal{L}}(s)}{e_{in,\mathcal{L}}(s)} = \frac{1}{RCs+1} = \frac{1/RC}{s+[1/RC]} \qquad 3.7.10.a.$$

Notice that the transfer functions for two different outputs of the same system are different, as you would expect.

Example 3.7.2. Distribution of a drug between two compartments.

When given intravenously, a certain drug, P, behaves as if it were instantaneously distributed throughout a well-stirred compartment, A, of volume $V_A = 10$ liters. P can diffuse between A and a second well-stirred compartment, B, of volume $V_B = 30$ liters. The rate of diffusion in mg/min is directly proportional to the concentration difference, $C_A(t) - C_B(t)$ mg/liter, between the two compartments, the constant of proportionality being 0.5 liter/min. The clearance of P from A by metabolism in the liver is 0.1 liter/min. If the input is $\dot{Q}_{in}(t)$, the rate of intravenous administration of P, and the output is defined as $C_B(t)$, what is the transfer function for the system?

The circuit diagram corresponding to the preceding description is shown in Figure 3.7.1. Since we want to use transfer function analysis, we must have zero initial conditions. We shall therefore assume that there is no P present in the system at $t = 0$, so that

$$C_A(0) = C_B(0) = 0 \text{ mg/L}$$

By Kirchhoff's T-variable law, at node $C_A(t)$,

$$\dot{Q}_{in}(t) = \frac{C_A(t) - C_B(t)}{2} + \frac{C_A(t)}{10} + 10\frac{dC_A(t)}{dt} \quad \text{mg/min} \qquad 3.7.11.$$

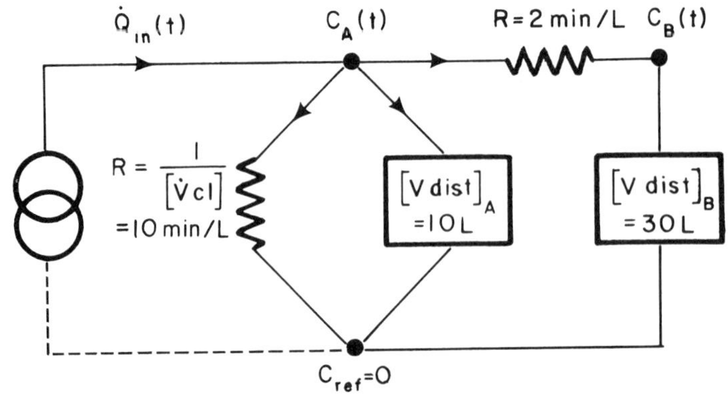

Figure 3.7.1. Circuit diagram for Example 3.7.2.

or,

$$100\frac{dC_A(t)}{dt} + 6C_A(t) - 5C_B(t) = 10\,\dot{Q}_{in}(t) \text{ mg/min} \qquad 3.7.11.a.$$

whose Laplace transform is

$$[100s + 6]\, C_{A,\mathcal{L}}(s) - 5C_{B,\mathcal{L}}(s) = 10\,\dot{Q}_{in,\mathcal{L}}(s) \text{ mg} \qquad 3.7.12.$$

At node $C_B(t)$,

$$\frac{C_A(t) - C_B(t)}{2} = 30\frac{dC_B(t)}{dt} \text{ mg/min} \qquad 3.7.13.$$

or,

$$60\frac{dC_B(t)}{dt} + C_B(t) = C_A(t) \text{ mg/min} \qquad 3.7.13.a.$$

whose Laplace transform is

$$[60s + 1]\, C_{B,\mathcal{L}}(s) = C_{A,\mathcal{L}}(s) \text{ mg} \qquad 3.7.14.$$

Combining equations 3.7.12 and 3.7.14. by eliminating $C_{A,\mathcal{L}}(s)$,

$$[100s + 6]\,[60s + 1]\, C_{B,\mathcal{L}}(s) - 5C_{B,\mathcal{L}}(s) = 10\dot{Q}_{in,\mathcal{L}}(s) \text{ mg} \qquad 3.7.15.$$

Solving equation 3.7.15. for the transfer function $C_{B,\mathcal{L}}(s)/\dot{Q}_{in,\mathcal{L}}(s)$,

$$\frac{C_{B,\mathcal{L}}(s)}{\dot{Q}_{in,\mathcal{L}}(s)} = \frac{1}{600\,s^2 + 46\,s + 0.1} = \frac{1/600}{[s + 0.00224]\,[s + 0.0744]} \text{ min/L}$$

3.7.15.a.

Having gotten the transfer function, we can easily find the drug concentration in compartment B at any time $t \geq 0^+$ for any pattern of drug administration which can be expressed as a Laplace-transformable function of time. For example, suppose we were to infuse drug P intravenously at a constant rate of $\dot{Q}_{in}(t) = 2$ mg per min starting at time zero. To find $C_B(t)$, we need only multiply the transfer function of equation 3.7.15.a. by $\dot{Q}_{in,\mathcal{L}}(s) = \mathcal{L}\{2\} = 2/s$ mg,

$$C_{B,\mathcal{L}}(s) = \frac{1}{300}\left\{\frac{1}{s\,[s + 0.00224]\,[s + 0.0744]}\right\} \frac{\text{mg}\cdot\text{min}}{\text{L}} \qquad 3.7.16.$$

break the right hand side of equation 3.7.16. into partial fractions (See Appendix),

$$C_{B,\mathcal{L}}(s) = \frac{20}{s} - \frac{20.6}{s + 0.00224} + \frac{0.621}{s + 0.0744} \quad \frac{\text{mg}\cdot\text{min}}{\text{L}} \qquad 3.7.17.$$

and take inverse Laplace transforms,

$$C_B(t) = 20 - 20.6\epsilon^{-0.00224t} + 0.621\epsilon^{-0.0744t} \text{ mg per liter} \qquad 3.7.18.$$

SECTION 3.8. Common types of nonperiodic inputs

Certain simple kinds of input forcing functions have been found particularly useful in the analysis of linear systems. These functions, their graphical representations and their Laplace transforms are summarized in Table 3.8.1. In dealing with such inputs, we shall make repeated use of the principle of superposition (Section 1.13.), that most endearing attribute of linear systems which assures us that the output produced by two different inputs superimposed upon each other (i.e., applied either simultaneously or seriatim) is simply the sum of the two separate outputs which we would get if we applied each of the two inputs separately.

The unit step, u(t), is defined as zero for all negative values of its argument, unity for zero or positive values of its argument. We have already used the function u(t) in the shifting theorem of Section 3.5. We shall regard u(t) as dimensionless. By multiplying any other function by u(t) we can suppress it for negative values of the argument of u, and "turn it on" when $t = 0$. This should be done explicitly for all input forcing functions whose Laplace transforms are to be multiplied by a transfer function, in order to ensure zero initial conditions. Often, however, the u(t) is taken for granted and is omitted. As a physiological example of how the unit step function is employed, suppose the input is $\dot{Q}_{in}(t)$, the rate of intravenous infusion of inulin (g per min). Then the infusion of 0.025 g of inulin per min starting at time zero and continuing (in theory at least!) forever would be called a step input of magnitude 0.025 g per min, and would be symbolized as $\dot{Q}_{in}(t) = 0.025u(t)$ g per min. But if the infusion were not started until some time after time zero, say at $t = t_1 = 20$ min, the input function would be $\dot{Q}_{in}(t) = 0.025u(t-20)$ g per min. This makes the input zero until $t = 20$ min because $t - 20$, the argument of u, is negative for all $t < 20$ min, and, by definition u( ) is zero for all negative arguments.

Table 3.8.1. Common kinds of non-periodic inputs

In the following table, the input function is $x(t)$. The significance of $u(t)$ and of $\delta(t)$ is explained in the text. k, A, b, t_1, and t_2 are arbitrary constants. Most commonly, $t_1 = 0$, so that the input begins at time zero.

Kind of Input	Graphical representation	$x(t) =$	$x_{\mathcal{L}}(s) =$	Duration
STEP unit step when $k = 1$	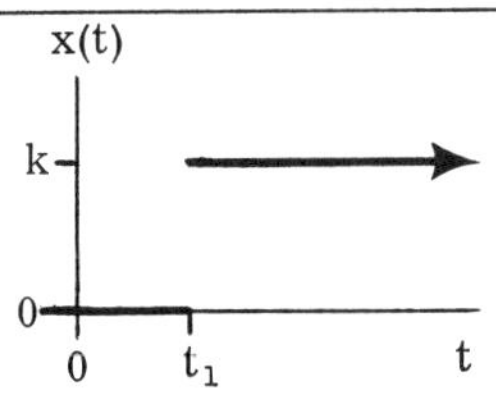	$ku(t-t_1)$	$\frac{k}{s}\epsilon^{-st_1}$	∞
PULSE	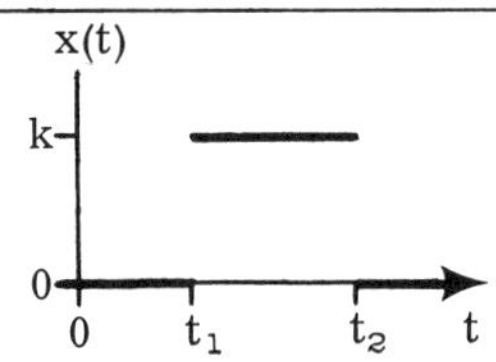	$k[u(t-t_1) - u(t-t_2)]$	$\frac{k}{s}[\epsilon^{-st_1} - \epsilon^{-st_2}]$	$t_2 - t_1$
IMPULSE unit impulse when $A = 1$	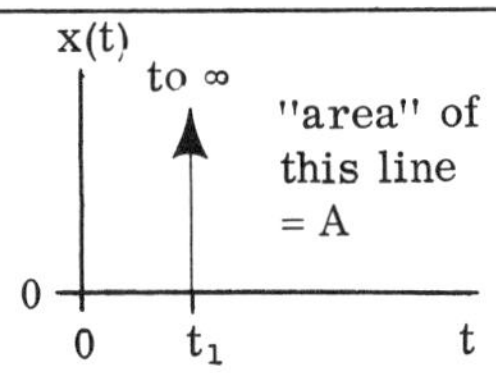	$A\delta(t-t_1)$	$A\epsilon^{-st_1}$	0
RAMP unit ramp when $b = 1$	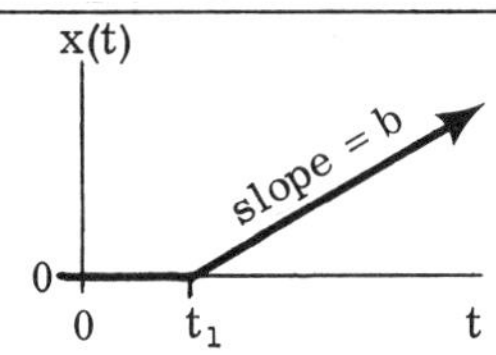	$b[t-t_1]u(t-t_1)$	$\frac{b}{s^2}\epsilon^{-st_1}$	∞

A pulse input is a step input of finite duration. The function $ku(t - t_1)$ jumps from zero to k at t_1 but continues to have the value k forever thereafter. If we want the input to return to zero at some particular instant, t_2, we simply add a second step input, $-ku(t - t_2)$, of equal magnitude but opposite sign, which starts at t_2 and continues forever thereafter, thus ensuring that the sum of the two inputs will be zero for all $t > t_2$. Continuing the example of the preceding paragraph, if the infusion of 0.025 g of inulin per min is started at $t_1 = 20$ min and discontinued at $t_2 = 60$ min, $\dot{Q}_{in}(t) = 0.025\,[u(t-20) - u(t-60)]$ g per min. Note that $\dot{Q}_{in,\mathcal{L}}(s) = [0.025/s]\,[\epsilon^{-20s} - \epsilon^{-60s}]$ g of inulin.

An impulse input is the limiting value which is approached by a pulse input as the duration of the pulse approaches zero while the integral of the input with respect to time (= the "area" of the pulse) remains finite and constant. As a concrete illustration, consider again the infusion of inulin. If we infuse 0.025 g per minute for a period of 80 minutes, starting at t_1, we shall have infused a total of two grams,

$$\text{Total inulin infused} = \int_{t_1}^{t_1+80} 0.025\,dt = 0.025t \Big|_{t_1}^{t_1+80} = 2 \text{ g}$$

But this same two-gram total could have been administered by using an input of 0.1 g per min for 20 min, or an input of 2/3 g per min for 3 min, or an input of 5 g/min for 0.4 min, or, in general, an input of 2/P g per min for P min where $P = t_2 - t_1$ is the duration of the pulse input,

$$\text{Total inulin infused} = \int_{t_1}^{t_1+P} \frac{2}{P}\,dt = \frac{2}{P}t \Big|_{t_1}^{t_1+P} = 2 \text{ g}$$

Now as P becomes shorter and shorter, we approach closer and closer to an input of infinite magnitude but zero duration, i.e., closer and closer to the instantaneous injection of 2 g of inulin at $t = t_1$. Of course it is physically impossible to inject anything truly instantaneously, but the idealized concept of such an impulse input is nonetheless of great theoretical and practical utility. For example, in transfer function analysis, impulse inputs, mathematically applied to the proper components of a system at $t_1 = 0$, can instantaneously establish the equivalent of initial conditions (see example 3.9.1.); though in a system with many components and several non-zero initial conditions this may introduce tedious complications due to multiple inputs. Even more important, it is often possible to identify an unknown transfer function by observing the response of a system to an actual input which is, for all practical purposes, equivalent to an impulse. (Section 3.11).

It was necessary to define the unit step function, $u(t - t_1)$, as a kind of "starting-time" function which specifies when any input function becomes non-zero but which does not otherwise affect the magnitude of that function. It is similarly necessary to define a "unit impulse function" (= Dirac delta function), $\delta(t - t_1)$, which specifies the instant, t_1, at which an impulse input is delivered, without affecting the "size" of the impulse as measured by the integral of the input with respect to time. But beware! A more misleading description than "unit impulse function" could hardly be imagined, for $\delta(t - t_1)$ is neither a proper function of time (as $u(t - t_1)$ is), nor does it ever have unit magnitude (as $u(t - t_1)$ does.) It is a "unit" function only in the peculiar sense that its integral with respect to time is unity, provided t_1 falls within the limits of integration. Moreover, unlike $u(t - t_1)$, $\delta(t - t_1)$ is not a dimensionless quantity but has the dimension of reciprocal time, so that its integral with respect to time is dimensionless. To quote the "International Dictionary of Applied Mathematics" (7),

"DELTA FUNCTION. Commonly written as $\delta(t)$, it lies outside the usual scope of function theory. It can be defined by the relation

$$\int_a^b \Phi(t)\,\delta(t-t_1)\,dt = \Phi(t_1)$$

where $\Phi(t)$ is continuous and $a \le t_1 \le b$. It can thus be thought of as a function which is zero for every value of t except the origin, where it is infinite in such a way that

$$\int_a^b \delta(t)\,dt = 1, \quad a < 0 < b"$$

(For uniformity with the present discussion, I have replaced "x" in the passage quoted by "t," and I have used t_1 in place of x_o.) The monograph goes on to say that the delta function can be considered as the derivative of the step function. The merit of this idea is debatable, but it is at least dimensionally sound; for, if $u(t-t_1)$ is dimensionless, $\frac{du}{dt}$ must have the dimension of reciprocal time.

Before discussing the use of the Dirac delta function, we shall investigate the nature of an impulse input more systematically. Let the input variable be x(t), and let the input forcing function be a pulse of magnitude A/P, starting at t_1, ending at t_2, and lasting for a period of $P = t_2 - t_1$ seconds. We shall further require the integral of the input with respect to time (= the area under the rectangular pulse) to remain constant with a value of A. Then the input will be,

$$x(t) = [A/P]\,[u(t-t_1) - u(t-t_2)] \qquad 3.8.1.$$

whose Laplace transform is

$$x_{\mathfrak{L}}(s) = [A/sP]\,[\epsilon^{-t_1 s} - \epsilon^{-t_2 s}] \qquad 3.8.2.$$

But since $P = t_2 - t_1$,

$$t_2 = P + t_1 \qquad 3.8.3.$$

Substituting this value in equation 3.8.2.,

$$x_{\mathfrak{L}}(s) = A\epsilon^{-t_1 s}\left[\frac{1-\epsilon^{-Ps}}{sP}\right] \qquad 3.8.4.$$

We now use L'Hospital's rule (See Appendix) to evaluate the bracketed expression as P approaches zero,

$$\lim_{P\to 0} \frac{1-\epsilon^{-Ps}}{sP} = \lim_{P\to 0} \frac{s\epsilon^{-Ps}}{s} = \lim_{P\to 0} \epsilon^{-Ps} = 1 \qquad 3.8.5.$$

Hence, combining equations 3.8.4. and 3.8.5.,

$$\lim_{P\to 0} x_{\mathcal{L}}(s) = A\epsilon^{-t_1 s} \qquad 3.8.6.$$

which is, therefore, the Laplace transform of an impulse input, delivered at t_1, whose integral with respect to time is A. By evaluating equation 3.8.6. at $t_1 = 0$, we find that the Laplace transform of the impulse input delivered at time zero is simply A, the integral of the input with respect to time.

We have just seen that the Laplace transform of an impulse input has a perfectly well-defined and finite value. Furthermore, this Laplace transform has the same dimensions as the Laplace transform of any other comparable input forcing function (see Exercise 7) so that it is dimensionally proper to add the Laplace transform of an impulse to the Laplace transforms of any other forcing functions put into the system at the same point. But however reasonable the properties of an impulse may be in the s-domain, in the time domain an impulse is nothing but a convenient mathematical fiction. Trying to "evaluate" the true magnitude of the input, $x(t) = A/P$, when x(t) has actually become an impulse, i.e., when $P = 0$, is exactly equivalent to trying to find the height of a rectangle of given finite area but with zero base. Since division by zero is forbidden, the height cannot be evaluated save by saying that it is infinite. But this tells us nothing because the height is "just as infinite" when the area is small as when the area is large. So if we insist upon regarding an impulse input as a function of time, the best we can do is to say that it has a perfectly well-defined value, zero, at every point in its domain except at the one point of interest, where its value is indefinable. Such a thing is a phantasm, not a function.

The delta "function" provides us with a clever and convenient way of sidestepping the uncomfortable ambiguity of talking about infinitely large inputs of zero duration. For if equation 3.8.1. is written

$$x(t) = A \left\{\frac{u(t-t_1) - u(t-t_2)}{t_2 - t_1}\right\} \qquad 3.8.7.$$

we can show (see exercise 3. 8) that for the limiting case where t_2 has not only approached t_1, but has actually reached it, so that x(t) is a bona fide impulse input, the bracketed expression may plausibly be considered to have precisely the characteristics of $\delta(t-t_1)$ which were defined above. Hence,

$$\lim_{t_2\to t_1} \frac{u(t-t_1) - u(t-t_2)}{t_2 - t_1} = \delta(t-t_1) \qquad 3.8.8.$$

Therefore, an impulse input, whose integrated value (= "area") is A, delivered at $t = t_1$, may be symbolized as $A\delta(t-t_1)$.

Fortunately we shall not encounter impulses when we take reverse transforms to obtain the time-domain outputs of real physical systems; real physical systems do not generate phantasms.

A ramp input has been included in table 3.8.1. as a simple example of an unbounded input. Its characteristics are too evident to need discussion. A variety of other non-periodic inputs are mentioned in exercise 3.2.

SECTION 3.9. Periodic inputs. Sine and cosine functions

In physiological as well as in physical systems, certain input forcing functions are periodic (= cyclic). Any function of t, f(t), is periodic with period P if $f(t) = f(t \pm nP)$ where P is a constant interval and n can have any integral value, positive or negative, consistent with the domain of t over which f has been defined. This simply means that the pattern of variation of f(t) with time, however complicated it may be within a given period is repeated without change again and again, period after period, so that by specifying f(t) explicitly for any single period we are, in effect, describing f(t) for all other periods as well.

The simplest and most fundamental periodic functions are the sine and cosine functions. The following example, previously solved by classical methods (Example 2.16.1.), will illustrate both transfer function analysis with a periodic input and the use of an impulse input to establish initial conditions.

Example 3.9.1. For the network shown in Figure 2.16.1. and described in example 2.16.1., use transfer function analysis to find i(t).

We have a series RC circuit in which the initial voltage across the capacitor is 2 volts. At time zero the circuit is activated by a voltage source of $e_{in}(t) = 10 + 3 \sin \omega t$ volts with a polarity such as to increase the charge on the capacitor.

The input is $e_{in}(t)$. The output is i(t). By equations 2.16.1. and 2.16.2. we have,

$$Ri(t) + \frac{1}{C}\int i(t)dt = e_{in}(t) \qquad 3.9.1.$$

whose Laplace transform with initial conditions of zero is,

$$Ri_{\mathcal{L}}(s) + \frac{1}{Cs}\, i_{\mathcal{L}}(s) = e_{in,\mathcal{L}}(s) \qquad 3.9.2.$$

Solving equation 3.9.2. for the transfer function, $i_{\mathcal{L}}(s)/e_{in,\mathcal{L}}(s)$.

$$\text{Transfer function} = \frac{1}{R + \frac{1}{Cs}} = \frac{\frac{1}{R}s}{s + \frac{1}{RC}} \qquad 3.9.3.$$

In order to establish the initial condition of 2 volts across the capacitor, we must include in the input an impulse voltage which will "instantaneously" deliver a charge of 2C coulombs through the resistance, R, to the capacitor. (Since this will require an "infinite" input voltage, acting at t = 0, we need not worry about the opposing voltage built up on the capacitor, and we can therefore treat the voltage across R as constant.) A voltage of e_R volts will cause a current, $i_R = e_R/R$ amperes, to flow through R into the capacitor. If this current lasts for P seconds, it will put a charge of $i_R P = [e_R/R]P$ coulombs on the capacitor. For a charge of 2C coulombs, we therefore require that,

$$[e_R/R]P = 2C \text{ coulombs, or } e_R P = 2RC \text{ volt}\cdot\text{sec} \qquad 3.9.4.$$

From the preceding discussion of impulses, it is clear that 2RC corresponds to the integral of the constant input voltage, e_R, over time, during the interval, P sec. Hence, if we allow e_R to become infinite and P to become zero, while keeping 2RC constant, we shall have the desired impulse input of $2RC\,\delta(t)$ volts. Adding this to the stated input of 10 + 3 sin ωt volts beginning at t = 0,

$$\text{total input} = 2RC\,\delta(t) + 10\,u(t) + 3\,u(t)\sin\omega t \quad \text{volts} \qquad 3.9.5.$$

whose Laplace transform is,

$$\mathcal{L}\{\text{input}\} = 2RC + \frac{10}{s} + 3\,\frac{\omega}{s^2 + \omega^2} \quad \text{volt}\cdot\text{sec} \qquad 3.9.6.$$

To obtain the Laplace transform of the corresponding output, we multiply the transfer function of equation 3.9.3. by $\mathcal{L}\{\text{input}\}$ of equation 3.9.6.,

$$\mathcal{L}\{\text{output}\} = i_{\mathcal{L}}(s) = \frac{2Cs}{s + \frac{1}{RC}} + \frac{10/R}{s + \frac{1}{RC}} + \frac{[\frac{3\omega}{R}]\,s}{[s + \frac{1}{RC}]\,[s^2 + \omega^2]} \qquad 3.9.7.$$

Breaking the last term of equation 3.9.7. into partial fractions,

$$\frac{[\frac{3\omega}{R}]\,s}{[s + \frac{1}{RC}]\,[s^2 + \omega^2]} = \frac{J}{s + \frac{1}{RC}} + \frac{Ks}{s^2 + \omega^2} + \frac{L}{s^2 + \omega^2} \qquad 3.9.8.$$

$$\text{where } J = -K = -\frac{3C\omega}{[RC\omega]^2 + 1}; \quad L = \frac{3RC^2\omega^3}{[RC\omega]^2 + 1}$$

For convenience, we shall now construct a table, (Table 3.9.1.), listing the five terms of $i_{\mathcal{L}}(s)$ on the left and the corresponding terms of the time-domain output, i(t), on the right.

Table 3.9.1. Partial fractions of Example 3.9.1. and the corresponding inverse transforms.

$i_{\mathcal{L}}(s)$	$i(t)$
$s2C\left[\frac{1}{s+\frac{1}{RC}}\right]$	$\frac{d}{dt}\left[2C\epsilon^{-\frac{1}{RC}t}\right] = -\frac{2}{R}\epsilon^{-\frac{1}{RC}t}$
$\frac{10}{R}\left[\frac{1}{s+\frac{1}{RC}}\right]$	$\frac{10}{R}\epsilon^{-\frac{1}{RC}t}$
$-\frac{3C\omega}{[RC\omega]^2+1}\left[\frac{1}{s+\frac{1}{RC}}\right]$	$-\frac{3C\omega}{[RC\omega]^2+1}\epsilon^{-\frac{1}{RC}t}$
$\frac{3C\omega}{[RC\omega]^2+1}\left[\frac{s}{s^2+\omega^2}\right]$	$\left[\frac{3C\omega}{[RC\omega]^2+1}\right]\cos\omega t$
$\frac{3R[C\omega]^2}{[RC\omega]^2+1}\left[\frac{\omega}{s^2+\omega^2}\right]$	$\frac{3R[C\omega]^2}{[RC\omega]^2+1}\sin\omega t$ $= \frac{3C\omega}{[RC\omega]^2+1}[RC\omega\sin\omega t]$

Collecting the terms on the right of Table 3.9.1.,

$$i(t) = \left[\frac{8}{R} - \frac{3C\omega}{[RC\omega]^2+1}\right]\epsilon^{-\frac{1}{RC}t} + \frac{3C\omega}{[RC\omega]^2+1}\left[\cos\omega t + RC\omega\sin\omega t\right] \quad \text{amp} \qquad 3.9.9.$$

which is exactly the solution previously obtained by classical methods (equation 2.16.8.).

SECTION 3.10. Representation of arbitrary periodic inputs by a Fourier series

Practically any periodic function--certainly any we are likely to encounter--can be approximated to any desired accuracy by a sum of sine and cosine functions in the form of a Fourier series. Although precise representation of an arbitrary periodic function would require summing an infinite number of terms, an approximate representation quite accurate enough for practical purposes can usually be achieved by using only the first few terms of the series, neglecting higher harmonics whose contribution to the approximation is relatively small. The actual procedure for calculating each term of the series is described in the Appendix and is there illustrated by two examples.

Approximation of an arbitrary periodic input by a Fourier series enables us to substitute easy-to-handle, well-behaved sine and cosine functions for the true input which is often more complicated and less tractable. If we then find the output corresponding to each individual term in the series, the principle of superposition (Section 1.13.) assures us that the total output will simply be the sum of the individual outputs. So if the arbitrary input function was reasonably well represented by a finite number of terms, the sum of the outputs due to those terms should be a reasonably good approximation of the true output. The advantage of using sine and cosine functions as inputs, particularly when we are interested only in the pure forced response, will become much clearer after we have discussed phasor analysis in the frequency domain (Section 3.15 and following.)

SECTION 3.11. The use of impulse inputs to find unknown transfer functions

Very frequently, especially in the biological sciences, the system which we want to study is a "black box." A black box contains any number of unknown components arranged in an unknown pattern. We cannot open the box to investigate the individual components directly, but we can observe the output of the system in response to some experimentally-controlled input. The black box problem is the problem of using these experimental observations to find the transfer function of the system hidden in the box. The transfer function in turn may enable us to formulate some more-or-less plausible hypothesis about what the components inside the box are. Until it is supported by additional experimental evidence, such an hypothesis can never be more than an intelligent guess, because, within the errors of our observations, an infinitely large number of systems might yield the same transfer function. But quite aside from any speculations which it may provoke about the contents of the black box, the transfer function itself may be useful because (ideally at least), it allows us to predict the output response to any input forcing function.

When it is experimentally possible to observe the response of the system to an impulse input, we have a particularly simple and direct way of finding the transfer function. Let $A\delta(t)$ be an impulse input, whose integral with respect to time is A, delivered at time zero. Let y(t) be the observed output, and let G(s) be the unknown transfer function. By definition,

$$G(s) = \frac{\mathcal{L}\{y(t)\}}{\mathcal{L}\{A\delta(t)\}} = \frac{y_{\mathcal{L}}(s)}{A} \qquad 3.11.1.$$

Equation 3.11.1. shows that the transfer function is the Laplace transform of the output response to an impulse whose time integral is unity, for if $A = 1$, $G(s) = y_{\mathcal{L}}(s)$. (If $A \neq 1$, we can still find G(s) merely by dividing the Laplace transform of the output by A.) As an immediate corollary, the inverse Laplace transform of a transfer function is simply the natural response to whatever initial conditions are established by a unit impulse input.

Example 3.11.1. 0.25 g of drug D was injected intravenously at t = 0 into a 15 Kg dog. Thereafter, the concentration of D in samples of blood plasma was determined at suitable intervals. These experimental observations could be described satisfactorily by a simple exponential curve (Figure 3.11.1.) whose equation is,

$$C(t) = 0.125\epsilon^{-0.084t} u(t) \quad g/L \qquad 3.11.2.$$

where C(t) is the concentration of drug D in plasma, and t is time in hours. If the input is defined as $\dot{Q}(t)$, the rate of intravenous administration of D in g/hr, and the output is defined as C(t).

a) What is the transfer function, G(s)?

b) Predict C(t) if, on another occasion, a single dose of 0.1 g of D were injected intravenously into the same dog every four hours for two days, the first such dose being given at t = 0.

c) Predict C(t) if, on a third occasion, D were infused continuously intravenously for two days, at a rate of 0.025 g/hr starting at t = 0.

a) The Laplace transform of the output response described by equation 3.11.2. is,

$$C_{\mathcal{L}}(s) = \frac{0.125}{s + 0.084} \quad [g/L] \quad hr \qquad 3.11.3.$$

The Laplace transform of $A\delta(t) = 0.25\delta(t)$ g/hr, (the impulse input), is,

$$\mathcal{L}\left\{A\delta(t)\right\} = A = 0.25 \quad g \qquad 3.11.4.$$

By equation 3.11.1, therefore, the transfer function is,

$$G(s) = \frac{C_{\mathcal{L}}(s)}{A} = \frac{0.5}{s + 0.084} \quad hr/L \qquad 3.11.5.$$

b) The input described consists of a series of 12 impulses, each of area 0.1 g, given every 4 hours. This input is

$$\dot{Q}(t) = 0.1[\delta(t) + \delta(t-4) + \delta(t-8) + \ldots + \delta(t-44)] \quad g/hr \qquad 3.11.6.$$

Its Laplace transform is

$$\dot{Q}_{\mathcal{L}}(s) = 0.1[1 + \epsilon^{-4s} + \epsilon^{-8s} + \ldots + \epsilon^{-44s}] \quad g \qquad 3.11.7.$$

Hence,

$$C_{\mathcal{L}}(s) = \dot{Q}_{\mathcal{L}}(s)\, G(s) = \frac{0.05}{s + 0.084}[1 + \epsilon^{-4s} + \epsilon^{-8s} + \ldots + \epsilon^{-44s}] \quad g \cdot hr/L \qquad 3.11.8.$$

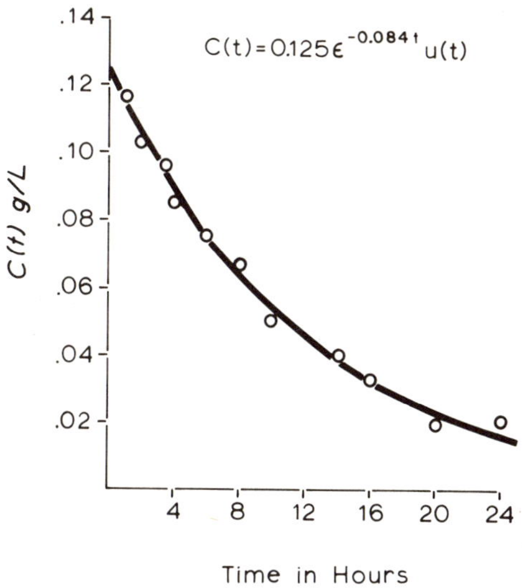

Figure 3.11.1. Exponential disappearance of a drug from the blood after an impulse input. (Example 3.11.1).

The inverse transform of equation 3.11.8. is,

$$C(t) = 0.05[\epsilon^{-0.084t}u(t) + \epsilon^{-0.084[t-4]}u(t-4) + \epsilon^{-0.084[t-8]}u(t-8) + \ldots +$$

$$\epsilon^{-0.084[t-44]}u(t-44)] = 0.05 \sum_{n=0}^{11} \epsilon^{-0.084[t-n]}u(t-n) \quad g/L \qquad 3.11.9.$$

Figure 3.11.2. shows C(t), calculated from equation 3.11.9., plotted against time. (Note: A series of regularly-spaced impulses might be regarded as a periodic function which could be approximated by a Fourier series. The fact that one can actually derive such a series, even for so bizarre a "function" as an impulse, is a remarkable example of the versatility of Fourier series representation. But so many terms would be needed for reasonable accuracy that approximating impulse inputs by a Fourier series is hardly a practical procedure.)

c) The input described is a pulse input,

$$\dot{Q}(t) = 0.025\,[u(t) - u(t-48)] \quad g/hr \qquad 3.11.10.$$

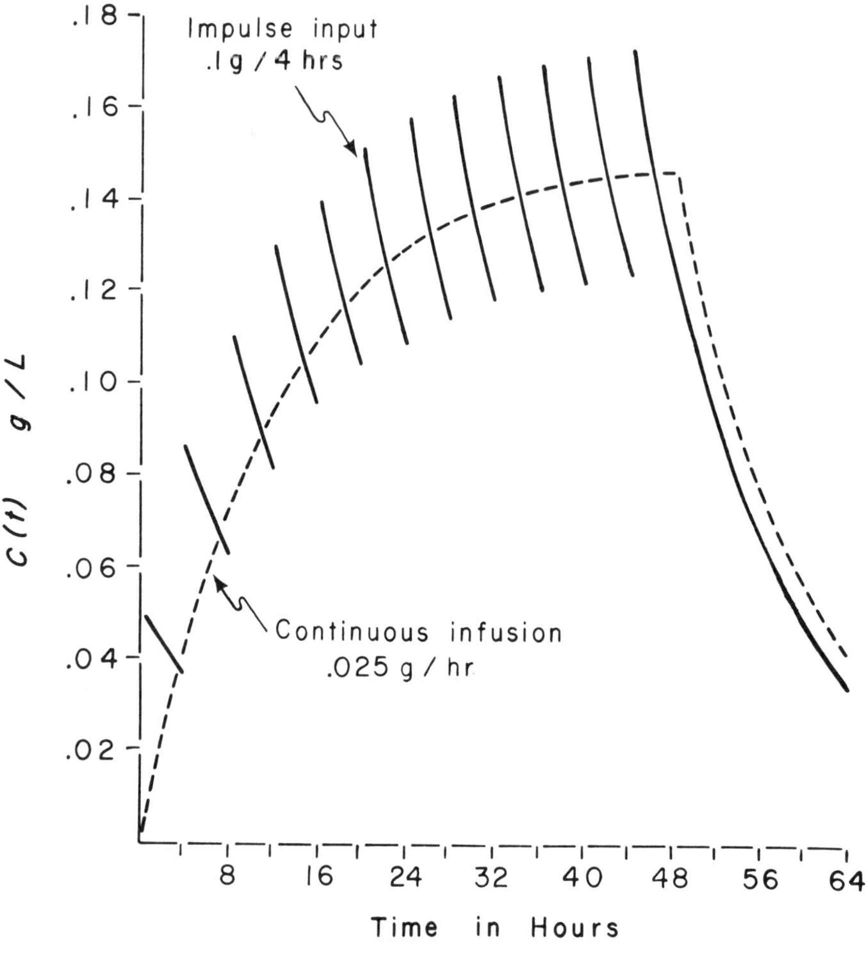

Figure 3.11.2. The plasma concentration of drug D (Example 3.11.1) as a function of time in response to two different inputs.

The curves shown here were calculated by using the transfer function obtained from the observed response to an inpulse input. The solid curves (equation 3.11.9) show how C(t), the plasma concentration of D, would behave if single doses (= impulse inputs) of 0.1 g of D were given intravenously every 4 hr for two days, starting at time zero. The broken curves (equation 3.11.14) show how C(t) would behave if the same total dose of D (i.e. 1.2 g) were administered as a continuous intravenous infusion at the rate of 25 mg/hr for 48 hr, starting at time zero. Notice that by the end of the two-day period of drug administration, the mean plasma concentration is nearly the same for both kinds of input, though with the intermittent impulse inputs, fluctuations above and below that mean continue to be prominent.

With a simple linear system like this, the behavior of the plasma concentration in response to any input function of time could be similarly calculated.

Hence

$$\dot{Q}_{\mathcal{L}}(s) = \frac{0.025}{s}\,[1 - \epsilon^{-48s}]\ \text{g} \qquad 3.11.11.$$

$$C_{\mathcal{L}}(s) = \dot{Q}_{\mathcal{L}}(s)G(s) = \frac{0.0125\,[1 - \epsilon^{-48s}]}{s\,[s + 0.084]}$$

$$= \frac{0.0125}{0.084}\left[\frac{1}{s} - \frac{1}{s + 0.084}\right][1 - \epsilon^{-48s}]\ \text{g}\cdot\text{hr/L} \qquad 3.11.12.$$

$$C(t) = \frac{0.0125}{0.084}\,[1 - \epsilon^{-0.084t}]u(t) - [1 - \epsilon^{-0.084(t-48)}]u(t-48)\ \text{g/L}$$

3.11.13.

or,

$$\left.\begin{aligned} C(t) &= \frac{0.0125}{0.084}\,[1 - \epsilon^{-0.084t}]\ \text{g/L} \quad \text{for } 0 \le t \le 48 \text{ hr} \\ C(t) &= C(48)\epsilon^{-0.084[t-48]}\ \text{g/L} \quad \text{for } t \ge 48 \text{ hr} \end{aligned}\right\} \qquad 3.11.14.$$

Figure 3.11.2. (broken lines) shows C(t), calculated from equation 3.11.13. or equation 3.11.14., plotted against time.

In the preceding example, the system itself remained a black box in the sense that the transfer function which we obtained could tell us nothing about whether D was being excreted in the urine, or in the exhaled air, or was being sequestered in bone, or was being metabolized in the liver or other tissues, or was being disposed of by two or more such pathways simultaneously. However, transfer function analysis did yield precisely the output function to be expected from the stated input functions for a drug which disappears exponentially. Unfortunately, matters are not always so straight-forward as will be evident from the following:

Horrible example 3.11.2. You enjoy the occasional cocktail? I thought so. Let me buy you a drink . . . on condition that you down it within five minutes. Agreed? Fine. Bartender! A quadruple martini for the reader! You didn't want that much? Sorry, a bargain's a bargain. Anyway, a mere 45 grams of ethanol isn't going to hurt you. . . . There! I knew you could do it. That's as close to an impulse input of alcohol as you're ever likely to get. Now just breathe into this alcohol analyzer every 15 min so that we can estimate your blood alcohol concentration. . . . Look! Your blood alcohol is decreasing as a practically perfect straight-line function of time (See Figure 3.11.3.). The observations are well fitted by the equation,

$$C(t) = [1.00 - 0.22t]\,[u(t) - u(t - \frac{1}{0.22})]\ \text{g/L} \qquad 3.11.15.$$

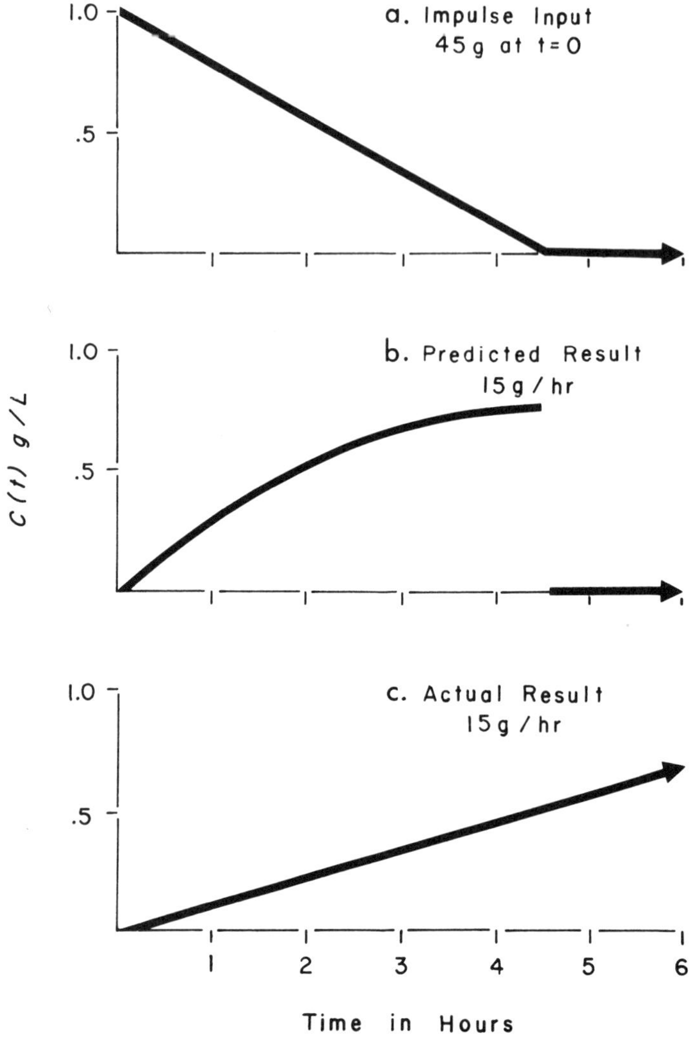

Figure 3.11.3. The blood ethanol fiasco. (Example 3.11.2)

After 45g of ethanol given as an impulse input at time zero, C(t), the concentration of ethanol in blood, decreases as a straight-line function of time as shown in graph a. (At very low concentrations, the disappearance would actually become exponential, but for simplicity this feature has been neglected here.) If a "transfer function" is calculated from graph a on the erroneous assumption that the system is linear, and if the transfer function is then used to predict the effect on C(t) of giving 15g of ethanol per hour continuously, we would obtain the ridiculous result shown in part b. For comparison, the actual result of continuously administering 15g of ethanol per hour is shown in part c.

where C(t) is blood ethanol concentration and t is time in hours. The Laplace transform of this output is,

$$C_{\mathcal{L}}(s) = \frac{s - 0.22}{s^2} \left[1 - \epsilon^{-\frac{1}{0.22}s}\right] \text{g·hr/L} \qquad 3.11.16.$$

The impulse input was 45δ(t) g/hr. Its Laplace transform is 45 g. Hence, by equation 3.11.1, the transfer function, G(s), should be,

$$G(s) = \frac{1}{45} \left[\frac{s - 0.22}{s^2}\right] \left[1 - \epsilon^{-\frac{1}{0.22}s}\right] \text{hr/L} \qquad 3.11.17.$$

Now let us use this transfer function, in the manner so successfully employed in the previous example, to predict how your blood ethanol concentration would change with time during a continuous intake of alcohol at a constant rate, Q(t) = 15 grams per hour, whose Laplace transform is 15/s grams. The product of this and the transfer function of equation 3.11.17. gives us,

$$C(s) = \frac{1}{3} \left[\frac{s - 0.22}{s^3}\right] \left[1 - \epsilon^{-\frac{1}{0.22}s}\right] \text{g·hr/L} \qquad 3.11.18.$$

from which, by taking inverse transforms, we obtain,

$$C(t) = \frac{1}{3}t\,[1 - 0.11t]\,[u(t) - u(t - \frac{1}{0.22})] \text{ g/L} \qquad 3.11.19.$$

which is flagrantly preposterous! (See Figure 3.11.3.)

I have deliberately led the slightly inebriated reader into an analytical ambush in order to stress a point which is absolutely fundamental. Transfer function analysis is applicable only to linear systems. Therefore we cannot safely calculate a transfer function for an unknown black box system unless we at least know that there is a linear relationship between input and output. For non-linear systems transfer functions do not exist. So although even in a non-linear system we can calculate the ratio between the Laplace transform of a particular output and the Laplace transform of the corresponding particular input, this ratio is not a valid transfer function and cannot be used to predict the output due to some other input.

Why is the ethanol system non-linear, the drug D system linear? After all, the disappearance of drug D may well be due to the same general kind of process, namely enzymatic degradation, which gets rid of ethanol. The answer is simply that whatever the process responsible for removing drug D, it is unsaturated so that the rate of disappearance of drug D is determined by the rate at which molecules of D are supplied to the process. The rate of disappearance is therefore directly proportional to drug concentration, as we see clearly if we differentiate equation 3.11.2. with respect to time,

$$\frac{dC(t)}{dt} = -0.084\ C(t)\ \ g/L\cdot hr \qquad 3.11.20.$$

or,

$$\frac{dC(t)}{dt} + 0.084C(t) = 0.\ \ g/L\cdot hr \qquad 3.11.20.a.$$

Note that equation 3.11.20. is a homogeneous linear differential equation, as it must be for a linear system which has been activated only by an impulse input. Since an impulse input to a linear system corresponds to establishing initial conditions, the response which it elicits must be the natural response of the system, and must therefore be described by a homogeneous equation (Section 2.13). In contrast, as is well known, the enzyme system responsible for the metabolism of ethanol behaves as if it is saturated even by such relatively small concentrations of ethanol as are sought after by moderate drinkers. Over a considerable range of concentrations, therefore, the rate of disappearance of ethanol remains constant. Had we differentiated equation 3.11.15. with respect to time,

$$\left.\begin{aligned}\frac{dC(t)}{dt} &= -0.22\ g/L\cdot hr \text{ for } 0 \le t < \frac{1}{0.22}\ hr\\ &= 0 \quad g/L\cdot hr \text{ for } t \ge \frac{1}{0.22}\ hr\end{aligned}\right\} \qquad 3.11.21.$$

we would have seen at once that the response of the system to an impulse input could not be described by a homogeneous equation, and that the system under investigation was therefore not linear. Saturation nonlinearities of this sort are very frequently encountered when physiological (or other) systems of limited capacity are presented with large inputs. Indeed, with living systems, non-linearity of one kind or another is the rule, not the exception. In addition to checking the homogeneity and linearity of the differential equation for the natural response, it would therefore be wise, when dealing with an unfamiliar system, to observe the response to two or more different inputs. If the same transfer function can be calculated from each, one may be reasonably confident that the system is behaving linearly, at least for the range of inputs being investigated.

SECTION 3.12. Obtaining inverse Laplace transforms by using the convolution theorem

When we have the transfer function, G(s), and the Laplace transform, $x_{\mathcal{L}}(s)$, of some particular input, x(t), which interests us, it is easy enough to find the Laplace transform, $y_{\mathcal{L}}(s) = G(s)\,x_{\mathcal{L}}(s)$ of the corresponding output function, y(t). But it is not always so easy to find y(t) itself by breaking $G(s)\,x_{\mathcal{L}}(s)$ into partial fractions and looking up the corresponding inverse transforms in a table. An alternative way of obtaining the inverse transform of $G(s)\,x_{\mathcal{L}}(s)$ is to use the convolution theorem (Theorem ix of Section 3.5.). Let $g(t) \doteq \mathcal{L}^{-1}G(s)$. Then, by equations 3.5.11. and 3.5.12.,

$$y(t) = \mathcal{L}^{-1}\left\{y_{\mathcal{L}}(s)\right\} = \mathcal{L}^{-1}\left\{G(s)x_{\mathcal{L}}(s)\right\} = \int_0^t g(t-\tau)x(\tau)d\tau = g(t) * x(t) \qquad 3.12.1.$$

Equation 3.12.1. shows that y(t) can be found by evaluating the convolution integral $\int_0^t g(t-\tau)x(\tau)d\tau$ in which τ is just a dummy variable of integration.

Notice that in equation 3.5.12, $f_1(s)$ and $f_2(s)$ can be <u>any</u> function of s. Hence, instead of identifying G(s) as $f_1(s)$ and $x_{\mathcal{L}}(s)$ as $f_2(s)$ we could just as well have let $x_{\mathcal{L}}(s)$ be $f_1(s)$ and G(s) be $f_2(s)$. Then instead of 3.12.1., we would have had

$$y(t) = \mathcal{L}^{-1}\left\{y_{\mathcal{L}}(s)\right\} = \mathcal{L}^{-1}\left\{x_{\mathcal{L}}(s)G(s)\right\} = \int_0^t x(t-\tau)g(\tau)d\tau = x(t) * g(t) \qquad 3.12.2.$$

In other words the convolution of x(t) with g(t) yields the same result as the convolution of g(t) with x(t)

$$g(t) * x(t) \equiv x(t) * g(t) \qquad 3.12.3.$$

<u>Example 3.12.1.</u> Using the convolution theorem, find the output, C(t), of the system of example 3.11.1. when the input is the pulse described by equation 3.11.10.

By inspection, the inverse Laplace transform of G(s), the transfer function in equation 3.11.5. is,

$$\mathcal{L}^{-1}\left\{G(s)\right\} = g(t) = 0.5\varepsilon^{-0.084t} \ L^{-1} \qquad 3.12.4.$$

The input (equation 3.11.10.) is

$$\dot{Q}(t) = 0.025u(t) - 0.025u(t-48) \ \ g/hr \qquad 3.12.5.$$

So,

$$C(t) = \mathcal{L}^{-1}\left\{\dot{Q}_{\mathcal{L}}(s)G(s)\right\} = \dot{Q}(t) * g(t) = \int_0^t g(t-\tau)\dot{Q}(\tau)d\tau$$

$$= \int_0^t 0.5\varepsilon^{-0.084(t-\tau)}[0.025u(\tau)-0.025u(\tau-48)]d\tau$$

$$= 0.0125\int_0^t \varepsilon^{-0.084t}\varepsilon^{0.084\tau}[u(\tau) - u(\tau - 48)]d\tau \quad g/L \qquad 3.12.6.$$

To evaluate the integral in equation 3.12.6, we note that $\epsilon^{-0.084t}$ is independent of τ, the variable of integration, so it can be taken outside the integral sign.

$$C(t) = 0.0125\epsilon^{-0.084t}\left\{\int_0^t \epsilon^{0.084\tau} u(\tau)d\tau - \int_0^t \epsilon^{0.084\tau} u(\tau-48)d\tau\right\} \text{ g/L} \qquad 3.12.7.$$

But $u(\tau-48) = 0$ for $\tau < 48$

$= 1$ for $\tau \geq 48$

Hence, in the second integral in equation 3.12.7, we can replace the lower limit of integration, zero, by 48 without affecting the value of the integral. Then,

$$C(t) = 0.0125\epsilon^{-0.084t}\left\{\int_0^t \epsilon^{0.084\tau} u(\tau)d\tau - \int_{48}^t \epsilon^{0.084\tau} u(\tau-48)d\tau\right\} \text{ g/L} \qquad 3.12.8.$$

Now the function u has the constant value of unity throughout the range of integration for both integrals, so we can take it outside the integral signs and integrate to obtain,

$$C(t) = \left[0.0125\epsilon^{-0.084t}u(t)\right] \frac{1}{0.084}\epsilon^{0.084\tau}\Big|_0^t - \left[0.0125\epsilon^{-0.084t}u(t-48)\right]\frac{1}{0.084}\epsilon^{0.084\tau}\Big|_{48}^t \text{ g/L} \qquad 3.12.9.$$

Whence,

$$C(t) = \frac{0.0125}{0.084}\left\{\left[1 - \epsilon^{-0.084t}\right] u(t) - \left[1 - \epsilon^{-0.084[t-48]}\right]u(t-48)\right\} \text{ g/L} \qquad 3.12.10.$$

which is exactly the result obtained previously (equation 3.11.13).

Sometimes the input, x(t), is an experimentally-observed function of time which is difficult or impossible to express mathematically but which can be described by plotting x(t) against time. Or g(t), the inverse transform of the transfer function may be available only as some complicated empirical curve obtained by plotting the observed response of the (known-to-be-linear!) system to an impulse input. (See Section 3.11.) Under these circumstances the convolution integral cannot be evaluated analytically, but it can still be estimated graphically by a method which is described, for example, in Reference 8.

SECTION 3.13. Conventional block diagrams

For simple systems, the circuit diagram is a convenient means of summarizing the interconnections between the individual elements and the direction of the T-variable when stored energy is exchanged between the elements or when energy is supplied to the system by one or more source elements. Similarly, with more general systems whose components may be activated by different forms of energy the interconnections between components and the direction of signal flow between them is commonly summarized in a block diagram. In this section, I shall describe what I shall term a conventional block diagram, while in the next section I shall mention briefly certain modifications and extensions of the block diagram concept.

Conventional block diagrams are used to describe stationary linear systems (i.e. those with constant coefficients) in which each component has a single input and a single output. Each such component is represented in the diagram by a rectangle (= block) containing either the actual transfer function or its symbol. The input variable (or its Laplace transform) is shown by a line with an arrowhead pointing to the block, the output variable (or its Laplace transform) is shown by a line with an arrowhead pointing away from the block. (Figure 3.13.1.a.) Such a block represents a complete subsystem in the sense that, given the transfer function in the block, the output variable is determined entirely by the input variable. Hence, as was previously emphasized for transfer functions themselves (See Section 3.7.), signal flow along each pathway in a block diagram must be unidirectional. In this respect, block diagrams differ from circuit diagrams where the T-variable can usually pass through the interconnecting leads from one passive element to another in either direction depending upon circumstances. In a block diagram, any group of passive elements interconnected in such a non-unidirectional way must be described by a single transfer function and must therefore be represented by a single block. For a concrete example, see Exercise 3.13.

When, as is usual, the variables depicted by the input and output pathways are represented in the time domain, the block and its entering and exiting lines symbolize all three of the following operations:

1. Take the Laplace transform of the input.

2. Multiply it by the transfer function in the block.

3. Take the inverse Laplace transform of the product to get the output.

Only two other features, splitting points and summing points, need be mentioned to complete the description of conventional block diagrams. A splitting point (Figure 3.13.1.b.) simply shows a single line branching into two or more lines, indicating that the single variable symbolized by the line must be considered at two or more points. Nothing happens to a variable at a splitting point; it follows each path unchanged. It is in no way comparable to a node, and Kirchhoff's node equation does not apply. A summing point, symbolized by a small circle, has two or more variables as inputs. Near (or in) the circle, the arrow for each of these inputs is labelled "+" or "-" to show whether

a.) A block. x(t) is the input, y(t) is the output, and G(s) is the transfer function.

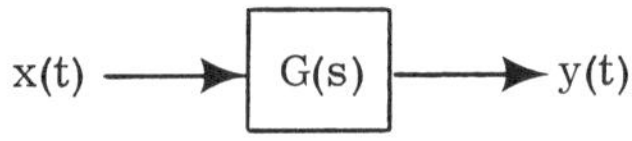

b.) A splitting point. Notice that the signal, y(t), follows each pathway unchanged. A splitting point is therefore not comparable to a node.

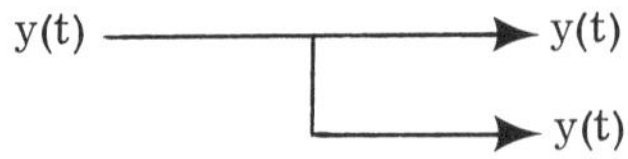

c.) A summing point.

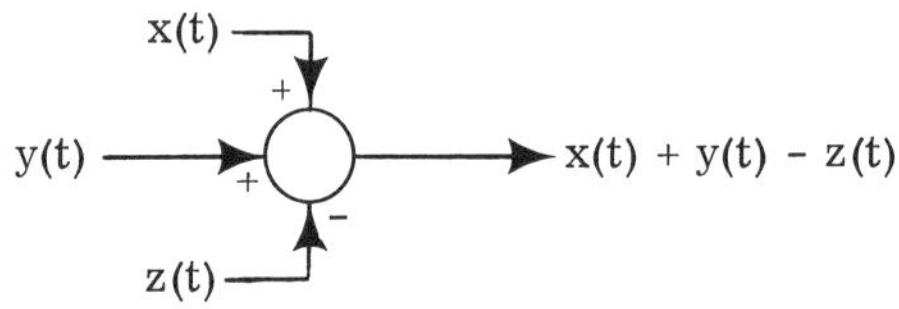

Figure 3.13.1. Features of a conventional block diagram.

---

the variable it represents is to be added to or subtracted from the others. The output from the summing point is then simply the algebraic sum of the inputs. (Figure 3.13.1.c.)

Example 3.13.1. Draw a block diagram for the system of tanks and the pressure gauge depicted in Figure 3.13.2.a. assuming that outflow is directly proportional to the hydrostatic pressure at the outflow valve and that $\theta(t)$, the angular displacement of the hand on the face of the pressure transducer gauge is also directly proportional to the hydrostatic pressure.

From the physical arrangement it is obvious that the two tanks and the pressure transducer can be considered as three separate subsystems. Each tank may be described by the circuit diagram in Figure 3.13.2.b. (C, the fluid capacitance, is directly proportional to the cross-sectional area of the tank, inversely proportional to the density of the fluid.) From the circuit diagram and Kirchhoff's node law, we have, for each tank,

$$\dot{V}_c(t) + \dot{V}_{out}(t) = \dot{V}_{in}(t) \qquad 3.13.1.$$

or

$$C\frac{dP(t)}{dt} + \frac{P(t)}{R} = \dot{V}_{in}(t) \qquad 3.13.2.$$

Taking Laplace transforms on both sides of equation 3.13.2.,

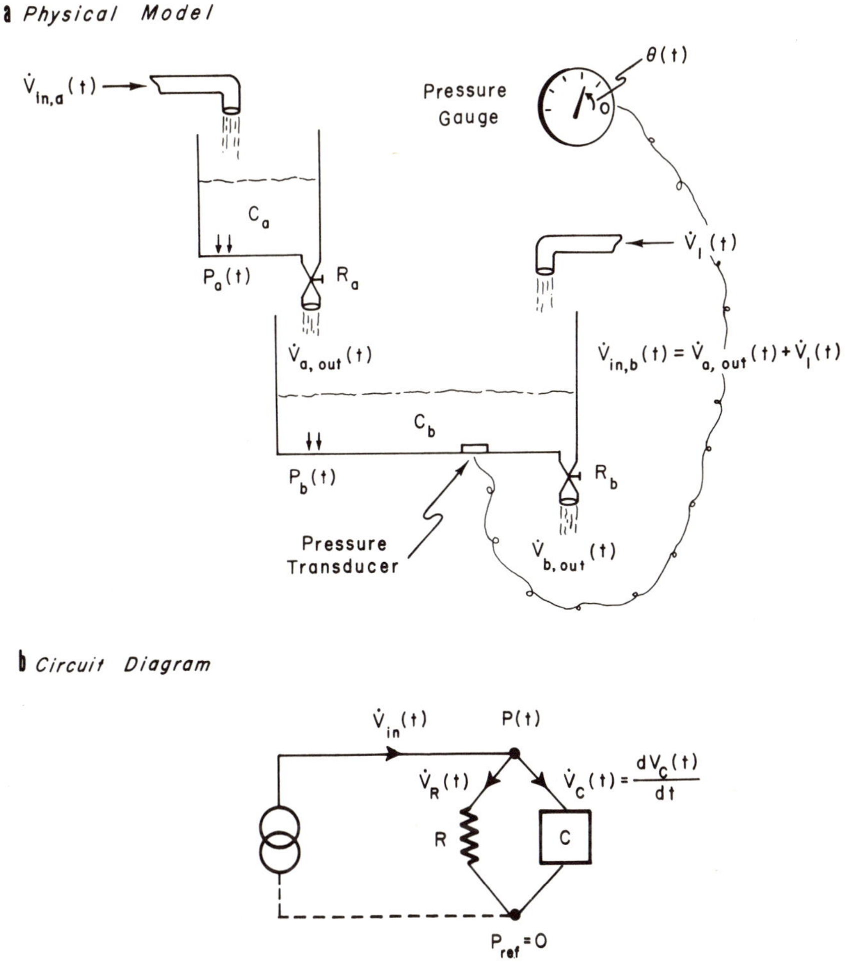

Figure 3.13.2. The two-tank system of example 3.13.1.

Part a. Physical Model. Subscript "a" refers to the upper, and subscript "b" to the lower tank. $\dot{V}$ = flow; $\dot{V}_{\ell}$ = load input (= external flow into tank b); C = fluid capacitance; P(t) = pressure on the bottom of the tank; $\dot{V}_R(t)$ = rate of flow out of the tank through the fluid resistor. $\theta(t)$ = angular displacement of hand on pressure gage.

Part b. Circuit diagram for either tank. $\dot{V}_{in}(t)$ = total flow into the tank; $\dot{V}_C(t) = dV_C(t)/dt$ = rate of increase of the volume in the tank; $\dot{V}_R(t)$ = flow out of the tank through the resistor.

It is important to understand this system thoroughly because a very similar one will be used to illustrate much of the discussion of feedback systems in Chapter 4.

$$[Cs + \frac{1}{R}]\ P_{\mathcal{L}}(s) = \dot{V}_{in,\mathcal{L}}(s) \qquad 3.13.3.$$

Now if we define the input as $\dot{V}_{in}(t)$ and the output as P(t), the transfer function is,

$$\frac{P_{\mathcal{L}}(s)}{\dot{V}_{in,\mathcal{L}}(s)} = \frac{1/C}{s + \frac{1}{RC}} = \frac{R}{RCs + 1} \qquad 3.13.4.$$

Equation 3.13.4. can be applied to either tank merely by appending appropriate subscripts. However, $P_a(t)$, the output from tank a, cannot be summed directly with $\dot{V}_{\ell}(t)$ to obtain the input, $\dot{V}_{in,b}(t)$, for tank b. It must first be multiplied by $\frac{1}{R_a}$ to convert it to $\dot{V}_{a,out}(t)$,

$$\dot{V}_{a,out}(t) = P_a(t)/R_a \qquad 3.13.5.$$

whence,

$$\dot{V}_{a,out,\mathcal{L}}(s)/\ P_{a,\mathcal{L}}(s) = \frac{1}{R_a} \qquad 3.13.6.$$

and we shall include a separate block* with this transfer function. Finally we shall treat the pressure transducer as a black box; all we need to know about it is that the output, $\theta(t)$, (the angular displacement of the dial hand), is proportional to the input, $P_b(t)$,

$$\theta(t) = kP_b(t) \qquad 3.13.7.$$

from which it is easy to see that the transfer function, $\theta_{\mathcal{L}}(s)/P_{b,\mathcal{L}}(s)$, is simply k, the constant of proportionality.

[*The reader should satisfy himself that inclusion of this block is permissible from the standpoint of unidirectionality of signal flow.]

The block diagram corresponding to this analysis is shown in Figure 3.13.3.a. Notice the following features which are characteristic of any conventional block diagram.

1. All of the parameters of the system are in the blocks. No variables (except for s) appear therein.

2. The wholly independent inputs (= sources) for the system (here $\dot{V}_{in,a}(t)$ and $\dot{V}_{\ell}(t)$) stand out clearly because they are always represented by lines coming from outside the system, not from any block or summing point within the system.

3. Two adjacent blocks can easily be combined into a single block whose transfer function is simply the product of the transfer functions of the

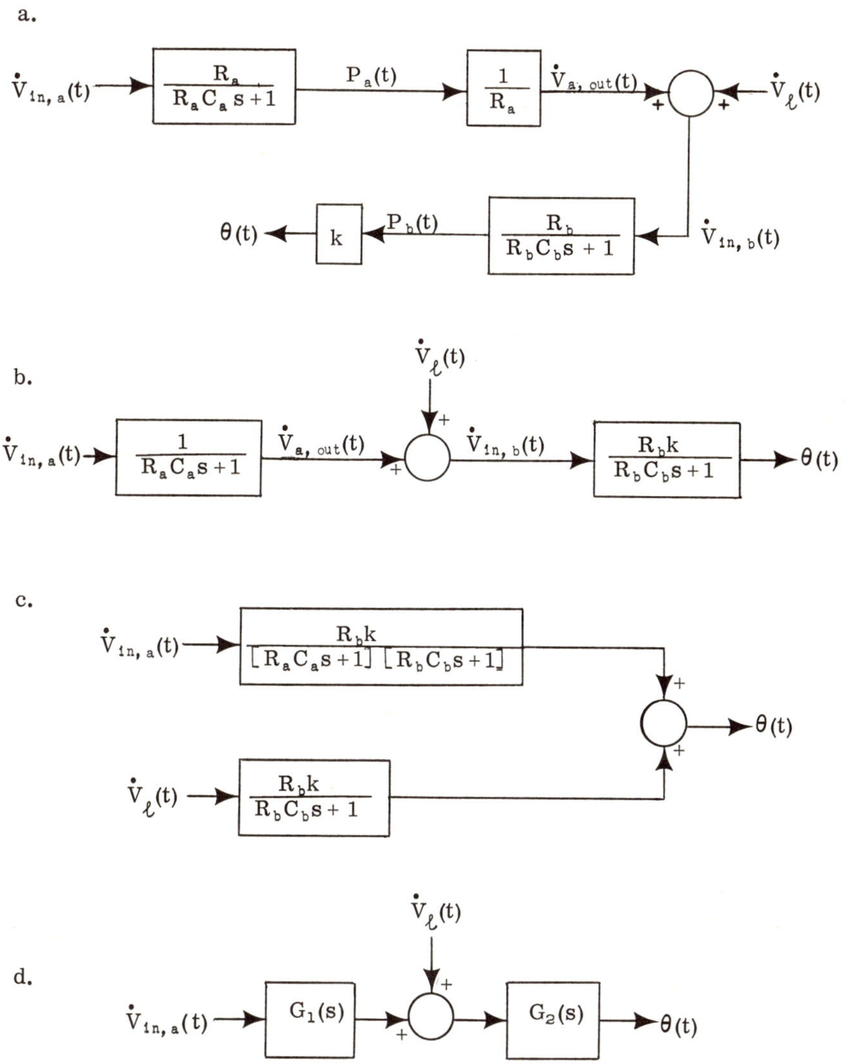

Figure 3.13.3. Block diagrams for the two-tank system of Figure 3.13.2.

The extended 4-block diagram of Panel a can be condensed to the 2-block system of Panel b by multiplying together the transfer functions in adjacent blocks, thereby deleting the intervening variables, here $P_a(t)$ and $P_b(t)$. Panel c shows another way of condensing the extended block diagram. Panel d shows that if we do not need to display the specific transfer functions, they can be replaced by suitably-defined functional symbols such as $G_1(s)$ and $G_2(s)$. Note that the general pattern of diagram b is faithfully preserved in diagram d.

original blocks. The intermediate variable is thereby eliminated. This procedure is precisely equivalent to combining two equations by eliminating a common variable. Repeatedly applied, it can reduce a complicated block diagram to one, or, at most, a few blocks which succinctly summarize the dependence of some final output variable on the wholly independent input variables of the system.

4. Just as is true of individual transfer functions, a block diagram is valid for <u>any</u> input functions of time which have Laplace transforms.

To illustrate how a block diagram can be simplified by combining adjacent blocks, the extended diagram of Figure 3.13.3.a. has been condensed to two blocks in Figure 3.13.3.b. Because inputs enter at two different points, further condensation is not possible if we are to preserve features 1 and 2 noted above. However, if we want to emphasize the different ways in which the two independent variables, $\dot{V}_{in,a}(t)$ and $\dot{V}_{\ell}(t)$, influence $\theta(t)$, we can rearrange the diagram so that the summation of the signals due to these two inputs is postponed until after each of them separately has been subjected to the operations symbolized by the appropriate blocks (Figure 3.13.3.c.).

When condensing a really complicated block diagram to simple form, it is helpful to begin by simplifying the notation. For example, with resistor-capacitor subsystems such as the one of Figure 3.13.2.b., it is customary to define a <u>time constant</u>, T,

$$T \equiv RC \qquad 3.13.8.$$

which is simply the inverse of the exponential rate constant characteristic of the RC system. (T is, in fact, the time needed for the exponential change to be $100\,[1 - 1/\epsilon]\ \% \approx 63.2\%$ complete). But the greatest economy in notation is achieved by defining a single symbol to represent a given transfer function. For example, if we define,

$$G_1(s) \equiv \frac{1}{R_a C_a s + 1} = \frac{R_a}{T_a s + 1} \qquad 3.13.9.$$

$$G_2(s) \equiv \frac{R_b k}{R_b C_b s + 1} = \frac{R_b k}{T_b s + 1} \qquad 3.13.10.$$

the block diagram of Figure 3.13.3.b. may be written as in Figure 3.13.3.d., from which it is easy by mere inspection to write the equation,

$$\theta_{\mathcal{L}}(s) = G_2(s)\left[\dot{V}_{\ell,\mathcal{L}}(s) + \dot{V}_{in,a\mathcal{L}}(s)\ \ G_1(s)\right] \qquad 3.13.11.$$

or,

$$\theta_{\mathcal{L}}(s) = G_2(s)\ \dot{V}_{\ell,\mathcal{L}}(s) + G_1(s)G_2(s)\dot{V}_{in,a\mathcal{L}}(s) \qquad 3.13.11.a.$$

which is the form corresponding to the block diagram in Figure 3.13.3.c.

With a system as simple as the one of example 3.13.1., the benefits of such economy of notation are not clearly evident. But suppose we need to analyze the behavior of a very complicated linear system consisting of many components interlinked in diverse ways and often including several feedback loops. If we know the relatively simple transfer functions of the individual components, and the input-output relations which link the components together, we can construct a block diagram such as the one in the upper part of Figure 3.13.4. With the diagram as a guide, it is then relatively easy to condense this complicated pattern to some simpler form such as the one in the lower part of Figure 3.13.4, and to find the overall transfer function for signal transfer from the initial input, x(t), to the final output, z(t). The legend of Figure 3.13.4. explains how to do this.* (See also Exercise 3.12.)

[*Still greater economy of notation may be achieved by dispensing with the blocks altogether, and expressing the interrelationships in the form of a pattern of directed lines and intersections technically known as a <u>graph.</u> One can then solve highly complicated problems of this sort in an efficient and orderly manner by applying a set of algorithms known as Mason's rules (9). These are well worth looking up if (as seems unlikely) the reader needs to solve many such problems of <u>linear</u> systems.]

<u>SECTION 3.14.</u> Unconventional block diagrams

The outstanding virtues of the conventional block diagram have prompted modifications both numerous and varied. (One almost gets the impression that putting <u>anything</u> into a little rectangle magically guarantees its scientific validity.) It is customary to refer to all of these modifications as "block diagrams."

When a system consists of both linear and non-linear components, each of the non-linear elements may be represented by a block containing a graphical sketch of the output variable as a function of the input variable (see Figure 6.20.2. as an example). The linear components are represented by their transfer functions as usual. This kind of diagram shows clearly where and how the non-linearities complicate the analysis, and identifies those parts of the system which can still be treated by the methods of linear analysis.

Linear systems with several inputs and several outputs are often best analyzed by vector and matrix techniques. (See Chapter 5). For such systems, a block has a vector input and a vector output, and may contain a symbol for the square matrix whose elements are the parameters of the system. Or it may contain a matrix equation for the time-derivative of the output vector. Such block diagrams must be combined according to the rules of matrix algebra.

Occasionally one sees diagrams with blocks containing non-algebraic operators such as $\int(\ \ )dt$ or $\frac{d(\ \ )}{dt}$. In such diagrams, blocks cannot be combined by algebraic rules, and for reasons previously discussed (Section 3.7.) Laplace transform representation is much to be preferred.

a.

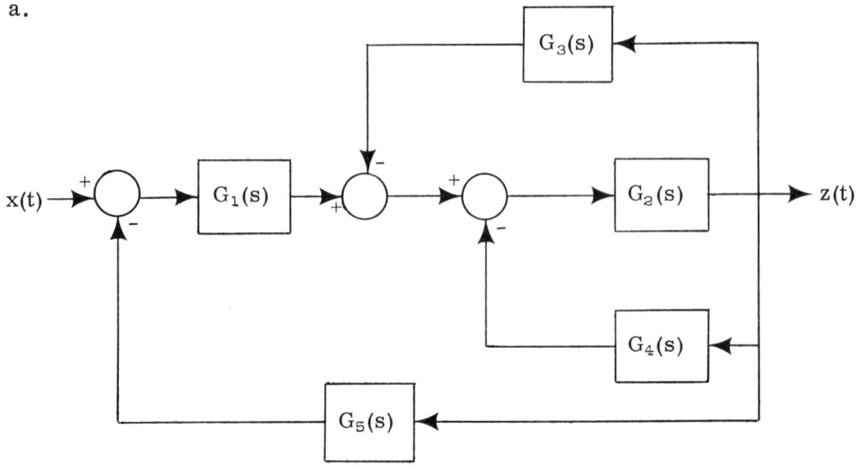

b.

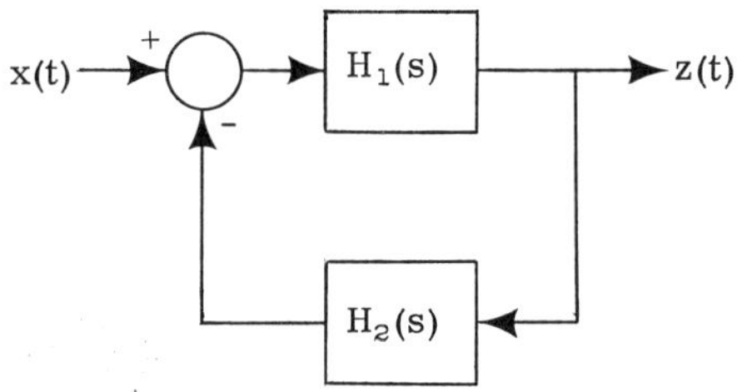

Figure 3.13.4. Reduction of a complicated block diagram to a simpler form.

In the upper diagram are shown the paths for signal flow from x(t) to z(t), including three feedback loops. If we assign symbols to each of the intermediate variables, we can write equations showing the relation between the Laplace transform of the input to, and the Laplace transform of the output from each of the summing points and blocks. By combining these equations so as to eliminate some of the intermediate variables, the complicated upper diagram can be condensed to the simpler lower one where $H_1(s)$ and $H_2(s)$ represent suitable combinations of the original G(s) transfer functions. A little practice makes it unnecessary to write down all of the single equations, since the block diagram will suggest how to combine some of the transfer functions and summing points "by inspection."

Finally, purely descriptive diagrams are often used to show the general pattern of signal flow between blocks containing no more than the name of the component, e.g. "controlled element " or "posterior pituitary" or "vice-president for medical affairs." Since the quantitative information summarized by such block diagrams is nil, they are better suited to a late-afternoon discussion by the Executive Committee of the proposed "Chain of Command for the New Medical Center" than to the serious scientific pursuits of our more wakeful hours. However, a biologist is in no position to scoff at such non-mathematical vagueness. Neatly-labelled, but quantitatively uncharacterized black boxes are likely to afflict our block diagrams for a long time to come.

SECTION 3.15. Frequency-domain analysis

In the preceding sections, the complex variable, s, has been used only as a convenient intermediate tool in our analysis of the time-domain behavior of physical systems, allowing us to summarize the characteristics of a system by means of one or more transfer functions and helping us find an unknown output from a known input by taking inverse transforms of partial fractions in s. We have thus found it convenient to manipulate s mathematically, but we have not yet made any serious use of its having been defined as a complex frequency. In fact, the methods thus far employed are usually considered to be part of time-domain analysis. We now turn to analysis in the frequency domain.

In the methods to be described here, we restrict attention to the non-transient response of stationary linear systems to sinusoidal inputs of constant magnitude and frequency. The two underlined restrictions are less confining than one might at first suppose: 1) very frequently the non-transient behavior of a system is the only response we are interested in; the complexities of its behavior just after a forcing input has been applied are then nothing but a nuisance. 2) Restriction of the input to sinusoidal functions is not severely limiting because, a) any reasonable periodic function may be approximated as closely as we please by a Fourier-series sum of sines and cosines (See Section 3.10.), and b) a constant input may be treated as the limiting value approached by a sinusoidal input as ω, the radian frequency, approaches zero.

The benefits which we reap by accepting these restrictions are quite astonishing, as the next few sections will show. Furthermore, as a considerable bonus it turns out that the formal structure of frequency-domain analysis is precisely the same as the formal structure of transfer function analysis. This makes it very easy to go from one to the other. However, some familiarity with complex numbers, and with their representation in exponential and in polar coordinate form (see Appendix) is a prerequisite for using the methods to be described.

The strategy underlying frequency domain analysis is as follows: Consider any stationary linear system which is being forced into activity by a long-continued sinusoidal input with a frequency of ω radians per unit of time. After the transient response has disappeared, any output (i.e. any dependent variable of the system) will be undergoing sinusoidal oscillations at precisely the same frequency, ω*. Furthermore, with a sinusoidal input, both the A-elements and the T-elements (as well as the D-elements) of the system can be completely characterized by the opposition or impedance they offer to the T-variable. These impedances can be treated much like the resistances of a simple dissipative circuit, except that the impedance offered by A- and T-elements will depend upon the imposed frequency, ω. (For example, the opposition of a capacitor to the flow of an alternating current through it will be less for a rapidly-oscillating high-frequency current than for a current of low frequency, while the opposite will be true of an inductor.) Finally, because of the monotonously regular recurrence of events cycle after cycle, a sinusoidal input of frequency ω radians per second can be fully characterized by its magnitude and by the phase of the sinusoidal cycle (expressed as the phase angle) which it has reached at some particular instant of time. Similarly, the non-transient output of the system at the same instant of time can be characterized by its magnitude and phase angle. Now although the frequency of any of the output responses of the system must be exactly the same as the frequency of the input signal, in general both the magnitude and the phase angle of the sinusoidal oscillations will change as the signal passes through the system from input to output. In frequency-domain analysis we are primarily interested in how these changes in magnitude and phase angle depend upon ω, the frequency of the sinusoidal oscillation which is being forced upon the system. These general ideas will soon be made quite specific as we develop some of the tactical weapons of the frequency domain.

[*NOTE WELL. This statement is not, in general, valid for nonlinear systems.]

SECTION 3.16. Phasors and impedances

Consider an electrical capacitor of capacitance, C, activated by an alternating current which is fluctuating sinusoidally about a mean value of zero. Any such alternating current with a frequency ω can be described by the equation,

$$i_{in}(t) = a \cos \omega t + b \sin \omega t \qquad 3.16.1.$$

where a and b are the magnitudes of the cosine and sine components respectively. Let the output be defined as $e_c(t)$, the voltage drop across the capacitor. To permit comparison between input and output, we must express both as the same kind of function, either sine or cosine, and not (as in equation 3.16.1.) as a mixture of the two. It is conventional to choose the cosine function. Accordingly, we rewrite equation 3.16.1. as,

$$i_{in}(t) = I \cos(\omega t + \theta) \qquad 3.16.2.$$

where $I = \sqrt{a^2 + b^2}$ is the magnitude of the now purely-cosine input, and $\theta = -\text{Arctan}[b/a]$ is the phase angle of the input. Similarly, the output must be written in the form,*

$$e_c(t, \omega) = E(\omega) \cos(\omega t + \Phi(\omega)) \qquad 3.16.3.$$

where $E(\omega)$ is the magnitude and $\Phi(\omega)$ is the phase angle of the non-transient output voltage. Notice that e_c, E, and Φ have all been written as functions of ω in recognition of the important role which ω plays as an independent variable in frequency-domain analysis.

[*As will be shown by Example 3.16.3, even when the input is a pure sinusoidal function with no non-zero constant term, a constant term may occur in the output. In phasor analysis, such terms are neglected.]

The next step, which is the key to the whole analysis, depends upon a clever application of the principle of superposition. Consider any stationary linear system. Let $y_1(t)$ be the output of the system when the input is $x_1(t)$, and let $y_2(t)$ be the output of the system when the input is $x_2(t)$. Now let α be any constant, real or complex. Because the system is linear, if we apply $x_1(t) + \alpha x_2(t)$ as an input we shall obtain $y_1(t) + \alpha y_2(t)$ as the output. In particular, let $\alpha = j \equiv \sqrt{-1}$, a perfectly legitimate constant. Then in response to the complex input, $x_1(t) + jx_2(t)$ we shall obtain the complex output, $y_1(t) + jy_2(t)$. We have thus shown that the real part of a complex output is the response to the real part of a complex input; the imaginary part of a complex output is the response to the imaginary part of a complex input. For the moment, we shall defer any explicit use of this theorem. But it does assure us that the real output of the system will in no way be obscured if we find it convenient, as we now do, to convert the real input of equation 3.16.2. to a complex input by adding a fictitious imaginary term. As we shall soon see, this innocent bit of mathematical make-believe will allow us to get rid of time as a variable so that we can study the magnitude and the phase angle of the output as functions of ω alone. Accordingly, let us add $jI \sin(\omega t + \theta)$ to the input function of 3.16.2, letting $i_{cplx}(t)$ symbolize the resulting complex function. Then, making use of Euler's identity, we have,

$$i_{cplx}(t) = I[\cos(\omega t + \theta) + j \sin(\omega t + \theta)]$$

$$= I\epsilon^{j[\omega t + \theta]} = I\epsilon^{j\omega t}\epsilon^{j\theta} \qquad 3.16.4.$$

Now the changes in magnitude and in phase angle of the imaginary sine wave signal as it traverses the system must be precisely the same as the changes in magnitude and in phase angle of the real cosine wave of the same frequency and magnitude; for the distinction between a sine wave and a like cosine wave depends solely upon where we choose to place the completely arbitrary origin of the time scale. Hence the complex output resulting from the complex input of equation 3.16.4 will be given by the following modification of equation 3.16.3,

$$e_{c,\,cplx}(t, \omega) = E(\omega)[\cos(\omega t + \Phi(\omega)) + j \sin(\omega t + \Phi(\omega))]$$

$$= E(\omega)\,\epsilon^{j[\omega t + \Phi(\omega)]} = E(\omega)\epsilon^{j\omega t}\epsilon^{j\Phi(\omega)} \qquad 3.16.5.$$

In a capacitor, the relationship between input current and output voltage is,

$$e_{c,\,cplx}(t, \omega) = \frac{1}{C}\int i_{cplx}(t)dt \qquad 3.16.6.$$

or,

$$\frac{de_{C,\,cplx}(t,\omega)}{dt} = \frac{1}{C}\, i_{cplx}(t) \qquad 3.16.7.$$

By differentiating equation 3.16.5. with respect to time, we obtain

$$\frac{de_{C,\,cplx}(t,\ \omega)}{dt} = j\omega\, E(\omega)\epsilon^{j\Phi(\omega)}\epsilon^{j\omega t} \qquad 3.16.8.$$

Substituting this value, and the value of $i_{cplx}(t)$ from equation 3.16.4. into equation 3.16.7.,

$$j\omega E(\omega)\epsilon^{j\Phi(\omega)}\epsilon^{j\omega t} = \frac{1}{C}\, I\epsilon^{j\omega t}\epsilon^{j\theta} \qquad 3.16.9.$$

Multiplying both sides of equation 3.16.9. by $\epsilon^{-j\omega t}$ to get rid of the time-dependent term, and rearranging,

$$\frac{E(\omega)\epsilon^{j\Phi(\omega)}}{I\epsilon^{j\theta}} = \frac{1}{j\omega C} \qquad 3.16.9.a.$$

or, as it is commonly written,

$$\frac{E(\omega)\underline{/\Phi(\omega)}}{I\underline{/\theta}} = \frac{1}{j\omega C} \qquad 3.16.9.b.$$

Equation 3.16.9.b. is a truly remarkable result. Notice first that the equation is no longer a relationship between functions of time, but is instead a relationship between functions of frequency. Next, notice the denominator on the left. It is just the magnitude and phase angle of the input. Think of the complex input function of equation 3.16.4. as corresponding to the counterclock-wise rotation of a vector in the complex plane. (See figure on p. 561 in the Appendix.) The absolute value of the length of the vector is the magnitude, I, of the input current, and the vector is rotating at a frequency of ω radians per second, passing repeatedly through all phases of its cycle. Suppose we suddenly halt the vector at some particular instant when $\omega t = 2\pi n$ radians, where n = any integer. Then the angle which the vector forms with the positive real axis will be the phase angle, θ radians, and the complex input can be fully characterized by stating the <u>PHAS</u>e angle and the vect<u>OR</u> length. Taken together, these two quantities constitute a <u>PHASOR</u>. The denominator on the left of equation 3.16.9.a., or 3.16.9.b., is thus called a <u>phasor current</u>. Similarly, the complex output may be represented by a vector rotating in the complex plane, and if we arrest its rotation at the same instant, it can be fully characterized by <u>its</u> magnitude and phase angle. Together these form the <u>phasor voltage</u> of the output, which is the numerator on the left of equation 3.16.9.a. or b. Finally, notice the term on the right of equation 3.16.9.a. or b. It depends partly on the capacitance, C, and partly on the radian frequency, ω. But for a given capacitor and a given frequency it is a constant. Furthermore it is equal to the ratio of a voltage and a current, so it behaves exactly like a resistance. In fact, if the phasor current,

$I\underline{/\theta}$ is in amperes, and the phasor voltage, $E(\omega)\underline{/\Phi(\omega)}$ is in volts, the quantity $\frac{1}{j\omega C}$ is in ohms. $\frac{1}{j\omega C}$ is called the impedance of the capacitor, and we shall symbolize it by $Z_c(j\omega)$,

$$Z_c(j\omega) = \frac{1}{j\omega C} \qquad 3.16.10.$$

Taken as a whole, equation 3.16.9.a. or b. is precisely analogous to Ohm's law. This means that, given any phasor voltage drop across any capacitative element we can calculate the phasor current thru the capacitor for any ω (or given the phasor current we can calculate the phasor voltage) almost as easily as we can calculate ordinary currents and voltages in ordinary resistors!

By using exactly the same sequence of steps, the reader should have no difficulty in showing that the impedance of an inductor whose inductance is L is

$$Z_L(j\omega) = j\omega L \qquad 3.16.11.$$

and that the impedance of a resistor whose resistance is R is simple R itself,

$$Z_R(j\omega) = R \qquad 3.16.12.$$

Furthermore, just as it is often convenient to work with the conductance (= 1/R) instead of the resistance when dealing with purely dissipative networks, so it is often convenient to work with the admittance, symbolized $Y(j\omega)$, and defined as

$$Y(j\omega) = \frac{1}{Z(j\omega)} \qquad 3.16.13.$$

when applying phasor analysis to more complicated electrical networks. Finally, with the end-paper table as a guide, it is a simple matter to extend the definition of phasors, impedances, and admittances to non-electrical systems. This means that we can equally well investigate the responses of mechanical, fluid, thermal, or concentrational systems to sinusoidal inputs at various frequencies merely by characterizing each element by its impedance or admittance, and by expressing the A-variables and T-variables as phasors. We can then treat the whole system as a quasi-dissipative network, amenable to the simple algebraic manipulations which were described in Section 2.12. This makes it comparatively easy to find out how the response of a complicated network is influenced by the frequency of a sinusoidal input. Such studies are crucially important in the analysis of electrical networks activated by AC inputs. But the frequency-dependent behavior of other kinds of systems is also of fundamental interest, as we shall see when we consider the stability of feedback systems (Section 4.12.).

In working with Laplace transforms, we found it convenient to define a special symbol, $x_{\mathfrak{L}}(s)$, for the Laplace transform of any suitable time-domain function, x(t). Similarly, we now define a special symbol, $\tilde{x}(j\omega)$, to represent the phasor of any suitable time-domain function, x(t). The superscript tilde reminds us that the "suitable functions" are sinusoidal; the argument, $j\omega$, reminds us that the phasor is no longer a function of time, but is a complex function of the input frequency, ω.

When actually using phasor analysis, it is of course unnecessary to repeat the steps employed in the derivation of equations 3.16.10, 3.16.11, and 3.16.12, because these equations are perfectly general and enable us to write down immediately the impedance or the admittance of any passive element whose parameter is known. The steps which are required in practice will now be developed and illustrated by means of three examples.

Example 3.16.1. Consider again the series RC circuit of Figure 2.16.1. Let the input be $e_{in}(t) = 10 + 3 \sin \omega t$ volts as in examples 2.16.1. and 3.9.1. Find the non-transient output, i(t), using phasor analysis.

The input consists of two terms: 10 volts, which may be regarded as a sinusoidal input where $\omega = 0$, and 3 $\sin\omega t$ volts where ω will not, in general, $= 0$. In phasor analysis, the responses to inputs with different frequencies must be calculated separately, because the impedance will, in general, have a different value for each different frequency. The separate responses must be added together only after each response has been translated back into time domain.

The impedance of a series RC circuit is,

$$Z(j\omega) = R + \frac{1}{j\omega C} = R - j\frac{1}{C\omega} = \frac{RC\omega - j}{C\omega} \qquad 3.16.14.$$

The admittance is,

$$Y(j\omega) = 1/Z(j\omega) = \frac{C\omega}{RC\omega - j} \qquad 3.16.15.$$

Applying Ohm's law, we can calculate $\tilde{i}(j\omega) = I(j\omega)\underline{/\Phi(j\omega)}$, the phasor output current, most simply by multiplying each phasor input voltage by $Y(j\omega)$. Taking the constant term first, we express it as a cosine function,

$$10 \text{ volts} = 10 \cos(0t + 0) \text{ volts} \qquad 3.16.16.$$

for which the phasor voltage is $10\underline{/0}$ volts.

Hence,

$$\tilde{i}(j0) = 10\underline{/0}\,[Y(j0)] = 10\underline{/0}\left[\frac{0}{0-j}\right] = 0 \text{ amp} \qquad 3.16.17.$$

which shows that there is no steady-state current in response to the constant term of the input voltage. This is, of course, just what we would expect for direct current flow through a capacitor. Next, we consider the sine-wave input.

We must convert it to a cosine function,

$$3 \sin \omega t \text{ volts} = 3 \cos(\omega t - \pi/2) \text{ volts} \qquad 3.16.18.$$

The corresponding phasor is $3\underline{/-\pi/2}$ volts so we have

$$\tilde{i}(j\omega) = 3\underline{/-\pi/2}\, Y(j\omega) = \frac{3C\omega\underline{/-\pi/2}}{RC\omega - j} \text{ amps} \qquad 3.16.19.$$

But

$$3C\omega\underline{/-\pi/2} = -j3C\omega \qquad 3.16.20.$$

Substitute this value into equation 3.16.19, and multiply both numerator and denominator by the complex conjugate of the denominator so as to get rid of j in the denominator,

$$\tilde{i}(j\omega) = \frac{-j3C\omega}{RC\omega - j}\left\{\frac{RC\omega + j}{RC\omega + j}\right\} = \frac{3C\omega - j3R[C\omega]^2}{[RC\omega]^2 + 1} \text{ amps} \qquad 3.16.21.$$

Now we shall spare ourselves a great deal of trouble if we recognize that, for any given value of ω, the right side of equation 3.16.21. is nothing but a complex number, say $\alpha + j\beta$, where

$$\alpha = \frac{3C\omega}{[RC\omega]^2 + 1} \text{ amps} \qquad 3.16.22.$$

$$\beta = -\frac{3R[C\omega]^2}{[RC\omega]^2 + 1} \text{ amps} \qquad 3.16.23.$$

So we may write

$$\tilde{i}(j\omega) = \alpha + j\beta = [\alpha^2 + \beta^2]^{1/2}\underline{/\Phi} = [\alpha^2 + \beta^2]^{1/2}\epsilon^{j\Phi} \text{ amps} \qquad 3.16.24.$$

where $\Phi = \text{Arctan}\, \beta/\alpha$. Next, we convert the frequency-domain phasor output of equation 3.16.24. to the corresponding complex output in time domain by multiplying both sides of the equation by $\epsilon^{j\omega t}$. Notice that this is just the reverse of what we did in passing from equation 3.16.9. to equation 3.16.9.a,

$$\tilde{i}(j\omega)\epsilon^{j\omega t} = i_{cplx}(t) = [\alpha^2 + \beta^2]^{1/2}\epsilon^{j\Phi}\epsilon^{j\omega t} = \sqrt{\alpha^2 + \beta^2}\; \epsilon^{j[\omega t + \Phi]}$$

$$= \sqrt{\alpha^2 + \beta^2}\cos(\omega t + \Phi) + j\sqrt{\alpha^2 + \beta^2}\sin(\omega t + \Phi) \qquad 3.16.25.$$

Finally, we make explicit use of the theorem which states that the real part of a complex output is the response to the real part of a complex input,

$$i(t) = \text{Re}\ i_{cplx}(t) = \sqrt{\alpha^2 + \beta^2}\ \cos(\omega t + \Phi)$$

$$= \alpha \cos \omega t - \beta \sin \omega t \quad \text{amps} \qquad 3.16.26.$$

Substituting the values of α and β from equations 3.16.22. and 3.16.23. into equation 3.16.26.,

$$i(t) = \frac{3C\omega}{[RC\omega]^2 + 1}\ [\cos\omega t + RC\omega \sin\omega t] \quad \text{amps} \qquad 3.16.27.$$

which is exactly what equation 3.9.9. becomes after its exponential terms have died of old age! But such complete agreement cannot be expected in general because the result obtained by evaluating the total response to a sinusoidal input as $t \rightarrow \infty$ sometimes includes a constant term which is missing from the solution obtained by phasor analysis. This phenomenon will be illustrated by Example 3.16.3.

From the present example, we see that there are really only five steps in phasor analysis:

1. Express the input as a pure cosine function.

2. Write its magnitude and phase angle in phasor form; i.e., as a complex number, using either rectangular or polar coordinates as convenience dictates.

3. Find the overall impedance (or, if more convenient, admittance) of the system.

4. Treating the impedance as if it were a simple resistance, calculate the output phasor from Ohm's law and express it as a complex number $\alpha + j\beta$.

5. Express the output either in the frequency domain as a magnitude, $\sqrt{\alpha^2 + \beta^2}$, and phase angle, Arctan β/α, or in the time domain as $\alpha \cos\omega t - \beta \sin\omega t$, or as $\sqrt{\alpha^2 + \beta^2}\ \cos(\omega t + \text{Arctan}\ (\beta/\alpha))$.

Example 3.16.2. Consider the parallel RLC network of Figure 3.16.1. If the output is e(t), find the non-transient response to an input current of $i(t) = 0.4 \cos\omega t$ amperes.

We follow the five steps given above.

1. The input is already expressed as a pure cosine function.

2. Its phasor form is $0.4\underline{/0} = 0.4 + j0$ amps.

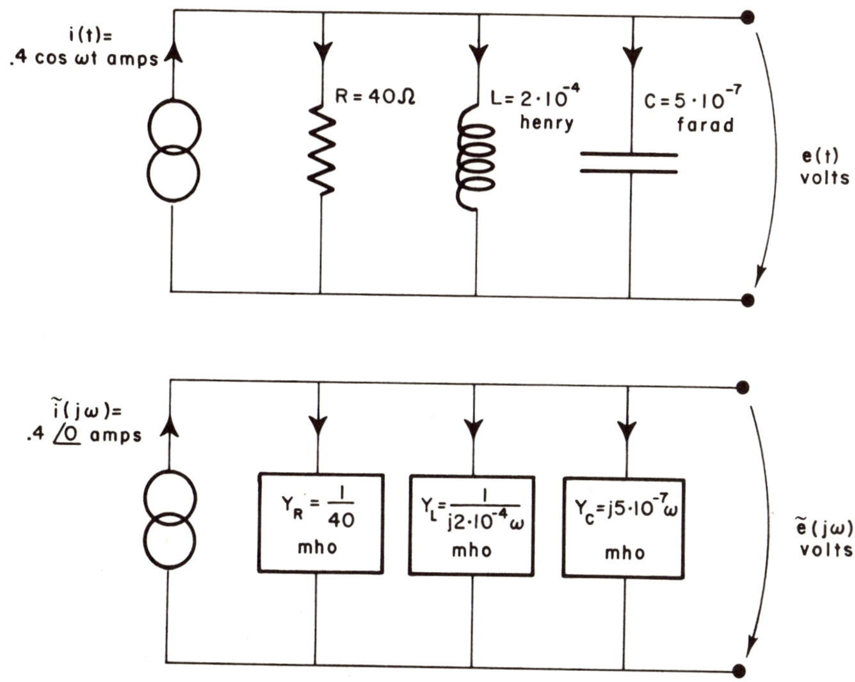

Figure 3.16.1. The parallel RLC system of example 3.16.2, activated by a cosinusoidal input. a.) Circuit diagram (time domain.) b.) Phasor representation (frequency domain).

3. The admittance of the network is

$$Y(j\omega) = \frac{1}{R} + j\omega C + \frac{1}{j\omega L} = \frac{1}{0.4 \cdot 10^2} + j5 \cdot 10^{-7}\omega - j\,\frac{1}{2 \cdot 10^{-4}\omega}$$

$$= 2.5 \cdot 10^{-2} + j5 \cdot 10^{-7}\omega - j\frac{5 \cdot 10^3}{\omega} = \frac{2.5 \cdot 10^{-2}\omega + j[5 \cdot 10^{-7}\omega^2 - 5 \cdot 10^3]}{\omega}$$

mhos 3.16.28.

4. By Ohm's law,

$$\tilde{e}(j\omega) = \tilde{i}(j\omega)Z(j\omega) = \tilde{i}(j\omega)\left[\frac{1}{Y(j\omega)}\right] \qquad 3.16.29.$$

whence:

$$\tilde{e}(j\omega) = \frac{0.4\omega}{2.5\cdot10^{-2}\omega + j[5\cdot10^{-7}\omega^2 - 5\cdot10^3]} \text{ volts} \qquad 3.16.30.$$

or, using the convenient rules given in the Appendix about multiples and reciprocals of complex numbers,

$$\tilde{e}(j\omega) = \frac{0.4\omega}{\sqrt{25\cdot10^{-14}\omega^4 - 43.75\cdot10^4\omega^2 + 25\cdot10^6}} \Bigg/ \tan^{-1} - \frac{5\cdot10^{-7}\omega^2 - 5\cdot10^3}{2.5\cdot10^{-2}\omega} \text{ volts} \qquad 3.16.31.$$

5. Therefore, in time domain,

$$e(t) = \frac{0.4\omega}{\sqrt{25\cdot10^{-14}\omega^4 - 43.75\cdot10^4\omega^2 + 25\cdot10^6}} \cos\left\{\omega t + \tan^{-1} - \frac{5\cdot10^{-7}\omega^2 - 5\cdot10^3}{2.5\cdot10^{-2}\omega}\right. \text{ volts} \qquad 3.16.32.$$

These steps become easier and easier as one becomes more and more familiar with complex numbers and the various ways of representing them.

In Figure 3.16.2, the magnitude and phase angle of the output voltage for this example have been plotted against frequency. When both a capacitor and an inductor are present in a network, we may observe the phenomenon of resonance. Resonance occurs when the frequency of the input current or voltage is the frequency at which energy can be exchanged without loss between the two storage elements. When this happens, the parallel RLC network behaves just as if only the dissipative element were present, because the admittances of the inductor and the capacitor at the resonant frequency, ω_{res}, are equal in magnitude but opposite in sign, so that they cancel each other. At the resonant frequency, therefore, the imaginary part of the admittance (or impedance) vanishes. In the present example, this will be true when

$$jC\omega + \frac{1}{jL\omega} = j5\cdot10^{-7}\omega - j\frac{1}{2\cdot10^{-4}\omega} = 0 \qquad 3.16.33.$$

Solving equation 3.16.33. for ω,

$$\omega = \omega_{res} = \frac{1}{\sqrt{LC}} = \frac{1}{\sqrt{10^{-10}}} = 10^5 \text{ radians per sec} \qquad 3.16.34.$$

At the resonant frequency, therefore, all of the current from the source must flow through the resistor. This accounts for the voltage peak at $\omega = \omega_{res} = 10^5$ rad/sec, in Figure 3.16.2. Notice also that the phase angle of the output voltage when $\omega = \omega_{res}$ is the same as the phase angle of the input (here, zero

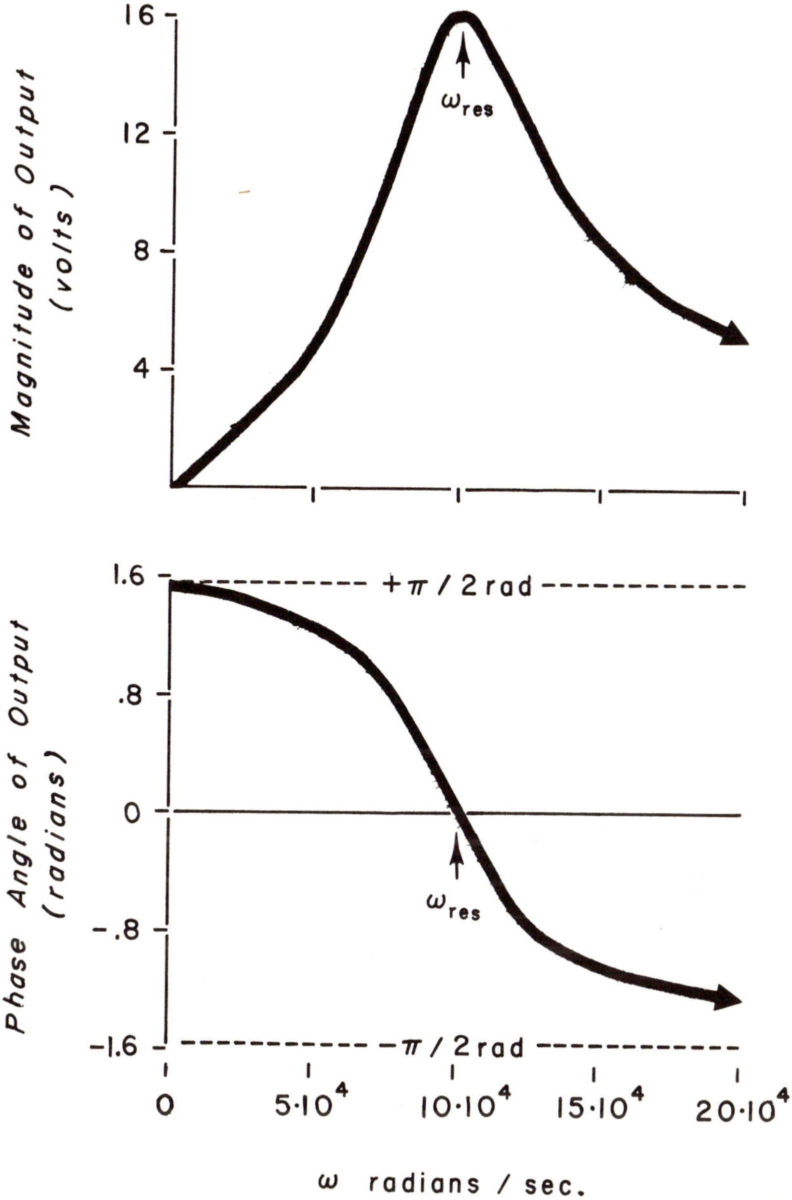

Figure 3.16.2. The output voltage of the system of Figure 3.16.1 showing that the magnitude peaks at ω_{res}, the resonant frequency where the phase shift is zero.

radians) as it always is for a purely dissipative network. In contrast, as $\omega \to 0$ the magnitude of the voltage output approaches zero while the phase difference, $\Phi_{output} - \theta_{input}$, approaches the $+\pi/2 \approx +1.571$ radians which is characteristic of alternating current flowing through a pure inductor. (The output voltage of a pure inductor is said to lead the input current by $+\pi/2$ radians or 90° because the output voltage reaches any particular phase of its cycle $\pi/2$ radians ($\frac{\pi}{2\omega}$ seconds) earlier than the input current does.) As $\omega \to \infty$, the magnitude of the voltage output again approaches zero, but the phase difference approaches the $-\pi/2$ radians which is characteristic of a pure capacitor. (The output is then said to lag the input by $\pi/2$ radians). The reader should be able to explain why the magnitude and phase angle of the output behave in this way as ω approaches extreme values of 0 or ∞. He may also wish to work out for himself the behavior of a series RLC circuit. (Exercise 3.9.)

While a detailed consideration of the properties of resonating circuits would take us too far afield, there is one common misconception which the present example can help clarify. The natural, or unforced frequency, ω_n, of a system is not the same as the forced resonant frequency ω_{res}, though in some systems they may differ from each other by a negligible amount.

The natural frequency may be found by writing the node equation for the system,

$$\frac{e(t)}{R} + C\frac{de(t)}{dt} + \frac{1}{L}\int_0^t e(\tau)d\tau = i_{in}(t) \qquad 3.16.35.$$

taking its Laplace transform with zero initial conditions,

$$\frac{1}{R}e_{\mathcal{L}}(s) + Cse_{\mathcal{L}}(s) + \frac{1}{Ls}e_{\mathcal{L}}(s) = i_{in,\mathcal{L}}(s) \qquad 3.16.36.$$

and solving for the transfer function $e_{\mathcal{L}}(s)/i_{in,\mathcal{L}}(s)$,

$$e_{\mathcal{L}}(s)/i_{in,\mathcal{L}}(s) = \frac{1}{\frac{1}{R} + Cs + \frac{1}{Ls}} = \frac{\frac{1}{C}s}{s^2 + \frac{1}{RC}s + \frac{1}{LC}} \qquad 3.16.37.$$

If we use a unit impulse input to establish initial conditions,

$$e_{\mathcal{L}}(s) = \frac{\frac{1}{C}s}{s^2 + \frac{1}{RC}s + \frac{1}{LC}} \qquad 3.16.38.$$

Now provided the damping ratio $\{[1/RC]^2/[4/LC]\}^{1/2}$ is less than unity (as it is in this example) the output will exhibit damped sinusoidal oscillations at the natural frequency, ω_n. Taking the inverse transform of equation 3.16.38,

$$e(t) = \frac{1}{C}\epsilon^{-\frac{1}{2RC}t}\cos\left\{\frac{1}{LC} - \left[\frac{1}{2RC}\right]^2\right\}^{1/2}_{t} \qquad 3.16.39.$$

so,

$$\omega_n = \left\{\frac{1}{LC} - \left[\frac{1}{2RC}\right]^2\right\}^{1/2} \qquad 3.16.40$$

while from equation 3.16.34,

$$\omega_{res} = \left\{\frac{1}{LC}\right\}^{1/2} \qquad 3.16.41.$$

Therefore if $\left[\frac{1}{2RC}\right]^2$, the square of the rate constant for exponential decay, is small compared with $\omega_{res}^2 = 1/LC$, ω_{res} will be very close to ω_n. In the present example, $[1/2RC]^2 = 6.25 \cdot 10^8$, while $\omega_{res}^2 = 1/LC = 10^{10}$. Hence,

$\omega_n = \sqrt{10^{10} - 0.0625 \cdot 10^{10}} = 9.68 \cdot 10^4$ radians per sec, compared with $\omega_{res} = 10.00 \cdot 10^4$ rad/sec. In radio circuits with accurate tuning, the agreement between ω_n and ω_{res} would be very much better than it is in this example.

<u>Example 3.16.3. Constant terms in the output when the input is a pure sinusoidal function.</u> A capacitor of capacitance C with zero initial charge is activated by an alternating current, $i(t) = I \sin \omega t$ starting at time zero. Compare $e_c(t)_{t\to\infty}$ calculated by finding the total response and then allowing t to $\to \infty$, with $e_c(t)$ calculated by using phasor analysis.

Applying the equation which describes the behavior of a capacitor,

$$e_c(t) = \frac{1}{C}\int_0^t I \sin\omega\tau\, d\tau \text{ with } e_c(0) = 0 \qquad 3.16.42.$$

Solving equation 3.16.42. by integration,

$$e_c(t) = \frac{I}{C}\left\{-\frac{1}{\omega}\cos\omega\tau\Big|_0^t\right\} = \frac{I}{C\omega} - \frac{I}{C\omega}\cos\omega t \qquad 3.16.43.$$

whence, letting time approach ∞,

$$e_c(t)_{t\to\infty} = \frac{I}{C\omega} - \frac{I}{C\omega}\cos\omega t \qquad 3.16.44.$$

Applying phasor analysis to the same problem,

$$i(t) = I \sin \omega t = I \cos(\omega t - \pi/2) \qquad 3.16.45.$$

$$\tilde{i}(j\omega) = I\underline{/ - \pi/2} \qquad 3.16.46.$$

$$Z_c(j\omega) = \frac{1}{j\omega C} = -j\frac{1}{C\omega} = \frac{1}{C\omega}\underline{/ - \pi/2} \qquad 3.16.47.$$

$$\tilde{e}_c(j\omega) = \tilde{i}(j\omega)Z_c(j\omega) = \frac{I}{C\omega}\underline{/ - \pi} \qquad 3.16.48.$$

$$e_c(t) = -\frac{I}{C\omega}\cos \omega t \qquad 3.16.49.$$

Equation 3.16.44. contains a constant term, $I/C\omega$, which equation 3.16.49 lacks. How can we account for this discrepancy?

The explanation is simple. Phasor analysis deals only with the magnitudes and phase angles of sinusoidal outputs as they oscillate about some mean value in response to a long-continued sinusoidal input. We have assumed (e.g. in equation 3.16.3.) that in response to an input fluctuating sinusoidally about a mean value of zero the output would also fluctuate about a mean value of zero. The present analysis shows that this assumption is not necessarily correct. For if at time zero a sine-wave input current starts to flow into an initially uncharged capacitor, the value of $e_c(t)$ will start to increase and will continue rising so long as the current is positive, i.e. during all of the first half of the cycle. During the second half of the cycle when the current is negative, current flowing out of the capacitor will remove the accumulated charge and bring the voltage back to zero again at the end of the cycle. So $e_c(t) \geq 0$ throughout the entire cycle, fluctuating sinusoidally about some non-zero mean value which in our example was $I/C\omega$, but which in general will depend upon the phase angle of the input current. Fortunately the mean value of the output in response to a sine or cosine input is often of no interest to us, so that we can still use phasor analysis as a tool for studying the oscillatory behavior of the output.

SECTION 3.17. The relation between transfer function analysis and phasor analysis

In the upper six rows of Table 3.17.1., transfer functions and impedances (or admittances), for the three basic elements are listed side by side, and in subsequent rows are shown transfer functions, and the corresponding phasor formulations for a few more complicated systems. On contemplating the upper part of Table 3.17.1., we are at once struck by the astonishing similarity between the transfer function linking the input and output of a single element and the corresponding impedance or admittance. The impedance (or admittance) of any system composed entirely of the three basic passive elements can be obtained directly from the appropriate transfer function simply by substituting $j\omega$ for s. One is at first tempted to view this remarkable interchangeability

Table 3.17.1.

Time-Domain Formulation	Transfer Function	Corresponding Phasor Formulation
D-element $e(t) = Ri(t)$ $i(t) = \frac{1}{R}e(t)$	$e_{\mathcal{L}}(s)/i_{\mathcal{L}}(s) = R$ $i_{\mathcal{L}}(s)/e_{\mathcal{L}}(s) = 1/R$	$\tilde{e}(j\omega)/\tilde{i}(j\omega) = Z_R(j\omega) = R$ $\tilde{i}(j\omega)/\tilde{e}(j\omega) = Y_R(j\omega) = 1/R$
A-element $e(t) = \frac{1}{C}\int i(t)dt$ $i(t) = C\frac{de(t)}{dt}$	$e_{\mathcal{L}}(s)/i_{\mathcal{L}}(s) = 1/Cs$ $i_{\mathcal{L}}(s)/e_{\mathcal{L}}(s) = Cs$	$\tilde{e}(j\omega)/\tilde{i}(j\omega) = Z_C(j\omega) = 1/Cj\omega$ $\tilde{i}(j\omega)/\tilde{e}(j\omega) = Y_C(j\omega) = Cj\omega$
T-element $e(t) = L\frac{di(t)}{dt}$ $i(t) = \frac{1}{L}\int e(t)dt$	$e_{\mathcal{L}}(s)/i_{\mathcal{L}}(s) = Ls$ $i_{\mathcal{L}}(s)/e_{\mathcal{L}}(s) = 1/Ls$	$\tilde{e}(j\omega)/\tilde{i}(j\omega) = Z_L(j\omega) = Lj\omega$ $\tilde{i}(j\omega)/\tilde{e}(j\omega) = Y_L(j\omega) = 1/Lj\omega$
Series RC circuit, Input = $e_{in}(t)$. Output = $e_c(t)$. $Ri(t) + \frac{1}{C}\int i(t)dt = e_{in}(t)$ $e_c(t) = e_{in}(t) - Ri(t)$	$\frac{e_{c,\mathcal{L}}(s)}{e_{in,\mathcal{L}}(s)} = \frac{1}{RCs+1}$	$\frac{\tilde{e}_c(j\omega)}{\tilde{e}_{in}(j\omega)} = \frac{1}{RCj\omega+1}$
Two-tank system of Figure 3.13.2 with $\dot{V}_{\ell}(t) = 0$	$\theta_{\mathcal{L}}(s)/\dot{V}_{in,a,\mathcal{L}}(s) =$ $\frac{R_bk}{[R_aC_as+1][R_bC_bs+1]}$	$\tilde{\theta}(j\omega)/\tilde{\dot{V}}_{in,a}(j\omega) =$ $\frac{R_bk}{[1+R_aC_aj\omega][1+R_bC_bj\omega]}$
GENERALIZATION $x(t) \rightarrow \boxed{G(s)} \rightarrow y(t)$	$y_{\mathcal{L}}(s)/x_{\mathcal{L}}(s) = G(s)$	$\tilde{y}(j\omega)/\tilde{x}(j\omega) = G(j\omega)$

of $j\omega$ and s as a kind of supernatural dispensation which the mathematical gods, in their infinite wisdom, have seen fit to bestow upon their human votaries. The steps used in the derivation of transfer functions seem so completely unrelated to the steps used in the derivation of impedances. Furthermore, the Laplace transform of a sinusoidal input and the phasor form of the same input do not even have the same dimensions. For example, if the input is current in amperes, the phasor form of the input will also be current in amperes. But the Laplace transform of the input will be charge in ampere-seconds (= coulombs). However, on further reflection one remembers that, after all, both transfer function analysis and phasor analysis are methods in which the behavior of a linear system can be divorced from time so that it can be analyzed by purely algebraic manipulations. So it is perhaps only natural that the algebraic description of comparable systems should turn out to be formally identical for the two methods. Nevertheless, we can hardly help admiring the ease of passage from transfer function analysis to frequency domain analysis (and vice versa) which this formal identity makes possible. Moreover, as with transfer function analysis, the utility of phasor analysis is by no means restricted to systems which deal with one particular form of energy. There is no reason why the phasor output of a given component in a suitable linear system cannot become the phasor input of the next component in line, even if the next component is a transducer which changes the signal from one form of energy to another. We shall make good use of this generality when we study the stability of feedback systems in the next chapter.

In summary, let G(s) be the transfer function for any stationary linear system which is stable (see Section 4.7.). If x(t) is the input, and y(t) is the output such that

$$y_{\mathcal{L}}(s) = G(s)x_{\mathcal{L}}(s) \qquad 3.17.1.$$

then

$$\tilde{y}(j\omega) = G(j\omega)\tilde{x}(j\omega) \qquad 3.17.2.$$

SECTION 3.18. The extension of phasor analysis to more general kinds of inputs

In the previous sections, phasor analysis has been restricted to undamped sinusoidal inputs. In transfer function analysis, undamped sinusoidal inputs correspond to partial fractions whose denominators are $s^2 + \omega^2$. Putting $s^2 + \omega^2 = 0$, we find that the corresponding roots are $s = \pm j\omega$, so that we have been limiting consideration to output terms of the general form $A\varepsilon^{j\omega t}$ and $B\varepsilon^{-j\omega t}$, i.e., to sums of sines and cosines. In fact, therefore, we have been using only the radian frequency or $j\omega$ part of the complex variable $s = \sigma + j\omega$, as, indeed, is implied by equation 3.17.2. This immediately raises the question of whether the phasor concept could be extended to inputs of the form $x(t) = K\varepsilon^{[\sigma + j\omega]t}$ (i.e., to inputs with a non-zero neper frequency) by generalizing equation 3.17.2. to,

$$\tilde{y}(\sigma + j\omega) = G(\sigma + j\omega)\,\tilde{x}\,(\sigma + j\omega) \qquad 3.18.1.$$

where we would have two independent variables, σ, and ω, instead of one. This is, indeed, possible. But although the extension of phasor analysis to neper frequencies and hence to exponential inputs is not difficult, and is occasionally convenient, I shall not describe it here. The results obtained must be interpreted rather carefully, because, when exponentially-changing inputs are permitted, the phasor response no longer corresponds to the sinusoidal part of the total response as $t \to \infty$, i.e., when the natural response has disappeared. What happens is that with exponential inputs, such an extension of phasor analysis yields terms corresponding to the forced part of the transient response but not terms corresponding to the simultaneous natural part of the transient response. While the natural part can be calculated separately and added to the forced part, it is doubtful whether this procedure offers any substantial advantage over transfer function analysis which gives us the total response directly.

SECTION 3.19. A comment upon the meaning of the variable, s

By way of overall summary, I would like to make a final comment upon the interpretation of the complex frequency, $s = \sigma + j\omega$. However vague s may seem to be when we use it as the Laplace transform variable, or in writing transfer functions, whenever we come to the point of wanting specific solutions, we have to give s specific values. These specific values are the values which make the denominators of the partial fractions in the Laplace transform of the output equal to zero (and they are therefore called "zeros" of the denominator; see Section 4.8). Each and every such specific value becomes a coefficient of time in the exponent of one (or, with repeated factors, more than one) of the exponential terms of the output, though we may prefer to express some of these terms as constants (when $s = 0$) or as sinusoidal functions (when $\omega \neq 0$). These specific values of s originate in two quite different ways: 1) When there is no input (or, more properly, only an impulse input to establish initial conditions) the specific values of s are the eigenvalues of the system; i.e., the roots of the characteristic equation obtained by equating to zero the denominator of the transfer function relating input to output.* For a given stationary linear system these eigenvalues are not independent variables but are constants, since they are determined by the parameters of the system itself. 2) When there is a true input, the input forcing function brings additional partial fractions into the Laplace transform of the output function. For these additional terms, the specific values of s are not eigenvalues but are the values of the exponents, σ, or the frequencies, ω, already present in the terms of the input forcing function. Whenever convenient, these imposed values of s can be treated as independent variables, and this is exactly how we treat ω in phasor analysis.

[*For a formal proof that the characteristic equation can be obtained by equating the denominator of the transfer function to zero, see Section 4.7.]

Exercises

1. Find the Laplace transform of $t^2\epsilon^{-2.7t}$ by applying Theorem iv instead of Theorem v as was done in the text.

2. Write the time-domain equations and the Laplace transforms for each of the following input functions. (Heavy lines.)

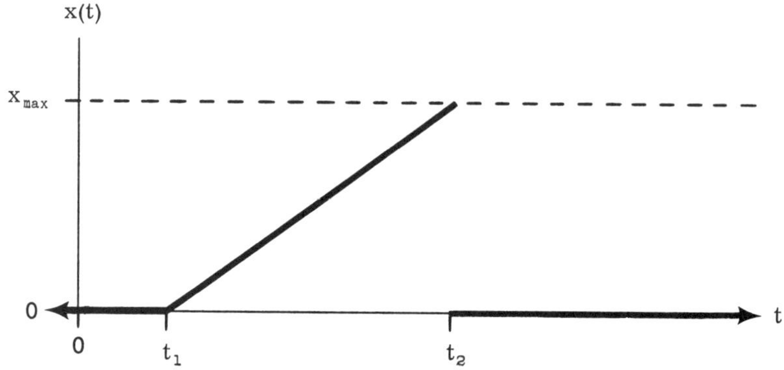

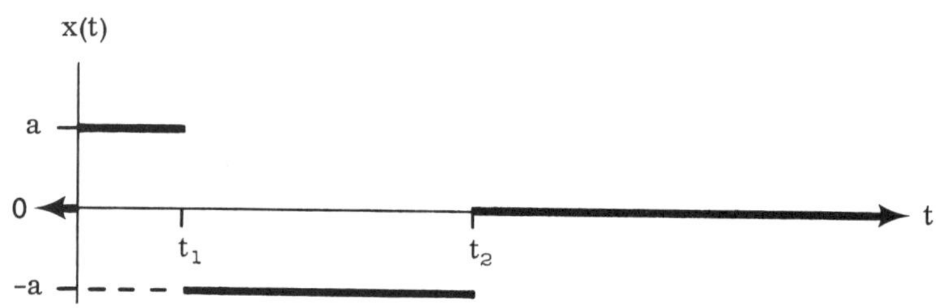

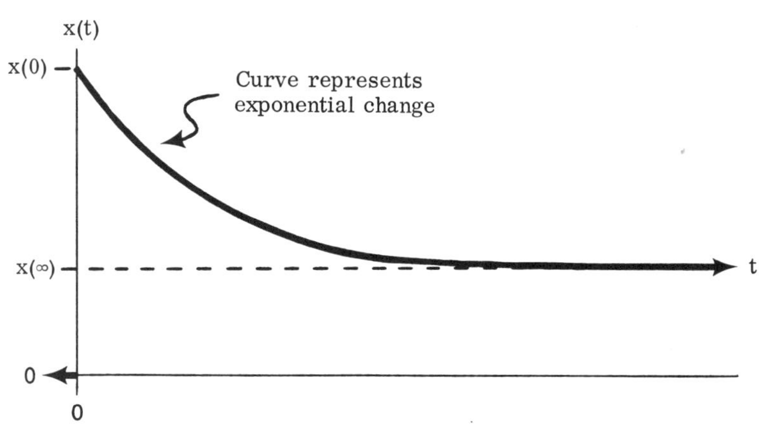

3. Calculate the Laplace transforms of the following time-domain functions by applying the Laplace operator, $\mathcal{L} = \int_0^{+\infty} \epsilon^{-st}[\ ]dt$ to each. Show all calculations.

(a.) $f_1(t) = 8\epsilon^{-2t} + 7$

(b.) $f_2(t) = 5\cos\omega t$; where ω is a constant radian frequency

4. If $x_{\mathcal{L}}(s) = \dfrac{9s^2 + 16s + 50}{s^3 + 6s^2 + 25s}$, find x(t). (Show necessary calculations, but feel free to use a table of Laplace transforms.)

5. Under the circumstances described in example 3.7.2, at what time will $C_B(t)$ be increasing most rapidly?

6. In the system of example 3.7.2, use transfer function analysis to find $C_A(t)$ for $t \geq 0$ when a single intravenous injection of 300 mg of P is given at t = 15 min.

7. Let $x_{\mathcal{L}}(s)$ be the Laplace transform of any input forcing function, x(t), which is not an impulse. Let 8 be the Laplace transform of $8\delta(t)$, an impulse input delivered at time zero. Show that 8 and $x_{\mathcal{L}}(s)$ have the same dimensions.

8. By requiring the integral of x(t) with respect to time in equation 3.8.7 to equal A even when $t_2 = t_1$, show that equation 3.8.8 is plausible, i.e., that $\lim_{t_2 \to t_1} \left\{ \dfrac{u(t - t_1) - u(t - t_2)}{t_2 - t_1} \right\}$ can be viewed as having the characteristics of $\delta(t - t_1)$.

9. Using phasor analysis find a general expression for the nontransient response of $e_R(t)$, the voltage drop across the resistor of a series RLC circuit activated by a voltage source, $e_{in}(t) = \cos \omega t$. Can this circuit exhibit resonance? If so, find an expression for ω_{res}.

10. Find the magnitude and phase angle of each of the following complex numbers. (Give the phase angle in actual radians, e.g. $\theta = 4.71$ radians, or $\theta = 3\pi/2$ radians, not θ = Arctan (0/-3) radians.)

a.) 14

b.) -14

c.) j

d.) $-|-1| + j2^{-2}$

e.) $-j^{-1}$

f.) 2 - j3

g.) -2 + j3

h.) $[-3+j2] + [2-j3]$

i.) $[-3+j2][2-j3]$

j.) $\dfrac{1}{[-3+j2][2-j3]}$

k.) $3\epsilon^{-j2}$

11. Reduce the following expression to a single complex number.

$$\frac{-1/2[4-j3\omega]}{[-2+j3\omega][1-j5\omega]j\omega}$$

12. Consider the block diagrams in Figure 3.13.4. Find expressions for $H_1(s)$ and $H_2(s)$ (part b.) in terms of the transfer functions (the G(s)'s) in part a of the figure.

13. Consider the two-tank model systems pictured below. The fluid is water at a density of 1g/ cm^3. For convenience, express pressure in cm H_20 (1 cm $H_20 \approx 980.6$ dynes/cm^2).

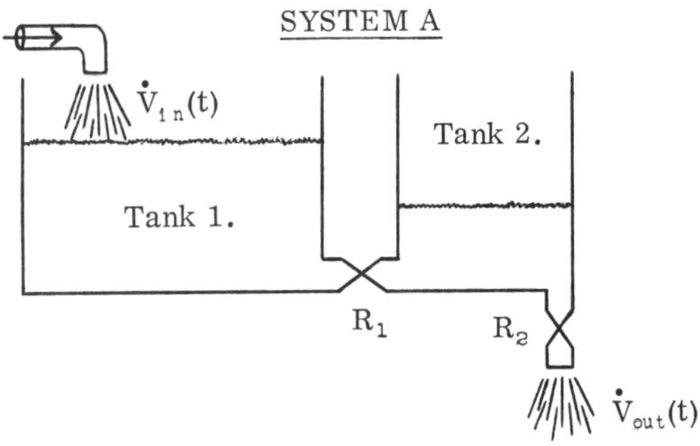

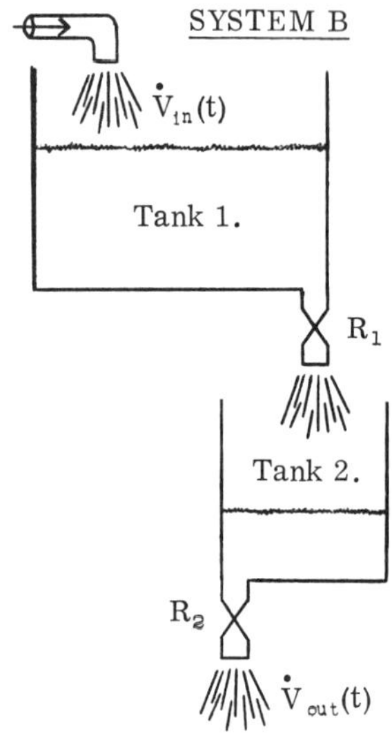

R_1 = 3 cm H_20/[L/min]
R_2 = 1.2 cm H_20/[L/min]

Floor Area, Tank 1, A_1 = 8000 cm^2
Floor Area, Tank 2, A_2 = 2000 cm^2

For each system:

a.) Draw a block diagram, showing the transfer functions of s, with $\dot{V}_{in}(t)$ as the input and $\dot{V}_{out}(t)$ as the output.

b.) Use phasor analysis to find the pure forced response of $\dot{V}_{out}(t)$ when $\dot{V}_{in}(t) = 20 + 5\cos\frac{1}{6}t + 10\sin\frac{1}{6}t$ L/min.

14. The <u>cup-of-coffee problem.</u>

When I get up in the morning, I want a drinkable cup of coffee <u>in a hurry.</u> I use soluble coffee, sugar, and boiling water to make 240 ml of black coffee at an initial temperature of 95° C. But I also want to add 30 ml of homogenized milk, kept in the refrigerator at 5° C. As I sit and stir my coffee (shivering slightly because the room temperature is only 15° C) I ponder the age-old question: should I add the milk at once to the 95° C brew and then wait for it to cool to 60° C, or should I wait until it cools enough so that stirring in the milk will instantaneously bring the mixture to 60° C? Which procedure will produce a potable 60° C cup of coffee faster, or does it make no difference? By actual measurement, I find that adding the milk increases the area for heat exchange by 8.00 per cent.

Make, and list, the simplest plausible assumptions which will allow you to arrive at reasonable estimates of the times of cooling. Then analyze the cup-of-coffee problem accordingly. Show all calculations to three significant figures, (i.e., slide-rule accuracy.)

15. The dehydrated camel.

When exposed to intense heat, most mammals defend their body temperature by evaporating large volumes of water, either from their lungs (panting) or from their skin (sweating). But when short of water, the desert camel is different. As Schmidt-Nielsen has shown (10), a dehydrated camel conserves water at the expense of a constant body temperature, in part relying upon its considerable bulk to "buffer" the temperature changes caused by diurnal exposure to extreme heat. It absorbs calories during the heat of the day, and unloads calories in the cool of the night, with comparatively little true temperature regulation.

Consider a camel, living in the Sahara Desert, deprived of water. We shall assume,

1. That his heat production is 250 Kcal per hour.

2. That the effective air temperature (i.e., the true air temperature plus an additional term representing radiant heat absorbed by the camel when exposed to sun during the daylight hours) varies sinusoidally in a 24-hour cycle with a maximum of 46.6° C at 2:00 P.M. and a mean of 33.3° C.

3. That the camel's body temperature also varies sinusoidally with a maximum of 40.6° C at 2.4 minutes past 7:00 P.M., and a minimum of 34.0° C.

4. That the dehydrated camel has no reflex mechanisms (such as sweating, panting, cutaneous vasodilation or vasoconstriction, shivering, etc.) for temperature regulation.

5. That we can treat the camel's body as a simple lumped-parameter heat reservoir.

6. That the conditions described are non-transient so that we can use phasor analysis.

a.) On a single graph of temperature vs. time, draw rough freehand curves showing body temperature and air temperature during two consecutive days.

b.) Draw a circuit diagram for the system.

c.) Write an equation for the effective air temperature, $T_{air}(t)$ and another for the body temperature, $T_{body}(t)$.

d.) Estimate the thermal conductance between body and air.

e.) Estimate the thermal capacitance of the camel.

f.) If the body weight of the camel is 1000 lbs (455Kg), is the estimated thermal capacitance reasonable? Comment briefly, bearing in mind the simplifying assumptions underlying your calculations.

NOTE: Contrary to what we have assumed in this example, the actual mean effective air temperature was roughly the same as, or even a bit higher than the mean body temperature. How can you account for this?

References

1. Machin, K.E.: Feedback theory and its application to biological systems. Sympos. Soc. exper. Biol., 18, 421-445, 1964.

2. Eckman, D.P.: Automatic Process Control. John Wiley and Sons, N.Y., 1966.

3. Handbook of Chemistry and Physics, Chemical Rubber Publishing Co., Cleveland. (Any recent edition.)

4. Korn, G.A., and Korn, T.M.: Mathematical Handbook for Scientists and Engineers. McGraw-Hill, N.Y., 1961.

5. Peterson, L.H.: Systems behavior, feedback loops, and high blood pressure research. Circulation Res., 12, 585-596, 1963.

6. Foerster, H. von: Basic concepts of homeostasis. Brookhaven Symp. Biol., 10, 216-242, 1957.

7. The International Dictionary of Applied Mathematics. D. van Nostrand Co., Princeton, 1960.

8. Aseltine, J.A.: Transform Method in Linear System Analysis. pp. 116-117, McGraw-Hill, N.Y., 1958.

9. Kuo, B.: Automatic Control Systems. pp. 48-53, Prentice-Hall Englewood Cliffs, N.J., 1962.

10. Schmidt-Nielsen, K.: The Physiology of the camel. Scient. Am., 201, 140-151, 1959.

Give me a fruitful error any time, full of seeds, bursting with its own corrections.

Vilfredo Pareto, COMMENT ON KEPLER

4

LINEAR FEEDBACK SYSTEMS

SECTION 4.1. Error-actuated feedback control systems

In Chapter 1, a general inquiry into how one variable can influence another led us to define feedback as one of the four kinds of working relationships which can exist between two variables. Historically, however, the study of feedback systems was not based upon any such broad theoretical considerations, but was the outgrowth of intensely practical demands for automatic control systems. As a consequence, investigations of feedback phenomena have been conducted largely by engineers whose primary aim was to design apparatus which would monitor the performance of some physical system and would then automatically adjust the controls of the system so as to reduce its error, which may be defined loosely as the difference between the actual performance and the desired or ideal performance. Self-regulating steel plants, automatic airplane pilots, and guidance systems for missiles and satellites are but three examples of how extraordinarily successful the control engineer has been in devising such error-actuated feedback control systems. Hence, even though physiological feedback systems are frequently not error-actuated, and are rarely linear, current discussions of feedback mechanisms usually assume that the reader is already familiar with the now-classical concepts which have guided the study of error-actuated linear control systems, and with the specialized jargon developed for their description. It is therefore imperative for anyone seriously interested in feedback mechanisms to become acquainted with the basic ideas and techniques discussed in the present chapter. We shall confine attention here to systems in which only a single input and a single output are considered at any one time, reserving for Chapter 5 consideration of multiple-input multiple-output systems.

When dealing with linear feedback systems in general, and with error-actuated systems in particular, we must bear in mind that transfer function analysis yields the correct output for a given input only when the system is inactive at $t = 0^-$, just prior to the initial application of the input signal. This means that all inputs and all initial conditions, and hence all variables, must be zero at $t = 0^-$. But in an error-actuated system the actuating error itself is an essential variable, and the error cannot be zero unless the system is already operating in the desired state as judged by the value of whatever output variable has been chosen as the most suitable criterion of performance. Yet how can the system be operating at all, if there are neither inputs nor initial conditions? We seem to be faced with an irreconcilable pair of requirements! In fact, however, the solution to this dilemma is simple enough. Since the "zero input, zero initial conditions" requirement is a purely mathematical, and not a physical requirement, all we need do is to define each variable which enters the analysis as the difference between its actual value and the "desired" or "ideal" value which it has when the error is zero. We then carry through the analysis in terms of these deviations-from-the-ideal. Nothing is lost, because we can always convert the output to its actual value, merely by adding back its error-free value. To emphasize the fact that the variables are defined as differences from their zero-error values, we might prefix Δ to the symbol for each variable. But this is not the usual practice, and the reader must simply take it for granted that every variable, whether an input or an output, has been normalized about its zero-error value.

In naturally-occurring systems, there is often no identifiable reference standard whose chosen value, the set point, is to be matched (for errorless performance) by the value of the corresponding output. However, even in such systems, where the term "error" makes no sense, it is often desirable to choose some "normal" steady-state value for the output variable of particular interest, and to carry out the analysis in terms of the deviations of each variable from its own corresponding "normal" value at the same steady state. This set of "normal" steady-state values, including the values of the inputs, can then be regarded as the "standard" operating point of the system. But in systems which lack a physically-real reference standard, the choice of such a standard operating point depends solely on convenience; as far as the mathematical analysis is concerned, any consistent set of steady-state values would serve equally well as a "standard." (See Section 4.6.)

SECTION 4.2. The behavior of error-actuated control systems

Consider the generalized feedback control system of Figure 4.2.1, where c(t) is the controlled variable which has been selected as the primary output. We shall define the output error as the actual value of c(t) minus its desired, or errorless value. In unity feedback systems (defined below) the reference standard, r(t), actually is the desired value of c(t), so that r(t) - c(t), the negative of the output error, is measured directly and used as the input to the controller. But it is often more convenient to compare some other variable, f(c(t)) (which must be a linear function of the output only), with a suitable reference standard, r(t). The difference between r(t) and f(c(t)) is defined as the actuating error, E(t), because it is the input which directly actuates the controlling mechanism.

$$E(t) = r(t) - f(c(t)) \qquad 4.2.1.$$

The actuating error is identical with the negative of the output error only for unity feedback systems where f(c(t)) = c(t), and H(s), the transfer function for the feedback path,is unity.

We start at $t = 0^-$ with the system physically operating in the desired errorless state, but mathematically quiescent, with both of the fully independent external inputs, r(t) and $\ell(t)$, (or, more precisely, their deviations) at zero, and with the deviations of all other variables at zero as well. We now

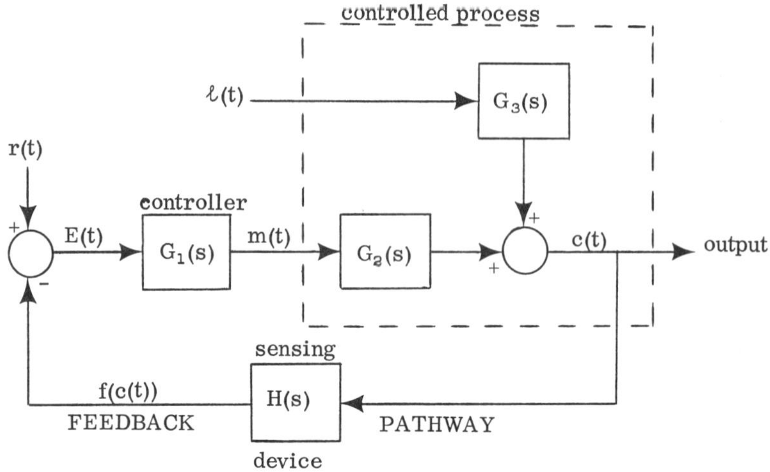

Figure 4.2.1. A generalized feedback control system.

r(t) = Reference standard or command signal (reference input)
f(c(t)) = Variable being compared with reference standard
E(t) = Actuating error
m(t) = Manipulated variable
ℓ(t) = Load variable (load input)
c(t) = Controlled variable (output of system)
$G_1(s)$, $G_2(s)$, $G_3(s)$, H(s), Transfer functions

---

ask two fundamental questions about the working behavior of the system: a,) With the load, $\ell(t)$, kept constant at zero, how effectively does the system respond to changes in the reference standard, r(t), by making the desired changes in the output variable, c(t); i.e., how good is the tracking ability of the system? b.) With r(t) kept at a constant set-point of zero, how effectively does the system minimize changes in c(t) when changes occur in the load, $\ell(t)$, i.e., how good is the stabilizing, or homeostatic ability of the system? Since both questions concern the response of the output to changes in one of the inputs, let us derive an equation for $c_{\mathcal{L}}(s)$ as a function of $r_{\mathcal{L}}(s)$ and $\ell_{\mathcal{L}}(s)$. By inspection of Figure 4.2.1, we have,

$$c_{\mathcal{L}}(s) = G_3(s)\ell_{\mathcal{L}}(s) + G_1(s)G_2(s)E_{\mathcal{L}}(s) \quad 4.2.2.$$

$$E_{\mathcal{L}}(s) = r_{\mathcal{L}}(s) - H(s)c_{\mathcal{L}}(s) \quad 4.2.3.$$

Combining equations 4.2.2. and 4.2.3. with the elimination of $E_{\mathcal{L}}(s)$, and solving for $c_{\mathcal{L}}(s)$,

$$c_{\mathcal{L}}(s) = \left\{\frac{1}{1 + G_1(s)G_2(s)H(s)}\right\} G_3(s)\ell_{\mathcal{L}}(s)$$

$$+ \left\{\frac{1}{1 + G_1(s)G_2(s)H(s)}\right\} G_1(s)G_2(s)r_{\mathcal{L}}(s) \qquad 4.2.4.$$

In equation 4.2.4, the first term on the right depends only upon the transfer functions which characterize the system and upon the load; it is therefore related to the stabilizing behavior of the system. The second term on the right depends only upon the transfer functions and the reference standard; it is therefore related to the tracking behavior of the system. Notice, however, that the expression in brackets which appears in both terms is the only part of the equation influenced by H(s), the transfer function for the feedback pathway. If H(s) were zero, the expression in brackets would be unity, and the system would be an open-loop system, because putting H(s) = 0 is equivalent to "opening" the feedback loop so that no signal can traverse it. (See Figure 4.2.2.a). It is clear, therefore, that in equation 4.2.4, the influence of an intact (= closed) feedback loop is determined by the quantity $G_1(s)G_2(s)H(s)$, regardless of whether we are interested in tracking behavior or stabilizing behavior. Yet the negative of this quantity, so obviously related to closed-loop behavior, is called the open-loop transfer function. The reason for this seemingly inappropriate term is explained by Figure 4.2.2.b. Consider only those components of the system which actually form the feedback loop. Let us open the loop at any point and "insert" at that point an appropriate input which we shall simply call the insert signal. After traversing the entire loop the signal will return to its starting point as the return signal which is the output of the open loop. By inspection of Figure 4.2.2.b,

$$\mathcal{L}\{\text{return signal}\} = - G_1(s)G_2(s)H(s)\,\mathcal{L}\{\text{insert signal}\} \qquad 4.2.5.$$

But the transfer function around the opened loop from input to output, (i.e., the open-loop transfer function) is, by definition, $\mathcal{L}\{\text{output}\}/\mathcal{L}\{\text{input}\}$, so

$$\text{Open-loop transfer function} = - G_1(s)G_2(s)H(s) \qquad 4.2.6.$$

We have gone as far as we profitably can in analyzing the generalized system of Figure 4.2.1. To show how specific answers to our fundamental questions about tracking behavior and stabilizing behavior can be obtained, we turn to a concrete example. In Figure 4.2.3 the two-tank system of Figure 3.13.2 has been modified so that the output of the pressure transducer is now a voltage, $e_p(t)$, which is directly proportional to the pressure, $P_b(t)$.

$$e_p(t) = \frac{1}{3}P_b(t) \quad \text{volt} \qquad 4.2.7.$$

The actuating error, E(t), is obtained by subtracting $e_p(t)$ from a reference voltage, $e_{ref}(t)$.

$$E(t) = e_{ref}(t) - e_p(t) \quad \text{volt} \qquad 4.2.8.$$

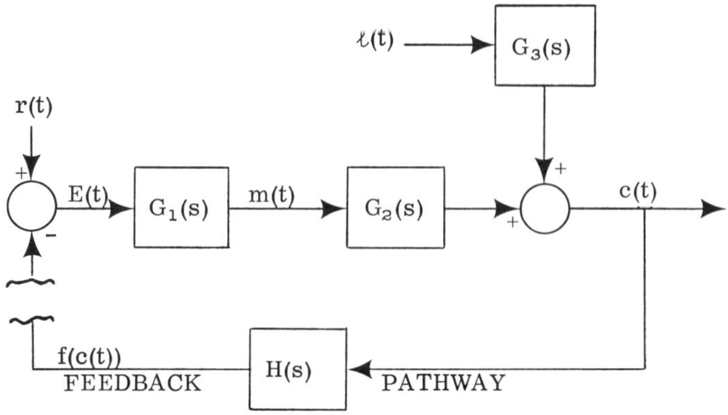

Figure 4.2.2.a. The feedback system of figure 4.2.1 with the feedback loop opened.

When the feedback pathway is interrupted (or, equivalently, when H(s) = 0) it becomes a dead end. Signals can no longer be "fed back" all the way around the loop to influence the behavior of the rest of the system. In such an "open-loop system," there really is no loop; there are only "forward paths" for direct signal flow from inputs to outputs.

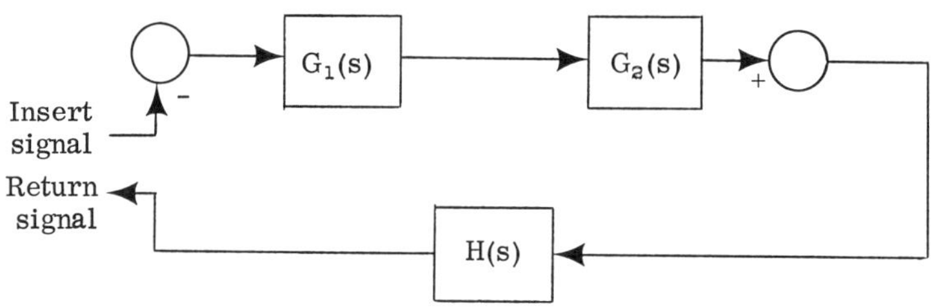

Figure 4.2.2.b. The meaning of "open-loop transfer function."

Consider only the components of the feedback loop, including summing points with their proper signs. Suppose we break the loop at some arbitrary point, and insert an input signal (= "insert signal") at the break. After traversing the opened loop, the signal will return to the point at which it was inserted, there appearing as the open-loop output (= "return signal"). So by definition, the open loop transfer function must be

$$\frac{\mathcal{L}\{\text{output}\}}{\mathcal{L}\{\text{input}\}} = \frac{\mathcal{L}\{\text{return signal}\}}{\mathcal{L}\{\text{insert signal}\}} = -\, G_1(s)G_2(s)H(s)$$

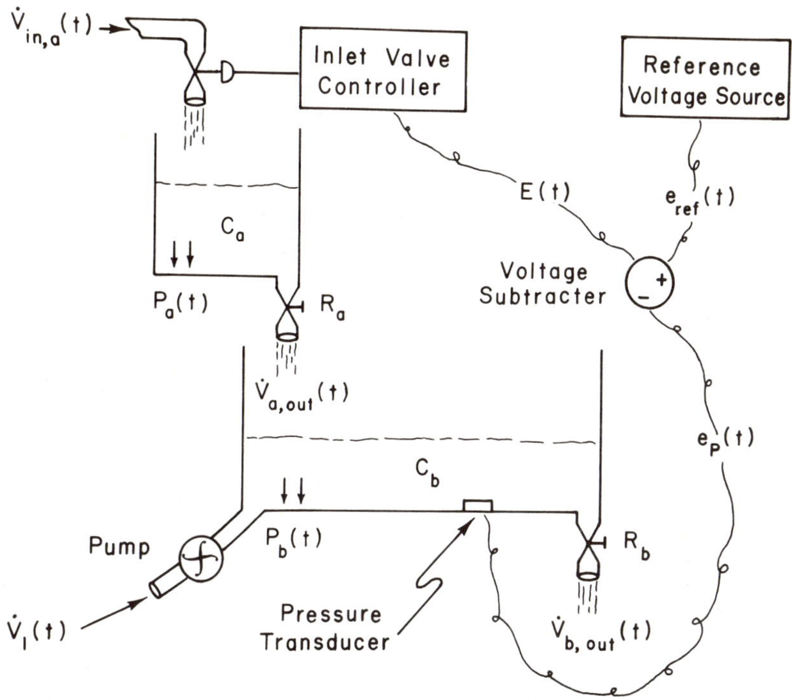

Figure 4.2.3. Physical model of a two-tank negative feedback system.

The following are the actual values at the "errorless" operating point. The values to be used in transfer function analysis are deviations from these.

INPUTS	Reference standard	$e_{ref}(0^-) = 128/3$ volt
	Load input	$\dot{V}_\ell(0^-) = -14$ L/min
PARAMETERS	Fluid resistances	$R_a = 2$; $R_b = 8$ cm H_2O/[L/min]
	Tank capacitances	$C_a = 5$; $C_b = 12.5$ L/cm H_2O
	Proportionality constant for pressure transducer	$k_{trans} = 1/3$ volt/cm H_2O (not shown in figure)
LOOP VARIABLES	Feedback voltage	$e_P(0^-) = 48/3$ volt
	Actuating error	$E(0^-) = 80/3$ volt
	Manipulated variable	$\dot{V}_{in,a}(0^-) = 20$ L/min
	Intermediate variables	$P_a(0^-) = 40$ cm H_2O
		$\dot{V}_{a,out}(0^-) = 20$ L/min
		$\dot{V}_{in,b}(0^-) = \dot{V}_{a,out}(0^-) + \dot{V}_\ell(0^-) = 6$ L/min
	Controlled output	$P_b(0^-) = 48$ cm H_2O
WHOLLY DEP. VAR.	Outflow from Tank b	$\dot{V}_{b,out}(0^-) = 6$ L/min

NOTES: 1.) A negative value of $\dot{V}_\ell(0^-)$ shows that fluid is being pumped out of tank b.

2.) The parameters for the three types of controllers are given in the legend for Figure 4.2.4.

E(t) is the input to a controller which controls a valve on the inlet pipe to tank a. The output of the controller is therefore $\dot{V}_{in,a}(t)$ (which is also the input to the first tank). In terms of Laplace transforms,

$$\dot{V}_{in,a,\mathcal{L}}(s) = G_1(s)\, E_{\mathcal{L}}(s) \text{ liters per min} \qquad 4.2.9.$$

Figure 4.2.4 shows a block diagram for the entire system.

The behavior of this system will be discussed in three sections, each illustrating one of the three basic types of linear control.

SECTION 4.3. Proportional control

The output of a proportional controller is directly proportional to the actuating error. For our generalized system (Figure 4.2.1),

$$m(t) = k_p E(t) \qquad 4.3.1$$

where k_p is a constant of proportionality with appropriate dimensions. In the present example, let $k_p = 3/4$ [liter/min] / volt, so that

$$G_1(s) = \frac{\dot{V}_{in,a,\mathcal{L}}(s)}{E_{\mathcal{L}}(s)} = \frac{3}{4}\ \frac{\text{liter/min}}{\text{volt}} \qquad 4.3.2.$$

To investigate tracking behavior, let $\dot{V}_{\ell}(t)$ remain at zero liters per min, and let $e_{ref}(t) = u(t)$ volts. In other words, at time zero we shall instantaneously change the set point of the reference standard from zero volt to one volt. Using the block diagram of Figure 4.2.4 as a guide, we apply equation 4.2.4 to this specific situation, thus obtaining

$$P_{b,\mathcal{L}}(s) = \left\{\frac{1}{1 + \frac{3}{4}\left[\frac{8}{[10s+1][100s+1]}\right]\frac{1}{3}}\right\}\left\{\frac{3}{4}\cdot\frac{8}{[10s+1][100s+1]}\right\}\frac{1}{s}$$

$$= \frac{2}{s} + \frac{10}{s+0.06} - \frac{12}{s+0.05} \text{ cm } H_2O\cdot\text{min} \qquad 4.3.3,$$

whence, taking inverse Laplace transforms,

$$P_b(t) = 2 + 10\epsilon^{-0.06t} - 12\epsilon^{-0.05t} \text{ cm } H_2O \qquad 4.3.4.$$

The actuating error is

$$E(t) = e_{ref}(t) - \frac{1}{3}P_b(t) = 1 - \frac{1}{3}P_b(t) \text{ volt} \qquad 4.3.5.$$

Combining equations 4.3.4 and 4.3.5, $E_{closed}(t)$, the closed-loop actuating error, is

$$P_{b,E=0}(t) = 3u(t) = 3 \text{ cm } H_2O \text{ for } t \geq 0^+ \qquad 4.3.6,$$

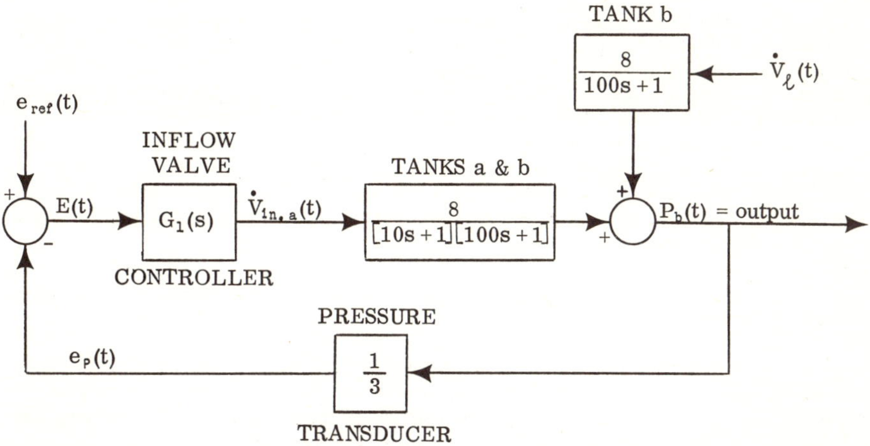

Figure 4.2.4. Conventional block diagram, in the pattern of Figure 4.2.1.a, for the two-tank system of Figure 4.2.2.a.

Illustrative values used in text

For proportional controller:	$k_p = \frac{3}{4} \cdot \frac{L/min}{volt}$	$G_1(s) = 3/4$
For integral controller:	$k_i = 0.00087 \frac{L/min}{volt \cdot min}$	$G_1(s) = 0.00087 \frac{1}{s}$
For derivative controller:	$k_d = 90 \frac{L/min}{volt/min}$	$G_1(s) = 90s$

$$E_{closed}(t) = 1 - \frac{2}{3} - \frac{10}{3}\epsilon^{-0.06t} + \frac{12}{3}\epsilon^{-0.05t}$$

$$= \frac{1}{3} - \frac{10}{3}\epsilon^{-0.06t} + \frac{12}{3}\epsilon^{-0.05t} \text{ volt} \qquad 4.3.6.$$

If the feedback loop were opened by interrupting the feedback pathway so that c(t) could no longer influence E(t), then $E_{open}(t)$, the open-loop actuating error would be

$$E_{open}(t) = e_{ref}(t) = 1 \text{ volt} \qquad 4.3.7.$$

The ratio of equation 4.3.6 to 4.3.7

$$\frac{E_{closed}(t)}{E_{open}(t)} = \frac{1}{3} - \frac{10}{3}\epsilon^{-0.06t} + \frac{12}{3}\epsilon^{-0.05t} \qquad 4.3.8$$

shows how much the negative feedback loop reduces the actuating error in response to a step increment in the reference standard. In particular, as $t \rightarrow \infty$ the ratio given by equation 4.3.8 approaches 1/3, showing that, in the steady state, the actuating error would be three times as large with no feedback as it is with feedback. This, then, is one answer to the first question posed in Section 4.2: "How effectively does the system respond to alterations in r(t) by making the desired changes in the output? " Discussion of a closely-related but somewhat more general answer will be postponed until we have studied the response of feedback system to sinusoidal inputs, and have carefully defined the much-abused term, "gain" (See Section 4.11).

In Figure 4.3.1, the values of $e_{ref}(t)$, E(t), and $P_b(t)$ obtained from this analysis have been plotted against time.

To answer the second question posed in Section 4.2, "How effectively does the system minimize changes in the output when changes occur in the load? ", we shall keep $e_{ref}(t)$ constant at zero volt while instantaneously increasing the load, $\dot{V}_{\ell}(t)$, from zero to one liter per min, starting at $t = 0^+$.

$$\dot{V}_{\ell}(t) = u(t) \text{ liters per min} \qquad 4.3.9$$

Using a derivation similar to the previous one, we shall compare the open-loop and closed loop responses of $P_b(t)$ to this step increment in the load. With the loop closed so that feedback can occur,

$$P_{b,\,closed,\,\mathcal{L}}(s) = \left\{\frac{1}{1 + \frac{3}{4}\left[\frac{8}{[10s+1][100s+1]}\right]\frac{1}{3}}\right\} \left\{\frac{8}{100s+1}\right\} \frac{1}{s}$$

$$= \frac{8/3}{s} - \frac{8}{s+0.05} + \frac{16/3}{s+0.06} \text{ cm } H_2O\cdot\text{min} \qquad 4.3.10.$$

$$P_{b,\,closed}(t) = \frac{8}{3} - \frac{24}{3}\varepsilon^{-0.05t} + \frac{16}{3}\varepsilon^{-0.06t} \text{ cm } H_2O \qquad 4.3.11.$$

Since the value of $P_b(t)$ for errorless performance is zero, equation 4.3.11 also represents the closed-loop error of the output.

With the loop opened by interrupting the feedback pathway,

$$P_{b,\,open,\,\mathcal{L}}(s) = \left[\frac{8}{100s+1}\right]\frac{1}{s} = \frac{8}{s} - \frac{8}{s+0.01} \text{ cm } H_2O\cdot\text{min} \qquad 4.3.12.$$

$$P_{b,\,open}(t) = 8 - 8\varepsilon^{-0.01t} \text{ cm } H_2O \qquad 4.3.13.$$

which also represents the open-loop error of the output. The ratio of equation 4.3.11 to equation 4.3.13 shows how much the output error has been reduced by the operation of the feedback loop.

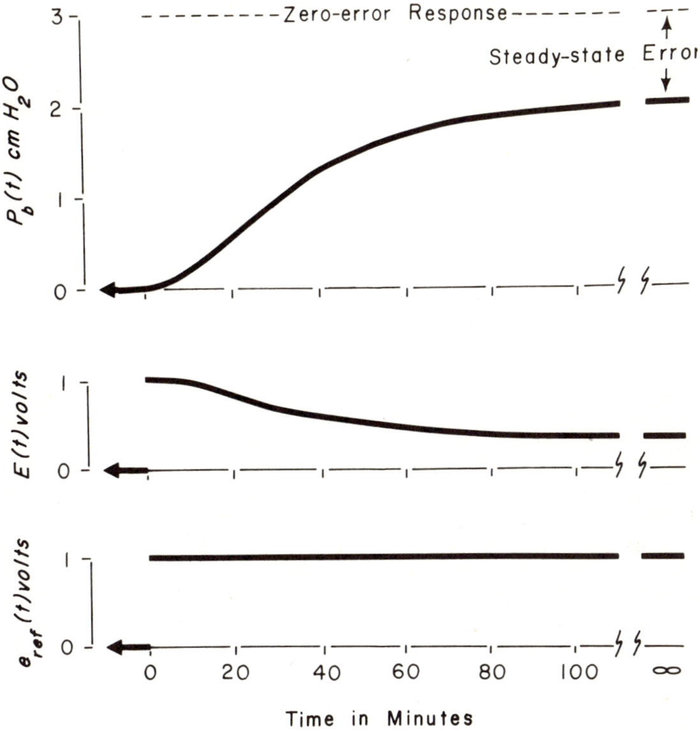

Figure 4.3.1. Response of the two-tank feedback control system with proportional control to a unit step increase in the set-point of the reference-standard voltage, $e_{ref}(t)$.

The arrowheads in this figure, and in some of the subsequent ones, merely indicate that the value of each of the variables is assumed to be zero for all $t \leq 0^-$.

If the system were able to follow the step change in the reference input (bottom graph) perfectly, the actuating error (middle graph) would remain unchanged at zero. This would require the output (top graph) to jump instantaneously to $P_b(t) = 3$ cm H_2O as indicated by the broken line for zero-error response. The actual ability of the system to track the step change in $e_{ref}(t)$ is far from ideal; the approach to the steady state is gradual, not instantaneous, and some error remains even when the steady state has been reached at time infinity.

$$\frac{\text{closed-loop error of output}}{\text{open-loop error of output}} = \frac{P_{b,\,closed}(t)}{P_{b,\,open}(t)} = \frac{1 - 3\varepsilon^{-0.05t} + 2\varepsilon^{-0.06t}}{3(1 - \varepsilon^{-0.01t})} \qquad 4.3.14$$

As $t \to \infty$ this ratio approaches 1/3, just as the ratio of actuating errors did when the input was a change in the set-point of the reference standard. So if we confine our attention to steady-state error, both of the fundamental questions posed in Section 4.2 have a similar answer: in the present system, with $G_1(s) = k_p = 3/4$ [L/min]/volt, feedback reduces the appropriate measure of error (= actuating error for a change in reference standard, output error for a change in load) to 1/3 of its open-loop value. (See, however, exercise 2).

If we try to achieve an even greater reduction in the steady-state error by increasing the constant of proportionality, k_p (now a mere 3/4 [liter/min] per volt) which determines how much the controlled variable will be altered for a given change in actuating error, we immediately encounter one of the problems of feedback control. It is not difficult to show that, for this system, the value $k_p = 3/4$ [liter/min] per volt is very close to the value which gives critical damping. (See exercise 4.3.). Hence, although we can reduce the steady-state error by increasing k_p we must then put up with an oscillatory approach to the steady state of the sort illustrated for $k_p = 3$ in Figure 4.3.2. (See also exercise 4.9.). (Whether such oscillations are undesirable will, of course, depend upon the particular system). Moreover, with systems of third or higher order, one cannot increase k_p to arbitrarily higher and higher values because at some point the system will become unstable. (See Section 4.7).

Figure 4.3.2 also illustrates another important feature of the transient response. The approach of $P_b(t)$ to its final steady-state value is considerably hastened by the feedback loop. For example, in the open-loop system with no feedback regulation (uppermost curve, Figure 4.3.2), $P_b(t)$ takes 230 minutes to undergo 90% of the total change needed to reach its new steady state, while in the closed-loop system when $G_1(s) = 3/4$, the time needed for $P_b(t)$ to undergo 90% of its total change is only 59 minutes. Note, however, that in this particular system once k_p has exceeded the value for critical damping, the exponential rate of change of the output (i.e., the neper frequency) becomes constant so that the approach to the new steady state is not further hastened by additional increases in k_p.

SECTION 4.4 Integral control

The output of an integral controller is proportional to the integral of the actuating error with respect to time. For our generalized system (Figure 4.2.1),

$$m(t) = k_i \int_0^t E(\tau)d\tau \qquad 4.4.1,$$

where τ is a dummy variable of integration and k_i is a proportionality constant with appropriate dimensions. In the present example, let $k_i = 0.00087$ [liters/min] per volt·min, so that

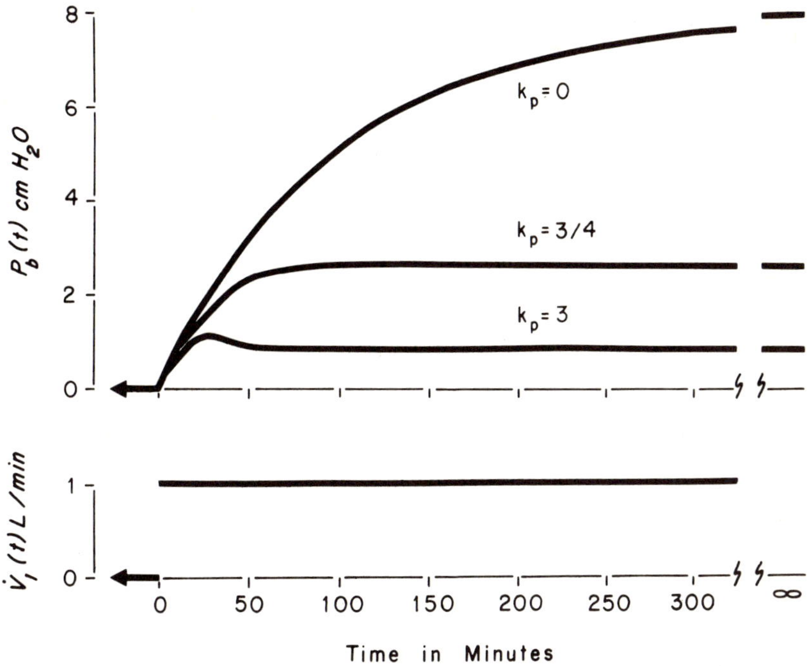

Figure 4.3.2 Response of the two-tank feedback control system with proportional control to a unit step increase in the load, $\dot{V}_\ell(t)$.

The approach to a new steady-state value of $P_b(t)$ is shown for $k_p = 0$ (open-loop system), $k_p = 3/4$ (system close to critical damping), and $k_p = 3$ (underdamped system). Note that negative feedback reduces both the steady-state error and the time needed to approach a new steady state.

$$\dot{V}_{in,a}(t) = 0.00087 \int_0^t E(\tau)d\tau \quad \text{liters/min} \qquad 4.4.2,$$

and

$$G_1(s) = \frac{\dot{V}_{in,a,\mathcal{L}}(s)}{E_{\mathcal{L}}(s)} = 0.00087 \left[\frac{1}{s}\right] \quad [\text{liters/min}] \text{ per volt} \qquad 4.4.3.$$

We shall now investigate the ability of the system to stabilize its output, $P_b(t)$, in the face of a unit step increment in load, using the same kind of analysis as was employed in the previous section for proportional control. Keeping $e_{ref}(t) = 0$, we let $\dot{V}_{\ell}(t) = u(t)$. Then

$$P_{b,\mathcal{L}}(s) = \left[\frac{1}{1 + \frac{0.00087}{s}\left\{\frac{8}{[10s+1]\,[100s+1]}\right\}\frac{1}{3}}\right]\left[\frac{8}{100s+1}\right]\frac{1}{s}$$

$$= \frac{0.00221}{s + 0.10025} + \frac{37.2}{s + 0.00380} - \frac{37.2}{s + 0.00595} \quad \text{cm } H_2O\cdot\text{min} \qquad 4.4.4.$$

$$P_b(t) = 0.00221\,\varepsilon^{-0.10025t} + 37.2\,\varepsilon^{-0.00380t} - 37.2\,\varepsilon^{-0.00595t} \quad \text{cm } H_2O \qquad 4.4.5,$$

or, since the first term in equation 4.4.5 is negligibly small compared with the other two,

$$P_b(t) \approx 37.2\left\{\varepsilon^{-0.00380t} - \varepsilon^{-0.00595t}\right\} \quad \text{cm } H_2O \qquad 4.4.6.$$

In Figure 4.4.1, $P_b(t)$ calculated from equation 4.4.6 has been plotted against time. With integral control the response to feedback is slow because the corrective adjustment of the inflow to tank a is proportional to the error integrated over time. Hence, no appreciable adjustment can occur until a considerable error has lasted for a considerable period of time. Consequently, the curve for pure integral control at first follows the open-loop, no-feedback curve very closely. Eventually, however, the integrated error becomes large enough to reverse the upward trend of the output. And thereafter as $t \to \infty$, the error approaches zero. The outstanding virtue of feedback with integral control is that, given enough time, it can completely eliminate steady-state error in the face of a constant input. Demonstration that this is as true when the input is a change in set point as it is when the input is a change in load has been left as an exercise (Exercise 4.5.)

As with proportional control, it is natural to ask whether we cannot improve the performance--i.e., hasten the response--of the integral controller by increasing the value of k_i. And again the answer is that we can do so only if we are willing to accept an oscillatory response, for the value $k_i = 0.00087$ [liter/min] per volt•min is very close to the value for critical damping (Exercise 4.9.) Moreover, even with this simple two-capacitance system, the value of k_i cannot be increased indefinitely, because at some point (see Exercise 4.8.) the system will become unstable. (Section 4.7).

SECTION 4.5 Derivative control

The output of a derivative controller is proportional to the derivative of

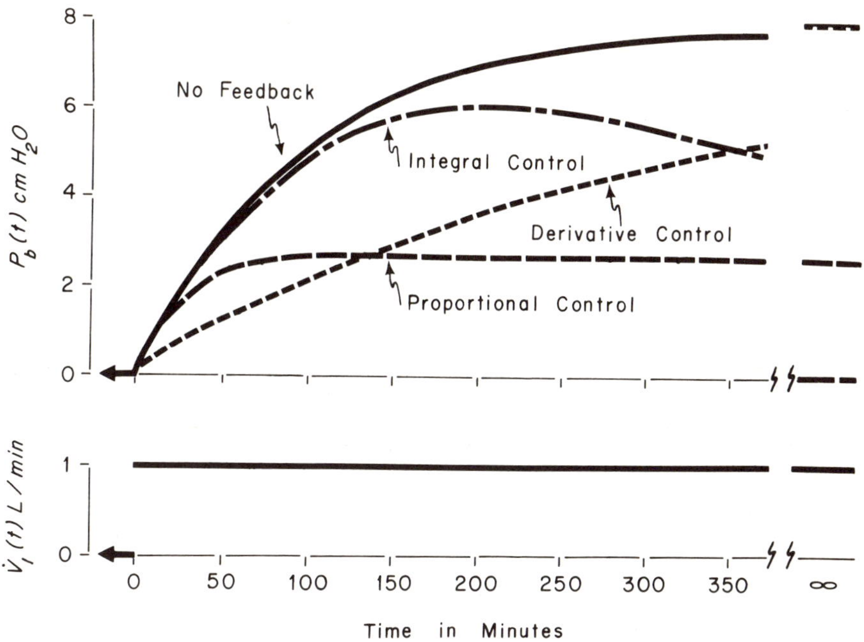

Figure 4.4.1. Response of the two-tank feedback system to a unit step increase in the load, $\dot{V}_\ell(t)$.

The curves were calculated using the illustrative values for $G_1(s)$ which are tabulated in the legend of Figure 4.2.4. The features which characterize each type of control are discussed in the text.

---

the actuating error with respect to time. For our generalized system (Figure 4.2.1),

$$m(t) = k_d \left[\frac{dE(t)}{dt}\right] \qquad 4.5.1,$$

where k_d is a proportionality constant with appropriate dimensions. In the present example, let $k_d = 90$ [liter/min] per [volt/min], so that

$$\dot{V}_{in,\,a}(t) = 90\,\frac{dE(t)}{dt} \qquad \text{liter/min} \qquad 4.5.2,$$

and

$$G_1(s) = \frac{\dot{V}_{in,\,a,\,\mathcal{L}}(s)}{E_{\mathcal{L}}(s)} = 90s \qquad [\text{liter/min}]/\text{volt} \qquad 4.5.3.$$

We again investigate the ability of the system to stabilize its output, $P_b(t)$, when a unit step load is imposed. Keeping $e_{ref}(t) = 0$, we let $\dot{V}_{\ell}(t) = u(t)$. Then

$$P_{b,\mathcal{L}}(s) = \left\{ \frac{1}{1 + 90s\left\{\frac{8}{[10s+1][100s+1]}\right\}\frac{1}{3}} \right\} \left\{\frac{8}{100s+1}\right\} \frac{1}{s}$$

$$= \frac{8}{s} - \frac{0.165}{s + 0.347} - \frac{7.835}{s + 0.0029} \text{ cm } H_2O \cdot \text{min} \qquad 4.5.4.$$

$$P_b(t) = 8 - 0.165\varepsilon^{-0.347t} - 7.835\varepsilon^{-0.0029t} \text{ cm } H_2O \qquad 4.5.5.$$

In Figure 4.4.1, $P_b(t)$ calculated from equation 4.5.5 has been plotted against time. With derivative control, the effect of feedback quickly becomes evident as a reduction in the rate at which $P_b(t)$ rises in response to an increased load. The outstanding virtue of feedback with derivative control is the rapidity with which it begins to make appropriate adjustments when changes occur in the input. However, as is evident from Figure 4.4.1, with pure derivative control, the effect of feedback regulation in the face of a steady input eventually becomes nil, for the error gradually approaches the open-loop error as $t \to \infty$.

In the system of the present example, (but not in general) the value of the proportionality constant, k_d, may be made arbitrarily large without danger of oscillations or instability. However, to make sure he understands how a system with derivative control would respond to a change in set point, the reader would do well to work exercise 4.4. In any event, he should certainly consider carefully how the derivative control system of the present example would behave if we had kept $\dot{V}_{\ell}(t)$ at zero and had let $e_{ref}(t) = u(t)$.

SECTION 4.6 Further remarks upon feedback systems with the three basic kinds of control

The preceding sections have illustrated some of the outstanding peculiarities of feedback systems with pure proportional, integral or derivative control; and the effect of each of these basic types of control upon the output of our illustrative system, disturbed by a unit step increment in load, has been summarized in Figure 4.4.1. I have deliberately concentrated upon load inputs because these are of the most universal interest in physiological systems. Each kind of control has obvious advantages and disadvantages. With proportional control, a new steady state can be approached rapidly, but some error always remains, even when the new steady state has been reached. Integral control is slow in getting started, but, in the fullness of time, eliminates steady-state error. Derivative control gets off to a fast start, but fades in the stretch. As one attempts to increase the precision of feedback regulation by increasing k_p, k_i, or k_d one may encounter problems due to oscillatory behavior or to downright instability. Although these problems are most prominent with integral controllers, and least prominent with derivative controllers, in more complex systems they may arise with any of the three types. It is small wonder that in designing practical control systems, engineers often use combinations of these pure types so as to achieve better overall results.

The physiologist's concern, however, is not usually with the design of new control systems, but with the analysis of naturally-occurring control systems, many of which have no real reference standard whose value serves as a set-point with which the output can be compared. Consider such a system. In order to start with zero initial conditions, we might (in accordance with the suggestion in the last paragraph of Section 4.1) define each variable as its deviation from the value which it has at some convenient steady-state "operating point." Then the difference between c_{op}, the value of the output variable at the chosen operating point, and c(t), its actual value at any time, t, will serve as a kind of quasi "error." (N.B. Throughout this section symbols such as c(t), m(t) etc. will always refer to actual values, not deviations.)

$$\text{"error"} = c_{op} - c(t) \qquad 4.6.1.$$

But c_{op} is not a part of the real system; it is no more than a numerical value which we have estimated as a "normal" value, or assumed as a "standard" value, or even chosen quite arbitrarily to suit our own convenience. Hence we must make sure that the value which we calculate for m(t), the actual output of our controller, is independent of the value chosen for c_{op}, our make-believe reference standard. To do this in a simple way, let us consider the engineer's error-actuated systems as described in the preceding sections; and let us see (separately for each kind of controller) whether $m_{calc}(t)$, calculated by using c_{op} as an arbitrary "reference standard," is the same as $m_{true}(t)$, calculated from c_{ref}, the true set-point of a physically-real reference standard. If it is the same, then as far as responses to load inputs are concerned the actual value of the set-point is irrelevant, and no physically-real reference standard is needed.* We shall assume unity feedback, so that $E(t) = c_{ref} - c(t)$.

[*Notice that this statement is valid only for the regulatory behavior of the system when disturbed by load inputs. Therefore, even when designing a system which could regulate perfectly well without a physically-real reference standard, the engineer may still want to include one as a convenient means of resetting the operating point. If a real reference standard is included, its actual value is by no means irrelevant to the tracking behavior of the system, since a change in the set-point will most certainly cause a change in the output.]

Consider derivative control. Its essential feature (equation 4.5.1) is that the value of the manipulated variable increases in proportion to the rate of increase of the error, regardless of what the actual value of the error is. Let us rewrite equation 4.5.1 as

$$m_{true}(t) - m_{ref} = k_d \frac{dE(t)}{dt} = k_d \frac{d[c_{ref} - c(t)]}{dt} \qquad 4.6.2,$$

where m_{ref} is the steady-state value of m when the set-point is c_{ref}. But since c_{ref} is a constant,

$$dc_{ref}/dt = 0 \qquad 4.6.3,$$

and equation 4.6.2 reduces to

$$m_{true}(t) - m_{ref} = - k_d [dc(t)/dt] \qquad 4.6.4,$$

which shows that even when we have a physically-real reference standard, its set-point, c_{ref}, does not influence $m_{true}(t)$. If we now rewrite equation 4.6.2 in terms of deviations from an arbitrary operating point, we obtain

$$m_{calc}(t) - m_{op} = k_d \frac{d[c_{op} - c(t)]}{dt} \qquad 4.6.5,$$

or, since c_{op} is a constant,

$$m_{calc}(t) - m_{op} = - k_d [dc(t)/dt] \qquad 4.6.6.$$

Like equation 4.6.4, equation 4.6.6 is independent of c_{ref}. But in fact, equation 4.6.6 is more general than equation 4.6.4 because it correctly states that the deviation of m(t) from any m_{op} is proportional to the rate of decrease of c(t). Equation 4.6.6 reduces to equation 4.6.4 for the particular case where the operating point value is m_{ref}. Then $m_{calc}(t) = m_{true}(t)$. Derivative control does not require a physically-real reference standard.

Consider proportional control. The essential feature of proportional control (equation 4.3.1) is that

$$m_{true}(t) - m_{ref} = k_p [c_{ref} - c(t)] \qquad 4.6.7,$$

or,

$$m_{true}(t) = m_{ref} + k_p c_{ref} - k_p c(t) \qquad 4.6.7.a.$$

Now suppose that we were to calculate m(t) using an arbitrary operating point, c_{op}, as a quasi set-point.

$$m_{calc}(t) - m_{op} = k_p [c_{op} - c(t)] \qquad 4.6.8,$$

or,

$$m_{calc}(t) = m_{op} + k_p c_{op} - k_p c(t) \qquad 4.6.8.a.$$

Using equation 4.6.7.a to calculate the true value of m_{op} from c_{op},

$$m_{op} = m_{ref} + k_p c_{ref} - k_p c_{op} \qquad 4.6.9.$$

Combining equations 4.6.8.a and 4.6.9 by eliminating m_{op},

$$m_{calc}(t) = m_{ref} + k_p c_{ref} - k_p c_{op} + k_p c_{op} - k_p c(t)$$

$$= m_{ref} + k_p c_{ref} - k_p c(t) \qquad 4.6.10.$$

By comparing equations 4.6.7.a and 4.6.10, we see that $m_{calc}(t) = m_{true}(t)$ regardless of what value we choose for c_{op}. Proportional control does not require a physically-real reference standard.

Consider integral control. The essential feature of integral control (Equation 4.4.1) is that

$$m_{true}(t) - m_{ref} = k_1 \int_0^t [c_{ref} - c(\tau)]\,d\tau = k_1 c_{ref} t - k_1 \int_0^t c(\tau)d\tau \qquad 4.6.11,$$

or,

$$m_{true}(t) = m_{ref} + k_1 c_{ref} t - k_1 \int_0^t c(\tau)d\tau \qquad 4.6.11.a.$$

Now suppose that we were to calculate m(t) using an arbitrary operating point, c_{op}, as a quasi set-point.

$$m_{calc}(t) - m_{op} = k_1 \int_0^t [c_{op} - c(\tau)]\,d\tau = k_1 c_{op} t - k_1 \int_0^t c(\tau)d\tau \qquad 4.6.12,$$

or,

$$m_{calc}(t) = m_{op} + k_1 c_{op} t - k_1 \int_0^t c(\tau)d\tau \qquad 4.6.12.a.$$

Using equation 4.6.11.a to calculate m_{op} by letting $c(\tau) = c_{op}$, a constant,

$$m_{op} = m_{ref} + k_1 [c_{ref} - c_{op}]\,t \quad \text{as } t \to \infty \qquad 4.6.13.$$

But now we are in trouble! For equation 4.6.13 shows that, in general, the absolute magnitude of m_{op} for an arbitrarily-selected constant value of c_{op} will be infinitely large, thus contradicting its definition as a steady-state operating-point value. In fact, equation 4.6.13 shows that m_{op} can assume a finite steady-state value (as it is required to do) if, and only if, $c_{op} = c_{ref}$ so that $m_{op} = m_{ref}$. Hence, with integral control, c_{op} is not arbitrary, but must equal c_{ref}. Integral control requires a physically-real reference standard with a definite set-point if it is to achieve a steady state (which, for integral control, is the errorless state) when the system is subjected to a constant load. With such a system, "having no set point" would be exactly equivalent to "having a set point whose value is zero."

Because integral control can eliminate steady-state error in the face of a constant load, we draw the following important conclusion: If the negative feedback loop of a linear control system contains a component which integrates its input with respect to time, the actual value of the output will tend towards zero unless the system has a physically-real, non-zero reference standard. Moreover, since only integral control can eliminate steady-state error, we may conclude that unless a negative feedback system has a physically-real reference standard, it cannot return to its original operating point when disturbed by a constant load. Because physically-real reference standards rarely exist within the moist and mobile world of living systems, integral control is uncommon, and most homeostatic mechanisms subjected to a constant load will show at least some residual steady-state deviation from the "normal" operating point. However, there are certain important biological feedback mechanisms which, via specialized sense-organs, can make use of physically-real set-points which are external to the body. When such feedback loops include the equivalent of an integral controller, they can practically eliminate steady-state error. Suppose, for example, that we want to maintain a constant speed of 60 mi/hr while driving along a level stretch of a limited-access highway. By adjusting the pressure of our foot on the accelerator, we can reduce the error, i.e., the difference between c(t), the actual position of the speedometer needle, and c_{ref}, the position marked "60" on the speedometer dial, to practically zero because we can actually see this error and use it as the input for the complicated neuromuscular system which controls our foot pressure and hence the speed of the car. The performance of such systems is often improved by anticipatory responses which have been learned from previous experience. For example, if we notice that the highway is starting to ascend a hill, we may increase the pressure on the accelerator even before we have seen any deviation of c(t) from c_{ref}. The tracking ability of such neuromuscular systems has been extensively studied (1,2) but their behavior is too complicated to consider in this book.

SECTION 4.7 The stability of linear systems

A particularly clear and simple definition of stability suffices for stationary linear systems: A stationary linear system is stable if its outputs are bounded for all bounded inputs. (A time-dependent variable is bounded if it remains finite as $t \to \infty$.) This means that for a system to be stable none of its output functions can contain any terms such as At, $B\epsilon^{+k_1 t}$, or $C\epsilon^{+k_2 t}\cos(\omega t + \theta)$ where A, B, C, k_1, k_2, ω, θ are constants and k_1 and k_2 are >0, because the absolute magnitude of such terms increases without bound as $t \to \infty$. I am sorry to say that, according to this definition, your savings-bank account is unstable.

Consider a stationary linear system whose input is x(t) and whose output is y(t). To prove that the system is stable, we must show that y(t) remains bounded for every bounded x(t) as $t \to \infty$. Let us assume (as will usually be true) that the transfer function can be written in the form

$$y_{\mathfrak{L}}(s)/x_{\mathfrak{L}}(s) = P_{num}(s)/P_{den}(s) \qquad 4.7.1,$$

where both the numerator, $P_{num}(s)$, and the denominator, $P_{den}(s)$, are polynomials in s. We first prove that the equation

$$P_{den}(s) = 0 \qquad 4.7.2.$$

is the characteristic equation of the system, so that its roots are eigenvalues. We then show that for the system to be stable, the real part of every eigenvalue must be negative.

Equation 4.7.1 may be rearranged in the form

$$P_{den}(s)y_{\mathcal{L}}(s) = P_{num}(s)x_{\mathcal{L}}(s) \qquad 4.7.1.a,$$

or, for an unforced system where $x_{\mathcal{L}}(s) = 0$,

$$P_{den}(s)y_{\mathcal{L}}(s) = 0 \qquad 4.7.3.$$

Now consider the "classical" solution of the homogeneous linear differential equation for such an unforced stationary system. (See Appendix.) The equation can always be written in the form

$$\frac{d^n y(t)}{dt^n} + a_{n-1}\frac{d^{n-1}y(t)}{dt^{n-1}} + \ldots + a_1\frac{dy(t)}{dt} + a_o = 0 \qquad 4.7.4,$$

or,

$$[D^n + a_{n-1}D^{n-1} + \ldots + a_1 D + a_o]\ y(t) = P(D)y(t) = 0 \qquad 4.7.5.$$

where $D^n \equiv d^n[\ \ ]/dt^n$. The characteristic equation is obtained by substituting λ for D in the differential operator which is enclosed in brackets in equation 4.7.5, and setting the resulting characteristic polynomial in λ equal to zero.

$$\lambda^n + a_{n-1}\lambda^{n-1} + \ldots + a_1\lambda + a_o = P(\lambda) = 0 \qquad 4.7.6.$$

The Laplace transform of equation 4.7.4 with zero initial conditions (as required for transfer function analysis) is

$$[s^n + a_{n-1}s^{n-1} + \ldots + a_1 s + a_o]\ y_{\mathcal{L}}(s) = 0 \qquad 4.7.7.$$

But in equation 4.7.7, the bracketted polynomial in s is of exactly the same form as is the characteristic polynomial in λ of equation 4.7.6. Hence equation 4.7.7 may be written as

$$P(s)y_{\mathcal{L}}(s) = 0 \qquad 4.7.8,$$

and the corresponding characteristic equation may be written as

$$P(s) = 0 \qquad 4.7.9,$$

which is the same characteristic equation as 4.7.6 with λ replaced by s. Combining equations 4.7.3 and 4.7.8,

$$P_{den}(s)y_{\mathcal{L}}(s) = P(s)y_{\mathcal{L}}(s) \qquad 4.7.10,$$

or,

$$P_{den}(s) = P(s) \qquad 4.7.10.a.$$

By substituting $P_{den}(s)$ for $P(s)$ in the characteristic equation 4.7.9, we obtain $P_{den}(s) = 0$, (equation 4.7.2). Hence equation 4.7.2 is the characteristic equation of the system, as was to be proved.

The roots of the characteristic equation may be distinct or multiple, real, imaginary, or complex. But from the standpoint of stability, the most important fact to remember is that the real part of any eigenvalue must appear in the corresponding term of the output function as a coefficient of time in the exponent of ϵ. We consider three general kinds of roots:

i) Roots with negative real parts (for example $\lambda_1 = -0.5$ or $\lambda_2 = -2 + j3$, $\lambda_3 = -2 - j3$) correspond to output terms with negative real exponents of ϵ. Such terms always decrease towards zero as time increases, even when (as may be true with multiple roots) some integral power of time also appears in the term as a coefficient of ϵ. (Cf. Example 3.3.1.) They are therefore compatible with a stable system.

ii) Roots with positive real parts (for example $\lambda_1 = 2$, or $\lambda_2 = 3 + j1.5$, $\lambda_3 = 3 - j1.5$) correspond to terms with positive real exponents of ϵ. Such terms increase in magnitude without bound as time increases and unequivocally indicate an unstable system.

iii) Roots whose real parts are zero. We distinguish two cases: A single conjugate pair of imaginary roots (for example $\lambda_1 = j8$, $\lambda_2 = -j8$) corresponds to sinusoidal functions of time which neither grow nor decay but maintain undamped oscillations in the value of the output. In unforced linear systems such as the one we are now considering, undamped oscillations can occur only in conservative systems which will never be encountered by the physiologist (See footnote Section 2.6). A single zero root corresponds to a constant term in the output. However, multiple zero roots bring in terms which are integral powers of time multiplied by some constant and which therefore increase in magnitude without bound as time increases. But this means that even a single zero eigenvalue is a sign of instability, for a constant forcing function (which is certainly a bounded input) would add to the output a second zero-root term, so that the output would grow in magnitude without limit.

In summary, even when the linear system is being forced by nothing more violent than a constant input, the system is stable only if all roots of the characteristic equation have negative real parts. Fortunately, however, this criterion applies equally well to any other bounded input. By insisting that all eigenvalues have negative real parts, we guarantee that, as time increases, all terms corresponding to the natural response of the system will eventually disappear from the output. And when the transfer function is multiplied by the Laplace transform of an input, the additional factors thus brought into the denominator retain their original form so that the inverse transforms of the corresponding partial fractions will be bounded as long as the terms of the input are themselves bounded. Even the special case where one of the roots corresponding to an input term is the same as an eigenvalue poses no threat to stability, because we are requiring the real part of every eigenvalue to be negative. At worst, then, multiplicity of roots can only bring into the solution terms such as $t^n \varepsilon^{-kt}$ [Acos ωt + Bsin ωt] which would still die away as $t \to \infty$.

We conclude that a stationary linear system whose transfer function can be written as a ratio of polynomials in s is stable if and only if all roots of its characteristic equation have negative real parts. When the transfer function cannot be written as a ratio of polynomials,* a different criterion of stability must be used. (See Section 4.13).

[*Strictly speaking, we need only require the denominator, $P_{den}(s)$ to be a polynomial in s, as long as we are sure that $P_{den}(s)$ is, in fact, the characteristic polynomial of the system. In practice, however, it is preferable to remember the ratio-of-polynomials rule, because exceptions to it are not only rare but are easily dealt with when they do arise. (See exercise 4.11.)]

SECTION 4.8 Poles, zeros and root loci in the complex plane

Because the roots, $\lambda_1, \lambda_2, \cdots \lambda_n$, of the characteristic equation 4.7.2 are values of s which make $P_{den}(s)$ zero, they are commonly called the zeros of $P_{den}(s)$. But a value of s which makes the denominator of the transfer function zero will make the transfer function itself indefinitely (= "infinitely") large. So the zeros of $P_{den}(s)$ are also called the poles of the transfer function. (The terms "zero" and "pole" are by no means restricted to transfer functions; they are used very generally to describe singularities of any function of a complex variable. For a graphical interpretation of the terms "pole" and "zero" see Figure 4.8.1).

It is often useful to plot the poles of the transfer function (= roots of the characteristic equation = eigenvalues) in the complex s-plane. The real part of each root is plotted as abscissa, the imaginary part as ordinate. From the previous discussion of stability, it is at once evident that for a system to be stable, all of the poles of its input-output transfer functions must lie in the left half-plane, because only then do the poles have negative real parts. Furthermore, the effect on stability of changing one of the parameters of the system (such as k_d, k_p, or k_i) may conveniently be summarized by plotting the loci of the roots in the s-plane as the parameter is varied.

Example 4.8.1 Plot in the complex plane the poles of the transfer function for the two-tank system of Section 4.4 (integral control) and sketch their loci as k_i is varied from zero [liters/min]/volt•min to values large enough to produce instability.

Leaving the proportionality constant, k_i, unevaluated, the transfer function with $e_{ref}(t)$ as input is easily calculated from figure 4.2.4 as

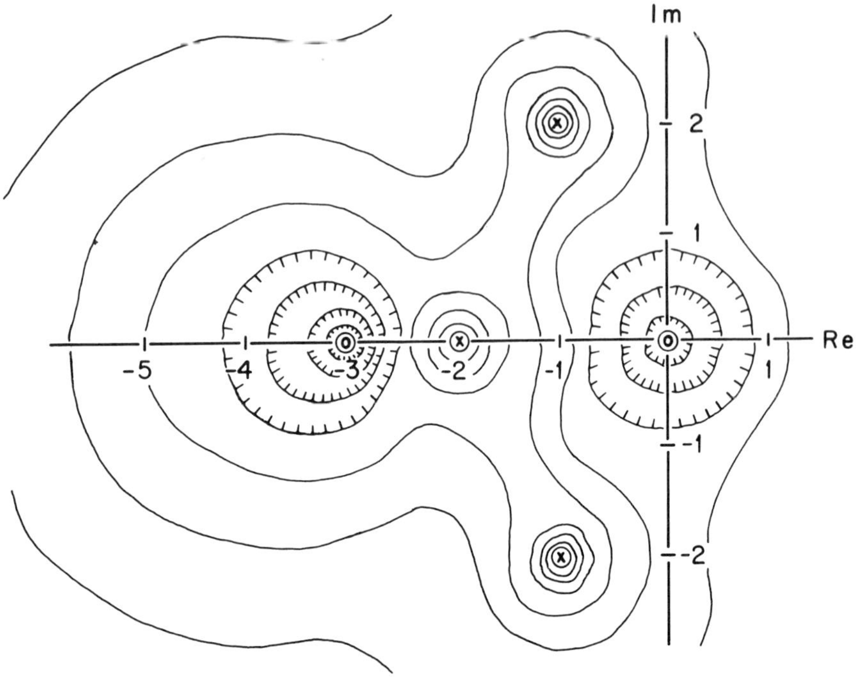

Figure 4.8.1. Poles and zeros in the complex plane.

If we evaluate the function

$$f(s) = \frac{s[s+3]}{[s+2][s^2+2s+5]}$$

for some particular value of the variable, $s = \sigma + j\omega$, we shall obtain a complex number whose magnitude we can plot in the third dimension directly above the point in the s-plane which represents the particular value of s. If we could do this for every point in the s-plane we would generate a curved surface which we can think of as a kind of tent pitched upon the s-plane. The tent is pegged down at "ground" or zero level at each zero of the function (marked as a small circle in the figure). The tent is stretched upwards, out of the plane of the page, by an infinitely long "tentpole" at each pole of the function (marked as a small cross in the figure). On a two-dimensional graph, the elevation of the curved surface of the tent could be indicated by contour lines of the sort shown in the figure.

(NOTE: the contour lines here drawn were not actually calculated, but were sketched freehand. They can therefore be regarded only as a crude representation of what the true surface would be like.)

$$\frac{P_{b,\mathcal{L}}(s)}{e_{ref,\mathcal{L}}(s)} = \frac{8k_1}{s[10s+1][100s+1]+\frac{8k_1}{3}} = \frac{0.008k_1}{s^3+0.11s^2+0.001s+\frac{8k_1}{3000}} \qquad 4.8.1.$$

Similarly, the transfer function for $\dot{V}_{\ell}(t)$ as input (Cf. equation 4.4.4) is

$$\frac{P_{b,\mathcal{L}}(s)}{\dot{V}_{\ell,\mathcal{L}}(s)} = \frac{[80\,s^2+8s]10^{-3}}{s^3+0.11s^2+0.001s+\frac{8k_1}{3000}} \qquad 4.8.2.$$

By inspecting these equations, we see that the denominator, $P_{den}(s)$, which is the characteristic polynomial of the transfer function, is exactly the same whether the input is the reference standard or the load. Actually we should have deduced this from the discussion in Section 4.7 which showed that the characteristic equation, $P_{den}(s) = 0$, can be obtained from the homogeneous equation which describes the system when all inputs and all initial conditions are zero. In fact, we can now show (exercise 4.12) that in a single-loop system where the output is a member of the loop,

$$1/[1 - \text{OLTF}] = N(s)/P_{den}(s) \qquad 4.8.3,$$

where N(s) is a polynomial in s, and OLTF is the open-loop transfer function (Section 4.2). Hence the characteristic equation for such a single-loop system can be calculated directly from the OLTF provided that 1/[1 - OLTF] can be arranged as a ratio of polynomials in s. In the present example,

$$\text{OLTF} = -\frac{8k_1}{3s[10s+1][100s+1]} \qquad 4.8.4$$

$$\frac{1}{1-\text{OLTF}} = \frac{s^3+0.11s^2+0.001s}{s^3+0.11s^2+0.001s+\frac{8k_1}{3000}} = \frac{N(s)}{P_{den}(s)} \qquad 4.8.5$$

From equation 4.8.5 with $k_1 = 0.00087$ we have,

$$P_{den}(s) = s^3 + 0.11s^2 + 0.001s + 2.32 \times 10^{-6} \qquad 4.8.6.$$

Using the graphical method of factoring described in the Appendix we find (Cf. the exponents in equation 4.4.5),

$$P_{den}(s) = [s + 0.10025]\,[s + 0.00380]\,[s + 0.00595] \qquad 4.8.6.a.$$

So the poles of the transfer function (= the zeros of $P_{den}(s)$ = the eigenvalues = the roots of the characteristic equation) are at $s = 1002.5 \cdot 10^{-4}$, $s = 38 \cdot 10^{-4}$, and $s = 59.5 \cdot 10^{-4}$. (Points "C" in Figure 4.8.2).

To find the loci of these roots as k_1 is varied, we write the characteristic equation with k_1 unevaluated.

$$P_{den}(s) = s^3 + 0.11s^2 + 0.001s + \frac{8k_1}{3000} = 0 \qquad 4.8.7.$$

Our objective is to find the roots of equation 4.8.7 which correspond to various values of k_1. A frontal attack, using selected values of k_1, would be tedious. So let us write,

$$s^3 + 0.11s^2 + 0.001s + \frac{8k_1}{3000} = [s + a][s^2 + bs + c] = 0 \qquad 4.8.8.$$

where a, b, and c are constants for a given k_1. It is now relatively easy to use selected values of a, and to calculate the corresponding values of k_1, b, and c. One root is, of course, -a; the other two are found from the quadratic term. For example, if we let a = 0.105, the computations would proceed as follows,

$$\begin{array}{r|l} & s^2 + 0.005s + 0.000475 \\ \hline s + 0.105 & s^3 + 0.11s^2 + 0.001s + \frac{8k_1}{3000} \\ & \underline{s^3 + 0.105s^2} \\ & \quad 0.005s^2 + 0.001s \\ & \quad \underline{0.005s^2 + 0.000525s} \\ & \qquad 0.000475s + \frac{8k_1}{3000} \\ & \qquad \underline{0.000475s + 0.000049875} \\ & \qquad\qquad \frac{8k_1}{3000} - 0.049875 \cdot 10^{-3} = \text{remainder} \end{array}$$

For the remainder to be zero, we must have,

$$k_1 = \frac{3}{8}[0.049875] = 0.018703 \qquad 4.8.9.$$

Putting the quadratic term equal to zero, and solving for its roots,

$$s = -0.0025 \pm j\,0.02165 \qquad 4.8.10.$$

So, for $k_1 = 0.0187$, the roots of $P_{den}(s) = 0$ are: $-1050 \cdot 10^{-4}$, $-25 \cdot 10^{-4} + j216.5 \cdot 10^{-4}$, and $-25 \cdot 10^{-4} - j216.5 \cdot 10^{-4}$ (points G in Figure 4.8.2). By repeating this procedure for various values of a, we can sketch the root loci in Figure 4.8.2. As is pointed out in the legend of that figure, such a root-locus diagram summarizes a considerable amount of information.

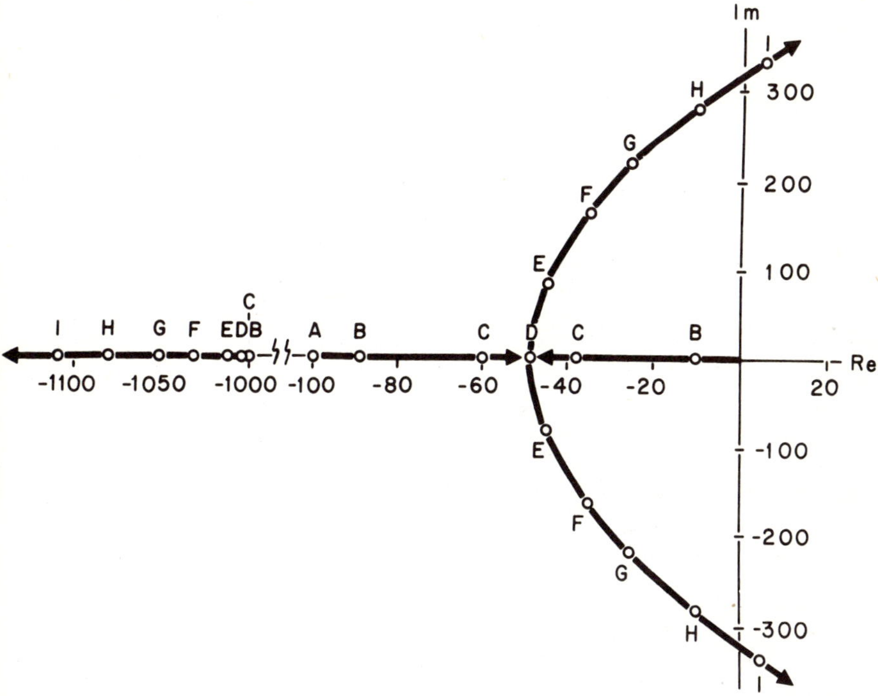

Figure 4.8.2. A root-locus plot for the two-tank feedback system with integral control.

NOTE: The scale values in this figure are 10^4 times the actual values.

Values of k_I at the lettered points

point	k_I $\frac{\text{liters}}{\text{min}}$ per volt• min
A	0.000000
B	0.000338
C	0.000870
D	0.000892
E	0.003447
F	0.010780
G	0.018700
H	0.031750
I	0.046250

When $k_I = 0$, the controller of Figure 4.2.4 is inoperative so that only the load input can influence the output. Consequently, the characteristic equation is simply $S + 0.01 = 0$ and only the single root $S = 100 \cdot 10^4$ exists (point A). But for any $k_I > 0$, there are three roots, since the characteristic equation (equation 4.8.7) is cubic. As k_I increases from near-zero values (points B) the root at the left of the graph moves along the negative real axis, becoming larger and larger in absolute magnitude. The other two roots move towards each other until they coincide at point D where k_I has the value corresponding to the critical damping ratio. With still further increases in k_I we have a conjugate pair of complex roots which move symmetrically towards the imaginary axis, finally crossing into the right half-plane between points H and I. When the imaginary axis has been reached (at $k_I \approx 0.04$) the system becomes unstable because the real parts of these complex roots are no longer negative. Higher values of k_I would carry the roots even farther into the right half-plane which unequivocally betokens instability.

SECTION 4.9 The Routh-Hurwitz method for testing stability

When the characteristic equation, $P_{den}(s) = 0$, is of fourth or higher order, it is still possible to find its roots, for example by applying the method of Lin described in the Appendix. But the labor involved is usually prohibitive unless a computer is available. The Routh-Hurwitz method is a technique for ascertaining whether all of the roots of a given characteristic equation have negative real parts without actually finding the roots themselves.

Since stability demands that all roots of $P_{den}(s) = 0$ have negative real parts, we need consider only polynomials whose factors are either of the form $[s + \alpha]$ or of the form $[s + \alpha + j\beta]\,[s + \alpha - j\beta] = s^2 + 2\alpha s + [\alpha^2 + \beta^2]$ where α is a real number greater than zero, and β is any real number. It is easy to see that any characteristic equation whose polynomial is the product of such factors must be of the form

$$s^n + a_{n-1}s^{n-1} + \ldots + a_1 s + a_0 = 0 \qquad 4.9.1,$$

where the constant coefficients (the a's) are all positive, the coefficient of the leading term (i.e., s^n) is unity, and all n + 1 terms are present. By multiplying equation 4.9.1 by a constant, c_1, we do not change the roots at all, but we obtain the slightly more general, and, for present purposes, more convenient form,

$$c_1 s^n + c_2 s^{n-1} + \ldots + c_n s + c_{n+1} = 0 \qquad 4.9.2,$$

where $c_2 = c_1 a_{n-1}$, $c_3 = c_1 a_{n-2}$, etc. Again, all n + 1 terms must be present, and all n + 1 c's must be of the same sign, usually positive. Hence, if any term is missing (e.g., $s^4 + 4s^2 + 6s + 18$, or $s^3 + 2s^2 + 3s$) or if the terms are not all of the same sign, we know by inspection that the system characterized by the equation is unstable. But these conditions, though necessary, are not sufficient. In order to prove unequivocally that all roots have negative real parts we use the n + 1 coefficients of equation 4.9.2 to calculate the entries in a Routh table as shown in Table 4.9.1. Once the pattern of calculation is understood, the table is easy to construct. Now inspect the n + 1 entries which comprise the first column of the table, beginning with c_1. If they are all strictly positive (or all strictly negative) the roots of equation 4.9.2 all have negative real parts. If there are any changes in sign as one passes from top to bottom, there are as many roots with positive real parts as there are sign changes. If, during construction of the table, a zero appears among the n + 1 entries of the first column, it also indicates instability, usually (but not always) due to complex conjugate poles on the imaginary axis. (When this happens, special technique are needed if one wants to complete the table. These are not difficult, but continuing the analysis once instability has been shown is not useful often enough to warrant description here.)

The examples in Table 4.9.2 will clarify the Routh-Hurwitz technique. For illustrative purposes they have been constructed from known roots whose actual locations in the complex plane are shown for comparison. In practice, of course, the Routh-Hurwitz method would not be used if the roots were actually known.

Table 4.9.1. Construction of the Routh table

First column				
c_1	c_3	c_5	c_7	$c_9 \ldots$
c_2	c_4	c_6	c_8	$c_{10} \ldots$
d_1	d_2	d_3	$d_4 \ldots$	$\vdots$
e_1	e_2	e_3	$e_4 \ldots$	
f_1	f_2	$f_3 \ldots$	$\vdots$	
g_1	g_2	$g_3 \ldots$		
h_1	$h_2 \ldots$	$\vdots$		
i_1	$i_2 \ldots$			
$j_1 \ldots$	$\vdots$			
$k_1 \ldots$				
$\vdots$				

The entries in the first two rows are the coefficients of equation 4.9.2.

There must be n + 1 rows in the table, n being the order of the equation.

Zeros can be inserted at the ends of rows if needed.

$$d_1 = \frac{c_2c_3 - c_1c_4}{c_2},\ d_2 = \frac{c_2c_5 - c_1c_6}{c_2},\ d_3 = \frac{c_2c_7 - c_1c_8}{c_2},\ d_4 = \frac{c_2c_9 - c_1c_{10}}{c_2},\ \text{etc.}$$

$$e_1 = \frac{d_1c_4 - c_2d_2}{d_1},\ e_2 = \frac{d_1c_6 - c_2d_3}{d_1},\ e_3 = \frac{d_1c_8 - c_2d_4}{d_1},\ e_4 = \frac{d_1c_{10} - c_2d_5}{d_1},\ \text{etc.}$$

$$f_1 = \frac{e_1d_2 - d_1e_2}{e_1},\ f_2 = \frac{e_1d_3 - d_1e_3}{e_1},\ f_3 = \frac{e_1d_4 - d_1e_4}{e_1},\ \text{etc.}$$

$$g_1 = \frac{f_1e_2 - e_1f_2}{f_1},\ g_2 = \frac{f_1e_3 - e_1f_3}{f_1},\ g_3 = \frac{f_1e_4 - e_1f_4}{f_1},\ \text{etc.}$$

$$h_1 = \frac{g_1f_2 - f_1g_2}{g_1},\ h_2 = \frac{g_1f_3 - f_1g_3}{g_1},\ \text{etc.}$$

$$i_1 = \frac{h_1g_2 - g_1h_2}{h_1},\ i_2 = \frac{h_1g_3 - g_1h_3}{h_1},\ \text{etc.}$$

$$j_1 = \frac{i_1h_2 - h_1i_2}{i_1},\ \text{etc.}$$

$$k_1 = \frac{j_1i_2 - i_1j_2}{j_1},\ \text{etc.}$$

Table 4.9.2 Examples of Routh tables

Example 4.9.1 $P_{den}(s) = s^5 + 5s^4 + 11.5s^3 + 17s^2 + 14.5s + 5 = 0$

$$= 2s^5 + 10s^4 + 23s^3 + 34s^2 + 29s + 10 = 0$$

2	23	29
10	34	10
$\frac{230 - 68}{10} = 16.2$	$\frac{290 - 20}{10} = 27$	0
$\frac{550.8 - 270}{16.2} = \frac{52}{3}$	$\frac{162 - 0}{16.2} = 10$	
$\frac{468 - 162}{52/3} = 17.65$	0	
$\frac{176.5 - 0}{17.65} = 10$		

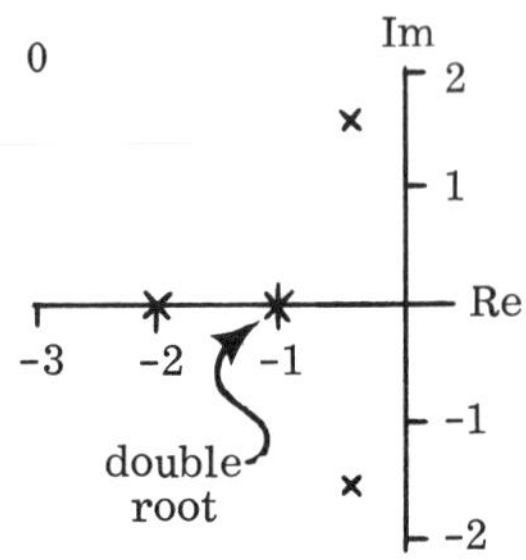

No zeros or sign changes in first column. All roots in left half-plane. Stable system.

Example 4.9.2 $P_{den}(s) = s^3 + 3s^2 + s + 20 = 0$

	1	1
plus	3	20
↓ minus	$\frac{3 - 20}{3} = -\frac{17}{3}$	0
↓ plus	$\frac{\left[\frac{-17}{3}\right] 20 - 0}{-17/3} = 20$	

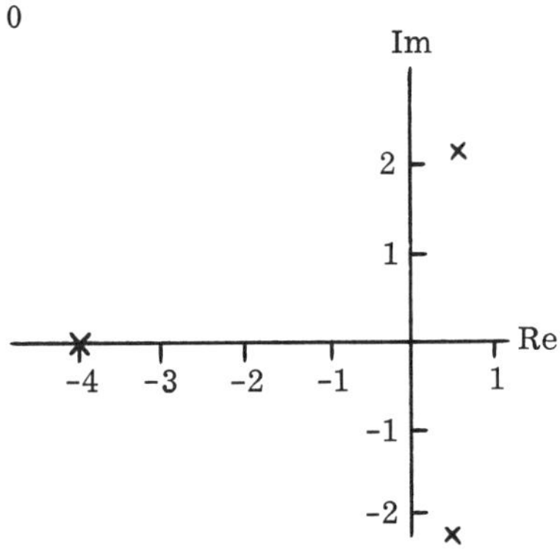

Two sign changes in first column. Two roots in right half-plane. Unstable system.

Table 4.9.2 continued

Example 4.9.3 $P_{den}(s) = s^4 + 5s^3 + 7s^2 + 5s + 6 = 0.$

1	7	6
5	5	0
$\frac{35 - 5}{5} = 6$	$\frac{30 - 0}{5} = 6$	
$\frac{30 - 30}{6} = 0$		
?		

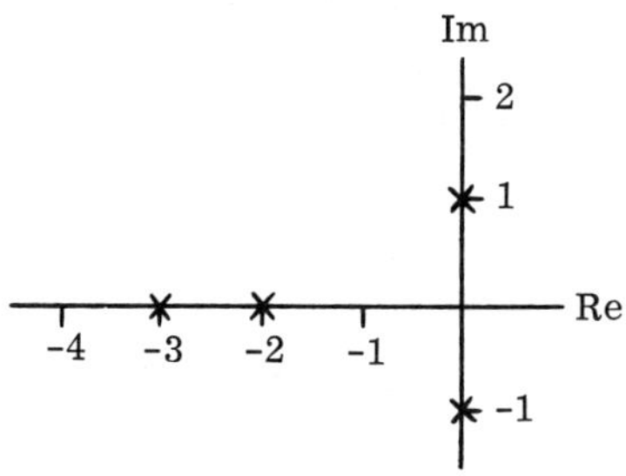

Zero in first column. Unstable system.

Example 4.9.4 (Cf. Figure 4.8.2)

$$P_{den}(s) = s^3 + 0.11s^2 + 0.001s + 0.00016 = 0.$$

	1	0.11
plus ↓	0.001	0.00016
minus ↓	$\frac{0.00011 - 0.00016}{0.001} = -0.05$	0
plus	$\frac{-0.05\,[0.00016]}{-0.05} = 0.00016$	

Two sign changes in first column. Two roots in right half-plane. Unstable system.

The Routh-Hurwitz technique may also be used to determine what values of a parameter (such as the k_p, k_d or k_i of Sections 4.3, 4.4, and 4.5) are compatible with stability.

Example 4.9.5 In the proportional control system of Section 4.3, what values of k_p are compatible with stability?

Leaving k_p unevaluated, we have as the characteristic equation, $P_{den}(s) = 3000s^2 + 330s + 3 + 8k_p = 0$. The corresponding Routh table is

$$\begin{array}{cc} 3000 & 3 + 8k_p \\ 330 & 0 \\ \dfrac{330\,[3 + 8k_p] - 0}{330} = 3 + 8k_p & 0 \end{array}$$

For stability, we require all terms in the first column to be positive, so we must have $3 + 8k_p > 0$. The critical value of k may therefore be found by setting

$$3 + 8k_{p,\,crit} = 0 \qquad 4.9.3,$$

whence $k_{p,\,crit} = -3/8$. This shows that we can have a stable system, even with negative values of k_p which make the system into a positive feedback system, provided $k_p > k_{p,\,crit} = -3/8$. Notice that when the value $k_p = -3/8$ is substituted into the characteristic equation, we have

$$P_{den}(s) = s^2 + 0.11s = 0 \qquad 4.9.4,$$

which, by inspection, indicates instability because the last term is missing.

For another example, work exercise 4.8.

SECTION 4.10 Frequency-domain analysis of feedback systems. Bode diagrams

The techniques of phasor analysis described in the last chapter are often used to investigate the response of a feedback system to sinusoidal inputs at various frequencies. Consider the generalized system of Figure 4.2.1. Let the reference standard, r(t), remain at zero. From equation 4.2.4, it is evident that the closed-loop transfer function for signal flow between the load input, $\ell(t)$, and the output, c(t) is

$$\frac{c_{\mathcal{L}}(s)}{\ell_{\mathcal{L}}(s)} = \frac{G_3(s)}{1 + G_1(s)G_2(s)H(s)} \qquad 4.10.1.$$

To investigate the response to phasor inputs we need only replace s in equation 4.10.1 by $j\omega$, thus obtaining

$$\tilde{c}(j\omega) = \frac{G_3(j\omega)}{1 + G_1(j\omega)G_2(j\omega)H(j\omega)}\ \tilde{\ell}(j\omega) \qquad 4.10.2.$$

For simplicity,* let the load input be cos ωt,

Then $\tilde{\ell}(j\omega) = 1$ and we have

$$\tilde{c}(j\omega) = \frac{G_3(j\omega)}{1 + G_1(j\omega)G_2(j\omega)H(j\omega)} = \alpha(\omega) + j\beta(\omega) \qquad 4.10.3.$$

where $\alpha(\omega)$ and $\beta(\omega)$ are, respectively, the real and imaginary parts of the complex phasor output. We could then plot the magnitude, $|\tilde{c}(j\omega)| = \sqrt{\alpha^2(\omega) + \beta^2(\omega)}$, and the phase angle, $\angle\tilde{c}(j\omega) = \varphi(j\omega) = \tan^{-1}(\beta\,\omega)/\alpha(\omega)$ against the forced frequency, ω, as was done in Figure 3.16.2 for the network of example 3.16.2. But it is often preferable, particularly when investigating a wide domain of frequencies, to plot** $\log_{10}|\tilde{c}(j\omega)|$ against $\log_{10}\omega$, and to plot $\angle c(j\omega)$ against $\log_{10}\omega$. Such plots are called Bode*** diagrams.

[*In phasor analysis we are usually interested in the difference between the phase angle of the input and the phase angle of the output, and in comparing the magnitudes of the outputs at various frequencies. Comparisons of this sort can be made whatever the actual values of the input magnitude and phase angle; so whenever we are free to choose the input (as we usually are since it is an external variable, independent of the system itself) we should select the simplest input possible, namely, cos ωt for which the magnitude is unity, the phase angle, zero.]

[**When making log-log or semilog plots, it is best to swallow our pure-mathematical pride and to use logarithms to the base 10 instead of to the base ϵ so that we can take advantage of the convenience of log-log or semi-log graph paper.]

[***Pronounced Bō•dē.]

Example 4.10.1. Construct Bode diagrams for the two-tank system with integral control (Section 4.4) when k_i = 0.03[L/min] per volt•min and $e_{ref}(t)$ = 0 volt.

The transfer function for signal flow from the load input, $\dot{V}_\ell(t)$, to the output, $P_b(t)$, (Figure 4.2.4) is

$$\frac{P_{b,\mathcal{L}}(s)}{\dot{V}_{\ell,\mathcal{L}}(s)} = \frac{8s[1 + 10s]}{s[1 + 10s][1 + 100s] + 0.08}\ \frac{\text{cm } H_2O}{\text{L/min}} \qquad 4.10.4.$$

Replacing s in equation 4.10.4 by $j\omega$, and solving for the phasor output, $\tilde{P}_b(j\omega)$,

$$\tilde{P}_b(j\omega) = \frac{j8\omega[1 + j10\omega]}{j\omega[1 + j10\omega][1 + j100\omega] + 0.08}\ \tilde{\dot{V}}_\ell(j\omega) \text{ cm } H_2O \qquad 4.10.5.$$

As usual, we shall let $\dot{V}_\ell(t) = \cos \omega t$ liter/min, so $\tilde{\dot{V}}_\ell(j\omega) = 1$ liter/min. Putting this value into equation 4.10.5 and rearranging,

$$\tilde{P}_b(j\omega) = \frac{\omega^2[800\omega^2 + 1.6]}{[0.08 - 110\omega^2]^2 + [\omega - 1000\omega^3]^2}$$

$$- j\left\{\frac{\omega[80,000\omega^4 + 800\omega^2 - 0.64]}{[0.08 - 110\omega^2]^2 + [\omega - 1000\,\omega^3]^2}\right\}$$

$$= \alpha(\omega) - j\beta(\omega) \text{ cm } H_20 \qquad 4.10.6.$$

where $\alpha(\omega)$ and $\beta(\omega)$ are, respectively, the real and imaginary parts of the phasor output $\tilde{P}_b(j\omega)$. We can now calculate the magnitude, $\sqrt{\alpha^2(\omega) + \beta^2(\omega)}$, and the phase angle, Arctan $\{\beta(\omega)/\alpha(\omega)\}$, for various values of ω so as to construct the Bode diagrams of Figure 4.10.1.a and b.

The most striking feature of these Bode diagrams is the pronounced peak of over 31 cm H_20 in the magnitude of the output. The peak occurs at the resonant frequency, ω_{res}, where the phase angle is zero. As was explained in Section 3.16, ω_{res} may be calculated by finding the frequency at which the imaginary part of the impedance (or admittance) vanishes. When (as is usual) the phase angle of the input is zero, ω_{res} is also the frequency at which the imaginary part of the output vanishes. Hence we may set $\beta(\omega) = 0$ and solve for ω_{res}. By inspection of equation 4.10.6, we see that for a non-zero $\tilde{P}_b(j\omega)$, $\beta(\omega) = 0$ implies that

$$80,000\,[\omega^2]^2 + 800[\omega^2] - 0.64 = 0 \qquad 4.10.7.$$

Solving this biquadratic for its positive root (we are not interested in imaginary values of ω) we obtain $\omega^2 = 7.45 \cdot 10^{-4}$, whence $\omega = \omega_{res} = 0.0273$ radians per min. Physically, this result means that if we were to vary the load sinusoidally over a domain of plus-or-minus one liter per minute, with a period of $2\pi/\omega_{res}$ = 115 minutes per cycle, the water pressure in tank b would be very poorly regulated indeed, varying (also sinusoidally and with a period of 115 min) between a low of 17 cm H_20 and a high of 79 cm H_20! So if there were any liklihood that the load would in fact fluctuate periodically at, or close to, this frequency, one or more of the parameters of the system would presumably have to be changed. The fluctuation of the input would not actually need to be sinusoidal, for any periodic input at a frequency of ω_{res} may be thought of,-- thanks to Fourier series representation--, as consisting of a sum of sines, or cosines, or both, of which the first (= fundamental) harmonic will be at the resonant frequency. (To keep a child happy in a swing, we need only deliver "pushes" (= intermittant pulses of force) at the proper frequency; we do not need to run back and forth continuously with the swing so as to deliver a continuous, sinusoidally-varying force throughout the entire cycle.)

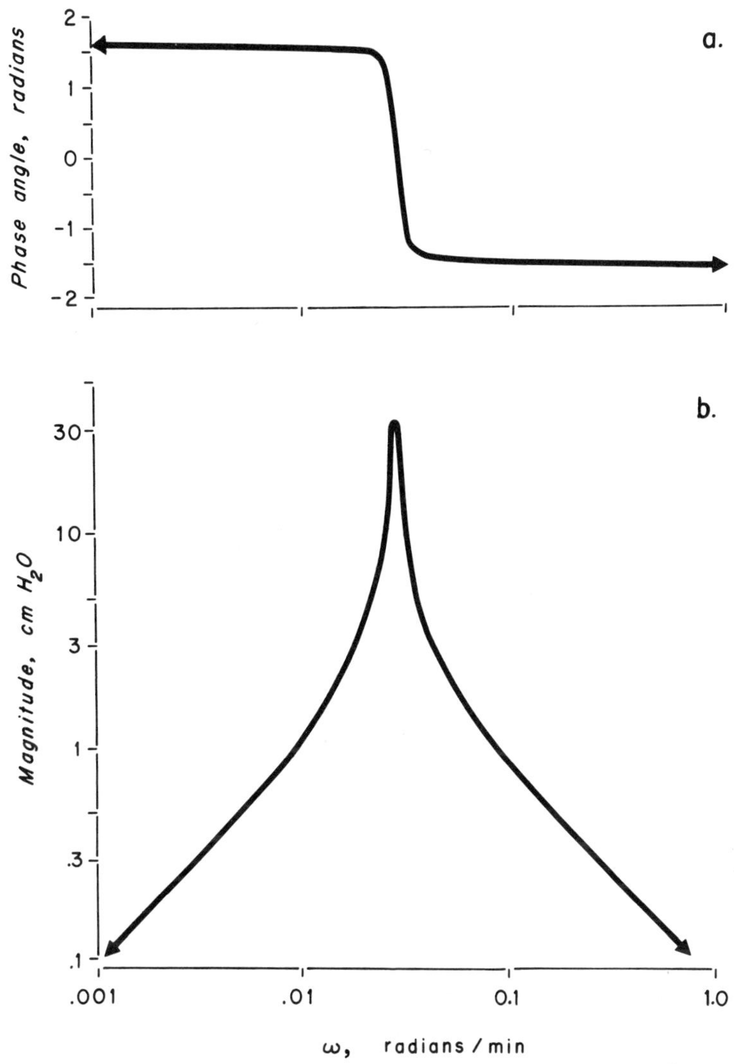

Figure 4.10.1 Bode diagrams for the two-tank feedback system with integral control.

a.) Phase angle, and b.) log magnitude of the phasor output, $\tilde{P}_b(j\omega)$, as a function of log ω when the input, $\dot{V}_\ell(t)$, is cos ωt liter/min. Since the angle of the corresponding input phasor is zero, the phase angle of the output phasor may also be called the phase difference (= phase shift) between input and output. And since the magnitude of the input phasor is unity, the magnitude of the phasor output may be called the gain between input and output provided that input and output have the same physical dimensions. (See Section 4.11). In this example, the input is flow, the output is pressure, so the term gain is not applicable here.

On both sides of the resonance peak, the magnitude of the output decreases sharply, approaching an asymptote of 100ω as ω becomes very small, and an asymptote of $0.08[1/\omega]$ as ω becomes very large. These asymptotes have practically been reached at the extremes of the Bode diagram in Figure 4.10.1.b, so the ends of the curve would continue downward indefinitely in the directions indicated by the arrows. As the magnitude declines, the phase angle also approaches asymptotic values of $+\pi/2$ radian/min for small ω and $-\pi/2$ radian/min for large ω. Bode diagrams characteristically make it easy to see what asymptotes are being approached, and how rapidly they are being approached. Incidentally, though the curves in Figure 4.10.1 may appear to be symmetrical about the resonant frequency, in fact, they are not, nor is there any theoretical reason to expect them to be.

SECTION 4.11. The meaning of the term "gain"

Consider a linear system activated by a sinusoidal input of such long duration that the system exhibits only a pure forced (= non-transient) response. Under these circumstances (See Section 3.15) although the output has the same frequency as the input, both the magnitude and the phase angle of the input signal may change as the signal traverses the system from input to output. Now a change in the phase angle can be calculated even when the input and the output do not have the same physical dimensions. It is just as meaningful to calculate the phase shift between input and output when the input is voltage and the output is current as when both are voltages. In contrast, a comparison of the magnitude of the output with the magnitude of the input is possible only when both signals have the same physical dimensions. It would be sheer nonsense to ask "How much larger is an output of five volts than an input of two amperes?" But if both output and input are voltages, such questions are perfectly legitimate: an output of five volts is two-and-one-half times as large as an input of two volts. Furthermore, since the magnitude of a sinusoidal signal is, by definition, always positive, any ratio of magnitudes is always positive. With these restrictions in mind, we shall define the term gain* as follows: When a sinusoidal signal passing from point A to point B of a system has the same physical dimensions at both points, the gain from A to B at a specified frequency, ω, is the ratio, [phasor magnitude at B]/[phasor magnitude at A].

[*I shall avoid the terms "amplitude" and "amplification." Although "amplitude" is commonly regarded as a synonym for the magnitude of a sinusoidal signal, both the International Dictionary of Applied Mathematics (3) and Webster's Third New International Dictionary (4) state that amplitude sometimes means the phase angle of a complex number. Could anything be more confusing! Amplification is sometimes used as a synonym for gain, but commonly refers to a gain in power. I shall also avoid the term "decibel," because I like to call a log a log.]

Example 4.11.1. What is the gain between the input, $\dot{V}_\ell(t)$, and the output, $P_b(t)$, in the system of Example 4.10.1?

$\dot{V}_\ell(t)$ and $P_b(t)$ do not have the same dimensions. The concept of gain is therefore inapplicable and the question is nonsense.

In feedback systems, the concept of gain is always applicable to the feedback loop itself because the insert signal and the return signal (See Section 4.2, and Figure 4.2.2.b), being measured at the same point in the loop, necessarily have the same dimensions. Hence we may always define the open-loop gain as the ratio of the magnitude of the return signal to the magnitude of the insert signal at a specified value of ω. Consider again the generalized feedback loop of Figure 4.2.2.b (opened loop), where the open-loop transfer function is

$$\frac{\mathcal{L}\{\text{Return signal}\}}{\mathcal{L}\{\text{Insert signal}\}} = -\,G_1(s)G_2(s)H(s) \qquad 4.11.1.$$

Converting equation 4.11.1 to phasor form,

$$\frac{\text{Phasor}\{\text{return signal}\}}{\text{Phasor}\{\text{insert signal}\}} = -\,G_1(j\omega)G_2(j\omega)H(j\omega) \qquad 4.11.2.$$

Taking magnitudes on both sides,

$$\text{Open-loop gain} \equiv \frac{|\text{Phasor}\{\text{return signal}\}|}{|\text{Phasor}\{\text{insert signal}\}|} = \left|-\,G_1(j\omega)G_2(j\omega)H(j\omega)\right| \qquad 4.11.3.$$

Equation 4.11.3 shows that the open-loop gain is the magnitude of the complex number obtained by evaluating the open-loop transfer function at $s = j\omega$. The next two sections will show how very important a quantity the open-loop gain is.

When we have an error-actuated, unity-feedback system (so that the error signal is simply the difference between the reference input and the output variable) we can always calculate a closed-loop gain for signal transfer between the reference input and the output (which, for unity feedback, necessarily have the same dimensions). Using the same kind of argument as for open-loop gain, the closed-loop gain for signal transfer between reference input and output in a unity feedback system is the magnitude of the complex number obtained by evaluating the corresponding transfer function at $s = j\omega$. But (as we have already seen from example 4.11.1) for load inputs, or for non-unity feedback systems it may or may not make any sense to speak of a closed-loop gain. The closed-loop gain, therefore, lacks the fundamental significance of the open-loop gain. When the unqualified term "loop gain" (or sometimes merely "gain") is used of a feedback system it should be taken to mean "open-loop gain."

SECTION 4.12. Open-loop gain and stability. Nyquist diagrams

The Routh-Hurwitz test (Section 4.9) is a way of discovering whether all of the roots of a system's characteristic equation have negative real parts; if not, the system is unstable. Another, and in some ways more instructive test for the stability of stationary linear feedback systems depends upon frequency-domain analysis.

Consider once more the open feedback loop of Figure 4.2.2.b. Equation 4.11.3 gives the gain for a sinusoidal signal traversing the elements of the open loop once and once only. The signal cannot jump the opening to re-enter the broken loop. But in actual operation, the loop is not open, and a given signal is free to circulate around the loop again and again. Now at most frequencies, after traversing the loop once, the sinusoidal signal will return to its point of entry more-or-less shifted in phase, so that the peaks of the return signal will not coincide with the peaks of the insert signal. Additional shifting in phase will occur the next time around, and the third time, and the nth time, so that the constantly shifting return signal will, on the average, oppose the insert signal as often as it reinforces the insert signal. Hence, even when the open-loop gain is large, the system remains stable, though as we have already seen (Figure 4.10.1.b) at or near the resonant frequency the magnitude of the output oscillations may become intolerably large. But now suppose that at some frequency the phase shift produced by one passage of a sinusoidal signal through the elements of the loop is exactly $-\pi$ radians. Then, if it were not subtracted from the insert signal, the return signal and the insert signal would be exactly opposite in phase. However, in a negative feedback loop, the loop signal is subtracted from the insert signal, and subtraction of a sinusoidal signal is equivalent to addition of the same signal shifted in phase by $-\pi$ radians. So the return signal will re-enter the closed loop with a phase shift of $-\pi$ radians (due to the components of the loop) plus $-\pi$ radians (due to the subtraction) making a total shift of -2π radians. But a total phase shift of -2π radians is equivalent to no phase shift at all! So the insert signal and the return signal will be precisely in phase, and this will continue to be true every time the return signal re-enters the closed loop. Hence, if the open-loop gain at this critical frequency equals or exceeds unity, the magnitude of the sinusoidal signals returning from the loop will be as great as or greater than the magnitude of the incoming train of sinusoidal signals to which (being precisely in phase) they will add so as to make an even larger insert signal the next time around. Once initiated, any train of sinusoidal signals at the critical frequency would thus increase in magnitude without limit and the system would be unstable. Hence, in a stable system the open-loop gain must be less than unity at the critical frequency where the total phase shift around the open loop is 2π radians.

The above argument is based upon the assumption that a train of sinusoidal impulses at the critical frequency has actually been inserted into the loop, presumably as an input. This suggests that open-loop gains equal to or greater than unity at the critical frequency might be compatible with a stable system if we could carefully shield the system from sinusoidal inputs at that frequency. In fact, however, such shielding is impossible: every system is subject to small random inputs ("noise") which behave as if they consisted of periodic inputs of all frequencies. The unstable system will inevitably select, and magnify without bound, any such random peaks of noisy input as happen to occur at the critical frequency. No deliberate or even identifiable input is needed.

The gains and phase lags produced at various frequencies by the components of a feedback loop are commonly displayed in the form of a Nyquist diagram. Let $-G_1(s)G_2(s)H(s)$ be the open-loop transfer function (equation

4.11.1). If we convert this to phasor form (equation 4.11.2) and choose the insert signal as cos ωt, the open-loop gain will be $|-G_1(j\omega)G_2(j\omega)H(j\omega)|$ (equation 4.11.3). But in calculating the phase shift for a Nyquist plot the negative sign is neglected. This means that the angle plotted is $\angle G_1(j\omega)G_2(j\omega)H(j\omega)$, not $\angle -G_1(j\omega)G_2(j\omega)H(j\omega)$, so the critical angle is $-\pi$ radians not -2π radians. These gains and phase angles are calculated for various ω from zero to ∞ and plotted on polar coordinates. Or, (and this is precisely equivalent), the real and imaginary parts of the complex number $G_1(j\omega)G_2(j\omega)H(j\omega)$ are plotted on rectangular coordinates.

It is worth noting that if we confine attention to open-loop transfer functions, the Nyquist diagram and the Bode diagrams represent alternative ways of displaying the same data; namely, the shifts in phase angle and the gains in magnitude which occur at various frequencies as a sinusoidal signal traverses the components of a feedback loop. Hence, if Bode diagrams have already been plotted, mere inspection of them may indicate clearly enough whether the open-loop gain is less than unity at the critical frequency. Under these circumstances, one would not usually bother to construct a Nyquist diagram, except perhaps as a brazenly vulgar display of erudition designed to impress the editors of a biological journal.

Example 4.12.1. Plot a Nyquist diagram for the two-tank feedback system of example 4.10.1, and calculate the critical frequency.

The open-loop transfer function is $-0.08/s[1 + 10s][1 + 100s]$. The corresponding phasor form is $-0.08/j\omega[1 + j10\omega][1 + j100\omega]$. For the Nyquist plot, the complex number to be plotted in the complex plane for various ω is,

$$\frac{0.08}{j\omega[1 + j10\omega][1 + j100\omega]} = \left\{\frac{0.08a}{a^2 + b^2}\right\} - j\left\{\frac{0.08b}{a^2 + b^2}\right\} \qquad 4.12.1,$$

where $a \equiv -110\omega^2$, and $b \equiv \omega[1 - 1000\omega^2]$.

If we choose to plot polar coordinates, the expression on the left of equation 4.12.1 is the easier to use, especially if we apply the rules (see Appendix) for products and reciprocals of complex numbers: In products, magnitudes multiply, phase angles add. In taking reciprocals magnitudes are straightforward reciprocals, phase angles simply change sign. The resulting polar plot* is shown in Figure 4.12.1.a.

[*Again, it may be best to abandon our lofty mathematical principles and to express angles in degrees so that we can profit from the convenience of polar coordinate graph paper.]

If we prefer to use rectangular coordinates, the expression on the right of equation 4.12.1 is the easier to work with. The resulting curve is shown in Figure 4.12.1.b. With an appropriate choice of scales these two Nyquist diagrams would be identical.

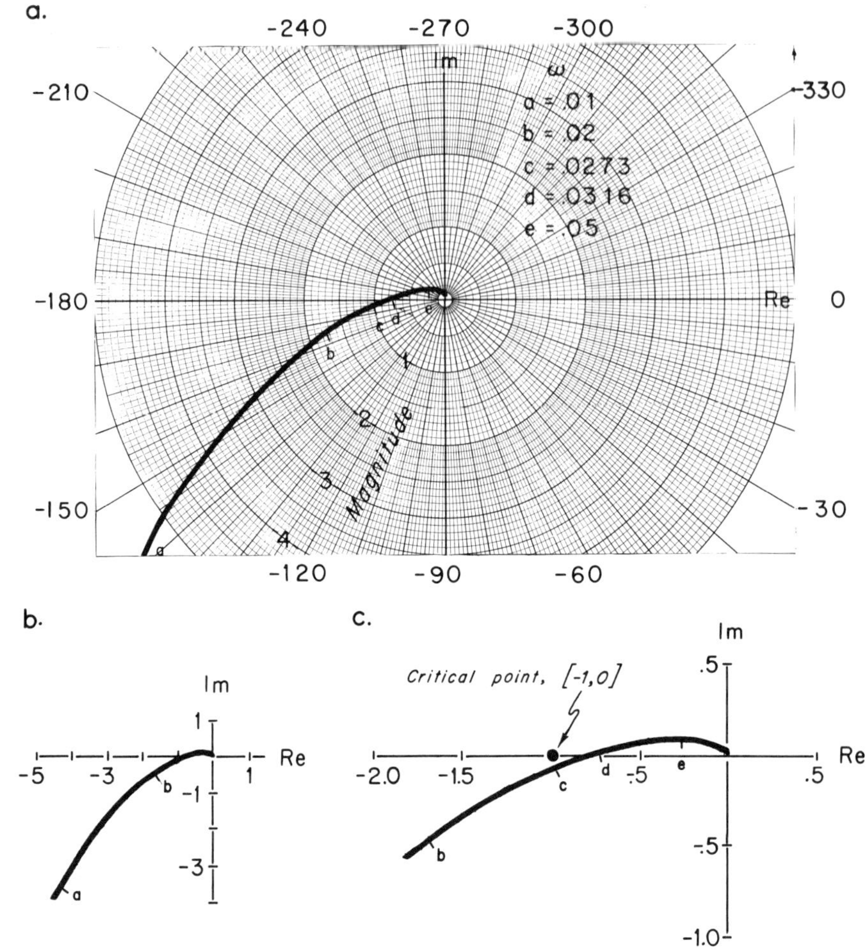

Figure 4.12.1. Nyquist diagrams for the two-tank feedback system with integral control.

The phase shift and gain due to transfer of a sinusoidal signal around the opened negative feedback loop (neglecting the sign inversion) are calculated as the phase angle and the magnitude of the complex number obtained by evaluating the negative of the open-loop transfer function at $s = j\omega$ for various values of ω between zero and plus infinity. In part a, the phase angle and magnitude have been plotted directly on polar coordinate graph paper. In part b, the same curve has been obtained by plotting the real and the imaginary parts of the complex number on the complex plane. In part c, the portion of the curve which is near the critical point at $-1 + j0$ is sketched in greater detail.

This system is stable, because the gain is less than unity when $\omega = 0.0316$ radian/min, the critical frequency at which the phase shift is $-\pi$ radians $(= -180^\circ)$.

In this example, the Nyquist plot crosses the negative real axis (which, of course, lies at the critical angle of $-\pi$ radians) to the right of the critical point for unit gain. (Figure 4.12.1.c). The system is therefore stable.

It is very easy to find the critical frequency, ω_{crit}, if we notice that at an angle of $-\pi$ radians the imaginary part of the complex number we are working with must be zero. Hence, from the second part of equation 4.12.1 we have,

$$0.08b = 0.08\omega_{crit}[1 - 1000\omega^2_{crit}] = 0 \qquad 4.12.2,$$

whence

$$1000\omega^2_{crit} = 1 \qquad 4.12.3,$$

and

$$\omega_{crit} = [1000]^{-1/2} = 0.0316 \text{ radian/min} \qquad 4.12.4.$$

Of course in certain other systems we might get several values of ω by setting the imaginary part equal to zero, for the Nyquist curve often originates (for $\omega = 0$) on the positive real axis, and it may cross the positive real axis or the negative real axis or both more than once. Common sense may show the proper meanings of these several values. However, certain complicated, multiple-loop, stable systems may yield Nyquist diagrams in which the curve traced out as ω increases from 0 to $+\infty$ not only crosses the negative real axis several times, but even includes some crossings at a gain equal to or greater than unity. Under these circumstances, a more general Nyquist criterion of stability than the one discussed above must be applied. (See, for example, references 5 and 6.)

We have now considered two apparently quite different stability criteria: the Routh-Hurwitz test, used to find out whether the characteristic equation has any roots with non-negative real parts, and the Nyquist diagram test to find out whether a sinusoidal signal at the critical frequency will be propagated around the loop with a gain in magnitude equal to or greater than unity. Yet whenever a system falls within the scope of both methods, they yield identical results.

At first glance, it might seem that the Routh-Hurwitz test would have the broader scope, since it could be applied to "open-loop" systems with no feedback. But on further reflection, we see that this advantage is illusory. For in any linear system composed entirely of the simple passive elements which we have been studying, the principle of superposition guarantees that an open-loop system must have a bounded output for any bounded input; in such systems instability can arise only through the operation of a closed feedback loop. Indeed, as we shall see in the next section, it is the Nyquist criterion which has the broader scope, since, unlike the Routh-Hurwitz test, it can take into account the effect of a delay in signal transmission.

SECTION 4.13. Delay (= Dead-time) elements and their effect on stability

In all of the previous discussion, we have made the tacit assumption that signals are transferred from one component of a system to another instantaneously. This assumption is by no means always valid. The transmission of a nerve impulse from point A to point B occupies a finite interval of time. So does the carriage of a chemical signal by the blood stream from one place to another. When the signal is not altered in transit but is merely delayed, the effect of the delay upon the behavior of the system can be represented by a delay element. If $x(t)u(t)$ is the input to a pure delay element, the output is $x(t - t_d)u(t - t_d)$ where t_d is the dead time or delay time. The transfer function for a delay element is therefore

$$\frac{\mathcal{L}\{x(t - t_d)u(t - t_d)\}}{\mathcal{L}\{x(t)u(t)\}} = \frac{x_{\mathcal{L}}(s)\varepsilon^{-st_d}}{x_{\mathcal{L}}(s)} = \varepsilon^{-st_d} \qquad 4.13.1.$$

Delay in the transmission of the signal has no effect upon the stability of an open-loop system. But if a delay element is one of the components in a negative feedback loop, it can drastically alter stability by increasing the phase lag of the return signal. Under these circumstances, the Routh-Hurwitz criterion for stability cannot be used, because the transfer function from input to output is no longer a simple ratio of polynomials but contains an exponential function of s. However, the Nyquist criterion is still applicable as the following argument will demonstrate.

Suppose a sinusoidal signal with a frequency of ω radian/sec is inserted into a feedback loop where transmission of the signal is delayed for t_d seconds without any change in its magnitude or wave form. It is obvious that the delay will cause the return signal to lag ωt_d radians behind the phase which would have been reached had there been no delay. (For example, if $\omega = 6\pi$ radians per sec, equivalent to 3 cycles per second, a "dead time" of 0.1 sec would delay the signals by 0.3 cycle, or $\omega t_d = 0.6\pi$ radians.) Now this is exactly the result which we would obtain by treating the transfer function of equation 4.13.1 in the usual way, replacing s by $j\omega$ to obtain the phasor form,

$$\frac{\text{Phasor}\{\text{return signal}\}}{\text{Phasor}\{\text{insert signal}\}} = \varepsilon^{-j\omega t_d} = \cos\omega t_d - j\sin\omega t_d \qquad 4.13.2.$$

The magnitude of this complex number is $\sqrt{\cos^2\omega t_d + \sin^2\omega t_d} = 1$; the phase angle is Arctan $(-\sin\omega t_d/\cos\omega t_d) = -\omega t_d$. In fact, if we are familiar with the exponential form of expressing complex numbers, (See Appendix), these values are evident by simple inspection of $\varepsilon^{-j\omega t_d}$.

SECTION 4.14. Open-loop gain and steady-state error with a constant input

In Section 4.3, the question, "How effectively does negative feedback reduce steady-state error when the set point or the load has been changed to some new but constant value?" was answered by numerical evaluation of the steady-state error with and without feedback for the specific two-tank system used as an example. By deriving a relationship between the magnitudes of the open-loop and closed-loop errors as $\omega \to 0$, we shall now show that the open-loop gain provides an analytical answer to this question. Consider again the generalized feedback system of Figure 4.2.1 which is described by equation 4.2.4. Let $r_{\mathcal{L}}(s)$, the reference input, = 0. We shall write the remaining part of equation 4.2.4 (the part describing the response to a load input) in phasor form, thereby automatically restricting our consideration to the non-transient response to sinusoidal inputs. Letting $\ell(t)$, the load input, $= \cos\omega t$ as usual, we have

$$\tilde{c}_{closed}(j\omega) = \left\{\frac{1}{1 + G_1(j\omega)G_2(j\omega)H(j\omega)}\right\} G_3(j\omega) \qquad 4.14.1.$$

To find the open-loop phasor, let $H(j\omega) = 0$

$$\tilde{c}_{open}(j\omega) = G_3(j\omega) \qquad 4.14.2.$$

Dividing equation 4.14.2 by equation 4.14.1,

$$\frac{\tilde{c}_{open}(j\omega)}{\tilde{c}_{closed}(j\omega)} = 1 + G_1(j\omega)G_2(j\omega)H(j\omega) \qquad 4.14.3.$$

Now for a <u>constant</u> input, $\omega = 0$, and equation 4.14.3 becomes

$$\frac{\tilde{c}_{open}(j\omega)}{\tilde{c}_{closed}(j\omega)} = 1 + G_1(0)G_2(0)H(0) = 1 - [\text{OLTF}]\,(0) \qquad 4.14.4,$$

where [OLTF] (0) is the open loop transfer function (See Figure 4.2.2.b) evaluated at $s = 0$. (Note that evaluating any function of s, say G(s), at $s = j\omega$ and then evaluating $G(j\omega)$ at $\omega = 0$ is equivalent to evaluating G(s) directly at $s = 0$.

Before proceeding further, let us consider equation 4.14.4 as it stands. If the controller represented by $G_1(s)$ is, at least in part, an integral controller (Cf. equation 4.4.3), $G_1(s)$ will approach infinity as $s \to 0$, so that by equation 4.14.1, $\tilde{c}_{closed}(0) = 0$. Therefore, as we have seen previously, there will be <u>no</u> steady-state error with a constant load. If the controller is a pure derivative controller, $G_1(s) = k_d s$, so $G_1(s)$ would equal zero when $s = 0$. Hence, by equation 4.14.4, $\tilde{c}_{open}(0)/\tilde{c}_{closed}(0)$ would be unity, so that, as we also saw previously, the steady-state error would not be decreased at all by feedback. We may therefore confine our further analysis of steady-state error to proportional control. Finally, we note that when $j\omega = 0$, the imaginary part of the complex number, $G_1(j\omega)G_2(j\omega)H(j\omega)$, vanishes so that both $G_1(0)G_2(0)H(0)$ and [OLTF] (0) must be real numbers.

We may now return to equation 4.14.4. Taking absolute values on both sides,

$$\frac{|\tilde{c}_{open}(0)|}{|\tilde{c}_{closed}(0)|} = |1 + G_1(0)G_2(0)H(0)| = |1 - [\text{OLTF}]\,(0)| \qquad 4.14.5.$$

In Chapter 1, we saw that in a symbol-and-arrow diagram, any negative feedback loop must contain an odd number of broken arrows. In a block diagram, the corresponding (in fact, precisely equivalent) requirement is that a signal passing around the entire negative feedback loop must undergo an odd number of sign changes, so that [OLTF] (0) is negative. Furthermore, the quantity $G_1(0)G_2(0)H(0)$ does not include the minus sign which shows that the feedback signal is to be subtracted from the reference input. Hence, $G_1(0)G_2(0)H(0)$ must be intrinsically positive. Therefore equation 4.14.5 may be written

$$\frac{|\tilde{c}_{open}(0)|}{|\tilde{c}_{closed}(0)|} = 1 + G_1(0)G_2(0)H(0) = 1 + \left|[\text{OLTF}]\ (0)\right| \qquad 4.14.6.$$

But by equation 4.11.3,

$$[\text{OLG}]\ (j0) = \left|-\ G_1(0)G_2(0)H(0)\right| = G_1(0)G_2(0)H(0) \qquad 4.14.7,$$

where we define [OLG] (j0) as the open loop gain for a constant input.* Combining equations 4.14.6 and 4.14.7,

$$\frac{|\tilde{c}_{open}(0)|}{|\tilde{c}_{closed}(0)|} = 1 + G_1(0)G_2(0)H(0) = 1 + \left|[\text{OLTF}]\ (0)\right| = 1 + [\text{OLG}]\ (j0) \qquad 4.14.8.$$

[*I prefer to include the "j" as part of the argument of [OLG] to remind us that "gain" is a concept which is derived from frequency-domain analysis.]

Finally, since we are assuming that $r(t) = 0$, the value of $c(t)$ for zero error is zero. Hence, $|\tilde{c}(0)|$ also represents the magnitude of

[actual value of $c(\infty)$] - [errorless value of $c(\infty)$],

which is the steady-state error for a constant load. So we may rewrite equation 4.14.8 as

$$[\text{OLG}]\ (j0) = \left|[\text{OLTF}]\ (0)\right| = G_1(0)G_2(0)H(0) = \frac{|\text{Open-loop error}|}{|\text{Closed-loop error}|} - 1 \qquad 4.14.9$$

Equation 4.14.9 shows that the open-loop gain at $\omega = 0$ measures the ability of the negative feedback system to minimize steady-state error (or, more generally, steady-state deviations from a given operating point) in the face of a constant load. A similar derivation may be used to show that equation 4.14.9 serves equally well for the steady-state error in the face of a step change in set-point. Notice that [OLG] (j0) is not, in fact, the ratio of the open-loop and closed-loop errors, but is the ratio of the absolute magnitudes of these errors, minus unity. The open-loop gain is thus a more sensible measure of regulatory effectiveness than the ratio of errors itself would be, because when the effectiveness of negative feedback is zero (corresponding to the open-loop situation) [OLG] (j0) is zero, while the ratio of errors would still be unity.

Example 4.14.1. How effective is negative feedback in minimizing the steady-state error in the two-tank system of Section 4.3 when the reference input is kept at zero but the load input is increased by a unit step?

With $k_p = 3/4$, numerical evaluation previously gave 3 as the ratio of the open-loop to the closed-loop error (Section 4.3). The magnitude of this ratio, minus 1, is 3 -1 = 2. This is exactly what we obtain by evaluating the absolute value of the open loop transfer function,

$$\left| - G_1(s)G_2(s)H(s) \right| = \frac{3}{4} \left\{ \frac{8}{[1 + 10s][1 + 100s]} \right\} \frac{1}{3} \qquad 4.14.9,$$

at s = 0,

$$[OLG]\,(j0) = G_1(0)G_2(0)H(0) = \frac{3}{4}[8]\frac{1}{3} = 2 \qquad 4.14.10.$$

SECTION 4.15. Open-loop gain in the steady state as a product of derivatives

If we confine our attention to the steady-state response of stationary linear systems with proportional (or proportional-derivative) control to constant inputs we can calculate the open-loop gain by another method which is important because it can be extended to non-linear systems.* (See Section 7.3).

[*With the italicized restrictions there is no longer any need to have zero initial conditions and we may therefore use the actual values of the variables, instead of their deviations from some operating point.]

In Chapter 1, feedback was simply defined as a closed unidirectional sequence of causal relationships linking two or more interdependent variables. Moreover, we found that to describe fully the working relationship between any two variables which are members of a feedback loop, we always had to have two different equations. We shall now use these equations to answer the question, "How effectively does negative feedback minimize steady-state deviations caused by a constant external disturbance?" While this is much the same question we have asked before, it is here worded in a somewhat more general way in order to show that we can study the ability of a negative feedback loop to minimize the influence of external changes without having to select a particular member of the feedback loop as the output and a particular independent variable as the input, and without having to talk in terms of "error."

Consider the generalized block diagram of Figure 4.15.1 which depicts a feedback system subject to the rather severe restrictions underlined in the first paragraph of this section. The $x(\infty)$'s and $y(\infty)$'s are variables which are actually members of the feedback loop; the U's are independent variables (= inputs). As in any conventional block diagram, the features of interest are the summing points and the blocks. Notice first that each y is merely the algebraic sum of the preceding loop variable and a constant input. In linear feedback loops, inputs always enter additively at a summing point; if any input multiplies or divides a loop variable, the system is not linear. Notice next that, with constant inputs, when $t \to \infty$ the blocks contain only parameters, for, as we

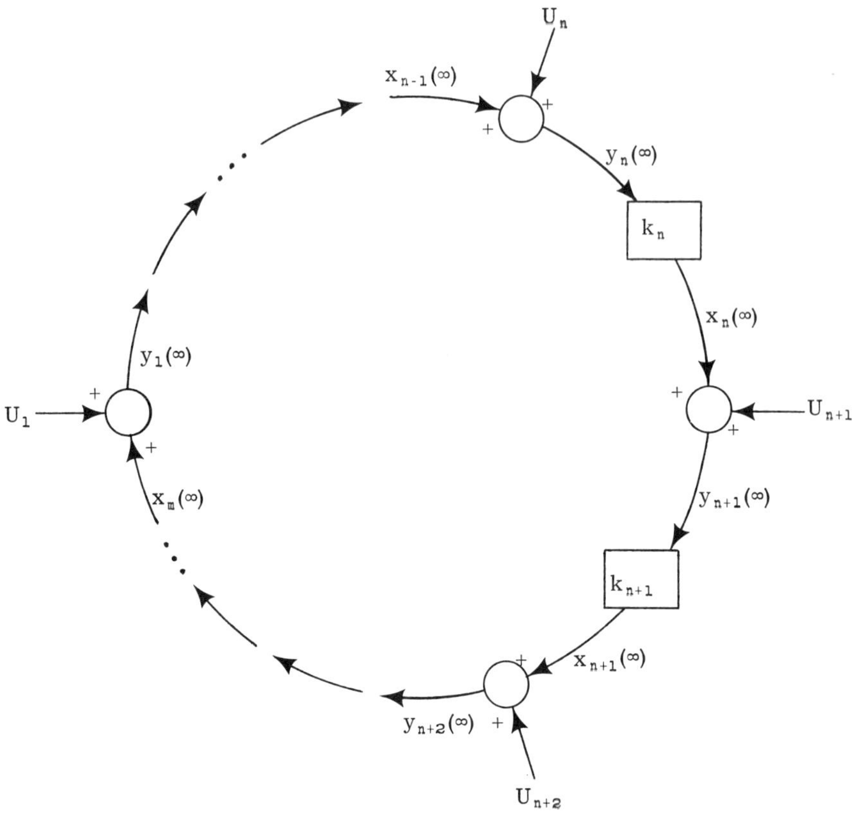

Figure 4.15.1. A generalized single feedback loop in the steady state.

Note: If the loop comprises m y's and m x's, equation 4.15.1 must become $y_1(\infty) = x_m(\infty) + U_1$ for the special case where the signal passes from the mth x to the first y. This special case is illustrated at the left of the diagram.

---

have already seen, s then equals zero. The entire feedback loop is therefore simply a closed chain of variables for each of which the working relationship is either of the form*

$$y_n(\infty) = x_{n-1}(\infty) + U_n \qquad 4.15.1,$$

or of the form

$$x_n(\infty) = k_n y_n(\infty) \qquad 4.15.2,$$

[*With the minor exception noted in the legend of Figure 4.15.1.]

where k_n is the transfer function of the nth block when $s = 0$. For convenience, we may combine these two types with no loss of generality by eliminating the y's, retaining only the x's as loop variables. Then the working relationship for $x_n(\infty)$ can be written in the general form

$$x_n(\infty) = k_n[x_{n-1}(\infty) + U_n] \qquad 4.15.3,$$

where $x_{n-1}(\infty)$ and k_n may be positive or negative but not zero, and U_n may be positive, zero, or negative.

Now let us concentrate upon <u>any</u> particular pair of loop variables, say $x_i(\infty)$ and $x_j(\infty)$, $i \neq j$. It should be obvious that by suitably combining equations of the 4.15.3 kind we can <u>eliminate all other loop variables</u> until we have only two working equations left: one for $x_i(\infty)$ as a function of $x_j(\infty)$,

$$x_i(\infty) = f(x_j(\infty)) \qquad 4.15.4,$$

and the other for $x_j(\infty)$ as a function of $x_i(\infty)$,

$$x_j(\infty) = g(x_i(\infty)) \qquad 4.15.5.$$

What will these functions, f and g, look like? Let us suppose, as a concrete example, that the loop contains five variables, $x_1(\infty)$, $x_2(\infty)$, $x_3(\infty)$, $x_4(\infty)$, and $x_5(\infty)$, so arranged that the specific versions of equation 4.15.3 are

$$x_1(\infty) = k_1[x_5(\infty) + U_1] \qquad 4.15.6,$$

$$x_2(\infty) = k_2[x_1(\infty) + U_2] \qquad 4.15.7,$$

$$x_3(\infty) = k_3[x_2(\infty) + U_3] \qquad 4.15.8,$$

$$x_4(\infty) = k_4[x_3(\infty) + U_4] \qquad 4.15.9,$$

$$x_5(\infty) = k_5[x_4(\infty) + U_5] \qquad 4.15.10.$$

Now suppose that we let $i = 2$ and $j = 5$. We can then get an equation corresponding to $x_i(\infty) = f(x_j(\infty))$ by combining equations 4.15.6 and 4.15.7 so as to eliminate $x_1(\infty)$.

$$x_2(\infty) = k_1k_2x_5(\infty) + k_1k_2U_1 + k_2U_2 = f(x_5(\infty)) \qquad 4.15.11.$$

Similarly, we can get an equation corresponding to $x_j(\infty) = g(x_i(\infty))$ by combining equations 4.15.8, 4.15.9 and 4.15.10, so as to eliminate $x_3(\infty)$ and $x_4(\infty)$.

$$x_5(\infty) = k_3k_4k_5x_2(\infty) + k_3k_4k_5U_3 + k_4k_5U_4 + k_5U_5 = g(x_2(\infty)) \qquad 4.15.12.$$

Finally, let us calculate the first derivative of equation 4.15.11 with respect to $x_5(\infty)$,

$$\frac{dx_2(\infty)}{dx_5(\infty)} = f'(x_5(\infty)) = k_1k_2 \qquad 4.15.13,$$

and the first derivative of equation 4.15.12 with respect to $x_2(\infty)$,

$$\frac{dx_5(\infty)}{dx_2(\infty)} = g'(x_2(\infty)) = k_3k_4k_5 \qquad 4.15.14,$$

and form the product of these two derivatives,

$$f'(x_5(\infty))\, g'(x_2(\infty)) = k_1k_2k_3k_4k_5 \qquad 4.15.15.$$

But what is $k_1k_2k_3k_4k_5$? It is nothing but the open-loop transfer function evaluated at $s = j0$. And since the corresponding open-loop gain is the magnitude of the open-loop transfer function evaluated at $s = j0$ (cf. equation 4.14.8) we have for $i = 2$, $j = 5$,

$$[OLG]\,(j0) = \left|f'(x_j(\infty))g'(x_i(\infty))\right| \qquad 4.15.16.$$

This result can easily be verified for all i and j such that $i \neq j$, however many variables there are in the feedback loop.

Example 4.15.1. Figure 4.15.2 is a symbol-and-arrow diagram showing the steady-state relationships among the variables of our well-worn, two-tank system (Figure 4.2.3). Each dependent variable is numbered, and in the legend an equation with the same number shows precisely how that variable depends upon other variables and upon the parameters of the system. Of the eight dependent variables, seven are members of the feedback loop. Since there are $\frac{n!}{p![n-p]!}$ different combinations of n variables taken p at a time (7), we have $7!/2!5! = 21$ different pairs of loop variables. By applying equation 4.15.16 to any one of these 21 pairs, we should obtain [OLG] (j0), which is already known (Cf. equation 4.14.10) to be 2 when the parameters are $k_{trans} = 1/3$ volt/cm H_2O, $k_p = 3/4$ [liter per min]/volt, $R_a = 2$ cm H_2O/ [liter per min], and $R_b = 8$ cm H_2O/[liter per min]. To illustrate, we shall perform the calculations for two of the 21 different pairs.

a) Let the two variables be $\dot{V}_{in,a}(\infty)$ and $\dot{V}_{in,b}(\infty)$. Combining equations 2, 1, 7, and 6 in Figure 4.15.2,

$$\dot{V}_{in,a}(\infty) = k_p e_{ref} - k_p k_{trans} R_b \dot{V}_{in,b}(\infty) \qquad 4.15.17,$$

whence

$$\dot{V}'_{in,a} = \frac{d\dot{V}_{in,a}(\infty)}{d\dot{V}_{in,b}(\infty)} = -k_p k_{trans} R_b \qquad 4.15.18.$$

Combining equations 5, 4, and 3 in Figure 4.15.2,

$$\dot{V}_{in,b}(\infty) = \frac{1}{R_a} \cdot R_a \dot{V}_{in,a}(\infty) + \dot{V}_\ell = \dot{V}_{in,a}(\infty) + \dot{V}_\ell \qquad 4.15.19,$$

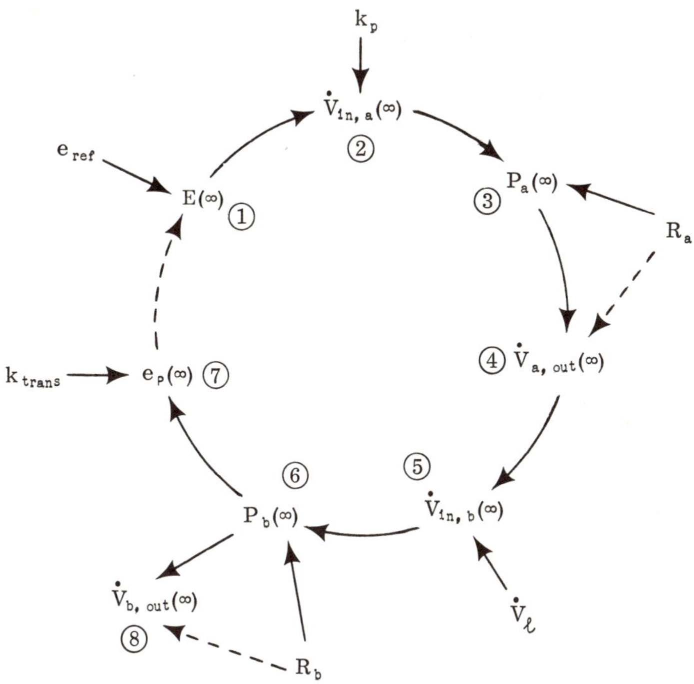

Figure 4.15.2. Symbol-and-arrow diagram for 2-tank feedback system in the steady state.

The symbols used here are the ones of Figure 4.2.3, with (∞) replacing (t) to show that the variables have reached their steady-state values. The number beside each dependent variable refers to the correspondingly numbered working equation in the following list:

1. $E(\infty) = e_{ref}(\infty) - e_p(\infty)$
2. $\dot{V}_{in,a}(\infty) = k_p E(\infty)$
3. $P_a(\infty) = R_a \dot{V}_{in,a}(\infty)$
4. $\dot{V}_{a,out}(\infty) = [1/R_a]\, P_a(\infty)$
5. $\dot{V}_{in,b}(\infty) = \dot{V}_{a,out}(\infty) + \dot{V}_\ell$
6. $P_b(\infty) = R_b\, \dot{V}_{in,b}(\infty)$
7. $e_p(\infty) = k_{trans} P_b(\infty)$
8. $\dot{V}_{b,out}(\infty) = [1/R_b]\, P_b(\infty)$

Notice that in the steady state, the behavior of the system is not influenced by the capacitances of the tanks, and that the remaining parameters of the system play much the same role as do the inputs. (See Section 7.1.)

Diagrams of this kind are particularly useful for displaying the steady-state relationships among the variables of nonlinear biological systems. They will therefore be used extensively in Part II of this book.

whence

$$\dot{V}'_{in,\,b} = \frac{d\dot{V}_{in,\,b}(\infty)}{d\dot{V}_{in,\,a}(\infty)} = 1 \qquad 4.15.20.$$

Taking the magnitude of the product of the two derivatives in equations 4.15.18 and 4.15.20 we have, by equation 4.15.16,

$$[\,OLG\,]\,(j0) = \left|\dot{V}'_{in,\,a}\ \dot{V}'_{in,\,b}\right| = k_p k_{trans} R_b = \frac{3}{4}\cdot\frac{1}{3}\cdot 8 = 2 \qquad 4.15.21.$$

b) Let the two variables be $e_p(\infty)$ and $P_a(\infty)$. Applying steps similar to those used in part a, we obtain

$$e_p(\infty) = k_{trans} R_b \dot{V}_\ell + [\,k_{trans} R_b / R_a\,]\ P_a(\infty) \qquad 4.15.22.$$

$$e'_p = \frac{de_p(\infty)}{dP_a(\infty)} = k_{trans} R_b / R_a \qquad 4.15.23.$$

$$P_a(\infty) = R_a k_p e_{ref} - R_a k_p e_p(\infty) \qquad 4.15.24.$$

$$P'_a = \frac{dP_a(\infty)}{de_p(\infty)} = -\,R_a k_p \qquad 4.15.25.$$

$$[\,OLG\,]\,(j0) = \left|e'_p P'_a\right| = k_p k_{trans} R_b = 2 \qquad 4.15.26.$$

as in part a.

An instructive graphical interpretation of this method for calculating the open-loop gain is shown in Figure 4.15.3. With linear systems, if the two loop equations describing the interrelationship between the variables chosen for study are plotted on the same sheet of rectangular coordinate graph paper, each equation yields a straight line. The coordinates of the point where the two lines intersect are the steady-state values of the two loop variables, and the first derivatives of the equations are the slopes of the two lines. However, since the independent variable for one equation is the dependent variable for the other, one of the two lines must necessarily be plotted in an unorthodox way with the independent variable as ordinate, the dependent as abscissa. The true slope of this line (which I shall always show as a broken line) is therefore just the inverse of the apparent slope as the line lies on the graph, so that when the slope of the broken line looks small, it is actually large. On such a diagram, it is easy to demonstrate how the two equations interact to reduce the effect of a change in one of the input variables. An example is fully explained in the legend of Figure 4.15.3.

The technique described in this section, though severely limited in scope by the restrictions specified in the opening paragraph, has the outstanding advantage that it can be used to calculate the open-loop gain, and thus to investigate the effectiveness of feedback regulation, when some of the loop variables are experimentally inaccessible and perhaps not even identifiable. This situation

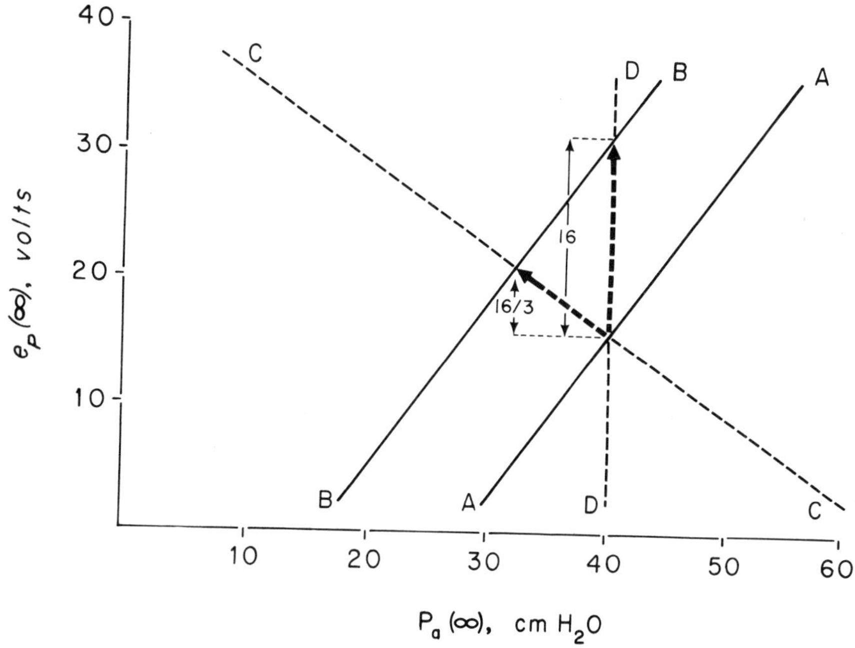

Figure 4.15.3. Steady-state changes in the two-tank system of Figure 4.2.2.a caused by a step increment in the load input.

(NOTE: The values in this figure are the actual values of the variables, not their deviations from some operating point.)

Line A, $e_p(\infty) = -[112/3] + [4/3]P_a(\infty)$ volt, shows the steady-state dependence of $e_p(\infty)$ on $P_a(\infty)$ when the inputs and the parameters have the values shown in Figure 4.2.3. Line B, $e_p(\infty) = -[64/3] + [4/3]P_a(\infty)$ volt, shows what this relationship would be if the load, $\dot{V}_\ell(t)$ were increased by a step from minus 14 liters/min to minus 8 liters/min. Line C, $P_a(\infty) = 64 - [3/2]e_p(\infty)$ cm H_2O, represents the steady-state dependence of $P_a(\infty)$ on $e_p(\infty)$ when the feedback loop is closed as it is in Figure 4.2.3. Line D, $P_a(\infty) = 40$ cm H_2O, (which has a slope of zero cm H_2O/volt) represents the open-loop condition produced by disconnecting the pressure transducer from the voltage subtractor, and substituting a constant signal of 16 volts.

For given conditions, the steady-state values of $e_p(\infty)$ and of $P_a(\infty)$ are the coordinates of the point at which the appropriate solid line intersects the appropriate broken line. Accordingly, when the feedback loop is open, the step increase of 6 liters/min in the load input will increase $e_p(t)$ by 16 volts (arrow along line D). But the same step increase with the loop closed will increase $e_p(\infty)$ by only 16/3 volts (arrow along line C). So in this system, negative feedback reduces the deviation caused by a load input to 1/3 of what it would be with no feedback.

is all too common in physiological investigations where we may count ourselves fortunate if we can obtain reliable quantitative information about the steady-state relationships between any pair of variables which are known to be members of the feedback loop. A further major advantage has already been mentioned, namely that the product-of-derivatives method can be adapted to the study of non-linear systems (See Section 7.3).

Penitential note: In a previous book, (8) which I wrote before I knew anything about control systems engineering, I invented the method described in this section and coined the term "Homeostatic Index" for the absolute value of the product of derivatives in equation 4.15.16. Coining a new synonym for a well-established term is a criminal act, especially in a field already distinguished by the richness and opacity of its jargon. In an attempt to purge myself of semantic sin, I hereby abandon "Homeostatic Index" in favor of "Open-loop gain," trusting that the context will make it clear when [OLG] (j0) (= Homeostatic Index), and not the more general [OLG] ($j\omega$) is intended.

SECTION 4.16. Positive feedback loops

Positive feedback loops are by no means uncommon in normal physiological systems, and they are encountered even more commonly in pathophysiology where "vicious cycles" (which are nothing but positive feedback loops) so often help to intensify or perpetuate disease processes. But the popular impression that positive feedback loops are necessarily unstable is quite wrong. Indeed, in discussing the Nyquist criterion we have already established the essential conditions for stability in any single feedback loop, be it positive or negative, namely that the open loop gain must not equal or exceed unity at the critical frequency where the insert signal and the return signal are precisely in phase. With negative feedback, we saw that this critical frequency was the frequency at which the phase lag due solely to the components of the loop (i.e., neglecting the minus sign at the reference input summing point) was π radians. But with positive feedback where no net sign inversion takes place around the loop, the insert signal and the return signal will be exactly in phase with each other when $\omega = 0$, that is when the insert signal is simply a constant. For when $\omega = 0$, the insert signal and the return signal are both real numbers whose phase angle is zero, so that they are necessarily "in phase" with each other. With positive feedback, therefore, if [OLG] (j0), the open-loop gain for a constant signal, equals or exceeds unity, the system will be unstable; we do not have to postulate a periodic input, because any deviation, however small, from a given steady state will be magnified without bound.

With positive feedback systems, the Routh-Hurwitz criterion for stability is just as valid as for negative feedback. In fact, we can now look back at Example 4.9.5 and see that the critical value of k_p for stability which, in that example, was minus 3/8[liters per min]/volt (thus making the system a positive feedback system) was precisely the value at which [OLG] (j0) = 1.

With negative feedback systems, whose business is to minimize the effects of an external disturbance, the open-loop gain and, more particularly, [OLG] (j0), is a sensible measure of regulatory effectiveness. But a positive feedback system is best regarded as a mechanism for magnifying the effect of an external change, and the value of [OLG] (j0) is not a suitable measure of this kind of activity. Let us therefore find an equation, comparable to equation 4.14.8, for the value of

$$[\text{Magnification}] - 1 = \frac{|\tilde{c}_{\text{closed}}(0)|}{|\tilde{c}_{\text{open}}(0)|} - 1 \qquad 4.16.1.$$

where [Magnification] is the factor by which positive feedback increases the open-loop magnitude of the output. For positive feedback, equation 4.14.5 and 4.14.6 become

$$\frac{|\tilde{c}_{\text{open}}(0)|}{|\tilde{c}_{\text{closed}}(0)|} = \left|1 - G_1(0)G_2(0)H(0)\right| = 1 - \left|G_1(0)G_2(0)H(0)\right| = 1 - [\text{OLG}]\,(j0) \qquad 4.16.2.$$

Combining equations 4.16.1 and 4.16.2 by eliminating the ratio $\tilde{c}_{\text{closed}}(0)/\tilde{c}_{\text{open}}(0)$,

$$[\text{Magnification}] - 1 = \frac{[\text{OLG}]\,(j0)}{1 - [\text{OLG}]\,(j0)} \qquad 4.16.3.$$

Hence [OLG](j0)/[1 - [OLG](j0)] measures the effectiveness of a positive feedback loop in the same sense that [OLG](j0) measures the effectiveness of a negative feedback loop.

For the engineer who is accustomed to dealing almost exclusively with error-actuated negative feedback systems, and who is thoroughly familiar with the techniques of frequency-domain analysis, the concept of gain is useful indeed. But for a biologist who must deal with both positive and negative feedback systems as they occur in nature, and who is often chiefly concerned with the steady state, the familiar idea of magnification is much more easily understood and applied. It is therefore unfortunate that the term "gain" has become so firmly entrenched in the vocabulary of control theory. A full discussion of various measures of how feedback influences the steady-state response to an external disturbance must be postponed to Section 7.4 and 5. At present, Table 4.16.1, summarizing the salient features of single-loop feedback systems, may help to display the forest to those readers who feel lost among the trees.

SECTION 4.17. Multiple feedback loops

Feedback systems often comprise several feedback loops. In particular, important physiological variables are usually regulated via multiple feedback pathways. While this tends to ensure that failure of any one feedback loop will not be disastrous, it complicates the job of the physiological analyst.

If two or more loops have only one variable in common, it is usually best to deal with each loop individually as if the others did not exist. This technique will be adopted for several of the systems described in Part II of this monograph.

Table 4.16.1. Fundamentals of single-loop feedback

	KIND OF FEEDBACK	
	NEGATIVE	POSITIVE
What the feedback loop does	Tends to decrease the effect of an external disturbance	Tends to increase the effect of an external disturbance
Magnification by feedback of the effect of an external disturbance	Between zero and plus unity	Between plus unity and plus infinity
How it is identified as positive or negative. (equivalent criteria)	[OLTF] (0) is negative Odd number of broken arrows in symbol-and-arrow diagram	[OLTF] (0) is positive Even number of broken arrows in symbol-and-arrow diagram
Single loop is not stable if	The open loop gain, [OLG] ($j\omega$), is (1.) equal to or greater than unity, when (2.) the insert signal and the return signal are exactly in phase.	
Instability will therefore occur when	[OLG] ($j\omega$) ≥ 1 where ω is the radian frequency for a phase lag of π radians neglecting sign inversion	[OLG] ($j0$) (i.e. the open-loop gain for a constant input) is ≥ 1
Range of [OLG] ($j0$) compatible with stability	Between zero and plus infinity	Between zero and plus unity
N.B. Damping has little to do with stability. A grossly underdamped system can still be stable provided its oscillations die away (however slowly) as $t \to \infty$.		

If two or more loops can be opened simultaneously by breaking the network at some one point, the open-loop gain for a signal injected at that point can be calculated in the usual manner. The value so obtained will represent the joint effectiveness of all the loops.

If two or more loops fall into some other pattern, references more advanced than the present volume should be consulted. However, when we confine attention to the steady state, we shall be able to deal with multiple loops almost as easily as with a single loop. (Section 7.4).

Exercises

1. One of the most important components of an electrical analog computer is the operational amplifier. Its circuit diagram is shown on the right. The voltages at the nodes are measured with respect to ground voltage.

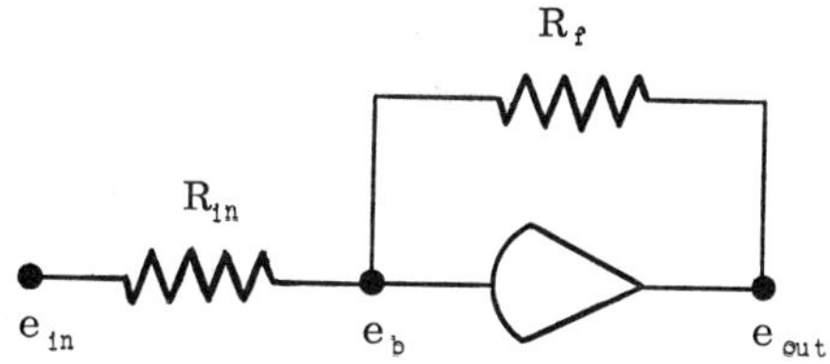

The sector between e_b and e_{out} symbolizes an amplifier whose gain, A, is very large. ($A \approx 10^8$.) The amplifier also causes a phase lag of π radians, so that it reverses the sign of the voltage.

$$e_{out} = -Ae_b$$

The current flow through the amplifier is negligible compared with the flow through the resistors. R_{in}, the input resistance, and R_f, the feedback resistance, do not usually differ from each other by more than a factor of 10. In this system, the input voltage, e_{in}, is the independent variable. It is usually restricted to a range of ± 10 volts.

A. Find an expression for e_b. It is sometimes said that e_b is "at virtual ground voltage." Is this statement justified?

B. With A, R_{in}, and R_f constant, e_{out}, the output, is a constant multiple (or fraction) of e_{in}. Find an expression for e_{out}/e_{in}.

C. Write a set of working equations for this system, and draw a block diagram accordingly.

D. To make an integrator, replace the feedback resistor by a capacitor of capacitance C. Let the input be $e_{in}u(t)$ where e_{in} is a constant voltage. Find an expression for $e_{out}(t)$ for $t \geq 0$. If we want $e_{out}(t)$ to equal $\int_0^t e_{in}u(\tau)\cdot d\tau$, what value must $R_{in}C$ have?

2. In Section 4.3 we found that in the two-tank feedback system with $k_p = 3/4$ [L/min]/volt the steady-state value of E(t) in response to a step change in $e_{ref}(t)$ was three times as large when the loop had been opened by interrupting the feedback pathway as it was in the closed-loop system.

A. Would we have obtained the same result if we had decided to compare the output error in the open-loop system with the output error in the closed-loop system in response to a unit step increase in $e_{ref}(t)$? (Note that by equation 4.3.5, the value of the output, $P_b(t)$, for errorless performance would be 3 cm H_2O.)

B. As a more general answer to the question in part A, consider a simplified version of the feedback system of Figure 4.2.1. Let $G_1(s) = k_1$, $G_2(s) = k_2$, $G_3(s) = k_3$, $H(s) = k_4$, where the k's are simply numerical constants. Let $c_{E=0}(t)$ be the value of c(t) if the system were able to track a change in the set-point of r(t) perfectly, so that $E(t) = 0$ when $r(t) = u(t)$. Find an expression for the ratio of open-loop error and closed-loop error, with the loop opened by letting $k_4 = 0$), when

i) $r(t) = u(t)$, $\ell(t) = 0$, and the error is $E(t)$, the actuating error.
ii) $r(t) = u(t)$, $\ell(t) = 0$, and the error is $c(t) - c_{E=0}(t)$, the output error.
iii) $r(t) = 0$, $\ell(t) = u(t)$, and the error is $E(t)$.
iv) $r(t) = 0$, $\ell(t) = u(t)$, and the error is $c(t) - c_{E=0}(t)$.

C. Repeat part B, but with the loop opened by letting $k_2 = 0$.

3. Consider the two-tank feedback system of Figure 4.2.3 whose block diagram is shown in Figure 4.2.4. Calculate to three significant figures the maximum value of k_p, the constant for a proportional controller, which will permit a non-oscillatory approach to a new steady state in response to a step input. (See the discussion in Section 4.3) You may choose to consider either a step increment in load or a step increment in set point, whichever you prefer.

4. A. Describe briefly how the two-tank feedback system with derivative control would respond to a step increment in the reference standard, $e_{ref}(t)$, with the load input remaining constant. (No calculations are needed!)
B. Use transfer function analysis to find the response of $P_b(t)$ to a unit ramp increase in $e_{ref}(t)$, starting at $t = 0$ and ending at $t = 1.5$ min.

5. Prove that with integral control, steady-state error will be eliminated after a change has been made in the set-point value of the input reference standard.

6. Sketch a root-locus plot in the manner of figure 4.7.2.b for the two-tank negative feedback system with proportional control.

7. Can a second-order negative feedback system whose characteristic equation is of the form

$$as^2 + bs + c = 0 \qquad ;\ a,\ b,\ c,\ \text{positive real numbers}$$

be unstable?

8. Consider the two-tank feedback system with integral control. Using the Routh tabulation as an aid, calculate the critical value of k_i which will just produce instability.

9. At what value of k_i will the two-tank feedback system be critically damped?

10. Write the characteristic equation for the two-tank feedback system with derivative control, leaving k_d unevaluated. Verify the assertion in the text that in this particular negative feedback system, k_d can be made arbitrarily large without producing instability. Can you find a value of k_d which will provide critical damping?

11. Write the transfer function for the following system where x(t) is the input and y(t) is the output.

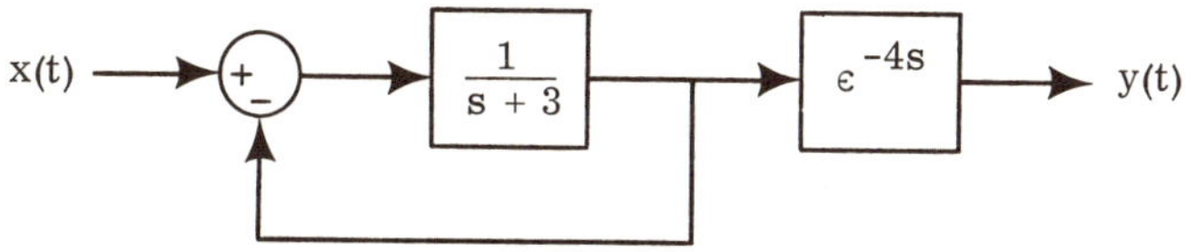

Is the transfer function a ratio of polynomials? What is the characteristic equation for the system? Find y(t) when x(t) = 2 u(t - 5).

12. Show that equation 4.8.3 is valid, provided that the output variable is a member of the feedback loop.

13. Sketch a Nyquist diagram for our two-tank negative feedback system with integral control when $k_i = 0.04125$. Is the system stable? Compare with Exercise 8.

14. If the open-loop transfer function for a linear feedback system comprising only the simple resistive, capacitative, and inductive elements which we have been studying is of nth order with respect to s, what is the greatest phase lag which you could possibly encounter when plotting a Nyquist diagram with ω ranging from 0 to ∞?

15. Occasionally an equilibrium among two or more variables, or even a unidirectional relationship between two variables is mistakenly identified as a feedback system, the "two feedback equations" being, in fact, the same equation (perhaps subtly disguised) solved first for one, then for the other variable.

A. Prove that if we stupidly insist upon treating such a system as a feedback system, the "feedback loop" will always be positive, and the "[OLG](j0)" will always be unity, so the system "cannot be stable!"

B. As a simple example of a feedback loop, Eckman (9) asks us to consider a rectangular tank. The inflow of fluid into the tank is m, and the outflow of fluid from the tank through an outlet valve in the floor of the tank is q. The head of pressure of fluid in the tank is h. Eckman draws a block diagram showing a two-component feedback loop. For one component, m - q is the input and h is the output; for the other, h is the input and q is the output. "The feedback occurs because an increase in outflow feeds back to decrease head, h. At the same time, an increase in inflow will increase the outflow. Thus the relationship of variables may be written as

$$q = f(h, t)$$

$$h = g(m - q, t)"$$

(NOTE that $m - q = C[dh(t)/dt]$, where C is the fluid capacitance of the tank)

i) Can you find explicit expressions for the functions f and g? If so, calculate $[OLG](j0)$. If not, try to find explicit expressions for $q(t) = f(h(t))$ and $h(t) = g(m - q(t))$. Again, calculate $[OLG](j0)$.

ii) Find an expression for h(t) as an explicit function of time, letting R = outflow resistance and C = the fluid capacitance of the tank.

iii) Draw a symbol-and-arrow diagram for the steady-state relationships in this system. Is there a feedback loop?

C. In Section 7.8, we shall consider a positive feedback system in which the product of an enzymatic reaction combines with an inactive enzyme precursor to form the active enzyme responsible for the conversion of substrate to the product. P(t), the concentration of product, and E(t) the concentration of the active enzyme are thus two members of the feedback loop. In the steady state,

$$P(\infty) = \frac{v + k_6 E(\infty)}{[v/P(\infty)] + k_5\{E_{tot} - E(\infty) - [SE](\infty)\}}$$

$$E(\infty) = \frac{v + k_5 P(\infty)\{E_{tot} - [SE](\infty)\}}{k_5 P(\infty) + [v/E(\infty)] + k_6}$$

where $[SE](\infty) = v/k_2$, and $v = \dot{Q}/V$. (For present purposes it is not necessary to define all of these symbols.)

Judging from the above equations, are $P(\infty)$ and $E(\infty)$ members of a feedback loop?

16. Study the first part (pages 1925-1934) of Reference 10 and answer the following questions briefly.

A. p. 1931, 9th from last line: Comment on the term: "the 'complex frequency' operator, s."
B. In equation 3, does the expression on the right really equal the function G(s)? (Compare with equation 4.). If not, what would you write instead of G(s) in equation 3?
C. Consider Figure 8. What does L_{ref} correspond to in the real biological system?
D. What experimental evidence leads the author to assume three "time-lag" elements (p. 1932, top left.)?
E. On p. 1933, middle right, the author states: "If the roots of the characteristic equation do not have a positive real component, then the response of the system to a bounded input will approach zero at some finite time: thus the system is stable." Is this statement correct? Comment briefly.

F. With the mathematical model being used, what will the phase lag approach as $\omega \to \infty$?

G. How does the author justify applying the techniques of linear systems analysis to his experimental data when he knows that the system is nonlinear?

References

1. Fender, D.H.: Techniques of systems-analysis applied to feedback pathways in the control of eye movements. Symp. Soc. exp. Biol., 18, 401-419, 1964.

2. Stark, L.: Neurological Control Systems: Studies in Bioengineering. Plenum Press, N.Y., 1968.

3. International Dictionary of Applied Mathematics. D. van Nostrand Co., Princeton, N.J., 1960.

4. Webster's Third New International Dictionary of the English Language. G & C Merriam Co., Springfield, Mass., 1961.

5. Truxall, J.G.: Automatic Feedback Control System Synthesis. McGraw-Hill, N.Y., 1955.

6. Bower, J.L., and Schultheiss, P.M.: Introduction to the Design of Servomechanisms. John Wiley and Sons, N.Y., 1958.

7. Handbook of Chemistry and Physics.p. A155, 45th edition. Chemical Rubber Publishing Co., Cleveland, 1964.

8. Riggs, D.S.: The Mathematical Approach to Physiological Problems. Williams and Wilkins Co., Baltimore, 1963.

9. Eckman, D.P.: Automatic Process Control. John Wiley and Sons, N.Y., 1958. pp. 9-10.

10. Stark, L.: Stability, oscillations, and noise in the human pupil servomechanism. Proc. Inst. Radio Engrs., 47, 1925-1939, 1959.

In nothing am I chang'd
But in my garments.

Shakespeare, KING LEAR

5

MATRIX METHODS IN THE ANALYSIS OF LINEAR SYSTEMS

SECTION 5.1. The advantages of matrix methods

In the preceding chapters, we have already covered the basic techniques used to analyze the behavior of the relatively simple linear systems which fall within the scope of this book; not much could be added without broadening that scope substantially. The reader is entitled to ask, therefore, why I have chosen to include a chapter on matrix techniques which, as here presented, will carry us no further than we have already gone and will be, in essence, only a redeployment of the same old mathematical troops for a fresh attack upon the same old linear systems. The reasons are: first, that matrix methods are practically indispensible for analyzing systems with multiple inputs and outputs, especially when the systems are of high order; second, that the behavior of a complicated system is often most easily described by a group of simultaneous first-order differential equations for the solution of which matrix methods are ideally suited; third, that when an nth- order differential equation is to be solved by using a digital computer, it must first be reduced to a series of n first-order differential equations of the sort analyzed by matrix techniques; and fourth, that because of its simplicity and generality, vector-matrix notation is now universally employed by control systems engineers when they are studying new problems and devising new techniques. Without some understanding of this notation therefore, one cannot hope to keep up with advances in the rapidly evolving field of feedback regulation.

There is another, purely didactic reason for discussing matrix methods. They offer an excellent opportunity to review many of the fundamental concepts which we have already studied, thereby helping to summarize and consolidate previous material by considering it from a different point of view.*

[*In accordance with the well-known rule for giving a good lecture: "First, tell them what you're going to tell them, then tell them, then tell them what you've told them."]

SECTION 5.2. State variables and phase variables

Consider an unforced system with m time-dependent variables. At any given instant of time each dependent variable will have a particular value, so the state of the system at the given instant is completely specified by the m particular values of the variables at that instant. But in general we do not need all m values in order to determine the state of the system because we may be able to calculate some of them from the remaining ones by using the equations which describe the system. Consider, then, any subset of the m variables so chosen that (1) no member of the subset can be calculated from the other members of the subset, and (2) every variable which is not in the subset can be calculated from members of the subset. Any set of n variables which conforms to these rules may be taken as a set of state variables for the system. The rules obviously ensure that n shall be the smallest number of variables from which it is possible to calculate all others, and hence to define the state of the system at any instant of time. The concept of state variables is very general, being just as valid for non-linear as for linear systems.

Example 5.2.1. Consider an ideal gas in a closed container. We shall regard it as a system consisting of the following m = 7 variables and two constants:

$P(t)$ = pressure exerted by the gas on the walls of the container, (atmospheres)
$V(t)$ = volume of the gas, (liters)
$n(t)$ = number of moles of gas
$T(t)$ = absolute temperature of the gas, (degrees Kelvin)
$C(t)$ = concentration of the gas, (moles per liter)
$\rho(t)$ = density of the gas, (grams per liter)
$v(t)$ = specific volume of the gas, (liters per gram)
M = gram molecular weight of the gas, (grams per mole)
R = ideal gas constant, (liter-atmospheres per mole-degree)

We know that the equation of state for an ideal gas is

$$P(t) = RC(t)T(t) \quad 5.2.1,$$

where, by definition,

$$C(t) = n(t)/V(t) \quad 5.2.2.$$

Also, by definition,

$$\rho(t) = MC(t) \quad 5.2.3,$$

$$v(t) = 1/\rho(t) \quad 5.2.4.$$

Now it is obvious that the value of any one variable in a given equation can be calculated when the values of all of the other variables in the same equation are known. Let us choose a set of 4 variables such that we can calculate the value of a different member of the set from each of the four independent* equations 5.2.1 through 5.2.4. (Such a choice will always be possible with independent equations.) To determine the state of the system at a given instant of time we shall then have to specify the values of the n = 3 remaining state variables. Any three variables from which we can calculate all the others by using the four equations may be selected as state variables. With 7 variables taken 3 at a time, we have altogether $7!/3![7-3]! = 35$

combinations of three different variables. Figure 5.2.1 makes it clear that 16 of these 35 combinations could, and 19 could not serve as state variables.

[*Independent in the sense that none of the equations can be derived from the others.]

In the preceding example, only algebraic equations were used to characterize the system. But exactly the same kind of reasoning is used to define state variables when at least one of the variables in the system is a first derivative (usually with respect to time) and where therefore at least one first-order differential equation is included among the equations describing the behavior of the system. Now to achieve a specific solution of a first-order differential equation, we must know the value of the differentiated variable at some instant of time; hence that variable must be included in our set of state variables. Furthermore, this restriction applies to each first-order differential equation used to describe the system. However, by making appropriate use of the algebraic relationships among the variables we can often replace the differential of one variable by the differential of another, so that we may still have considerable freedom to choose which variables are to be defined as state variables.

Before we can apply this reasoning to systems described by one or more differential equations of order higher than first, we must learn how to break up any n^{th}-order differential equation into n first-order differential equations; for when $n > 1$, what appears to be a single relationship is really a composite of n independent relationships.

Consider the general n^{th}-order homogeneous linear differential equation

$$a_n x_1^{(n)}(t) + a_{n-1} x_1^{(n-1)}(t) + \cdots + a_2 \ddot{x}_1(t) + a_1 \dot{x}_1(t) + a_0 x_1(t) = 0 \qquad 5.2.5,$$

where $x_1(t)$ is one of the variables of the system and $\dot{x}_1(t)$, $\ddot{x}_1(t)$, $\cdots x_1^{(n)}(t)$ symbolize its first, second, $\cdots$, n^{th} derivatives with respect to time. (The a's are coefficients which may or may not vary with time.) We now define a set of new variables each of which is the first time-derivative of the preceding one.

$$x_2(t) = \dot{x}_1(t) \qquad 5.2.6.$$

$$x_3(t) = \dot{x}_2(t) = \ddot{x}_1(t) \qquad 5.2.7.$$

$$\vdots$$

$$x_{n-1}(t) = \dot{x}_{n-2}(t) = x_1^{(n-2)}(t) \qquad 5.2.8.$$

$$x_n(t) = \dot{x}_{n-1}(t) = x_1^{(n-1)}(t) \qquad 5.2.9.$$

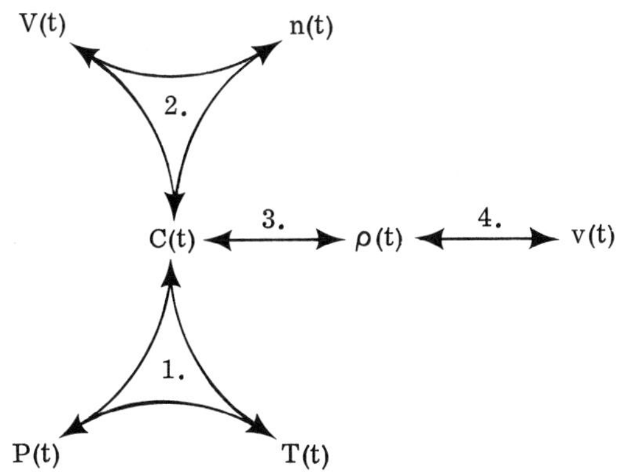

Figure 5.2.1. Identification of state variables. (See Section 5.2).

This is not a symbol-and-arrow diagram showing the working relationships among variables. Rather, the numbered groups of arrows, corresponding to equations 5.2.1, 5.2.2, 5.2.3, and 5.2.4 show how any variable can be calculated from others. In this diagram, if one arrowhead of a numbered group points to a variable, the value of that variable can be calculated from the variables which are pointed to by all other arrowheads in the same numbered group.

This figure makes it clear that to obtain a set of three state variables we must either choose

C(t) or ρ(t) or v(t)	plus	n(t) and P(t) or n(t) and T(t) or V(t) and P(t) or V(t) and T(t)	(Twelve combinations)

or we must choose any three of the following four variables: n(t), V(t), P(t), T(t) (Four combinations). None of the other 19 combinations of three variables, namely

C(t) or ρ(t) or v(t)	plus	n(t) and V(t) or P(t) and T(t)	or	C(t) plus ρ(t) or C(t) plus v(t) or ρ(t) plus v(t)	plus	any of the other five variables
(Six combinations)				(Thirteen different combinations)		

could serve as state variables, since one cannot calculate all remaining variables from such sets of three.

Taking the first derivative of $x_n(t)$ in equation 5.2.9,

$$\dot{x}_n(t) = x_1^{(n)}(t) \qquad 5.2.10.$$

Combining equations 5.2.5 through 5.2.10 to eliminate all derivatives of $x_1(t)$,

$$a_n\dot{x}_n(t) + a_{n-1}x_n(t) + a_{n-2}x_{n-1}(t) + \cdots + a_2x_3(t) + a_1x_2(t) + a_0x_1(t) = 0 \qquad 5.2.11.$$

Now let us solve equations 5.2.6, 5.2.7, 5.2.8, 5.2.9, and 5.2.11 for the first derivatives which they contain.

$$\dot{x}_1(t) = x_2(t) \qquad 5.2.6.a$$

$$\dot{x}_2(t) = x_3(t) \qquad 5.2.7.a$$

$$\vdots$$

$$\dot{x}_{n-2}(t) = x_{n-1}(t) \qquad 5.2.8.a$$

$$\dot{x}_{n-1}(t) = x_n(t) \qquad 5.2.9.a$$

$$\dot{x}_n(t) = -\frac{1}{a_n}\left[a_{n-1}x_n(t) + a_{n-2}x_{n-1}(t) + \cdots + a_2x_3(t) + a_1x_2(t) + a_0x_1(t)\right] \qquad 5.2.11.a.$$

These equations represent a set of n independent first-order differential equations which, taken together, convey the same information as the original n^{th}-order equation, 5.2.5. When the time-dependent behavior of a system can be characterized by such a set of equations in which each of the variables (save the first) is defined as the first derivative of the preceding variable, the variables are commonly called phase variables. It should be clear from the previous discussion that each phase variable must be one of the state variables; so if the first-order differential equations completely describe the system, the set of phase variables is the set of state variables. Indeed we may look upon these first-order equations as a set of n equations in m = 2n variables: the n phase variables and their n first derivatives. To solve the n equations, we must know the values of all n phase variables at some instant, usually time zero when they are the initial conditions. Of course if the system is forced, we must also know the forcing functions.

The reader may well feel that however useful it may prove to be for ease of mathematical manipulation, defining a series of phase variables is a pretty artificial business. In fact, however, it often turns out to be the most natural and informative way to approach a problem, as the following example will show:

Example 5.2.2. Consider the simple pendulum shown in Figure 5.2.2.a. Any torque, τ, tending to move the mass m around its circular path of radius ℓ will either store energy in the mass by virtue of Ω, the angular velocity which

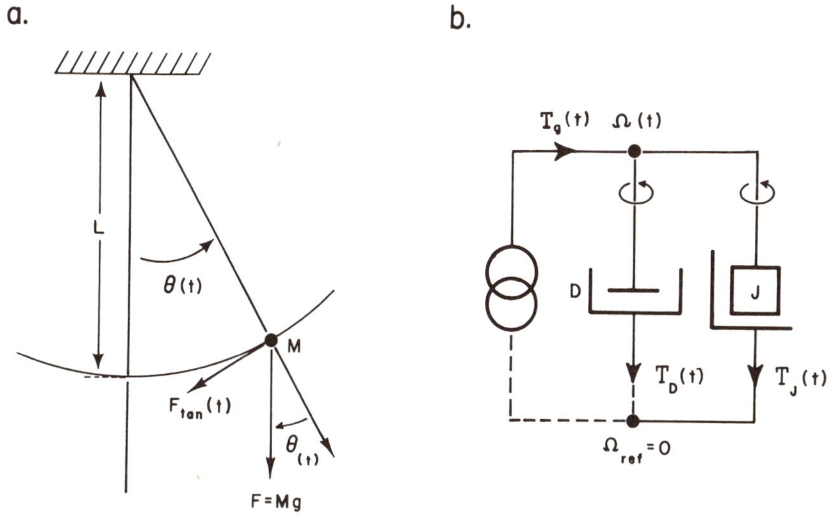

Figure 5.2.2. The pendulum system of Example 5.2.2.

Figure 5.2.2.a. A simple pendulum of point mass, m, swings back and forth in accordance with the tangential component of the gravitational force on m. But because of damping due to air resistance etc., the magnitude of $\theta(t)$, the angular displacement, gradually decreases with time, approaching zero asymptotically.

Figure 5.2.2.b. The circuit diagram for the pendulum. Notice that this system is nonlinear because the torque, $\tau_g(t) = -\ell mg \sin\theta(t)$, provided by the T-variable source is itself a function of $\theta(t)$, the angular displacement.

---

is the across variable for this system, or will dissipate energy as heat (damping). We can therefore represent the system by the circuit diagram of Figure 5.2.2.b. The torque, τ_g, due to gravity is equal to F_{tan}, (the tangential component of the force of gravity acting on the displaced mass, m) multiplied by ℓ, the length of the pendulum.

$$\tau_g = \ell\ F_{tan} = -\ell\ mg \sin\theta(t) \qquad 5.2.12,$$

where g is the acceleration due to gravity. (The torque is negative when θ, the angular displacement, is positive, because the gravitational force then tends to decrease θ.) The node equation is

$$m\ell^2\ \ddot{\theta}(t) + D\dot{\theta}(t) = -\ell mg \sin\theta(t) \qquad 5.2.13,$$

which is a nonlinear equation because the forcing function is a trigonometric function of $\theta(t)$ the unknown variable, and not just a function of time. However for <u>small</u> angular displacements (e.g. $-0.1 \le \theta \le +0.1$ radians) we shall make only a small error by using the approximation,

$$\sin \theta(t) \approx \theta(t) \qquad 5.2.14.$$

Combining equations 5.2.13 and 5.2.14, and re-arranging,

$$m\ell^2 \ddot{\theta}(t) + D\dot{\theta}(t) + m\ell g\theta(t) = 0 \qquad 5.2.15,$$

which is a second-order, linear, homogeneous equation.*

[*Fortuitously homogeneous because the forcing function merely happens to be of the form which another <u>passive</u> element, namely a torsional spring, would introduce!]

In accordance with the previous discussion, we now define the <u>phase variables</u> as follows:

$$x_1(t) = \theta(t) \qquad 5.2.16,$$

$$x_2(t) = \dot{x}_1(t) = \dot{\theta}(t) = \Omega(t) \qquad 5.2.17,$$

so that in this system the phase variables are the angular displacement and the angular velocity of the pendulum. Combining equations 5.2.15, 5.2.16, and 5.2.17, we obtain

$$m\ell^2 \dot{x}_2(t) + Dx_2(t) + m\ell gx_1(t) = 0 \qquad 5.2.18.$$

Solving equations 5.2.17 and 5.2.18 for the first derivatives which they contain,

$$\dot{x}_1(t) = x_2(t) \qquad 5.2.17.a,$$

$$\dot{x}_2(t) = -[D/m\ell^2]\, x_2(t) - [g/\ell]\, x_1(t) \qquad 5.2.18.a.$$

We shall see presently how these two simultaneous first-order differential equations can be solved using matrix methods. For the present, let $\ell = 98$ cm (roughly the length of the pendulum on a grandfather's or perhaps by this time great-grandfather's clock.) Let $M = 1041$ grams, $g = 980$ cm/sec^2, and $D = 2 \times 10^6$ dyne·cm·sec. With these values, equation 5.2.15 is

$$10^7 \ddot{\theta}(t) + 2\cdot 10^6 \dot{\theta}(t) + 10^8 \theta(t) = 0 \ \text{dyne}\cdot\text{cm} \qquad 5.2.19,$$

or

$$\ddot{\theta}(t) + 2\cdot 10^{-1} \dot{\theta}(t) + 10\, \theta(t) = 0 \qquad \sec^{-2} \qquad 5.2.19.a.$$

Let the initial conditions be $\theta = 0.1$ radian, $\dot{\theta} = 0$ rad/sec. Taking Laplace transforms of equation 5.2.19.a,

$$s^2\theta_{\mathcal{L}}(s) - s[0.1] - 0 + 2\cdot 10^{-1}[s\theta_{\mathcal{L}}(s) - 0.1] + 10\theta_{\mathcal{L}}(s) = 0 \ \ \text{sec}^{-1} \qquad 5.2.20.$$

Solving equation 5.2.20 for $\theta_{\mathcal{L}}(s)$,

$$\theta_{\mathcal{L}}(s) = \frac{0.1s + 0.02}{s^2 + 2\cdot 10^{-1}\,s + 10} \qquad \text{sec} \qquad 5.2.20.a.$$

Taking inverse transforms,

$$\theta(t) = x_1(t) = \epsilon^{-0.1t}[0.1\cos\omega t + \frac{10^{-2}}{\omega}\sin\omega t] \ \ \text{rad} \qquad 5.2.21.$$

where $\omega = \sqrt{9.99} \approx \sqrt{10}$ rad/sec.

Differentiating equation 5.2.21,

$$\dot{\theta}(t) = \dot{x}_1(t) = x_2(t) = -\epsilon^{-0.1t}[\frac{0.001}{\omega} + 0.1\,\omega]\sin\omega t \ \ \text{rad/sec} \qquad 5.2.22.$$

But with $\omega \approx \sqrt{10}$ rad/sec, we see that the first term within brackets in equation 5.2.22 is negligibly small compared with the second term. Furthermore, since $\omega \approx \sqrt{10}$ rad/sec, the second term may be approximated by $1/\omega$, and equation 5.2.22 may be simplified to

$$\dot{\theta}(t) = x_2(t) = -\frac{1}{\omega}\epsilon^{-0.1t}\sin\omega t \ \ \text{rad/sec} \qquad 5.2.23.$$

Differentiating equation 5.2.23,

$$\ddot{\theta}(t) = \dot{x}_2(t) = \epsilon^{-0.1t}[\frac{0.1}{\omega}\sin\omega t - \cos\omega t] \ \ \text{rad/sec}^2 \qquad 5.2.24.$$

Far from being artificial, in this example the two phase variables, plus the first derivative of $x_2(t)$, are precisely the quantities we might be interested in at any time, namely, the angular displacement, the angular velocity and the angular acceleration of the pendulum. Furthermore, it is often instructive to plot the phase variables, $x_1(t)$, $x_2(t)$, or their first derivatives, $\dot{x}_1(t)$ $(= x_2(t))$ and $\dot{x}_2(t)$ against each other at various instants of time (Figure 5.2.3.a and 5.2.3.b). If the system is stable, the curve (= trajectory) obtained by plotting the first derivatives must approach the origin as $t \to \infty$, for in a stable unforced system, whatever the initial conditions (i.e., wherever the curve starts in the plane) the system must eventually approach a state of complete quiescence where all time derivatives are zero. In the present example, this happens to

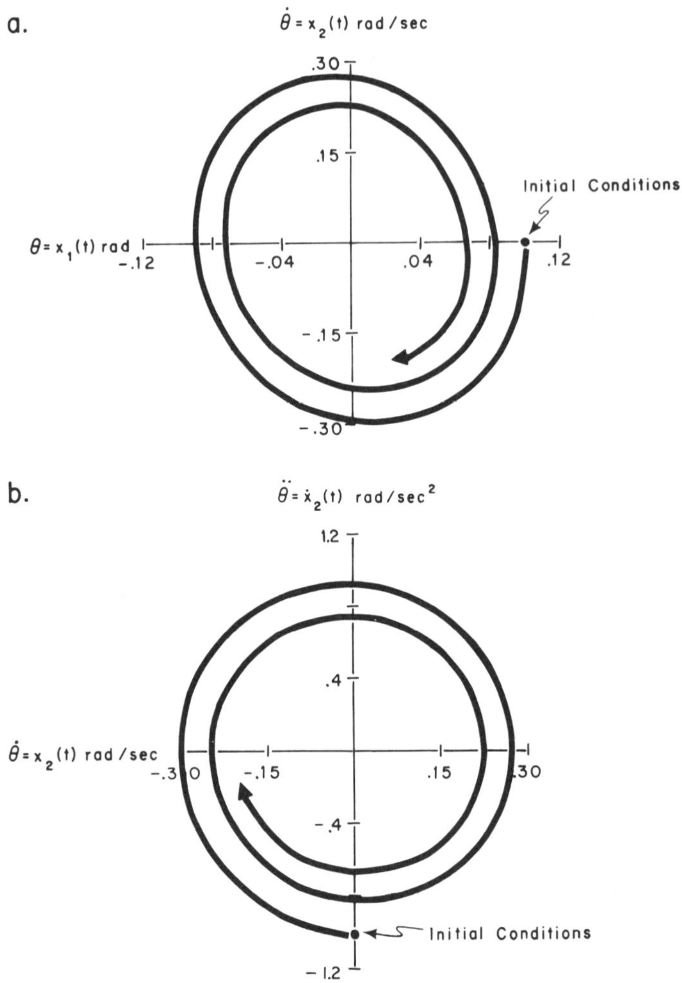

Figure 5.2.3. Trajectories traced out in phase space by the phase variables of the pendulum system (Example 5.2.2).

Figure 5.2.3.a. The trajectory of $\theta(t)$ and $\dot{\theta}(t)$ as time increases from zero at the initial-conditions starting point. In this damped system, both $\theta(t)$ and $\dot{\theta}(t)$ approach zero as $t \to \infty$, so that the trajectory spirals inward towards the origin. Notice that the magnitude of the angular velocity reaches its maximum slightly before the pendulum attains the vertical position where $\theta(t) = 0$ and the torque is zero.

Figure 5.2.3.b. A similar plot showing the trajectory of $\dot{\theta}(t)$ and $\ddot{\theta}(t)$ as time increases. Unlike the trajectory of Figure 5.2.3.a, this trajectory is not tilted with respect to the axes of coordinates because $|\dot{\theta}|$ must always be maximum when $|\ddot{\theta}|$ is zero, and $|\ddot{\theta}|$ must always be maximum when $|\dot{\theta}|$ is zero.

be true not only for the angular velocity and the angular acceleration but also for the angle itself.

Unfortunately, it is not practical to try to plot three, and impossible to plot four or more such variables against each other using Cartesian coordinates. However, it is of course quite legitimate to talk about a phase space, or, more generally, a state space of more than three dimensions, and this is frequently done. Furthermore, there is no reason why time cannot be added explicitly as another coordinate, and the resulting system of coordinates may then be called the solution space. Certain very general methods for investigating the stability of non-linear as well as linear systems (e.g. Liapunov's second method, Section 6.19) are based upon studying the trajectory traced by the state variable vector in solution space as time increases. These methods lie beyond (but not far beyond!) the scope of this book. For further discussion, see Rosen (1).

SECTION 5.3. Matrix techniques. General remarks

I shall assume that the reader is already familiar with certain elementary rules for the algebraic manipulation of vectors and matrices. (See Appendix). In this book, vectors are denoted by underlined small letters, matrices by underlined capitals. Vectors are defined as column vectors.

Consider any unforced linear system which requires at least one non-algebraic equation to describe its behavior. In preparation for matrix analysis, we convert any integral or integro-differential equations to differential form. By defining phase variables, we change any higher-order differential equations into a series of first-order equations. As state variables, we choose any appropriate set for which differential equations can be written. Finally, we eliminate from the differential equations all the variables which are not state variables by combining the algebraic equations with each other and with the differential equations. We thus obtain a series of n first-order differential equations in the n state variables and their first derivatives. In addition, we may have some purely algebraic equations showing how variables not included in the differential equations can be calculated from those which are included. Such algebraic relationships pose no analytical problems. For simplicity, therefore, in the following discussion we shall assume that the system is fully described by the n differential equations. Each such equation gives the rate of change, $\dot{x}_i(t) = dx_i(t)/dt$ of one of the n state variables as a function of the state variables themselves and the parameters of the system. This set of n equations may be written as

$$\left.\begin{aligned} \dot{x}_1(t) &= a_{11}x_1(t) + a_{12}x_2(t) + \cdots + a_{1n}x_n(t) \\ \dot{x}_2(t) &= a_{21}x_1(t) + a_{22}x_2(t) + \cdots + a_{2n}x_n(t) \\ &\vdots \\ \dot{x}_n(t) &= a_{n1}x_1(t) + a_{n2}x_2(t) + \cdots + a_{nn}x_n(t) \end{aligned}\right\} \quad 5.3.1,$$

or, equivalently, in matrix form, as

$$\underline{\dot{x}}(t) = \underline{A}\,\underline{x}(t) \qquad 5.3.2,$$

where

$$\underline{x}(t) \equiv \begin{bmatrix} x_1(t) \\ x_2(t) \\ \cdot \\ \cdot \\ \cdot \\ x_n(t) \end{bmatrix} ; \quad \underline{\dot{x}}(t) \equiv \begin{bmatrix} \dot{x}_1(t) \\ \dot{x}_2(t) \\ \cdot \\ \cdot \\ \cdot \\ \dot{x}_n(t) \end{bmatrix} ; \quad \text{and}$$

$$\underline{A} \equiv \begin{bmatrix} a_{11} & a_{12} \cdots & a_{1n} \\ a_{21} & a_{22} \cdots & a_{2n} \\ \cdot & \cdot & \cdot \\ \cdot & \cdot & \cdot \\ \cdot & \cdot & \cdot \\ a_{n1} & a_{n2} \cdots & a_{nn} \end{bmatrix}$$

Notice that since we always have n differential equations in n state variables, $\underline{A}$ must be an n by n square matrix. Notice also that many of the a_{ij}'s are likely to be zero, especially if we work with phase variables. So much the better for ease of computation!

Example 5.3.1. Using the numerical values, $m = 1041$ gram, $\ell = 98$ cm, $D = 2 \cdot 10^6$ dyne·cm·sec and $g = 980$ cm/sec^2, write equations 5.2.17.a and 5.2.18.a in matrix form. We have

$$\left.\begin{aligned} \dot{x}_1(t) &= 0x_1(t) + x_2(t) \\ \dot{x}_2(t) &= -10x_1(t) - 0.2x_2(t) \end{aligned}\right\} \qquad 5.3.3,$$

whence

$$\underbrace{\begin{bmatrix} \dot{x}_1(t) \\ \dot{x}_2(t) \end{bmatrix}}_{\underline{\dot{x}}(t)} = \underbrace{\begin{bmatrix} 0 & 1 \\ -10 & -0.2 \end{bmatrix}}_{\underline{A}} \underbrace{\begin{bmatrix} x_1(t) \\ x_2(t) \end{bmatrix}}_{\underline{x}(t)} \qquad 5.3.4,\ 5.3.2.$$

We have dealt thus far only with unforced systems. If the system is forced, any or all of the original equations describing its behavior may have a nonhomogeneous term which specifies an input forcing function. With n state variables, we might have as many as n different forcing functions, each of which might affect every one of the state variables. For a general forced system, therefore, we must add to equation 5.3.2 a second term to represent the effect of the various possible inputs upon $\underline{x}(t)$;

$$\dot{\underline{x}}(t) = \underline{A}\,\underline{x}(t) + \underline{B}\underline{f}(t) \qquad 5.3.5,$$

where $\underline{B}$ is an n by m matrix, and $\underline{f}(t)$ is a vector with m elements which are the various forcing inputs.

Example 5.3.2. Suppose that the pendulum described by equation 5.2.15 is subjected to a small* (non-gravitational) torque, $\tau(t)$.

Equation 5.2.15 will then become

$$m\ell^2\,\ddot{\theta}(t) + D\,\dot{\theta}(t) + m\ell g\,\theta(t) = \tau(t) \qquad 5.3.6.$$

As before, we shall define $x_1(t) = \theta(t)$, $x_2(t) = \dot{x}_1(t) = \dot{\theta}(t)$. This gives us the two equations

$$\left.\begin{aligned} \dot{x}_1(t) &= x_2(t) \\ \dot{x}_2(t) &= -\frac{g}{\ell}x_1(t) - \frac{D}{m\ell^2}x_2(t) + \frac{\tau(t)}{m\ell^2} \end{aligned}\right\} \qquad 5.3.7,$$

which can be written in matrix form as

$$\underbrace{\begin{bmatrix} \dot{x}_1(t) \\ \dot{x}_2(t) \end{bmatrix}}_{\dot{\underline{x}}(t)} = \underbrace{\begin{bmatrix} 0 & 1 \\ -\frac{g}{\ell} & -\frac{D}{m\ell^2} \end{bmatrix}}_{\underline{A}} \underbrace{\begin{bmatrix} x_1(t) \\ x_2(t) \end{bmatrix}}_{\underline{x}(t)} + \underbrace{\begin{bmatrix} 0 \\ \frac{1}{m\ell^2} \end{bmatrix}}_{\underline{B}} \underbrace{[\tau(t)]}_{\underline{f}(t)} \qquad 5.3.8,\ 5.3.5.$$

[*Small because it must not force the pendulum to exceed the very limited range of angular displacements for which equation 5.2.15 is reasonably valid.]

Before proceeding further, it is worth remarking upon the great generality of equation 5.3.5. $\dot{\underline{x}}(t)$ represents any number of first derivatives; $\underline{A}$ and $\underline{B}$ together comprise all of the parameters of the system; $\underline{f}(t)$ represents all inputs to the system; $\underline{x}(t)$ represents all of the outputs chosen as state variables. Furthermore, equation 5.3.5 is general enough to cover any linear system, forced or unforced, with or without initial conditions, with many feedback loops, or with none. A method for solving equation 5.3.5 will therefore really be a method for solving all of the linear systems previously discussed. Yet it should equally be emphasized that equation 5.3.5 is no more than a shorthand way of writing a number of simultaneous first-order differential equations. These could be solved without recourse to matrix methods, for example by taking Laplace transforms of each, solving the resulting algebraic equations simultaneously, and taking inverse transforms. Using matrix methods we do no more than this, but we do it in the most orderly and efficient way.

In the following sections a general method of solving equation 5.3.5 using Laplace transforms will be presented and illustrated by several examples. The relation of matrix notation to transfer function notation will then be discussed. Finally, a biological system of some complexity will be analyzed using the matrix technique.

SECTION 5.4. Matrix techniques. Laplace transform solution

We consider the solution of

$$\underline{\dot{x}}(t) = \underline{A}\,\underline{x}(t) + \underline{B}\,\underline{f}(t) \qquad 5.4.1,$$

with $\underline{x}(0)$, the initial conditions, known. Bearing in mind that equation 5.4.1 simply represents a set of n simultaneous equations, the reader should have no difficulty in taking Laplace transforms on both sides of equation 5.4.1,

$$s\underline{x}_{\mathcal{L}}(s) - \underline{x}(0) = \underline{A}\,\underline{x}_{\mathcal{L}}(s) + \underline{Bf}_{\mathcal{L}}(s) \qquad 5.4.2,$$

or

$$s\underline{x}_{\mathcal{L}}(s) - \underline{A}\,\underline{x}_{\mathcal{L}}(s) = \underline{x}(0) + \underline{Bf}_{\mathcal{L}}(s) \qquad 5.4.2.a,$$

or

$$[s\underline{I} - \underline{A}]\,\underline{x}_{\mathcal{L}}(s) = \underline{x}(0) + \underline{Bf}_{\mathcal{L}}(s) \qquad 5.4.2.b.$$

where $\underline{I}$ is the n by n identity matrix. Next, we assume that $[s\underline{I} - \underline{A}]$ is non-singular, so that there exists the inverse matrix, $[s\underline{I} - \underline{A}]^{-1}$, such that $[s\underline{I} - \underline{A}]^{-1}[s\underline{I} - \underline{A}] = [s\underline{I} - \underline{A}][s\underline{I} - \underline{A}]^{-1} = \underline{I}$. Premultiply both sides of equation 5.4.2.b by $[s\underline{I} - \underline{A}]^{-1}$.

$$\underline{x}_{\mathcal{L}}(s) = [s\underline{I} - \underline{A}]^{-1}\,\underline{x}(0) + [s\underline{I} - \underline{A}]^{-1}\underline{Bf}_{\mathcal{L}}(s)$$

$$= \underline{\Phi}_{\mathcal{L}}(s)\,\underline{x}(0) + \underline{\Phi}_{\mathcal{L}}(s)\underline{Bf}_{\mathcal{L}}(s) \qquad 5.4.3,$$

where,

$$\underline{\Phi}_{\mathcal{L}}(s) \equiv [s\underline{I} - \underline{A}]^{-1} \qquad 5.4.4.$$

$\underline{\Phi}(t)$, the inverse transform of $\underline{\Phi}_{\mathcal{L}}(s)$, is called the fundamental matrix of the system. Taking inverse transforms of equation 5.4.3,* we get

$$\underline{x}(t) = \mathcal{L}^{-1}\left\{\underline{\Phi}_{\mathcal{L}}(s)\,[\underline{x}(0) + \underline{Bf}_{\mathcal{L}}(s)]\right\} \qquad 5.4.5,$$

which is the desired solution. In calculating the inverse transform, the vector on the right of equation 5.4.5 is dealt with element by element, so that in this final step we are really finding the solutions for $x_1(t)$, $x_2(t) \cdots x_n(t)$ separately.

[*If we take inverse transforms of equation 5.4.3 using the convolution theorem for the product on the right, we get

$$\underline{x}(t) = \underline{\Phi}(t)\,\underline{x}(0) + \int_0^t \underline{\Phi}(t-\tau)\,\underline{B}\,\underline{f}(\tau)d\tau$$

Now the general solution of a single first-order differential equation with constant coefficients, $[dx(t)/dt] = \lambda x(t) + f(t)$, can be written as

$$x(t) = \epsilon^{\lambda t}x(0) + \int_0^t \epsilon^{\lambda(t-\tau)}f(\tau)\,d\tau$$

By comparing these two equations we find that $\underline{\Phi}(t)$ in the first plays the same role as $\epsilon^{\lambda t}$ in the second. This is no idle analogy, for the fundamental matrix, $\underline{\Phi}(t)$ can actually be defined as $\epsilon^{\underline{A}t}$, where $\underline{A}$ is the same matrix as in equation 5.4.1. $\epsilon^{\underline{A}t}$ is a matrix exponential with properties quite analogous to those of ordinary exponentials. By using these properties, it is possible to carry out the entire matrix analysis in the time domain. (See, for example, reference 2.) Mathematically, such an analysis is interesting and instructive, but the manipulations required, particularly the element-by-element integrations of the convolution integral, $\int_0^t \underline{\Phi}(t-\tau)\,f(\tau)\,d\tau$, are likely to be more tedious than the corresponding Laplace transform maneuvers. This is the primary reason for not discussing the pure time-domain analysis here.]

Example 5.4.1. As an easy introductory example, consider the unforced series RLC circuit of Figure 5.4.1. The switch is closed at t = 0. Thereafter, by Kirchoff's voltage law, the sum of the voltage drops around the circuit must be zero,

$$e_L(t) + e_R(t) + e_C(t) = 2\frac{di(t)}{dt} + 20i(t) + 32\int i(t)dt = 0 \text{ volt} \qquad 5.4.6.$$

Differentiating equation 5.4.6 with respect to time, and multiplying the result by 1/2,

$$\frac{d^2i(t)}{dt^2} + 10\frac{di(t)}{dt} + 16\,i(t) = 0 \qquad 5.4.7.$$

At $t = 0^-$, $i(0^-) = 0$. Since current through an inductor cannot change instantaneously, $i(0^+) = i(0^-) = 0$. With zero current $e_R(0^+) = 0$. Hence, at $t = 0^+$, equation 5.4.6 gives us,

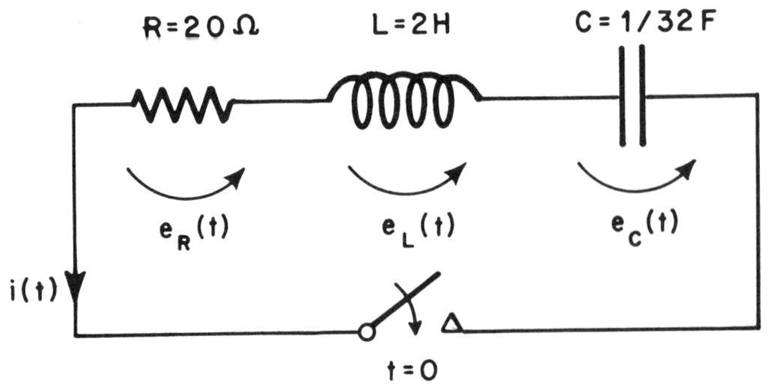

$e_c(0) = -30$ volt

Figure 5.4.1. The series RLC circuit of Example 5.4.1.

$$e_L(0^+) = 2\left.\frac{di(t)}{dt}\right|_{t=0^+} = -\ e_c(0^+) = -\ (-30) = 30 \text{ volt} \qquad 5.4.8,$$

or

$$\left.\frac{di(t)}{dt}\right|_{t \doteq 0^+} = 15 \text{ amp/sec} \qquad 5.4.8.a.$$

We define phase variables as

$$x_1(t) = i(t) \qquad 5.4.9,$$

$$x_2(t) = \dot{x}_1(t) \qquad 5.4.10.$$

Substituting these in equation 5.4.7,

$$\dot{x}_2(t) + 10\,x_2(t) + 16\,x_1(t) = 0 \qquad 5.4.11.$$

Solving equations 5.4.10 and 5.4.11 for their first derivatives,

$$\dot{x}_1(t) = x_2(t) \qquad 5.4.10.a,$$

$$\dot{x}_2(t) = -16\,x_1(t) - 10\,x_2(t) \qquad 5.4.11.a.$$

Writing equations 5.4.10.a and 5.4.11.a in matrix form,

$$\underbrace{\begin{bmatrix} \dot{x}_1(t) \\ \dot{x}_2(t) \end{bmatrix}}_{\dot{\underline{x}}(t)} = \underbrace{\begin{bmatrix} 0 & 1 \\ -16 & -10 \end{bmatrix}}_{\underline{A}} \underbrace{\begin{bmatrix} x_1(t) \\ x_2(t) \end{bmatrix}}_{\underline{x}(t)} \qquad 5.4.12.$$

Subject to the initial conditions $x_1(0^+) = 0$ amp, $x_2(0^+) = 15$ amp/sec.

$$[s\underline{I} - \underline{A}] = s\begin{bmatrix} 1 & 0 \\ 0 & 1 \end{bmatrix} - \begin{bmatrix} 0 & 1 \\ -16 & -10 \end{bmatrix} = \begin{bmatrix} s & -1 \\ 16 & s+10 \end{bmatrix} \qquad 5.4.13.$$

$$[s\underline{I} - \underline{A}]^{-1} = \underline{\Phi}_{\mathcal{L}}(s) = \frac{1}{|s\underline{I} - \underline{A}|}\begin{bmatrix} s+10 & 1 \\ -16 & s \end{bmatrix} \qquad 5.4.14,$$

where $|s\underline{I} - \underline{A}|$ = the determinant of the matrix $[s\underline{I} - \underline{A}]$.

$$|s\underline{I} - \underline{A}| = s^2 + 10s + 16 = [s+2][s+8] \qquad 5.4.15.$$

Check,

$$[s\underline{I} - \underline{A}][s\underline{I} - \underline{A}]^{-1} = \begin{bmatrix} s & -1 \\ 16 & s+10 \end{bmatrix} \frac{1}{s^2 + 10s + 16}\begin{bmatrix} s+10 & 1 \\ -16 & s \end{bmatrix} = \begin{bmatrix} 1 & 0 \\ 0 & 1 \end{bmatrix} = \underline{I}$$

Using equation 5.4.3,

$$\underline{x}_{\mathcal{L}}(s) = \underline{\Phi}_{\mathcal{L}}(s)\,\underline{x}(0) =$$

$$\frac{1}{[s+2][s+8]}\begin{bmatrix} s+10 & 1 \\ -16 & s \end{bmatrix}\begin{bmatrix} 0 \\ 15 \end{bmatrix} = \begin{bmatrix} \dfrac{15}{[s+2][s+8]} \\ \dfrac{15s}{[s+2][s+8]} \end{bmatrix} \qquad 5.4.16.$$

Taking inverse Laplace transforms after breaking both elements of the $\underline{x}_{\mathcal{L}}(s)$ vector in equation 5.4.16 into partial fractions,

$$\underline{x}(t) = \begin{bmatrix} x_1(t) = i(t) = 2.5\epsilon^{-2t} - 2.5\epsilon^{-8t} & \text{amp} \\ x_2(t) = \dfrac{di(t)}{dt} = -5\epsilon^{-2t} + 20\epsilon^{-8t} & \text{amp/sec} \end{bmatrix} \qquad 5.4.17.$$

Notice that by setting the polynomial in s of equation 5.4.15 equal to zero, we obtain the characteristic equation for the system. This will always be true. $|s\underline{I} - \underline{A}| = 0$ is the characteristic equation* whose roots are the eigenvalues which become the coefficients of time in the exponents of ϵ.

[*When the analysis is carried out entirely in time domain, the characteristic equation is commonly expressed as $|\underline{A} - \lambda\underline{I}| = 0$ whose roots, $\lambda_1, \lambda_2 \cdots \lambda_n$, are the eigenvalues. But this equation is precisely equivalent to $|s\underline{I} - \underline{A}| = 0$; for when we confine attention to the roots of the characteristic equation, $s = \lambda$. After replacing s by λ, the second expression can be changed into the first by multiplying it by -1.]

Example 5.4.2. Solve the pendulum problem (equation 5.3.8) using the values $m = 1041$ grams, $\ell = 98$ cm, $D = 2 \cdot 10^6$ dyne•cm•sec and $g = 980$ cm/sec^2 as before, with $\tau = -4 \cdot 10^6$ dyne•cm. (i.e., τ is a torque tending to rotate the pendulum clockwise.) Let the initial conditions be $x_1(t) = \theta(t) = 0.1$ radian and $x_2(t) = \dot{x}_1(t) = 0$ radian/sec. We then have

$$\begin{bmatrix} \dot{x}_1(t) \\ \dot{x}_2(t) \end{bmatrix} = \begin{bmatrix} 0 & 1 \\ -10 & -0.2 \end{bmatrix} \begin{bmatrix} x_1(t) \\ x_2(t) \end{bmatrix} + \begin{bmatrix} 0 \\ 10^{-7} \end{bmatrix} [-4 \cdot 10^6] \qquad 5.4.18,$$

$$\text{with } \underline{x}(0) = \begin{bmatrix} 0.1 \\ 0 \end{bmatrix}$$

$$[s\underline{I} - \underline{A}] = \begin{bmatrix} s & -1 \\ 10 & s + 0.2 \end{bmatrix} \qquad 5.4.19.$$

$$[s\underline{I} - \underline{A}]^{-1} = \underline{\Phi}_{\mathcal{L}}(s) = \frac{1}{s^2 + 0.2s + 10} \begin{bmatrix} s + 0.2 & 1 \\ -10 & s \end{bmatrix} \qquad 5.4.20.$$

$$\underline{x}(0) + \underline{B}\underline{f}_{\mathcal{L}}(s) = \begin{bmatrix} 0.1 \\ 0 \end{bmatrix} + \begin{bmatrix} 0 \\ -0.4/s \end{bmatrix} = \begin{bmatrix} 0.1 \\ -0.4/s \end{bmatrix} \qquad 5.4.21.$$

$$\underline{x}_{\mathcal{L}}(s) = \frac{1}{s^2 + 0.2s + 10} \begin{bmatrix} s + 0.2 & 1 \\ -10 & s \end{bmatrix} \begin{bmatrix} 0.1 \\ -0.4/s \end{bmatrix} = \begin{bmatrix} \dfrac{0.1s^2 + 0.02s - 0.4}{s[s^2 + 0.2s + 10]} \\ \dfrac{-1.4}{s^2 + 0.2s + 10} \end{bmatrix} \qquad 5.4.22.$$

$$x_1(t) = \theta(t) = \mathcal{L}^{-1}\left\{\frac{0.1s^2 + 0.02s - 0.4}{s[s^2 + 0.2s + 10]}\right\} = \mathcal{L}^{-1}\left\{\frac{-0.04}{s} + \frac{0.14s + 0.028}{s^2 + 0.2s + 10}\right\}$$

$$= -0.04 + 0.14\epsilon^{-0.1t}\cos\omega t + \frac{0.014}{\omega}\epsilon^{-0.1t}\sin\omega t \text{ rad} \qquad 5.4.23.$$

$$x_2(t) = \dot{\theta}(t) = \mathcal{L}^{-1}\left\{\frac{-1.4}{s^2 + 0.2s + 10}\right\} = -\frac{1.4}{\omega}\epsilon^{-0.1t}\sin\omega t \text{ rad/sec} \qquad 5.4.24,$$

where $\omega = \sqrt{9.99}$ rad/sec.

(Check: equation 5.4.24 can be obtained directly from equation 5.4.23 by differentiation.)

Notice once more that the determinant $|s\underline{I} - \underline{A}|$ (the denominator in equation 5.4.20) is the characteristic polynomial of the system.

Example 5.4.3. Consider the two-tank feedback system of Section 4.2 with proportional control. Let the initial state of the system be as specified by the values in Figure 4.2.3. At $t = 0$, $e_{ref}(t)$ is increased from 128/3 volt to 144/3 volt, and $\dot{V}_\ell(t)$ is decreased from -14 liter/min to -20 liter/min. Find $P_a(t)$ and $\dot{V}_{b,\,out}(t)$.

The basic equation describing the behavior of each tank is equation 3.13.2. Applying this equation to each tank separately, we have

$$\frac{dP_a(t)}{dt} + \frac{1}{R_a C_a} P_a(t) = \frac{1}{C_a}\dot{V}_{in,\,a}(t) \qquad 5.4.25,$$

$$\frac{dP_b(t)}{dt} + \frac{1}{R_b C_b} P_b(t) = \frac{1}{C_b}\dot{V}_{in,\,b}(t) \qquad 5.4.26.$$

In addition, the system is characterized by the following algebraic equations (Cf. Figure 4.15.2).

$$E(t) = e_{ref}(t) - e_p(t) \qquad 5.4.27.$$

$$\dot{V}_{in,\,a}(t) = k_p E(t) \qquad 5.4.28.$$

$$\dot{V}_{a,\,out}(t) = P_a(t)/R_a \qquad 5.4.29.$$

$$\dot{V}_{in,\,b}(t) = \dot{V}_{a,\,out}(t) + \dot{V}_\ell(t) \qquad 5.4.30.$$

$$e_p(t) = k_{trans} P_b(t) \qquad 5.4.31.$$

$$\dot{V}_{b,\,out}(t) = P_b(t)/R_b \qquad 5.4.32.$$

This is a system of 8 equations in 10 variables, not counting the input forcing functions, $e_{ref}(t)$ and $\dot{V}_{\ell}(t)$. As state variables, we must therefore select two of the ten which cannot be calculated from each other. Their values, the forcing functions, and the values of the parameters (see Table 5.4.1) must be known at some instant if we are to determine the state of the system at all times. Since we are asked to solve for $P_a(t)$ and $\dot{V}_{b,\,out}(t)$, let us choose them as state variables. (They are a possible choice because neither of them can be calculated from the other). Accordingly, we define

$$x_1(t) \equiv P_a(t) \qquad 5.4.33,$$

$$x_2(t) \equiv \dot{V}_{b,\,out}(t) \qquad 5.4.34.$$

(Notice that with this choice of state variables, phase variables cannot be defined. Suppose we tried to define $x_1(t) = P_a(t)$ and $x_2(t) = \dot{x}_1(t) = \dot{P}_a(t)$ as phase variables. Then we would need to write two first-order differential equations, one for $\dot{x}_1(t) = \dot{P}_a(t)$, and one for $\dot{x}_2(t) = \ddot{x}_1(t) = \ddot{P}_a(t)$. But $\ddot{P}_a(t)$ would be a new eleventh variable, not one of the original ten. Since we have only eight equations, we would then have to define three state variables instead of two, which gets us nowhere! The same thing would be true if we tried to choose $x_1(t) = P_b(t)$ and $x_2(t) \equiv \dot{P}_b(t)$ as phase variables. Once we have chosen the variables for whose first derivatives we are going to write equations, those variables must be our state variables!)

Combine* equations 5.4.27, 5.4.28, 5.4.31 and 5.4.32 to eliminate E(t), $e_p(t)$, and $P_b(t)$,

$$\dot{V}_{in,\,a}(t) = k_p e_{ref}(t) - k_p k_{trans} R_b \dot{V}_{b,\,out}(t) \qquad 5.4.35.$$

Combine equations 5.4.25, 5.4.33, and 5.4.35 to eliminate $\dot{V}_{in,\,a}(t)$, $P_a(t)$, and $\dot{V}_{b,\,out}(t)$.

$$\dot{x}_1(t) = -\frac{1}{R_a C_a} x_1(t) - \frac{k_p k_{trans} R_b}{C_a} x_2(t) + \frac{k_p}{C_a} e_{ref}(t) \qquad 5.4.36.$$

Combine equations 5.4.26 and 5.4.32, to eliminate $P_b(t)$.

$$\frac{d\dot{V}_{b,\,out}(t)}{dt} = -\frac{1}{R_b C_b} \dot{V}_{b,\,out}(t) + \frac{1}{R_b C_b} \dot{V}_{in,\,b}(t) \qquad 5.4.37.$$

Combine equations 5.4.37, 5.4.29, 5.4.30 and 5.4.33 to eliminate $\dot{V}_{in,\,b}(t)$, $\dot{V}_{a,\,out}(t)$, and $P_a(t)$.

$$\frac{d\dot{V}_{b,\,out}(t)}{dt} = -\frac{1}{R_b C_b} \dot{V}_{b,\,out}(t) + \frac{1}{R_a R_b C_b} x_1(t) + \frac{1}{R_b C_b} \dot{V}_{\ell}(t) \qquad 5.4.38.$$

Combine equations 5.4.38 and 5.4.34 to eliminate $\dot{V}_{b,\,out}(t)$.

$$\dot{x}_2(t) = \frac{1}{R_a R_b C_b} x_1(t) - \frac{1}{R_b C_b} x_2(t) + \frac{1}{R_b C_b} \dot{V}_{\ell}(t) \qquad 5.4.39.$$

[*Footnote is at end of section.]

Table 5.4.1. Known and unknown quantities in two-tank system, Example 5.4.3.

Known Parameters	Known Forcing Functions (= Inputs)
R_a = 2 cm H_2O/ [liter/min]	$e_{ref}(t)$ = 128/3 volts $-\infty < t < 0^+$
C_a = 5 liter/cm H_2O	$e_{ref}(t)$ = 144/3 volts $0^+ \le t$
R_b = 8 cm H_2O/ [liter/min]	$\dot{V}_\ell(t)$ = -14 liter/min $-\infty < t < 0^+$
C_b = 12.5 liter/cm H_2O	$\dot{V}_\ell(t)$ = -20 liter/min $0^+ \le t$
k_{trans} = 1/3 volt/cm H_2O	**Known Initial Conditions**
k_p = 3/4 [liter/min] /volt	$P_a(0)$ = 40 cm H_2O
	$\dot{V}_{b,\,out}(0)$ = 6 liter/min

Unknown Variables (= Outputs)

$dP_a(t)/dt$, $dP_b(t)/dt$, $P_a(t)$, $P_b(t)$, $\dot{V}_{in\,a}(t)$, $\dot{V}_{a,\,out}(t)$, $\dot{V}_{in\,b}(t)$, $\dot{V}_{b,\,out}(t)$, $e_p(t)$ and $E(t)$.

The preceding steps have eliminated all variables but the forcing functions, the state variables, and their first derivatives, so we now have two equations 5.4.36 and 5.4.39, in the four unknowns $x_1(t)$, $\dot{x}_1(t)$, $x_2(t)$ and $\dot{x}_2(t)$, the parameters of the system, and the inputs. Substituting into these equations the known numerical values from Table 5.4.1,

$$\left.\begin{aligned} \dot{x}_1(t) &= -0.1\,x_1(t) - 0.4\,x_2(t) + [0.15]\,[144/3] \\ \dot{x}_2(t) &= 0.005x_1(t) - 0.01x_2(t) + [0.01]\,[-20] \end{aligned}\right\} \quad 5.4.40,$$

subject to the initial conditions,

$$x_1(0) = 40 \text{ cm } H_2O; \quad x_2(0) = 6 \text{ liter/min}$$

In matrix notation, we have

$$\underline{\dot{x}}(t) = \underbrace{\begin{bmatrix} -0.1 & -0.4 \\ 0.005 & -0.01 \end{bmatrix}}_{\underline{A}} \underline{x}(t) + \underbrace{\begin{bmatrix} 0.15 & 0 \\ 0 & 0.01 \end{bmatrix}}_{\underline{B}} \underbrace{\begin{bmatrix} 48 \\ -20 \end{bmatrix}}_{\underline{f}(t)} \quad 5.4.41,$$

whence

$$[s\underline{I} - \underline{A}] = \begin{bmatrix} s + 0.1 & 0.4 \\ -0.005 & s + 0.01 \end{bmatrix} \qquad 5.4.42.$$

$$[s\underline{I} - \underline{A}]^{-1} = \frac{1}{s^2 + 0.11s + 0.0003} \begin{bmatrix} s + 0.01 & -0.4 \\ 0.005 & s + 0.1 \end{bmatrix} \qquad 5.4.43.$$

$$\underline{x}(0) + \underline{Bf}_{\mathcal{L}}(s) = \begin{bmatrix} 40 \\ 6 \end{bmatrix} + \begin{bmatrix} \frac{7.2}{s} \\ \frac{-0.2}{s} \end{bmatrix} = \begin{bmatrix} \frac{40s + 7.2}{s} \\ \frac{6s - 0.2}{s} \end{bmatrix} \qquad 5.4.44.$$

$$\underline{x}_{\mathcal{L}}(s) = \begin{bmatrix} \frac{s + 0.01}{s^2 + 0.11s + 0.003} & \frac{-0.4}{s^2 + 0.11s + 0.003} \\ \frac{0.005}{s^2 + 0.11s + 0.003} & \frac{s + 0.1}{s^2 + 0.11s + 0.003} \end{bmatrix} \begin{bmatrix} \frac{40s + 7.2}{s} \\ \frac{6s - 0.2}{s} \end{bmatrix} \qquad 5.4.45.$$

$$x_{1,\mathcal{L}}(s) = \frac{40s^2 + 5.2s + 0.152}{s[s + 0.05][s + 0.06]} = \frac{152/3}{s} + \frac{48/3}{s + 0.05} - \frac{80/3}{s + 0.06} \qquad 5.4.46.$$

$$x_1(t) = P_a(t) = \frac{152}{3} + 16\epsilon^{-0.05t} - \frac{80}{3}\epsilon^{-0.06t} \text{ cm } H_2O \qquad 5.4.47.$$

$$x_{2,\mathcal{L}}(s) = \frac{6s^2 + 0.6s + 0.016}{s[s + 0.05][s + 0.06]} = \frac{16/3}{s} - \frac{6/3}{s + 0.05} + \frac{8/3}{s + 0.06} \qquad 5.4.48.$$

$$x_2(t) = \dot{V}_{b,\,out}(t) = \frac{16}{3} - 2\epsilon^{-0.05t} + \frac{8}{3}\epsilon^{-0.06t} \text{ liter/min} \qquad 5.4.49.$$

Checks:

1. At $t = 0$, equations 5.4.47 and 5.4.49 give the proper initial conditions.
2. The original differential equations, 5.4.40, are satisfied by substituting into them the values of $x_1(t)$ and $x_2(t)$ from equations

5.4.47 and 5.4.49, and the values of $\dot{x}_1(t)$ and $\dot{x}_2(t)$ obtained by differentiating equations 5.4.47 and 5.4.49.

These first two checks validate the mathematical solution of equations 5.4.40. The following additional checks show that the solutions make sense in terms of the physical system itself, thus providing evidence that the derivation of equations 5.4.40 was also correct.

3. At $t = \infty$, equations 5.4.47 and 5.4.49 give the steady-state values which are correct for the new inputs.

4. At $t = 0^+$, any changes in tank 1 will be due entirely to the increased value of $e_{ref}(t)$ since the simultaneous decrease in $\dot{V}_\ell(t)$ cannot change $e_p(t)$ instantaneously. But the increase of 16/3 volt in $e_{ref}(t)$ will instantaneously increase $\dot{V}_{in,a}(t)$ by [3/4] [16/3] = 4 liter / min, which, in a tank whose capacitance is 5 liter/cm H_2O, corresponds to an initial value of 4/5 cm H_2O per min for $\dot{x}_1(0^+) = [d\,P_a(t)/dt]_{t\,=\,0^+}$. This is just what we get by differentiating equation 5.4.47 and evaluating at $t = 0^+$. A similar check can be used for the initial effect of $\dot{V}_\ell(t)$ on the first derivative of equation 5.4.49.

This third example begins to reveal some of the outstanding advantages of matrix analysis. Notice first that, in contrast to ordinary transfer function analysis, we do not have to define our variables as deviations from some operating point in order to establish zero initial conditions. Instead, we can work with the actual values of the variables and with the actual initial conditions. Second, notice that the characteristic polynomial automatically emerges during the analysis as the determinant of $[s\underline{I} - \underline{A}]$; if we want to use it for testing stability, there it is! Finally, notice how easily we can cope with two simultaneous inputs, and how easily we get two different outputs.

[*An orderly way of combining equations by eliminating unwanted variables is described in Chapters 12 and 13 of reference 3.]

SECTION 5.5. The relation between transfer functions and the matrix product, $\underline{\Phi}_{\mathcal{L}}(s)\underline{B}$

A comparison between the matrix product $\underline{\Phi}_{\mathcal{L}}(s)\underline{B}$ (Cf. equation 5.4.3) for a given system and the transfer functions for the same system is instructive. To make such a comparison at all, we shall have to accept the restriction peculiar to transfer function analysis that the initial conditions are zero.

Consider a generalized first-order system with a single input, $f_1(t)$ and a single output $x_1(t)$. The behavior of the system can be described by the equation

$$\dot{x}_1(t) + a_{11}x_1 = b_{11}f_1(t) \qquad 5.5.1,$$

whence

$$sx_{1,\mathcal{L}}(s) + a_{11}x_{1,\mathcal{L}}(s) = b_{11}f_{1,\mathcal{L}}(s) \qquad 5.5.2,$$

so the transfer function is

$$\frac{x_{1,\mathcal{L}}(s)}{f_{1,\mathcal{L}}(s)} = \frac{b_{11}}{s + a_{11}} \qquad 5.5.3.$$

Stupid though it sounds, let us apply our technique of matrix analysis to this single equation. (There is nothing at all illegitimate mathematically in working with a 1 x 1 matrix!) Solving equation 5.5.1 for its first derivative,

$$\dot{x}_1(t) = -a_{11}x_1 + b_{11}f_1(t) \qquad 5.5.4,$$

or,

$$\underline{\dot{x}}(t) = \underline{A}\,\underline{x}(t) + \underline{Bf}(t) \qquad 5.5.5.$$

$$\underline{A} = [-a_{11}] \quad ; \quad \underline{B} = [b_{11}] \qquad 5.5.6.$$

$$[s\underline{I} - \underline{A}] = [s + a_{11}] \qquad 5.5.7.$$

$$[s\underline{I} - \underline{A}]^{-1} = \underline{\Phi}_{\mathcal{L}}(s) = \left[\frac{1}{s + a_{11}}\right] \qquad 5.5.8.$$

$$\underline{x}_{\mathcal{L}}(s) = \underline{\Phi}_{\mathcal{L}}(s)\,[\underline{x}(0) + \underline{Bf}_{\mathcal{L}}(s)] \qquad 5.5.9.$$

But we are assuming that $\underline{x}(0) = 0$; hence equation 5.5.9 becomes,

$$\underline{x}_{\mathcal{L}}(s) = [\underline{\Phi}_{\mathcal{L}}(s)\,\underline{B}]\,\underline{f}_{\mathcal{L}}(s) = \left[\frac{b_{11}}{s + a_{11}}\right]\underline{f}_{\mathcal{L}}(s) \qquad 5.5.10.$$

Now compare equations 5.5.3 and 5.5.10. With a first-order system, the product of $\underline{\Phi}_{\mathcal{L}}(s)$, the Laplace transform of the fundamental matrix, and $\underline{B}$, the matrix of forcing function coefficients, is the transfer function. Notice again that $\underline{\Phi}_{\mathcal{L}}(s)\,\underline{B}$ contains all of the parameters of the system.

Consider next a generalized second-order system described by

$$\dot{x}_1(t) + a_{11}x_1(t) = a_{12}x_2(t) + b_{11}f_1(t) + b_{12}f_2(t) \qquad 5.5.11,$$

$$\dot{x}_2(t) + a_{22}x_2(t) = a_{21}x_1(t) + b_{21}f_1(t) + b_{22}f_2(t) \qquad 5.5.12.$$

A block diagram for this generalized system is shown in Figure 5.5.1. With two inputs, $f_1(t)$ and $f_2(t)$, and two outputs, $x_1(t)$ and $x_2(t)$, we have four possible input-output pairs, each of which will have a different transfer function. As an example, let us calculate the transfer function for signal transfer between $f_1(t)$ and $x_1(t)$. To do this, we must take Laplace transforms of equations 5.5.11 and 5.5.12 with $f_2(t) = 0$, and with $x_1(0) = x_2(0) = 0$.

$$[s + a_{11}]\,x_{1,\mathcal{L}}(s) = a_{12}x_{2,\mathcal{L}}(s) + b_{11}f_{1,\mathcal{L}}(s) \qquad 5.5.13,$$

$$[s + a_{22}]\,x_{2,\mathcal{L}}(s) = a_{21}x_{1,\mathcal{L}}(s) + b_{21}f_{1,\mathcal{L}}(s) \qquad 5.5.14.$$

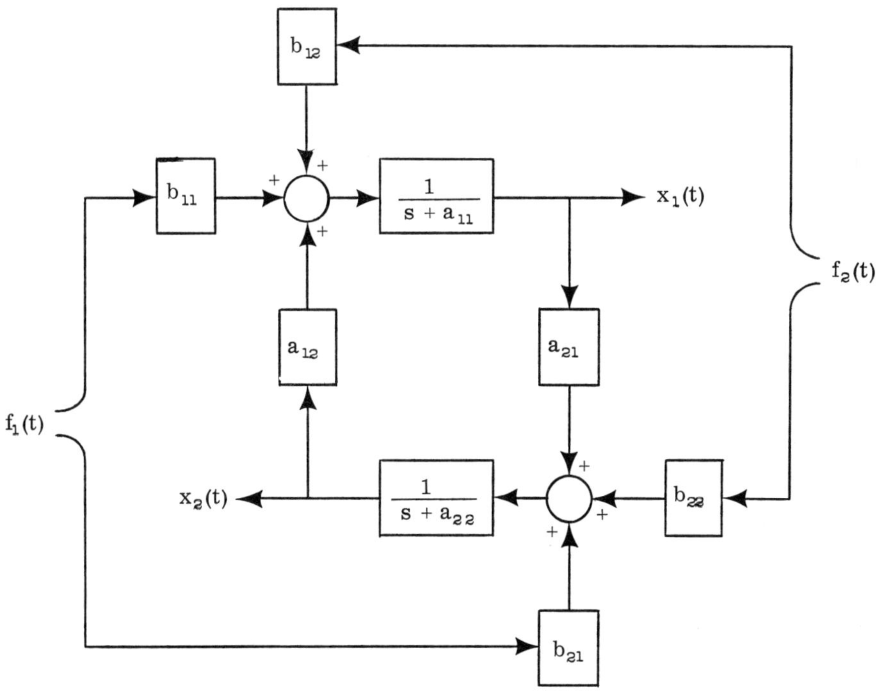

Figure 5.5.1. A block diagram for the generalized second-order system described by equations 5.5.11 and 5.5.12.

Combining equations 5.5.13 and 5.5.14 with the elimination of $x_{2,\mathcal{L}}(s)$,

$$x_{1,\mathcal{L}}(s) = \left\{\frac{[s+a_{22}]\,b_{11} + a_{12}b_{21}}{s^2+[a_{11}+a_{22}]\,s+[a_{11}a_{22}-a_{12}a_{21}]}\right\} f_{1,\mathcal{L}}(s) \qquad 5.5.15.$$

Solving equation 5.5.15 for $[TF]_{f_1,x_1}(s) \equiv x_{1,\mathcal{L}}(s)/f_{1,\mathcal{L}}(s)$, the transfer function between $f_1(t)$ and $x_1(t)$,

$$[TF]_{f_1,x_1}(s) = \frac{[s+a_{22}]\,b_{11}+a_{12}b_{21}}{s^2+[a_{11}+a_{22}]\,s+[a_{11}a_{22}-a_{12}a_{21}]} \qquad 5.5.16.$$

Similarly, we could calculate

$$[TF]_{f_1,x_2}(s) = \frac{a_{21}b_{11}+[s+a_{11}]\,b_{21}}{s^2+[a_{11}+a_{22}]\,s+[a_{11}a_{22}-a_{12}a_{21}]} \qquad 5.5.17,$$

$$[TF]_{f_2, x_1}(s) = \frac{[s + a_{22}]b_{12} + a_{12}b_{22}}{s^2 + [a_{11} + a_{22}]s + [a_{11}a_{22} - a_{12}a_{21}]} \quad 5.5.18,$$

$$[TF]_{f_2, x_2}(s) = \frac{[a_{21}b_{12}] + [s + a_{11}]b_{22}}{s^2 + [a_{11} + a_{22}]s + [a_{11}a_{22} - a_{12}a_{21}]} \quad 5.5.19.$$

Now let us use matrix analysis to solve equations 5.5.11 and 5.5.12. We have

$$\dot{\underline{x}}(t) = \begin{bmatrix} -a_{11} & a_{12} \\ a_{21} & -a_{22} \end{bmatrix} \begin{bmatrix} x_1(t) \\ x_2(t) \end{bmatrix} + \begin{bmatrix} b_{11} & b_{12} \\ b_{21} & b_{22} \end{bmatrix} \begin{bmatrix} f_1(t) \\ f_2(t) \end{bmatrix} \quad 5.5.20.$$

$$[s\underline{I} - \underline{A}] = \begin{bmatrix} s + a_{11} & -a_{12} \\ -a_{21} & s + a_{22} \end{bmatrix} \quad 5.5.21.$$

$$[s\underline{I} - \underline{A}]^{-1} = \underline{\Phi}_{\mathcal{L}}(s) = \frac{1}{s^2 + [a_{11} + a_{22}]s + [a_{11}a_{22} - a_{12}a_{21}]} \begin{bmatrix} s + a_{22} & a_{12} \\ a_{21} & s + a_{11} \end{bmatrix} \quad 5.5.22.$$

and, with zero initial conditions,

$$\underline{x}_{\mathcal{L}}(s) = \underline{\Phi}_{\mathcal{L}}(s)\underline{B}\underline{f}_{\mathcal{L}}(s) =$$

$$\frac{1}{|s\underline{I} - \underline{A}|} \begin{bmatrix} s + a_{22} & a_{12} \\ a_{21} & s + a_{11} \end{bmatrix} \begin{bmatrix} b_{11} & b_{12} \\ b_{21} & b_{22} \end{bmatrix} \begin{bmatrix} f_{1,\mathcal{L}}(s) \\ f_{2,\mathcal{L}}(s) \end{bmatrix}$$

$$= \frac{1}{|s\underline{I} - \underline{A}|} \begin{bmatrix} [s + a_{22}]b_{11} + a_{12}b_{21}, & [s + a_{22}]b_{12} + a_{12}b_{22} \\ a_{21}b_{11} + [s + a_{11}]b_{21}, & a_{21}b_{12} + [s + a_{11}]b_{22} \end{bmatrix} \begin{bmatrix} f_{1,\mathcal{L}}(s) \\ f_{2,\mathcal{L}}(s) \end{bmatrix} \quad 5.5.23,$$

where

$$|s\underline{I} - \underline{A}| = s^2 + [a_{11} + a_{22}]s + [a_{11}a_{22} - a_{12}a_{21}]$$

whence we have

$$\underline{x}_{\mathcal{L}}(s) = \begin{bmatrix} x_{1,\mathcal{L}}(s) \\ \\ x_{2,\mathcal{L}}(s) \end{bmatrix} = \begin{bmatrix} \dfrac{\{[s+a_{22}]b_{11}+a_{12}b_{21}\}f_{1,\mathcal{L}}(s) + \{[s+a_{22}]b_{12}+a_{12}b_{22}\}f_{2,\mathcal{L}}(s)}{s^2+[a_{11}+a_{22}]s+[a_{11}a_{22}-a_{12}a_{21}]} \\ \\ \dfrac{\{a_{21}b_{11}+[s+a_{11}]b_{21}\}f_{1,\mathcal{L}}(s) + \{a_{21}b_{12}+[s+a_{11}]b_{22}\}f_{2,\mathcal{L}}(s)}{s^2+[a_{11}+a_{22}]s+[a_{11}a_{22}-a_{12}a_{21}]} \end{bmatrix}$$

5.5.24.

If we now compare equations 5.5.16 through 5.5.19 with equation 5.5.23, we see at once that each element of the matrix $\underline{\Phi}_{\mathcal{L}}(s)\underline{B}$ is one of the four transfer functions. Indeed, by substituting equations 5.5.16 through 5.5.19 into equation 5.5.23, we can express the product $\underline{\Phi}_{\mathcal{L}}(s)\underline{B}$ as

$$\underline{\Phi}_{\mathcal{L}}(s)\underline{B} = \begin{bmatrix} [TF]_{f_1,x_1} & [TF]_{f_2,x_1} \\ \\ [TF]_{f_1,x_2} & [TF]_{f_2,x_2} \end{bmatrix} \qquad 5.5.25.$$

The product $\underline{\Phi}_{\mathcal{L}}(s)\underline{B}$ may thus be viewed as a kind of universal transfer function for the system.

SECTION 5.6. Matrix analysis of a multicompartmental biological feedback system

Linear biological systems are few and far between. But there is one general kind of biological problem, namely, the distribution of solutes between compartments, which is usually treated as a linear problem. I have discussed the significance of the term "compartment" elsewhere (3). In essence, a solute (which may be a drug or a metabolite) is pictured as being distributed between a number of more-or-less imaginary but convenient "pools" or compartments in the body. The basic assumption which lets us treat such systems as linear is that the rate at which the solute leaves a given compartment to go to another compartment, or to get out of the system via storage, metabolism or excretion, is directly proportional to its effective concentration in the compartment. When this assumption is valid, we can treat the system as a concentrational system where each compartment behaves like a capacitance and the pathways for flow of solute behave like conductances. However, the distribution of solute between two compartments is by no means always a simple passive approach to equilibrium as the distribution of electric charge between two capacitors via a resistor would be. In compartmental analysis therefore, even when there is a two-way interchange of solute between two compartments, it is customary to treat transfer in each direction separately, the net transfer of solute then being the algebraic sum of the transfers in opposite directions. For example, suppose compartment A containing $Q_A(t)$ moles of solute in a volume of V_A, and compartment B, containing $Q_B(t)$ moles of solute in a volume of V_B, can exchange solute with each other. According to the basic assumption underlined above, $\dot{Q}_{A,B}(t)$, the gross quantity of solute passing unidirectionally from A to B is proportional to $C_A(t)$.

$$\dot{Q}_{A,B}(t) = [\dot{V}cl]_{A,B} C_A(t) \qquad 5.6.1,$$

where the proportionality constant, $[\dot{V}cl]_{A,B}$, (analogous to a conductance) is defined as the clearance of A via the pathway to B.* But it is often more convenient to work with quantities rather than concentrations. Accordingly, let us multiply the right side of equation 5.6.1 by V_A/V_A,

$$\dot{Q}_{A,B}(t) = \frac{[\dot{V}cl]_{A,B}}{V_A} C_A(t)V_A = k_{A,B}Q_A(t) \qquad 5.6.2,$$

where $k_{A,B}$, the rate constant for transfer from A to B, is defined as

$$k_{A,B} = \frac{[\dot{V}cl]_{A,B}}{V_A} \qquad 5.6.3.$$

Similarly, for transfer from B to A,

$$\dot{Q}_{B,A}(t) = k_{B,A}Q_B(t) \qquad 5.6.4.$$

The net change in the quantity of solute in a given compartment will be the difference between the rate of entry of solute into, and the rate of exit from the compartment via all pathways. For this example,

$$\frac{dQ_A(t)}{dt} = \dot{Q}_{B,A}(t) - \dot{Q}_{A,B}(t) = k_{B,A}Q_B(t) - k_{A,B}Q_A(t) \qquad 5.6.5.$$

For each compartment in a system of compartments we can write a first-order differential equation similar to equation 5.6.5, showing the additions to, and the losses from the compartment as products of rate constants and quantities. (If it is more convenient to work with concentrations, it is easy enough to divide the equation for a given compartment by the volume of that compartment.)

[*"Clearance of A via the pathway to B" may be thought of as the volume of A from which the solute is, in effect, completely removed, i.e., "cleared," per unit of time via that pathway. It may equally well be defined (Cf. equation 5.6.1) as the rate of removal of solute from A via the pathway to B, divided by the concentration in A. It is important to understand that although a clearance can be defined for any single unidirectional pathway, it will represent net removal of solute from A only when there is no return of solute from B to A via any route. In practice, therefore, the term "clearance" is rarely used except for the irreversible removal of solute by metabolism, storage, or excretion. Problems such as the present one which involve the interchange of a solute between two or more compartments are better formulated in terms of rate constants than in terms of clearances.]

As a specific example, we shall consider the distribution of iodine in the human body, using a somewhat antiquated three-compartment model which I proposed many years ago. (4). While my choice of this particular model is no doubt influenced by nostalgia, the model is about as complex as one can handle easily without a computer, and it still gives a reasonably satisfactory picture of some of the main features of iodine metabolism. In the present context, it has the additional interest of being a positive feedback system. This model is depicted in Figure 5.6.1.a, and the corresponding block diagram is shown in Figure 5.6.1.b. The three first-order differential equations for the quantities of iodine in the three compartments are

$$\frac{dQ_I(t)}{dt} = -[k_1 + k_4]Q_I(t) + k_3 Q_H(t) + \dot{Q}_{in,I}(t) \qquad 5.6.6,$$

$$\frac{dQ_G(t)}{dt} = k_1 Q_I(t) - k_2 Q_G(t) \qquad 5.6.7,$$

$$\frac{dQ_H(t)}{dt} = k_2 Q_G(t) - [k_3 + k_5]Q_H(t) + \dot{Q}_{in,H}(t) \qquad 5.6.8.$$

(The symbols used in these equations are defined in Figure 5.6.1). We naturally choose $x_1(t) = Q_I(t)$, $x_2(t) = Q_G(t)$, $x_3(t) = Q_H(t)$ as our state variables. Furthermore, to simplify notation in the following analysis, let,

$$k_I = k_1 + k_4 \qquad 5.6.9,$$

$$k_H = k_3 + k_5 \qquad 5.6.10.$$

Each of these new symbols is the total rate constant for exit of iodine from the compartment indicated by the subscript. With these symbols, we may write equations 5.6.6, 5.6.7, and 5.6,8 in matrix form as

$$\dot{\underline{x}}(t) = \begin{bmatrix} -k_I & 0 & k_3 \\ k_1 & -k_2 & 0 \\ 0 & k_2 & -k_H \end{bmatrix} \begin{bmatrix} x_1(t) \\ x_2(t) \\ x_3(t) \end{bmatrix} + \begin{bmatrix} \dot{Q}_{in,I}(t) \\ 0 \\ \dot{Q}_{in,H}(t) \end{bmatrix} \qquad 5.6.11.$$

$$\dot{\underline{x}}(t) = \underline{A} \qquad \underline{x}(t) \quad + \quad \underline{Bf}(t)$$

$$[s\underline{I} - \underline{A}] = \begin{bmatrix} s + k_I & 0 & -k_3 \\ -k_1 & s + k_2 & 0 \\ 0 & -k_2 & s + k_H \end{bmatrix} \qquad 5.6.12.$$

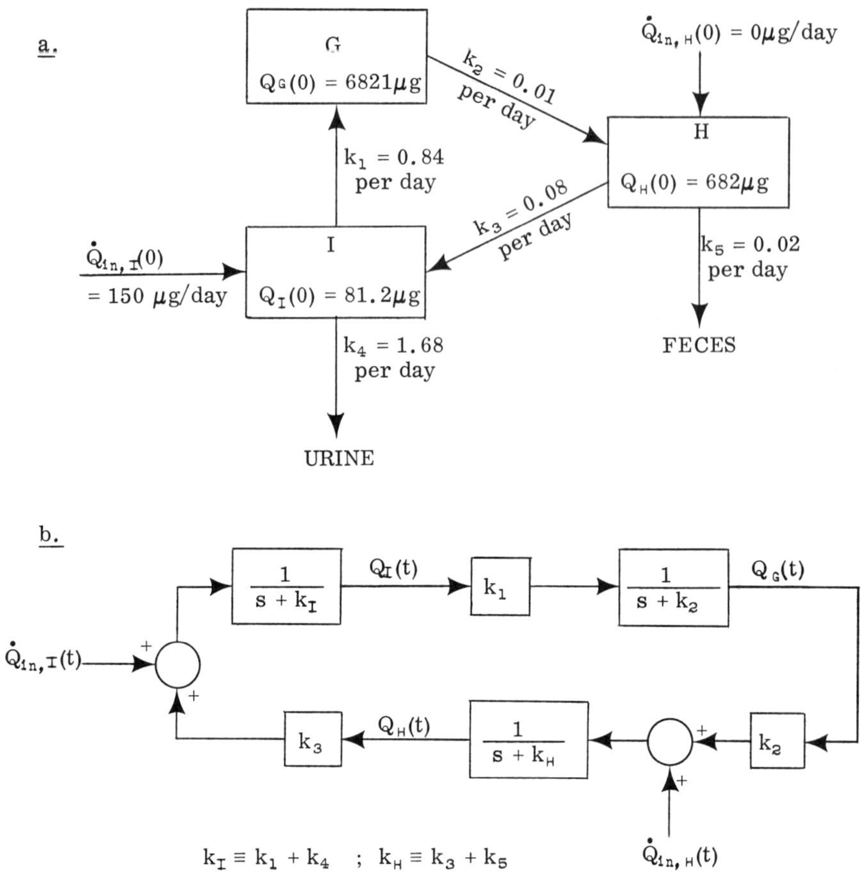

Figure 5.6.1. Compartmental analysis of the distribution of iodine in man.

Figure 5.6.1.a. A physical model showing three compartments and the pathways by which iodine enters and leaves them. Q_I is the quantity of iodide ion in the inorganic iodide compartment, I; Q_G is the quantity of organic iodine stored in the thyroid gland compartment, G; Q_H is the quantity of hormonal iodine in the extrathyroidal tissue compartment, H. $\dot{Q}_{in,I}$ is the rate of entry of exogenous iodide; $\dot{Q}_{in,H}$ is the rate of entry of exogenous hormonal iodine. Each pathway for transfer of iodine is labelled with a rate constant, k, showing what proportion per day of the iodine in the compartment of origin of that pathway leaves the compartment by the pathway. In each compartment is shown the initial quantity assumed to be present for the analysis described in the text. These quantities are the steady-state values determined by the constant inputs and rate constants shown in the figure.

Figure 5.6.1.b. The block diagram for the three-compartment model of Figure 5.6.1.a.

To save time and trouble, we shall use the general expression given in the Appendix for the inversion of a 3 by 3 matrix. We thus obtain

$$[s\underline{I} - \underline{A}]^{-1} = \underline{\Phi}_{\mathcal{L}}(s)$$

$$= \frac{1}{|s\underline{I} - \underline{A}|} \begin{bmatrix} [s+k_2][s+k_H] & k_2k_3 & k_3[s+k_2] \\ k_1[s+k_H] & [s+k_I][s+k_H] & k_1k_3 \\ k_1k_2 & k_2[s+k_I] & [s+k_I][s+k_2] \end{bmatrix} \quad 5.6.13,$$

where

$$|s\underline{I} - \underline{A}| = \frac{1}{[s+k_I][s+k_2][s+k_H] - k_1k_2k_3}$$

Equation 5.6.13 gives the Laplace transform of the fundamental matrix for the system.

To illustrate how a numerical solution is obtained, let $k_1 = 0.84$ per day, $k_2 = 0.01$ per day, $k_3 = 0.08$ per day, $k_4 = 1.68$ per day and $k_5 = 0.02$ per day. While these values have been selected partly for computational convenience, they are of the correct magnitude for a normal adult human whose diet contains adequate iodine ($\dot{Q}_{in,I}(t) \approx 150$ µg per day). For these values,

$$\underline{\Phi}_{\mathcal{L}}(s) = \frac{1}{|s\underline{I} - \underline{A}|} \begin{bmatrix} s^2 + 0.11s + 0.001 & 0.0008 & 0.08s + 0.0008 \\ 0.84s + 0.084 & s^2 + 2.62s + 0.252 & 0.00672 \\ 0.0084 & 0.01s + 0.0252 & s^2 + 2.53s + 0.0252 \end{bmatrix} \quad 5.6.14,$$

where the determinant, $|s\underline{I} - \underline{A}|$ (= the characteristic polynomial), is

$$|s\underline{I} - \underline{A}| = s^3 + 2.63s^2 + 0.2782s + 0.001848 \quad 5.6.15.$$

We shall take as initial conditions the steady-state values corresponding to inputs of $\dot{Q}_{in,I}(t) = 150\,\mu g$ per day and $\dot{Q}_{in,H}(t) = 0\,\mu g$ of exogenous hormonal iodine per day.

$$\underline{x}(0) = \begin{bmatrix} Q_I(0) \\ Q_G(0) \\ Q_H(0) \end{bmatrix} = \begin{bmatrix} 81.2\,\mu g \\ 6821\,\mu g \\ 682.1\,\mu g \end{bmatrix} \quad 5.6.16.$$

With these initial conditions, what will happen if the daily intake of iodine, $\dot{Q}_{in,I}(t)$, falls from 150 μg per day to 15 μg per day, as it might if the person represented by this example moved to a region of intense iodine deficiency? (5). We want to find

$$\underline{x}(t) = \mathcal{L}^{-1}\left\{ \underline{\Phi}_{\mathcal{L}}(s)\ [\underline{x}(0) + \underline{B}\underline{f}_{\mathcal{L}}(s)] \right\} = \mathcal{L}^{-1}\left\{ \underline{\Phi}_{\mathcal{L}}(s) \begin{bmatrix} 81.2 & + & \frac{15}{s} \\ 6821 & + & 0 \\ 682.1 & + & 0 \end{bmatrix} \right\} \qquad 5.6.17.$$

For ease in computation, let us replace the actual values in the matrix of equation 5.6.14 by the conventional algebraic symbols, a_{ij}. We then have

$$\underline{x}_{\mathcal{L}}(s) = \frac{1}{|s\underline{I} - \underline{A}|} \begin{bmatrix} a_{11} & a_{12} & a_{13} \\ a_{21} & a_{22} & a_{23} \\ a_{31} & a_{32} & a_{33} \end{bmatrix} \begin{bmatrix} \frac{81.2s + 15}{s} \\ 6821 \\ 682.1 \end{bmatrix}$$

$$= \begin{bmatrix} \frac{[81.2s + 15]\,a_{11} + 6821sa_{12} + 682.1sa_{13}}{s\,|s\underline{I} - \underline{A}|} \\ \frac{[81.2s + 15]\,a_{21} + 6821sa_{22} + 682.1sa_{23}}{s\,|s\underline{I} - \underline{A}|} \\ \frac{[81.2s + 15]\,a_{31} + 6821sa_{32} + 682.1sa_{33}}{s\,|s\underline{I} - \underline{A}|} \end{bmatrix} \qquad 5.6.18.$$

We can no longer postpone the somewhat tedious task of finding the roots of the characteristic equation, (Cf. equation 5.6.15),

$$s^3 + 2.630s^2 + 0.2782s + 0.001848 = 0 \qquad 5.6.19.$$

Lin's method (see Appendix) with three iterations, gives the following approximate values for the roots,

$$\left.\begin{aligned} s &= -2.52 \\ s &= -0.103 \\ s &= -0.00712 \end{aligned}\right\} \qquad 5.6.20.$$

The factors of the characteristic polynomial should therefore be $[s + 2.52]$, $[s + 0.103]$, and $[s + 0.00712]$.

Check: by direct multiplication, we find

$[s + 2.52][s + 0.103][s + 0.00712] = s^3 + 2.630s^2 + 0.2782s + 0.001848$, which happens to agree exactly with the original polynomial. With as few as three iterations, Lin's method cannot always be expected to yield roots so nearly correct!*

[*When I made these calculations, I was staying at "The Brasenose Inn" near Banbury, an English market town noted for its cakes and ale. Since I do not enjoy cakes, the astonishing accuracy of the results is no doubt due to the ability of ale to prevent those tiresome numerical errors which are likely to creep into any lengthy manual computation.]

We are now ready to find $\underline{x}(t)$ by taking the inverse transforms indicated by equation 5.6.17. To obtain $x_1(s)$, we perform the multiplications in the numerator of the first element on the right of equation 5.6.18 and collect terms.

$$x_{1,\mathcal{L}}(s) = \frac{81.2s^3 + 78.5s^2 + 7.734s + 0.015}{s[s + 2.52][s + 0.103][s + 0.00712]} =$$

$$\frac{8.12}{s} + \frac{53.6}{s + 2.52} - \frac{1.58}{s + 0.103} + \frac{21.08}{s + 0.00712} \qquad 5.6.21.$$

Therefore, taking inverse transforms,

$$x_1(t) = Q_I(t) = 8.12 + 53.6\varepsilon^{-2.52t} - 1.58\varepsilon^{-0.103t} + 21.08\varepsilon^{-0.00712t}\ \mu g \qquad 5.6.22.$$

Before going on to evaluate $x_2(t)$ and $x_3(t)$, let us check equation 5.6.22 at the extremes of time. As $t \to \infty$, $x_1(t) \to 8.12\,\mu g$, exactly one-tenth the original amount. In a linear system this makes sense because the input, which here is the daily intake of iodine, is one-tenth the original. And by equation 5.6.22, $x_1(0) = 81.22\,\mu g$ which agrees with the known initial condition of $81.2\,\mu g$. But if the reader has actually carried through all of the above calculations, he may well be wondering if there is not some easier way to obtain equation 5.6.22 and its two companions, still uncalculated. And indeed there is! For while we are not <u>obliged</u> to have zero initial conditions when using matrix analysis, there is nothing against our defining the variables so as to take advantage of zero initial conditions when, as in this example, it simplifies the computations. Accordingly, let us define all variables (just as in transfer function analysis) as <u>deviations</u> from their values at their original operating point,

$$\Delta x_i(t) = x_i(t) - x_i(0) \qquad i = 1, 2, 3. \qquad 5.6.23,$$

and let us also define the forcing function as its deviation from its original value,

$$\Delta\dot{Q}_{in,I}(t) = \dot{Q}_{in,I}(t) - \dot{Q}_{in,I}(0^-) \qquad 5.6.24.$$

We can now write equation 5.6.18 in the much simpler form,

$$\frac{1}{|s\underline{I}-\underline{A}|}\begin{bmatrix} a_{11} & a_{12} & a_{13} \\ a_{21} & a_{22} & a_{23} \\ a_{31} & a_{32} & a_{33} \end{bmatrix}\begin{bmatrix} -135/s \\ 0 \\ 0 \end{bmatrix} = \begin{bmatrix} \frac{-135a_{11}}{s\,|s\underline{I}-\underline{A}|} \\ \frac{-135a_{21}}{s\,|s\underline{I}-\underline{A}|} \\ \frac{-135a_{31}}{s\,|s\underline{I}-\underline{A}|} \end{bmatrix} \quad 5.6.25,$$

whence,

$$\Delta x_{1,\mathcal{L}}(s) = \frac{-135s^2 - 14.85s - 0.135}{s\,[s+2.52]\,[s+0.103]\,[s+0.00712]} =$$

$$\frac{-73.1}{s} + \frac{53.6}{s+2.52} - \frac{1.58}{s+0.103} + \frac{21.0}{s+0.00712}\ \mu g\cdot sec$$

5.6.26.

$$\Delta x_1(t) = -73.1 + 53.6\varepsilon^{-2.52t} - 1.58\varepsilon^{-0.103t} + 21.0\varepsilon^{-0.00712t}\ \mu g \quad 5.6.27.$$

To find $x_1(t)$, we simply add $x_1(0) = 81.2\,\mu g$ to both sides of equation 5.6.26.

$$x_1(t) = \Delta x_1(t) + x_1(0) = 8.1 + 53.6\varepsilon^{-2.52t} - 1.58\varepsilon^{-0.103t} + 21.0\varepsilon^{-0.00712t}\ \mu g$$

5.6.22,

as before. Equations for $x_2(t)$ and $x_3(t)$ are similarly derived.

$$\Delta x_{2,\mathcal{L}}(s) = \frac{-11.34\,[10s+1]}{s\,[s+2.52]\,[s+0.103]\,[s+0.00712]} =$$

$$-\frac{6138}{s} - \frac{17.9}{s+2.52} + \frac{14.3}{s+0.103} + \frac{6141}{s+0.00712}\ \mu g\cdot sec$$

5.6.28.

$$\Delta x_2(t) = -6138 - 17.9\varepsilon^{-2.52t} + 14.3\varepsilon^{-0.103t} + 6141\varepsilon^{-0.00712t}\ \mu g \quad 5.6.29.$$

$$x_2(t) = 683 - 17.9\varepsilon^{-2.52t} + 14.3\varepsilon^{-0.103t} + 6141\varepsilon^{-0.00712t}\ \mu g \quad 5.6.30.$$

$$\Delta x_{3,\mathcal{L}}(s) = \frac{-1.134}{s\,[s+2.52]\,[s+0.103]\,[s+0.00712]} =$$

$$-\frac{614}{s} + \frac{0.0741}{s+2.52} - \frac{47.5}{s+0.103} + \frac{661.2}{s+0.00712}\ \mu g\cdot sec$$

5.6.31.

$$\Delta x_3(t) = -614 + 0.074\epsilon^{-2.52t} - 47.5\epsilon^{-0.103t} + 661.2\epsilon^{-0.00712t}\ \mu g \quad 5.6.32.$$

$$x_3(t) = 68.1 + 0.074\epsilon^{-2.52t} - 47.5\epsilon^{-0.103t} + 661.2\epsilon^{-0.00712t}\ \mu g \quad 5.6.33.$$

In Figure 5.6.2, the values of $x_1(t) = Q_I(t)$, $x_2(t) = Q_G(t)$ and $x_3(t) = Q_H(t)$ have been plotted against time for a period of more than a year. We have here completely neglected the very important negative feedback loop, involving the rate of secretion of thyrotropic hormone from the pituitary, which helps to minimize the slow decrease in $Q_H(t)$ during iodine deficiency (See Part II). But Figure 5.6.2 shows that even without that loop, the huge amount of hormone stored in the thyroid is able to prevent abrupt changes in $Q_H(t)$, the physiologically important variable which, loosely speaking, represents the amount of thyroid hormone available to the tissues upon which it acts.

The system which we have been analyzing is a positive feedback system because the thyroid reutilizes some of the iodine released by the breakdown of hormone in the tissues. The positive feedback loop thus helps the body to conserve iodine which, as a trace element, is hard to come by. How effective is this positive feedback loop? The open-loop gain for $s = j\omega = 0$ is the absolute magnitude of the open-loop transfer function when $s = 0$. (Section 4.14).

$$[OLG]j(0) = \left|\frac{k_1k_2k_3}{k_1k_2k_H}\right| = 0.2667 \quad 5.6.34.$$

Hence, by equation 4.16,3,

$$\text{Magnification} - 1 = \frac{[OLG](j0)}{1 - [OLG](j0)} = \frac{0.2667}{0.7333} = 0.364 \quad 5.6.35,$$

or, if you prefer,

$$\text{Magnification} = 1.364 \quad 5.6.36.$$

This means that for a given steady input of iodide, the steady-state quantity of iodine in each compartment is 1.364 times as great as it would be without feedback, i.e., with k_H still 0.1 per day but with $k_3 = 0$. For example, if $Q_{in,I}(t) = 150\ \mu g$ per day and the loop is closed, $Q_H(\infty) = 682\ \mu g$ (Cf. equation 5.6.16). But if the loop were opened, so that none of the iodine in compartment H could return to compartment I, $Q_H(\infty)$ would be only [682/Magnification] = 500 μg. A similar, but even more important, positive feedback loop operates within the thyroid gland itself where the proteolysis of thyroglobulin releases not only hormone for secretion but also the hormonally inactive precursors, monoiodotryosine and diiodotyrosine. The iodine in these amino acids is enzymatically removed and returned to the pool of inorganic iodine within the thyroid gland itself for immediate reutilization in hormone synthesis. That pool actually constitutes a fourth compartment which, to keep the discussion simple, we have deliberately disregarded. (See Exercise 5.9.)

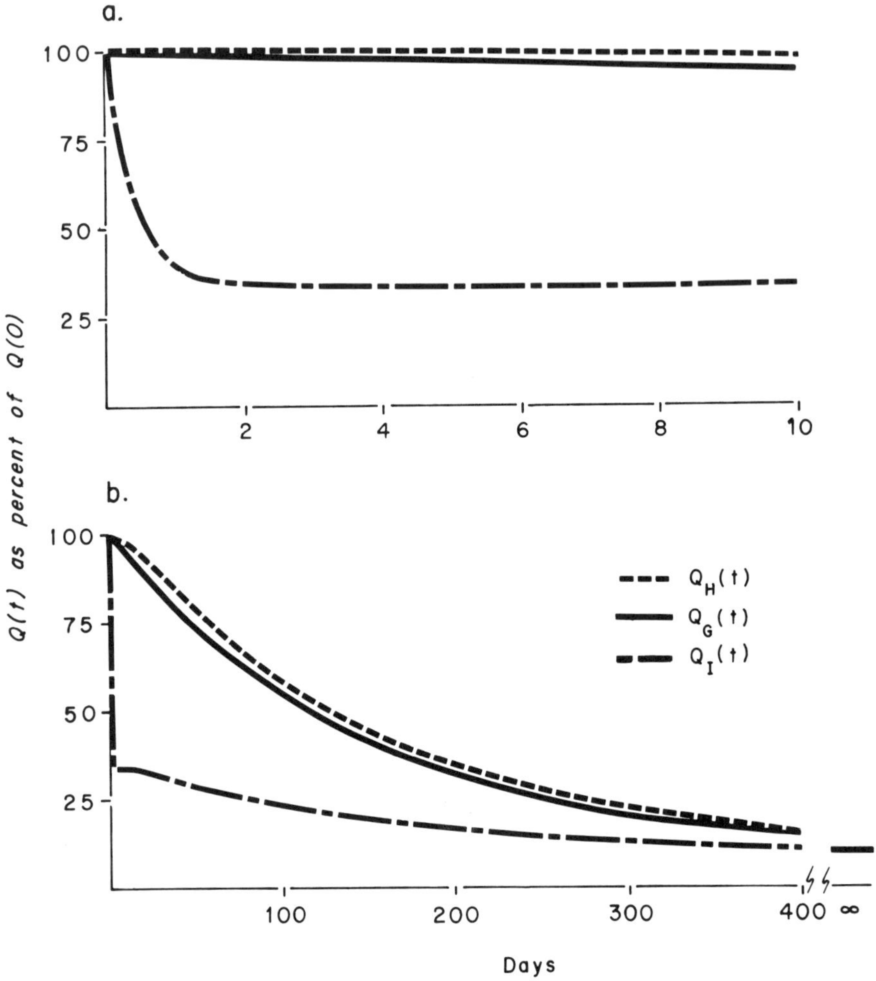

Figure 5.6.2. The quantity of iodine in the various compartments of Figure 5.6.1 as per cent of the quantity at time zero, plotted against time.

At time zero, $\dot{Q}_{in,I}(t)$, the intake of iodide, was abruptly decreased from 150 ug/day to 15 ug/day, and was maintained constant thereafter. Panel a shows the changes during the first ten days. Because of the rapid turnover in compartment I, $Q_I(t)$ (lowest curve) falls rapidly towards a nearly constant value sustained in large part by the iodide being returned to the compartment by the metabolism of hormone. In contrast, turnover in the thyroid gland (middle curve) and in the compartment of extrathroidal hormone (top curve) is so slow that little decrease is evident in ten days. However, as panel b shows, the amount of iodine in each compartment does eventually decline towards a constant value of 10% as required by the decrease of the intake to 10% of its original value.

Let the reader not underestimate what we have learned to do. The explicit solution of a three-compartment problem is no trivial exercise. A number of years ago, I published an extensive review of iodine metabolism in man, (4) based upon the very three-compartment model described by equations 5.6.6, 5.6.7, and 5.6.8. But I could not solve the equations. Now it seems so easy! I mention this not to prove that an old dog can learn new tricks, but to point out that multicompartmental problems are of genuine biological importance, and that anyone who has mastered the techniques presented here need never be humiliated, as I was, by having to describe the behavior of a linear system without knowing how to solve the relevant differential equations.

Exercises

1. Consider a system described by the equations

$$\frac{dx_1(t)}{dt} = a_{11}x_1(t) + a_{12}y(t)$$

$$\frac{dy(t)}{dt} = a_{21}x_1(t) + a_{22}y(t)$$

Show that it can also be described in terms of the phase variables $x_1(t)$ and $x_2(t) \equiv \dot{x}_1(t)$. [HINT. Start by taking Laplace transforms and eliminating $y_{\mathcal{L}}(s)$]

2. Use matrix techniques to solve Exercise 13, part b, Chapter 3.

3. Solve

$$2\frac{d^2x(t)}{dt^2} + 14\frac{dx(t)}{dt} + 34x(t) + \int 30x(t)\cdot dt = 3t + 2$$

$$x(0) = 0;\ \dot{x}(0) = 1;\ \ddot{x}(0) = 0$$

by matrix methods.

4. Construct a diagram like that of Figure 5.2.1 and use it to decide which pairs of variables in the two-tank system of Example 5.4.3 could, and which pairs could not serve as state variables.

5. Perform a matrix analysis of the two-tank feedback system like that of Example 5.4.3, but with $\dot{V}_{in,a}(t)$ and $\dot{V}_{a,out}(t)$ as the state variables.

6. Consider a linear, one-compartment system. The volume of the compartment is [Vdist]. C(t) is the concentration of a solute S, in the compartment. [$\dot{V}$cl] is the clearance of S from the compartment by all routes. $C(0^-) = 0$. At time zero, a quantity of S, Q_s, is added instantaneously to [Vdist], so as to produce an initial concentration of $C(0^+) > 0$. Thereafter

$$C(t) = C(0^+)\,\varepsilon^{-kt}$$

Prove that

$$[\dot{V}cl] = \frac{Q_s}{\int_0^\infty C(t)dt}$$

7. Matthews (6) discusses the distribution of plasma proteins, labelled with radioactive iodine, between compartment 1, a central compartment of intravascular protein, and n -1 "mammillary" compartments communicating with compartment 1 but not with each other (Figure 1, p. 37 of Matthews' paper). Write the appropriate differential equations for a four-compartment system of this kind, and use it to verify Matthews' equation 1 which is the characteristic equation for such a system.

8. Let A and B be the two compartments in a general two-compartment system through which a solute, S, is distributed. We assume that A and B are well-stirred at all times. At time 0^- when the concentration of S in A is $C_A(0) = C_{A,0}$, and the concentration of S in B is $C_B(0) = C_{B,0}$, we start to infuse S into A at a rate of $\dot{Q}_{in,A}$ and to infuse S into B at a rate of $\dot{Q}_{in,B}$. The clearance of S from A by renal excretion is $[\dot{V}cl]_A$, the clearance of S from B by metabolic degradation is $[\dot{V}cl]_B$. The rate constant for transfer from A to B is k_{AB}; the rate constant for transfer from B to A is k_{BA}.

A. Draw a diagram showing a physical model of this system.
B. Draw the corresponding circuit diagram.
C. Use matrix analysis to find $C_A(t)$, the concentration of S in A at any tim and $C_B(t)$, the concentration of S in B at any time t.

9. Consider the following four-compartment system in which we do not neglect IG, the pool of inorganic iodide within the thyroid gland.

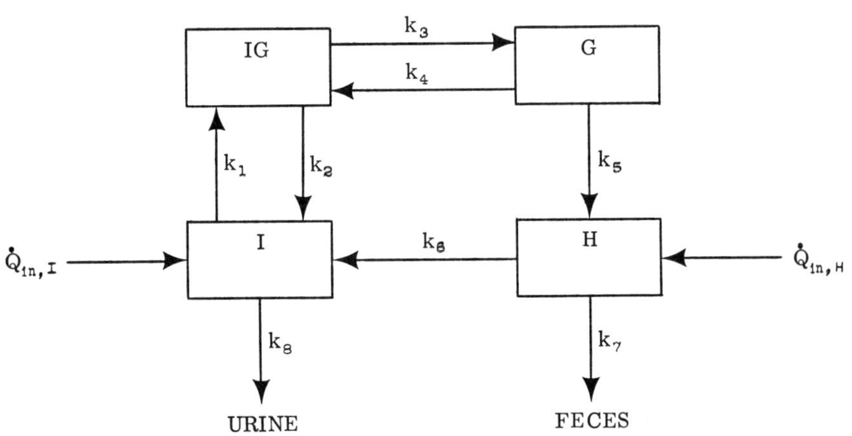

A. Write the differential equations describing this system.
B. Draw a block diagram for this system in the manner of Figure 5.6.1.b. How many feedback loops are there? Are they positive or negative?
C. Write equations analogous to 5.6.11 and 5.6.12 for this system.

(Don't try to carry the analysis further unless you have intense masochistic tendencies.)

References

1. Rosen, R.: Dynamical Systems Theory in Biology. Vol. I. Stability Theory and its Applications. John Wiley and Sons, N.Y. 1970.

2. DeRusso, P.M., Roy, R.J., and Close, C.M.: State Variables for Engineers. John Wiley and Sons, N.Y., 1965.

3. Riggs, D.S.: The Mathematical Approach to Physiological Problems. The Williams and Wilkins Co., Baltimore, 1963.

4. Riggs, D.S.: Quantitative aspects of iodine metabolism in man. Pharmacol. Rev., 4, 284-370, 1952.

5. Stanbury, J.B., Brownell, G.L., Riggs, D.S., Perinetti, H., Itoiz, J., and DelCastillo, E.B.: Endemic Goiter. Harvard Univ. Press, Cambridge, 1954.

6. Matthews, C.M.E.: The theory of tracer experiments with ^{131}I-labelled plasma proteins. Phys. Med. Biol., 2, 36-53, 1957.

You must not. . .be too precise or scientific about birds and trees and flowers...

Walt Whitman, SPECIMEN DAYS

6

NONLINEAR SYSTEMS

SECTION 6.1. Linear vs. nonlinear systems

In the previous chapters we have surveyed some of the outstanding characteristics of linear systems. Linear systems analysis is like a city founded upon solid mathematical rock and laid out on an orderly plan. Its avenues are wide and straight; its buildings beautiful and useful;--a tidy place, yet limited in scope, and essentially man-made. We have by no means exhausted its charms. In particular, we have not studied distributed-parameter systems, those upper floors to be reached only via partial differential equations. But we have exhausted its surprises. We know that any new skyscrapers, any new subdevelopments, will be built along the same lines. It is time for us to leave the city and to start exploring the strange nonlinear world which lies beyond. If you are uneasy about the prospect, you had best abandon biology in favor of something less wayward, --pure mathematics for example.

In the broadest sense, both our objectives and our strategy in the analysis of nonlinear systems are the same as in the analysis of linear systems. Our major objective is still to specify the state of the system (i.e. the values of all dependent variables) at every instant of time. To achieve this objective we must have the same information as before. (See Section 2.18). We must know the parameters of, and the input-output relationships for each component in the system; we must know how the several components are related to each other; and we must know the values of a set of state variables at some instant of time. Our general strategy is still to study the transfer of a given signal from input to output, via the unidirectional pathways which link the several components. But in a nonlinear system, there is at least one component whose operator, which tells us how to get the output from the input, is a nonlinear operator. In such a component (often called simply a nonlinearity) the output is no longer proportional to the input over the range of interest, superposition no longer holds, and we can no longer write a transfer function which is valid for all inputs. We are therefore obliged to study different inputs, or even different ranges of the same input, separately, as the next section will show.

SECTION 6.2. The thermostat as a nonlinear device

Before attempting a systematic approach, I shall illustrate some of the peculiarities which may be encountered in a nonlinear system by the familiar example of heat exchange in a water bath warmed by a thermostatically-controlled heater. We shall assume that the load input, which here is T_{ext}, the external temperature, remains constant during any given calculation. For simplicity we shall also assume that the heater warms up instantaneously when turned on, and cools off instantaneously when turned off, and that the heat it supplies is instantaneously and uniformly distributed throughout the water bath. Under these circumstances, the bath will behave like a parallel RC network (see Figure 6.2.1.a).

$$C\frac{d\Delta T(t)}{dt} + \frac{\Delta T(t)}{R} = \dot{q}(t) \qquad 6.2.1,$$

where $\Delta T(t)$ is the temperature drop across either element,

$$\Delta T(t) = T_{bath}(t) - T_{ext} \qquad 6.2.2,$$

and

$T_{bath}(t)$ = bath temperature ($^\circ$C)

C = heat capacitance of the bath (cal/$^\circ$C)

R = resistance to flow of heat out of bath ($^\circ$C•min/cal)

$\dot{q}(t)$ = rate of heat production by the heater (cal/min)

The transfer function for this component of our system is

$$\frac{\Delta T_{\mathcal{L}}(s)}{\dot{q}_{\mathcal{L}}(s)} = \frac{1/C}{s + \frac{1}{RC}} \qquad 6.2.3,$$

which is shown in a block in Figure 6.2.1.b. The figure also shows a summing point corresponding to equation 6.2.2. The only other feature specified in Figure 6.2.1.b, is subtraction of the bath temperature from the set-point or reference temperature input, $T_{ref}(t)$, to produce an error signal, $E(t)$.

$$E(t) = T_{ref}(t) - T_{bath}(t) \qquad 6.2.4.$$

(Actually, $T_{ref}(t)$ is not a true reference temperature but is a kind of pseudo-temperature, i.e., a point chosen on the temperature scale of the thermostat. However, it does no harm to pretend that this set-point is a true temperature.)

Suppose first that the thermostat acted as a proportional controller (Section 4.3), as shown in Figure 6.2.1.c. Then

$$\dot{q}(t) = k_p E(t) \qquad 6.2.5,$$

where k_p is the slope of the line in the graph which has been put in the "thermostat + heater" block (instead of a transfer function) to show the relationship between its input, $E(t)$ and its output $\dot{q}(t)$, (Figure 6.2.1.c). By solving equations 6.2.1, 6.2.2, 6.2.4, and 6.2.5 simultaneously, we would find that for a constant set-point, T_{ref},

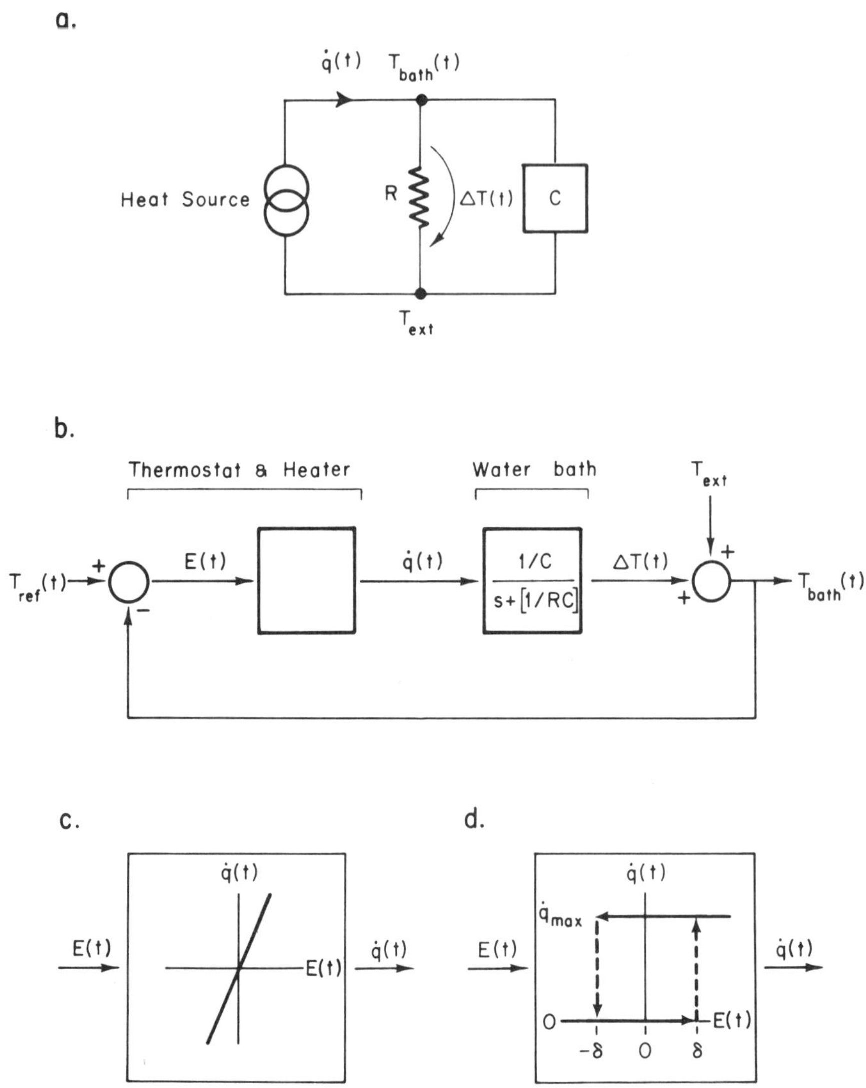

Figure 6.2.1.a, Circuit diagram, and b, c, and d, block diagram for the thermostat-water bath system of Section 6.2.

In part b, the empty block for the heater whose input is E(t) and whose output is $\dot{q}(t)$ is to be replaced by block c for proportional control, or by block d for the more usual nonlinear, on-off thermostat with hysteresis which is fully discussed in the text.

$$T_{bath}(t) = \{T_{bath}(0) - T_{bath}(\infty)\}\, \epsilon^{-\left[\frac{1 + k_pR}{RC}\right] t} + T_{bath}(\infty) \qquad 6.2.6,$$

where $T_{bath}(\infty)$, the steady-state bath temperature approached as $t \to \infty$ can be calculated from Figure 6.2.1.b and c by letting $s = 0$.

$$T_{bath}(\infty) = \frac{T_{ext} + k_pRT_{ref}}{1 + k_pR} \qquad 6.2.7.$$

Furthermore, the open-loop gain for $s = j\omega = j0$ (which, as was pointed out in Section 4.14, is a measure of the steady-state effectiveness of a negative feedback loop) is

$$[OLG]\,(j0) = k_pR \qquad 6.2.8.$$

In this simple linear first-order system there is no danger of instability, so in theory, by increasing k_p more and more (particularly if we also insulate the bath more effectively so as to increase R) we should be able to achieve any desired precision of temperature control. But in practice there are several things wrong with this scheme. In the first place, according to equation 6.2.5 and Figure 6.2.1.c, when E(t) is negative (as it will be if the external temperature is higher than the set-point temperature) q(t), the heat supplied to the bath, must also be negative. In other words, for high outside temperatures, our bath heater must become a bath cooler! Devices of this sort exist, but they are expensive. In the second place, even if we confine our attention to external temperatures below the desired set-point temperature, to achieve a large k_p with ordinary insulation we would have to install a huge heater equipped with an expensive mechanism for varying its energy supply (presumably electrical) according to the difference between the reference temperature and the bath temperature.

Now actually, as the reader knows perfectly well, satisfactory regulation of bath temperature can be obtained by using a very inexpensive thermostat which acts as a relay, turning the heater full on when the bath temperature falls to some specified value, say $T_{ref}-\delta$, below the desired temperature, and turning it full off when the bath temperature rises to some specified value, say $T_{ref}+\delta$, above the desired temperature. To keep the bath temperature within a reasonable range, the relay is usually so constructed that δ is small. The behavior of this nonlinear device is shown in Figure 6.2.1.d.

However successful such a thermostat may be in practice, it is a nasty thing to deal with mathematically. Properly speaking, the output is not even a function of the input! For with a function, each value of the independent variable (here E(t)), must correspond to one and only one value of the dependent variable (here q(t)). But in the present example (Figure 6.2.1.d), every value of E(t) between $-\delta$ and $+\delta$ corresponds to two different values of q(t): zero when the bath is cooling off (so that E(t) is rising); maximum when the bath is heating up (so that E(t) is falling). This kind of phenomenon is known as hysteresis, and a nonlinear component showing hysteresis is called a nonlinearity with

memory. Furthermore, as already mentioned, heaters are not usually reversible devices, able to cool as well as to heat. This is why the plot in Figure 6.2.1.d is not symmetrical about the origin.* In spite of these difficulties, do not despair! For with a system as simple as this, we can conduct our analysis piecewise. After all, at any particular instant the heater is either on, delivering $\dot{q}_{max}$ cal per min to the bath, or it is off delivering zero cal per min to the bath. If we confine our attention to non-transient states, so that the influence of any initial conditions has disappeared, there are only three possible situations:
(1.) T_{ext} is so low that the heater is on all the time; $\dot{q}_{max}$ just balances the loss of heat from the bath, and $T_{bath}(t)$ remains constant. Then by equations 6.2.1 and 6.2.2, with $\dot{q}(t) = \dot{q}_{max}$,

$$T_{bath}(t) = T_{ext} + R\dot{q}_{max} \qquad 6.2.9.$$

(2.) T_{ext} is so high (above $T_{ref} + \delta$) that the heater remains off all the time. Then $\dot{q}(t) = 0$, and

$$T_{bath}(t) = T_{ext} \qquad 6.2.10.$$

(3.) Between these two extremes, the heater keeps turning on and off as the bath temperature fluctuates cyclicly between $T_{ref} - \delta$, the turn-on temperature, and $T_{ref} + \delta$, the turn-off temperature. Consider any one cycle, starting when the bath has just cooled to $T_{ref} - \delta$. The cycle will consist of two distinct periods: a heating period (when $\dot{q}(t) = \dot{q}_{max}$) which lasts until $T_{bath}(t)$ has risen to $T_{ref} + \delta$, and a cooling period (when $\dot{q}(t) = 0$) which lasts until $T_{bath}(t)$ has fallen to $T_{ref} - \delta$ again. We shall treat these two periods separately. In particular, for each period we shall redefine time zero so that it occurs at the beginning of the period. In the heating period, from equations 6.2.1 and 6.2.2,

$$T_{bath}(t) = T_{ext} + R\dot{q}_{max} + \{T_{bath}(0) - T_{ext} - R\dot{q}_{max}\}\, \epsilon^{-[1/RC]t} \qquad 6.2.11,$$

where $T_{bath}(0)$ is, of course, $T_{ref} - \delta$, the bath temperature at $t = 0$, the start of the period we are considering. Hence, during the heating portion of the cycle,

$$T_{bath}(t) = T_{ext} + R\dot{q}_{max} + \{T_{ref} - \delta - T_{ext} - R\dot{q}_{max}\}\, \epsilon^{-[1/RC]t};\ 0 \le t \le t_{off} \qquad 6.2.12.$$

As soon as $T_{bath}(t)$ reaches $T_{ref} + \delta$ at $t = t_{off}$, the heater turns off. We reset the time scale to $t = 0$ as the start of the cooling period which lasts until $T_{bath}(t) = T_{ref} - \delta$, thus ending the cycle at $t = t_{on}$. The same kind of derivation as was used to obtain equation 6.2.12 shows that during the cooling period

$$T_{bath}(t) = T_{ext} + \{T_{ref} + \delta - T_{ext}\}\, \epsilon^{-\frac{1}{RC}t};\ 0 \le t \le t_{on} \qquad 6.2.13.$$

[*Unsymmetrical behavior of this sort is shunned by mathematical engineers, who would avoid it by defining the output of the nonlinearity as $\dot{q}_{act}(t) - [\dot{q}_{max}/2]$ where $\dot{q}_{act}(t)$ is the actual output and $\dot{q}_{max}$ is the maximum output.

For our present purposes, such a readjustment of the origin of the output scale is unnecessary]

As a numerical example, consider a 4-liter water bath where C = 4kg-cal/°C, R = 5° C·min/kg-cal, T_{ref} = 37° C, δ = 1° C, q_{max} = 4 kg-cal/min. (For illustrative purposes, the range through which the bath temperature is allowed to fluctuate, ± 1.0° C, has been chosen rather large.) In Figure 6.2.2, one or more cycles of bath temperature have been plotted against time for two external temperatures. When T_{ext} = 27° C, each "on" period and each "off" period lasts about 4 min, a time which is rather short compared with the exponential time constant, RC = 20 min. This is why each segment of the curve for T_{ext} = 27° C looks like a straight line, though it is actually a small portion of an exponential curve. In contrast when T_{ext} = 35.5° C, the rate of cooling (now much slower than the rate of heating) is obviously exponential.

Let us find $\overline{T}_{bath}$, the mean temperature of the bath during the cycle, by dividing the integral of bath temperature with respect to time over the whole cycle by the duration of the cycle.

$$\overline{T}_{bath} = \frac{\int_0^{t_{off}} T_{bath}(t)dt + \int_0^{t_{on}} T_{bath}(t)dt}{t_{off} + t_{on}} \qquad 6.2.14,$$

where the limits of integration are defined separately for each integral. To simplify notation, we shall express both equation 6.2.12 for the heating period and equation 6.2.13 for the cooling period in the form

$$T_{bath}(t) = a + b\epsilon^{-\frac{1}{RC}t} \qquad 6.2.15,$$

where

$$a = T_{ext} + R\dot{q}(t)$$

$$b = \begin{cases} T_{ref} - \delta - a \text{ for the heating period} \\ T_{ref} + \delta - a \text{ for the cooling period.} \end{cases}$$

However, we shall continue to deal with each period separately. To find t_{off}, solve equation 6.2.15 for t when $T_{bath}(t) = T_{ref} + \delta$.

$$[T_{ref} + \delta - a]/b = \epsilon^{-\frac{t}{RC}t_{off}} \qquad 6.2.16.$$

Inverting both sides of equation 6.2.16, and taking logarithms,

$$\ln\left\{\frac{T_{ref} - a - \delta}{T_{ref} - a + \delta}\right\} = \frac{1}{RC}t_{off} \qquad 6.2.17,$$

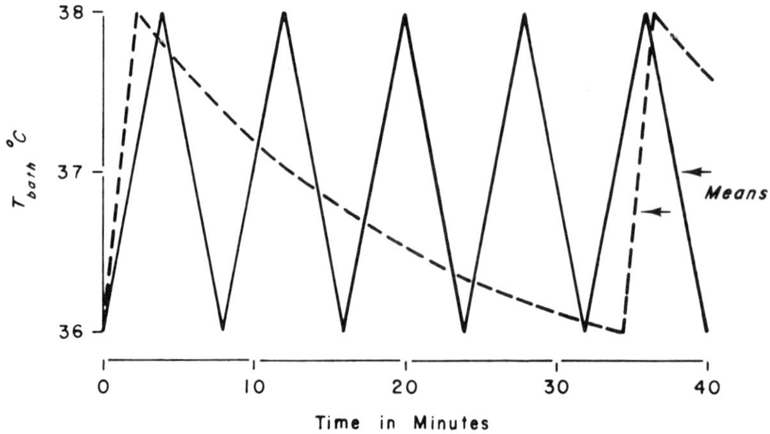

Figure 6.2.2. Temperature of the water bath of Section 6.2 as a function of time at external temperatures of 27° C (solid curve) and 35.5° C (broken curve).

The numerical values used to calculate the curves are given in the text. The mean bath temperature at each external temperature is indicated by an arrow at the right of the graph.

---

whence,

$$t_{off} = RC \ \ln \left\{ \frac{T_{ref} - T_{ext} - R\dot{q}_{max} - \delta}{T_{ref} - T_{ext} - R\dot{q}_{max} + \delta} \right\} \qquad 6.2.17.a.$$

The same series of steps applied to the cooling period gives

$$t_{on} = RC \ \ln \left\{ \frac{T_{ref} - T_{ext} + \delta}{T_{ref} - T_{ext} - \delta} \right\} \qquad 6.2.18.$$

For example, using the same numerical values as before, when $T_{ext} = 35.5°C$, equation 6.2.17.a gives

$$t_{off} = 20 \ln \left\{ \frac{19.5}{17.5} \right\} = 2.17 \text{ min} \qquad 6.2.19,$$

$$t_{on} = 20 \ln \left\{ \frac{2.5}{0.5} \right\} = 32.19 \text{ min} \qquad 6.2.20,$$

so the total duration of the cycle (the denominator in equation 6.2.14) is

$$t_{off} + t_{on} = 2.17 + 32.19 = 34.36 \text{ min} \qquad 6.2.21.$$

We next evaluate the integrals in equation 6.2.14. Using equation 6.2.15 for the heating period,

$$\int_0^{t_{off}} T_{bath}(t)\,dt = \int_0^{t_{off}} a + b\epsilon^{-\frac{1}{RC}t}\,dt = at - RCb\epsilon^{-\frac{1}{RC}t}\Big|_0^{t_{off}}$$

$$= at_{off} + RCb - RCb\epsilon^{-\frac{1}{RC}t_{off}} \qquad 6.2.22.$$

Substituting the value of $b\epsilon^{-\frac{1}{RC}t_{off}}$ from equation 6.2.16 into equation 6.2.22, next replacing b by $T_{ref} - \delta - a$, and finally replacing a by $T_{ext} + R\dot{q}(t)$,

$$\int_0^{t_{off}} T_{bath}(t)\,dt = at_{off} - 2RC\delta = [T_{ext} + R\dot{q}_{max}]\,t_{off} - 2RC\delta \qquad 6.2.23.$$

Similarly, for the cooling period,

$$\int_0^{t_{on}} T_{bath}(t)\,dt = T_{ext}t_{on} + 2RC\delta \qquad 6.2.24.$$

Substituting the expressions from equations 6.2.23 and 6.2.24 into equation 6.2.14,

$$\overline{T}_{bath} = T_{ext} + \frac{R\dot{q}_{max}t_{off}}{t_{on} + t_{off}} \qquad 6.2.25,$$

where t_{off} and t_{on} are to be calculated from equations 6.2.17.a and 6.2.18. Continuing with our previous numerical example,

$$\overline{T}_{bath} = 35.5 + \frac{20\,[2.17]}{34.36} = 36.76^\circ C$$

In Figure 6.2.3, $\overline{T}_{bath}$, regarded as the output of the system, has been plotted against the load input, T_{ext}, for our numerical example. The broken line illustrates proportional control. For $T_{ext} \le 37^\circ C$, it was calculated from equation 6.2.7 with $k_p = 5$[kg-cal per min]/°C. As you would expect, there is a simple linear relationship between the deviation of T_{bath} from 37°C and the deviation of T_{ext} from 37°C: When T_{ext} is 13°C below 37°C, T_{bath} is 0.5°C below 37°C; when T_{ext} is 26° C below 37°C, T_{bath} is 1.0°C below 37°C, etc. The slope of the line is thus 1/26 which of course is the value of the slope, $1/[1+k_pR]$, in equation 6.2.7 when $k_p = 5$[kg-cal per min]/°C and R = 5°C/[kg-cal per min]. For $T_{ext} > 37^\circ C$, there is no feedback control because the heater is continuously off. Hence $T_{bath} = T_{ext}$, and the slope of the line is unity.

Now look at the solid curve in Figure 6.2.3 which shows the peculiar relationship between mean bath temperature and external temperature when the bath temperature is regulated via the nonlinear on-off relay with memory.

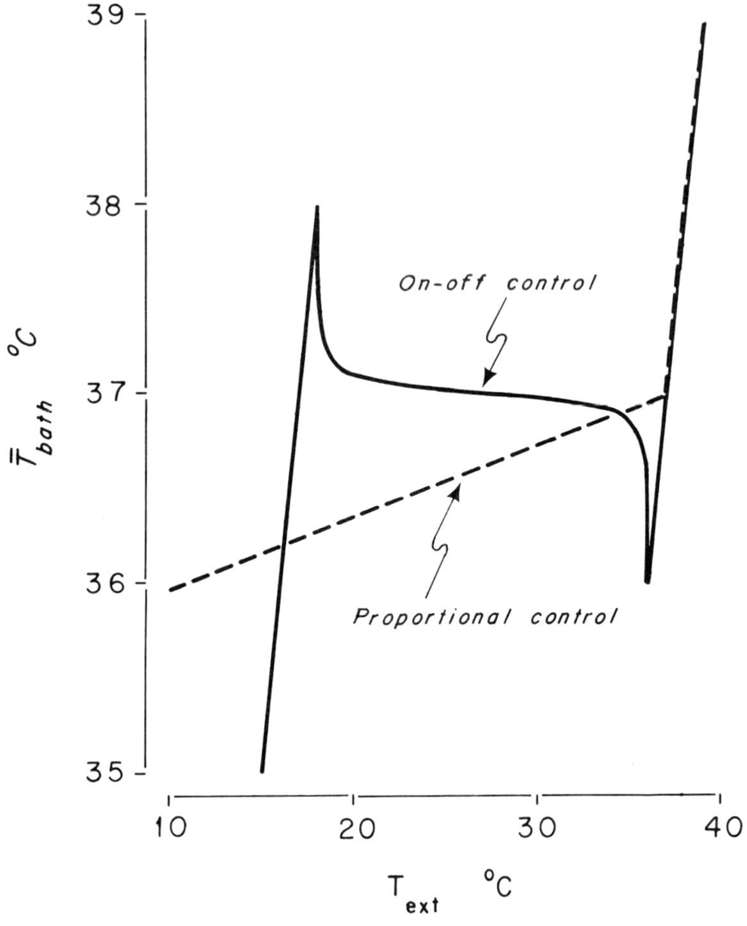

Figure 6.2.3. Mean water bath temperature as a function of the external temperature. (Section 6.2).

Broken lines: The result of using a heater with proportional control (equation 6.2.7.) for external temperatures below 37°C. For higher temperatures, where the heater is continuously off, the mean bath temperature equals the external temperature.

Solid lines: The peculiar result of using an on-off relay with hysteresis to control the heater. At very high external temperatures ($T_{ext} > 36$°C), when the heater is continuously off, or at very low external temperatures ($T_{ext} < 18$°C) when the heater is continuously on, regulation fails. But in the intermediate range where 18°C $< T_{ext} <$ 36°C, the mean bath temperature decreases as the external temperature increases.

The three portions of the curve correspond to the three situations previously analyzed: (1) When $T_{ext} \leq 18°$, $\overline{T}_{bath}$ is determined by equation 6.2.9. The heater can only supply 4 kg-cal/min, which, with $\overline{R} = 5$ [°C•min]/kg-cal, is just enough to maintain a temperature difference of $\overline{T}_{bath} - T_{ext} = q_{max}R = 20°C$ (Equation 6.2.9). So when T_{ext} is < 18°C, $\overline{T}_{bath}$ is < 38°C, the turn-off temperature of 38°C is never reached, and the heater stays on continuously. Under these circumstances, any further decrease in external temperature causes an equal decrease in bath temperature, so there is no regulation of bath temperature. (2) When $T_{ext} > 36°C$, $\overline{T}_{bath}$ is determined by equation 6.2.10. T_{bath} never falls low enough for the heater to turn on, and the mean bath temperature equals the external temperature, again showing no regulation. (3) When $18°C \leq T_{ext} \leq 36°C$, $\overline{T}_{bath}$ is regulated within $\pm\,\delta = \pm\,1°C$ of 37°C, the set-point temperature. But what a paradoxical relationship! Throughout this entire range, but most noticably at its extremes, <u>the mean bath temperature rises as the external temperature falls!</u> By considering Figure 6.2.2 again and imagining what the curves in that figure would look like for a T_{ext} <u>very</u> slightly less than 36°C or <u>very</u> slightly more than 18°C, the reader should have no difficulty in explaining why mean bath temperature behaves in such an odd way.

The preceding analysis of the thermostat highlights some of the difficulties commonly encountered when dealing with nonlinear systems. (1) In a nonlinear system a change in the input no longer causes a proportional change in the output. Therefore the principle of superposition, upon which all of linear analysis is founded, no longer holds and we cannot apply any of the previous methods, such as matrix analysis, or transfer function analysis, or phasor analysis directly to non-linearities. (2) There are no truly <u>general</u> methods for nonlinear systems which can take the place of the ones we have to abandon. As in our example, nonlinear problems often have to be attacked piece by piece, and our choice of weapons for the attack will depend heavily upon the nature of the individual problem. (3) In stable linear systems, sustained oscillatory behavior never occurs except in response to an oscillatory input. In nonlinear systems, sustained oscillations (Cf. Figure 6.2.2) may occur in response to a <u>constant</u> input. This suggests (and quite correctly) that the simple criteria of stability used for linear systems are not directly applicable to nonlinear systems. (4) Though not illustrated by the present example, when we encounter a system whose behavior must be described by a nonlinear differential equation, we shall usually have to content ourselves with an approximate solution.

As a consequence of these difficulties, we shall often have to resort to tedious numerical methods when dealing with nonlinear systems. The wise investigator will therefore make as much use of computers,--both analog and digital--, as he possibly can.

<u>SECTION 6.3.</u> General ways of studying nonlinear systems

The first step in analyzing a nonlinear system is to identify the nonlinearities and to specify their relationships to other components in the system, for example by drawing a block diagram of the sort shown in Figure 6.2.1. If there is but one nonlinearity, the remainder of the system may be treated as a linear subsystem, but if there are two or more nonlinearities interspersed among linear elements, only an uninterrupted sequence of linear elements can be analysed together as a linear subsystem. The reason is that shifting the position of a nonlinearity with respect to another element may alter the output completely. An example is shown in Figure 6.3.1.

Next, consider any particular nonlinearity. Its input-output relationship is either algebraic or it is not. If it is algebraic, then for any known input we can calculate the output exactly, (just as we did in the thermostat example.) We may then use the output of the nonlinearity as the input for the next component. You might think therefore that when a system contains only this kind of nonlinearity we should always be able to construct a complete and exact solution even though we might have to do it bit by bit. Unfortunately this is not true. In fact, the reason we could solve the thermostat problem completely was that we were able to break it into a finite number of parts each part, by itself, being a strictly linear problem. When this is impossible, we are in deeper trouble!

Example 6.3.1. Consider again the regulation of temperature in the water-bath problem of Section 6.2. As before, let the two inputs, T_{ref} and T_{ext}, remain constant during any particular calculation, and for convenience let all temperatures be measured as $^{\circ}$C above T_{ext} so that $T_{ext} = 0$. Then equation 6.2.1 becomes,

$$\frac{dT_{bath}(t)}{dt} + \frac{T_{bath}(t)}{RC} = \frac{\dot{q}(t)}{C} \qquad 6.3.1.$$

We shall again define the error by equation 6.2.4.

$$E(t) = T_{ref} - T_{bath}(t) \qquad 6.3.2.$$

But now, instead of having a simple on-off relay controlling $\dot{q}(t)$, let us suppose that $\dot{q}(t)$ is proportional to the square of the error.

$$\dot{q}(t) = kE^2(t) \quad ; \qquad E(t) \geq 0 \qquad 6.3.3.$$

Combining equations 6.3.2 and 6.3.3,

$$\dot{q}(t) = k[T_{ref}^2 - 2T_{ref}\,T_{bath}(t) + T_{bath}^2(t)] \qquad 6.3.4,$$

an equation which will certainly allow us to calculate the exact heat input to the bath for any bath temperature. However, when we put the expression for $\dot{q}(t)$ from equation 6.3.4 into equation 6.3.1, we get

$$\frac{dT_{bath}(t)}{dt} + \frac{T_{bath}(t)}{RC} = \frac{kT_{ref}^2}{C} - \left[\frac{2kT_{ref}}{C}\right] T_{bath}(t) + \left[\frac{k}{C}\right] T_{bath}^2(t) \qquad 6.3.5,$$

or,

$$\frac{dT_{bath}(t)}{dt} + \left[\frac{1 + 2kT_{ref}R}{RC}\right] T_{bath}(t) - \left[\frac{k}{C}\right] T_{bath}^2(t) = \left[\frac{kT_{ref}^2}{C}\right] \qquad 6.3.5.a,$$

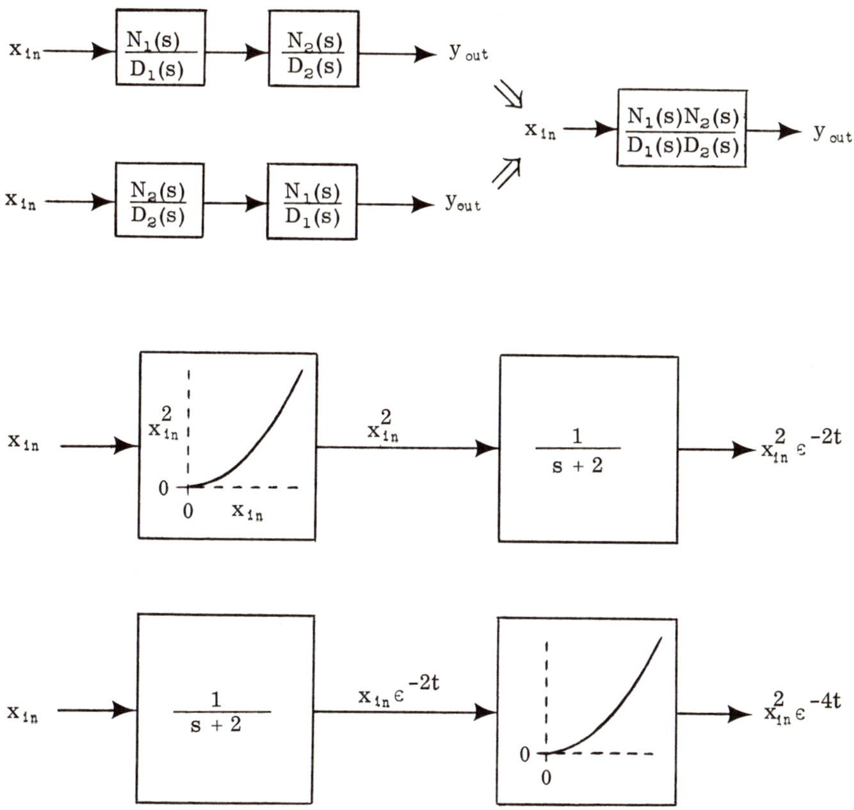

Figure 6.3.1. Two adjacent linear elements in an unbranched series can be interchanged without altering the output (upper part of figure), but this is not true when one of the elements is nonlinear, as exemplified in the lower part of the figure.

---

which is of the form

$$\frac{dT_{bath}(t)}{dt} + aT_{bath}(t) - bT_{bath}^2 = c \qquad 6.3.6,$$

where a, b, c are constants.

Now equation 6.3.6 is a nonlinear differential equation which presents us with the very same difficulties we would have faced ab initio had the nonlinearity itself needed a differential equation to specify how its output depended upon its input. Nonlinear systems will frequently confront us with equations such as 6.3.6, and several of the subsequent sections will be devoted to methods for handling nonlinear differential equations.

Finally, suppose that through ignorance (or indolence!) we cannot (or do not care to) analyze the time-dependent behavior of a nonlinear feedback system. All need not be lost; for it may still be possible to undertake an analysis of the steady-state behavior of the system which will provide useful information about [OLG] (j0) as an index of how well the system copes with external disturbances (See Section 7.3)

SECTION 6.4. Nonlinear differential equations: General remarks

Consider a general first-order differential equation describing a stationary linear system,

$$\dot{y}(t) + ay(t) = f(t) \qquad 6.4.1,$$

together with the initial condition,

$$y(0) = y_0 \qquad 6.4.2.$$

With the constants a and y_0 known, and with the forcing function, f(t) also known, we can use familiar methods to find the unknown variables, y(t) and $\dot{y}(t)$ at any time t. In other words, we can find the locus of y(t) on the t, y(t) plane as t increases. But the basic reason we are able to do this is that equation 6.4.2 tells us exactly where to start on the t, y(t) plane, while equation 6.4.1 tells us how to get to the next point, and from there to the next, and thence to the next, and so on forever.

Now consider a general first-order differential equation describing a stationary nonlinear system,

$$\Sigma\, a_{ijk}\dot{y}^{i}(t)y^{j}(t)f^{k}(t) = 0 \quad ; \quad \begin{matrix} i = 0,\ 1,\ 2,\ 3\ldots \\ j = 0,\ 1,\ 2,\ 3\ldots \\ k = 0,\ 1,\ 2,\ 3\ldots \end{matrix} \qquad 6.4.3,$$

where the indices may occur in all possible permutations, with one term in the sum for each permutation. By algebraic manipulation, we can usually express equation 6.4.3 in the form,

$$\dot{y}(t) = \varphi(y(t),\ f(t)) \qquad 6.4.4.$$

As always, we must know both f(t) the input forcing function, and the initial condition,

$$y(0) = y_0 \qquad 6.4.5.$$

Now, just as with the first-order linear equation, equation 6.4.5 tells us where to start on the t, y(t) plane, and equation 6.4.4 tells us how to get to the next point. The major difference between the linear case and the nonlinear case is that with a linear equation we have methods for obtaining an explicit expression for y(t) which is not only exact at every instant of time but which can also be written in symbolic form so that it can be applied to any system of the same general sort, whatever its parameters may be. In contrast, to solve most nonlinear equations we really do have to start at the given point [t = 0, y(t) = y_o] and move slowly along, point by laborious point. If we could actually calculate every point this way, our solution would be as exact as is the analytical solution of a linear equation. But with an infinite number of points to calculate, this is hardly practical. Instead, we seek an approximate (but acceptably accurate) solution by taking a succession of small finite steps starting with the initial condition and predicting one additional point at each step. Several methods for doing this will be explained and illustrated in the following sections. Unfortunately, the solution obtained with any such numerical method is valid only for the particular parameters which were used in the calculation; the entire sequence of steps has to be repeated if the numerical values of the parameters are changed.

SECTION 6.5. Nonlinear differential equations. Solution by direct integration

In this section and in most of the following ones we shall confine attention to first-order nonlinear differential equations.

Whenever we are confronted by such an equation, the first thing to do is to see whether the variables can be separated. If so, we should seek an analytical solution, either by actually carrying out the integrations on both sides of the equation or by looking them up in a table of integrals. For example, in equation 6.3.6, the variables $T_{bath}(t)$ and t can be separated.

$$\int \frac{dT_{bath}(t)}{c - aT_{bath}(t) + bT^2_{bath}(t)} = \int dt \qquad 6.5.1.$$

The integral on the left can be found in a table:

$$\int \frac{dT_{bath}(t)}{c - aT_{bath}(t) + bT^2_{bath}(t)} = \frac{2}{\sqrt{v}} \tan^{-1}\left(\frac{2bT_{bath}(t) - a}{\sqrt{v}}\right) + K \qquad 6.5.2,$$

where $v = 4bc - a^2$, and K is a constant of integration to be evaluated from the initial condition. We therefore have

$$\frac{2}{\sqrt{v}} \tan^{-1}\left(\frac{2bT_{bath}(t) - a}{\sqrt{v}}\right) + K = t \qquad 6.5.3.$$

where

$$K = -\frac{2}{\sqrt{v}} \tan^{-1}\left(\frac{2bT_{bath}(0) - a}{\sqrt{v}}\right)$$

an equation which allows us (with the aid of a table of Arctans) to calculate $T_{bath}(t)$ for any t.

SECTION 6.6. Nonlinear differential equations. Linear approximation

If the variables are not separable, we shall rarely be able to obtain a precise analytical solution, and we shall probably have to content ourselves with some kind of an approximation.*

[*In Chapter 4 of his text, Cunningham (1) lists various kinds of differential equations, including nonlinear ones, for which exact solutions can be found analytically, and he illustrates the solution of each type by an example.]

The easiest (therefore the most tempting!) approximation is to assume that within our range of interest the deviation of the system from linearity is small enough to neglect. We can then replace the nonlinear differential equation with an appropriate linear differential equation, thereby reducing the problem to one we can solve with our familiar linear tools. The dangers of this head-in-the-sand, "there-really-isn't-any-appreciable-nonlinearity" approach are obvious enough. But when it can be justified by objective arguments, or by comparison with the results of a trial computation by a method which does not neglect the nonlinearity, it would be folly not to exploit the tremendous advantages of linear analysis. As an example which has already been discussed, by confining consideration to small displacements it was possible to treat the simple pendulum of Section 5.2 as a linear system. (Compare equations 5.2.13 and 5.2.15.)

Even when the nonlinearity is too great to be neglected throughout the entire range of interest, it may still be possible to treat the system as linear within some sufficiently small region in the immediate neighborhood of a given operating point. Linearization about a point is, in one way or another, the most important tool of the nonlinear analyst! Several of the methods soon to be described are merely different ways of exploiting this basic technique.

SECTION 6.7. Numerical approximation. General remarks

In the next few sections, several approximate methods of solution will be presented. Following Cunningham (1) I shall illustrate each method by applying it to the solution of the same very simple linear equation, namely,

$$\frac{dx(t)}{dt} = -2x(t) + 3t \qquad 6.7.1,$$

$$x(0) = 1/4$$

whose exact solution,

$$x(t) = \varepsilon^{-2t} + [3/2]t - [3/4] \qquad 6.7.2,$$

is easily obtained analytically. The simplicity of the equation will make it easy to illustrate the steps used in each computation without wasting too much time with tedious arithmetic. And since the exact solution for any t can be obtained from equation 6.7.2, we can see just how accurate each approximation is.

In dealing with discrete points, the following symbols are useful:

t_n = the value of the independent variable for the nth calculated point.

$h = \Delta t$ = the amount by which the independent variable is increased when calculating a new point from a previous point.

x_n = the value of the dependent variable for the nth calculated point.

$x_o = x(0)$ = the given value of x(t) at t = 0.

$x_n^{(m)} = \left.\frac{d^m x(t)}{dt^m}\right|_{t = t_n}$ = the mth derivative of x(t) evaluated at $t = t_n$.

Note that (m) is not an exponent; it simple denotes the mth derivative. Note also that when m = 0, the function x(t) is "differentiated zero times," so $x^{(0)}(t) = x(t)$.

We shall assume that our starting point is given in the form of an initial condition, so that its coordinates on the t, x(t) plane are $[0, x_o]$. If some other point is given, say at $t = t_1$, it is easy enough to move the origin of the t scale to t_1 by working with $\tau \equiv t - t_1$ instead of t itself.

SECTION 6.8. The point-slope, or Euler method

This is the simplest possible method, and the least accurate. Each value of x(t) is calculated from the preceding point in the t, x(t) plane by assuming that between the two points, x(t) changes at the constant rate which is given by evaluating the differential equation at the preceding point. For our example, namely,

$$\frac{dx(t)}{dt} = -2x(t) + 3t \; ; \; x(0) = 1/4 \qquad 6.8.1,$$

the relation assumed is

$$x_{n+1} = x_n + x_n^{(1)}h = x_n + h[-2x_n + 3t_n] = x_n[1 - 2h] + 3ht_n \qquad 6.8.2.$$

If we choose h = 0.1, equation 6.8.2 becomes

$$x_{n+1} = 0.8x_n + 0.3t_n, \qquad 6.8.3,$$

and the computation would proceed, row by row, as shown in Table 6.8.1.

Table 6.8.1. The point-slope method for equation 6.7.1.

			Column: A	B	A + B
n	t_n	x_n	$0.8x_n$	0.3t	x_{n+1}
		given			
0	0	0.2500	0.2000	0.0000	0.2000
1	0.1	0.2000	0.1600	0.0300	0.1900
2	0.2	0.1900	0.1520	0.0600	0.2120
3	0.3	0.2120	0.1696	0.0900	0.2596
4	0.4	0.2596	etc.		

Notice that if h is kept constant, $t_n = nh$.

In a computation such as this, it is wise to carry one or even two more significant figures than are wanted in the final result. This counteracts the tendency for rounding errors to accumulate during an iterative computation.

Figure 6.8.1. shows the results of this simple straight-line extrapolation from point to point. When h = 0.2 the estimated points are very far from the true curve. With so long an interval, extrapolation from t = 0 to t = 0.2 according to the known negative slope at t = 0 completely neglects the rapid change to a positive slope and gets the whole computation off to a miserable start. While a choice of h = 0.1 gives somewhat better results, the estimated curve still lies regrettably far from the true one. Indeed, the rapid divergence of the line which is tangent at t = 0 from the true curve suggests that for a really good approximation we might have to choose h = 0.025 which would mean forty steps of calculation to cover the range in Figure 6.8.1,--a heavy price to pay for the simplicity of a single calculation!

SECTION 6.9. Series approximations. Taylor's series

A function, x(t), is an analytic function throughout a given domain of t if it has a finite derivative at every point in that domain. If we know the equation for the derivative of such a function (i.e., our differential equation) and if we know any one point [a, x(a)] in the domain (e.g. [0, x(0)] given by the initial conditions) we can calculate the value of x(t) at any other point [a + h, x(a + h)] to any desired precision by evaluating a sufficient number of terms in Taylor's series,

$$x(a+h) = \sum_{n=0}^{\infty} \left[\frac{x^{(n)}(a)}{n!}\right] h^n$$

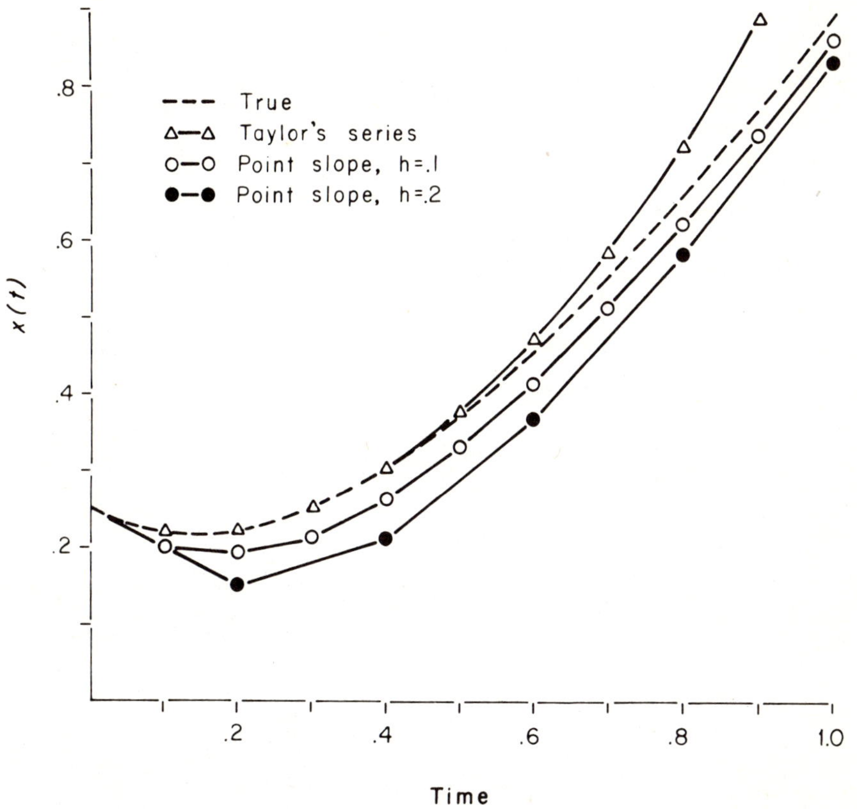

Figure 6.8.1. The function $x(t) = \epsilon^{-2t} + [3/2]t - [3/4]$ (broken line) and three numerical approximations of it calculated from equation 6.7.1.

For the Taylor's series approximation, the first five terms were used to estimate x(a + h) with $a = t_o = 0$ throughout, and h = 0.1, 0.2, 0.3, etc.

$$= \frac{x^{(0)}(a)}{0!}h^0 + \frac{x^{(1)}(a)}{1!}h^1 + \frac{x^{(2)}(a)}{2!}h^2 + \frac{x^{(3)}(a)}{3!}h^3 + \cdots + \frac{x^{(n)}(a)}{n!}h^n + \cdots \quad 6.9.1,$$

where

$$\frac{x^{(0)}(a)}{0!}h^0 = x(a) \text{ because } 0! \equiv 1.$$

In preparation for using Taylor's series, let us obtain the first three higher derivatives by successive differentiation,* starting with the first derivative which is given by our differential equation.

$$\left.\begin{aligned} x^{(1)}(t) & & &= -2x(t) + 3t \quad \text{(given)} \\ x^{(2)}(t) &= -2x^{(1)}(t) + 3 & &= 4x(t) - 6t + 3 \\ x^{(3)}(t) &= 4x^{(1)}(t) - 6 & &= -8x(t) + 12t - 6 \\ x^{(4)}(t) &= -8x^{(1)}(t) + 12 & &= 16x(t) - 24t + 12 \end{aligned}\right\} \quad 6.9.2.$$

[*or, alternatively, by calculating the partial derivatives in the equation,

$$x^{(n+1)}(t) = \frac{\partial x^{(n)}(t)}{\partial t} + \left[\frac{\partial x^{(n)}(t)}{\partial x}\right] x^{(1)}(t)$$

e.g. $x^{(4)}(t) = 12 + [-8][-2x(t) + 3t] = 12 + 16x(t) - 24t$]

Two ways of using equation 6.9.1 will be exemplified. In the first, we extend the solution to any desired point from the given initial point by keeping $a = 0$ throughout and letting h vary. For $a = 0$, equations 6.9.2 becomes

$$\left.\begin{aligned} x^{(1)}(0) &= -2x_o & &= -0.5 \\ x^{(2)}(0) &= 4x_o + 3 & &= 4 \\ x^{(3)}(0) &= -8x_o - 6 & &= -8 \\ x^{(4)}(0) &= 16x_o + 12 & &= 16 \end{aligned}\right\} \quad 6.9.3.$$

Substituting equations 6.9.3 into equation 6.9.1,

$$x(h) \approx 0.25 - 0.5h + 2h^2 - \frac{4}{3}h^3 + \frac{2}{3}h^4 \qquad 6.9.4,$$

which can be evaluated for any desired value of h. For example,

$$x(0.2) \approx 0.25 - 0.5[0.2] + 2[0.04] - \frac{4}{3}[0.008] + \frac{2}{3}[0.0016] =$$

$$0.2500 - 0.1000 + 0.0800 - 0.0107 + 0.0011 = 0.2204 \qquad 6.9.5,$$

$$x(0.8) \approx 0.25 - 0.5\,[0.8] + 2\,[0.64] - \frac{4}{3}\,[0.512] + \frac{2}{3}\,[0.4096] =$$

$$0.2500 - 0.4000 + 1.2800 - 0.6827 + 0.2737 = 0.7210 \qquad 6.9.6.$$

The results of such calculations are shown in Figure 6.8.1. It is obvious from the figure that the five terms of Taylor's series which we used above give a good approximation to the true value for x(0.2) because the last term is already small compared to three-figure accuracy. This is not true for x(0.8). We could take more terms as illustrated in Figure 6.9.2 but that would increase the labor of computation considerably. This suggests the second way of using equation 6.9.1, namely to let h remain constant and to extend the solution from each newly calculated point instead of anchoring it upon the point given by the initial conditions. For example, with h = 0.2, our first calculation would be the same as in equation 6.9.5. But for the second step, we would re-evaluate equations 6.9.2 using the value x(0.2) = 0.221 from equation 6.9.5.

$$\left.\begin{aligned} x^{(1)}(0.2) &= -2\,[0.221] + 3\,[0.2] = 0.158 \\ x^{(2)}(0.2) &= 4\,[0.221] - 6\,[0.2] + 3 = 2.684 \\ x^{(3)}(0.2) &= -8\,[0.221] + 12\,[0.2] - 6 = -5.368 \\ x^{(4)}(0.2) &= 16\,[0.221] - 24\,[0.2] + 12 = 10.736 \end{aligned}\right\} 6.9.7.$$

Substituting equations 6.9.7 into equation 6.9.1 with h = 0.2,

$$x(0.4) \approx 0.221 + 0.158\,[0.2] + 1.342\,[0.04] - 0.8947\,[0.008] + 0.4473\,[0.0016]$$

$$= 0.2210 + 0.0316 + 0.0537 - 0.0072 + 0.0007$$

$$= 0.2998$$

which is very close to the true value. Further pursuit of this method would, in fact, continue to yield points lying practically upon the theoretical line. But at what a cost in labor! Even for a simple equation, the point-by-point re-evaluation of the derivatives, and their subsequent use in equation 6.9.1 is a tiresome business, and serves chiefly to make us properly grateful for the more efficient methods of approximation described in Section 6.11 and Section 6.12.

Note in passing that if we use only the first two terms of Taylor's series in the method of point-by-point extension just described, this method reduces to the simple point-slope method of Section 6.8.

SECTION 6.10. Series approximations. Power series representation

For analytic functions, there is another, and in some respects more instructive method of obtaining the successive terms in Taylor's series. By

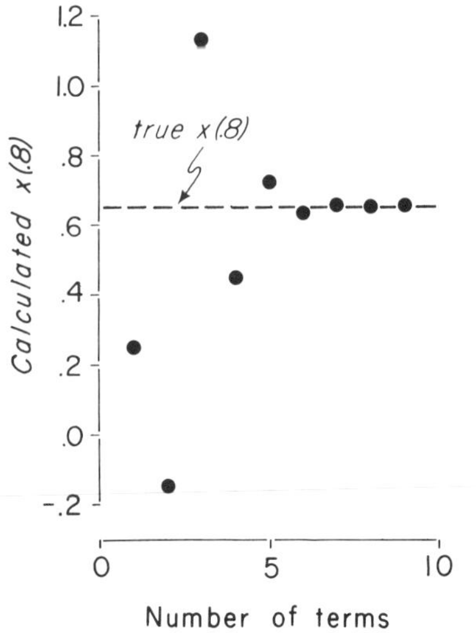

Figure 6.9.2. How a Taylor's series approximation approaches the true value of x(0.8) as the number of terms in the series increases.

inspection of equation 6.9.1, it is evident that when a = 0, h = t, and the equation is then of the general form,

$$x(t) = b_0 + b_1 t + b_2 t^2 + b_3 t^3 + \cdots + b_n t^n \qquad 6.10.1,$$

where the b's are constants depending only upon the known value,

$$x(0) = x_0 = b_0 \qquad 6.10.2,$$

and upon the values of the various derivatives at t = 0. Direct differentiation of equation 6.10.1 gives

$$\frac{dx(t)}{dt} = b_1 + 2b_2 t + 3b_3 t^2 + \cdots + nb_n t^{n-1} + (n+1)b_{n+1} t^n + \cdots \qquad 6.10.3.$$

By writing our differential equation,

$$\frac{dx(t)}{dt} = -2x(t) + 3t \qquad 6.10.4,$$

in terms of these infinite series, we obtain

$$b_1 + 2b_2 t + 3b_3 t^2 + \cdots + nb_n t^{n-1} + (n+1)\, b_{n+1} t^n + \cdots =$$

$$-2b_0 - [2b_1 - 3]\, t - 2b_2 t^2 - \cdots - 2b_{n-1} t^n - 2b_n t^n - \cdots \qquad 6.10.5.$$

To evaluate the constant b's, we use the theorem (2) that two infinite power series are equal if and only if they are equal term by term. (Or, equivalently, we may note that since equation 6.10.5 must be true for all values of t, the coefficients of like powers of t must be equal.) Hence,

$$\left.\begin{aligned}
b_1 &= -2b_o = -2x_o \\
b_2 &= -\frac{1}{2}[2b_1 - 3] = +\frac{1}{2}\cdot 2^2 x_o + \frac{1}{2}\cdot 3 \\
b_3 &= -\frac{1}{3}[2b_2] \quad = -\frac{1}{2}\cdot\frac{1}{3}2^3 x_o - \frac{1}{2}\cdot\frac{1}{3}\cdot 2.3 \\
b_4 &= -\frac{1}{4}[2b_3] \quad = \frac{1}{2}\cdot\frac{1}{3}\cdot\frac{1}{4}2^4 x_o + \frac{1}{2}\cdot\frac{1}{3}\cdot\frac{1}{4}\cdot 2^2.3 \\
&\quad\vdots \\
b_n &= -\frac{1}{n}2b_{n-1} \quad = [-1]^n \left\{\frac{2^n x_o}{n!}\right\} + [-1]^n \left\{\frac{3\cdot 2^{n-2}}{n!}\right\} \\
&= \frac{[-1]^n 2^{n-2}}{n!}[4x_o + 3] = [x_o + \frac{3}{4}]\,\frac{[-2]^n}{n!}
\end{aligned}\right\} \quad 6.10.6,$$

or, for $b_o = x_o = 0.25$,

$$\left.\begin{aligned}
b_1 &= -0.5 \\
b_2 &= 0.5 + 1.5 = 2 \\
b_3 &= -1/3 - 1 = -4/3 \\
b_4 &= \frac{1}{6} + \frac{1}{2} \quad = 2/3
\end{aligned}\right\} \quad 6.10.7.$$

Substituting equations 6.10.7 into equation 6.10.1,

$$x(t) \approx 0.25 - 0.5t + 2t^2 - \frac{4}{3}t^3 + \frac{2}{3}t^4 \qquad 6.10.8.$$

But since $a = 0$, $h = t$. Therefore (mirabile dictu!) equation 6.10.8, derived without obtaining any higher derivatives, is identical to equation 6.9.4.

Before turning to other methods of approximation, we may use the present example to show that with a truly infinite number of terms the series representation actually does yield the exact solution. Substituting equations 6.10.6 into equation 6.10.1, and rearranging terms,

$$x(t) = x_o - \frac{[2t]}{1!}x_o + \frac{[2t]^2}{2!}x_o - \frac{[2t]^3}{3!}x_o + \frac{[2t]^4}{4!}x_o - \cdots + \frac{[-1]^n[2t]^n}{n!}x_o + \cdots$$
$$+ \frac{t^2}{2!}\cdot 3 - \frac{2t^3}{3!}\cdot 3 + \frac{2^2t^4}{4!}\cdot 3 - \cdots + \frac{[-1]^n 2^{n-2} t^n}{n!}\cdot 3 + \cdots$$

6.10.9.

But it is well known that any exponential, ϵ^x, may be expressed by the infinite series

$$\epsilon^x = 1 + \frac{x}{1!} + \frac{x^2}{2!} + \frac{x^3}{3!} + \cdots + \frac{x^n}{n!} + \cdots \qquad 6.10.10.$$

Therefore the first group of terms on the right of equation 6.10.9 can be written as

$$x_0 \left[1 + \frac{[-2t]}{1!} + \frac{[-2t]^2}{2!} + \frac{[-2t]^3}{3!} + \cdots + \frac{[-2t]^n}{n!} + \cdots\right]$$

$$= x_0\,\epsilon^{-2t} = \frac{1}{4}\epsilon^{-2t} \qquad 6.10.11,$$

which suggests trying to make a similar complete series out of the second group of terms so that it, too, will define an exponential. Notice that in the second group the numerator of each term contains $[-2]^{n-2}[t^n]$ as a factor. Let us make it $[-2t]^n$ by multiplying each term by 2^2, at the same time dividing by 4 so as to leave the expression as a whole unchanged.

$$3\left[\frac{t^2}{2!} - \frac{2t^3}{3!} + \frac{2^2t^4}{4!} - \cdots + \frac{[-2]^{n-2}t^n}{n!}\right] =$$

$$\frac{3}{4}\left[\frac{[-2t]^2}{2!} + \frac{[-2t]^3}{3!} + \frac{[-2t]^4}{4!} + \cdots + \frac{[-2t]^n}{n!} + \cdots\right] \qquad 6.10.12.$$

Next, we complete the series on the right by adding $3/4 + [3/4][-2t/1!]$, subtracting these same terms outside the brackets so that again we leave the value of the expression as a whole unchanged.

$$\frac{3}{4}\left[1 + \frac{[-2t]}{1!} + \frac{[-2t]^2}{2!} + \frac{[-2t]^3}{3!} + \cdots + \frac{[-2t]^n}{n!} + \cdots\right] - \frac{3}{4} + \frac{3}{2}t =$$

$$\frac{3}{4}\epsilon^{-2t} - \frac{3}{4} + \frac{3}{2}t \qquad 6.10.13.$$

Substituting the values obtained in equations 6.10.11 and 6.10.13 into equation 6.10.9,

$$x(t) = \frac{1}{4}\epsilon^{-2t} + \frac{3}{4}\epsilon^{-2t} + \frac{3}{2}t - \frac{3}{4} = \epsilon^{-2t} + \frac{3}{2}t - \frac{3}{4} \qquad 6.10.14,$$

which is precisely what we obtained (equation 6.7.2) by ordinary analytical methods!

For an easy equation like this, obtaining the solution by means of an infinite series is like travelling from New York to Washington via San Francisco. But for certain less tractable equations, an infinite series may be the only practical way to express the solution exactly. For this purpose, it is convenient to use the general formula for the recursive relationship which tells how to calculate any additional coefficient, b_n, (except irregular ones at the beginning of the series) from those already calculated. For our example, the last of equations 6.10.6 gives the recursive relationship, so the full solution in series form could be written compactly as

$$x(t) = x_0 - 2tx_0 + \left[x_0 + \frac{3}{4}\right] \sum_{n=2}^{\infty} \frac{[-2t]^n}{n!} \qquad 6.10.15.$$

Certain important functions which are not analytic can also be expressed by infinite series of a form more complicated than a simple power series. But a discussion of such series, in fact of infinite series in general, lies beyond the scope of this book, and the interested reader must consult more advanced texts. (2, 3, 4.)

SECTION 6.11. The modified Euler method, with a "predictor-corrector" feature

In several of the methods for solving differential equations by numerical approximation, substantial improvement in accuracy is obtained without too must cost in computational labor by calculating each new point from two or more known points instead of one. Since only one point (the initial condition) is known to begin with, one or more additional points have to be calculated initially be some starting formula, usually Taylor's series. The simplest of these methods is the modified Euler method which needs only one additional point. If h is chosen judiciously small, this method should yield results quite accurate enough for most physiological problems, and the arithmetical steps can be performed fairly rapidly, even when no computer more elaborate than a slide rule is available. I would suggest, however, a quiet evening at home with a stack of your favorite LP records to provide background music.

We begin by using Taylor's series to calculate x_1 at $t = h$ from our known initial condition at $t = 0$. Let $h = 0.1$. Then four terms should more than suffice. By equation 6.9.4,

$$x_1 = x(0.1) \approx 0.25 - 0.5[0.1] + 2[0.01] - \frac{4}{3}[0.001] = 0.2187 \qquad 6.11.1.$$

Next, we note that Taylor's series can be used just as legitimately to extrapolate backwards as forwards. For an increment, h, of t, by equation 6.9.1,

$$x_{n+1} = x(a+h) = x(a) + \left[\frac{x^{(1)}(a)}{1!}\right] h + \left[\frac{x^{(2)}(a)}{2!}\right] h^2 + \left[\frac{x^{(3)}(a)}{3!}\right] h^3 + \cdots \qquad 6.11.2.$$

For a decrement, -h, of the same magnitude,

$$x_{n-1} = x(a-h) = x(a) - \left[\frac{x^{(1)}(a)}{1!}\right] h + \left[\frac{x^{(2)}(a)}{2!}\right] h^2 - \left[\frac{x^{(3)}(a)}{3!}\right] h^3 + \cdots \qquad 6.11.3.$$

Subtract equation 6.11.3 from equation 6.11.2.

$$x_{n+1} - x_{n-1} = 2h\,\frac{x^{(1)}(a)}{1!} + 2h^3\,\frac{x^{(3)}(a)}{3!} + \cdots \qquad 6.11.4.$$

Neglect all terms of order h^3 or higher in equation 6.11.4, and solve for x_{n+1}.

$$x_{n+1} \approx x_{n-1} + 2h\,x^{(1)}(a) = x_{n-1} + 2h\,x_n^{(1)} \qquad 6.11.5,$$

which is the predictor equation for the modified Euler method. Applying this to our illustrative problem with h = 0.1, we have

$$\left.\begin{aligned}
x_0 &= 0.2500 && \text{(given)}\\
x_1 &= 0.2187 && \text{(From Taylor's series, equation 6.11.1)}\\
x_2 &= 0.2500 + 0.2\{-2[0.2187] + 3[0.1]\} = 0.2225\\
x_3 &= 0.2187 + 0.2\{-2[0.2225] + 3[0.2]\} = 0.2497\\
x_4 &= 0.2225 + 0.2\{-2[0.2497] + 3[0.3]\} = 0.3026\\
&\text{etc.}
\end{aligned}\right\} \qquad 6.11.6,$$

The results of these calculations are compared with the true curve in Figure 6.11.1.

Notice that this is a simple point-slope method of extending the solution, just as the unmodified Euler method was. But in the unmodified method all terms of Taylor's series in powers of h higher than the first were neglected while in the modified method all terms in powers of h higher than the second are neglected. For small values of h this means a very substantial increase in accuracy (compare Figure 6.11.1 with Figure 6.8.1). Geometrically, the unmodified method predicts the second of two points from the location of the first point and the slope at the first point. The modified method predicts the third of three points from the location of the first point and the slope at the middle point.

Once the third point has been estimated, the estimate can usually be improved by calculating the slope at the estimated third point, averaging it with the slope at the middle point, and using this average slope and the location of the middle point to obtain a corrected value for the third point. If we

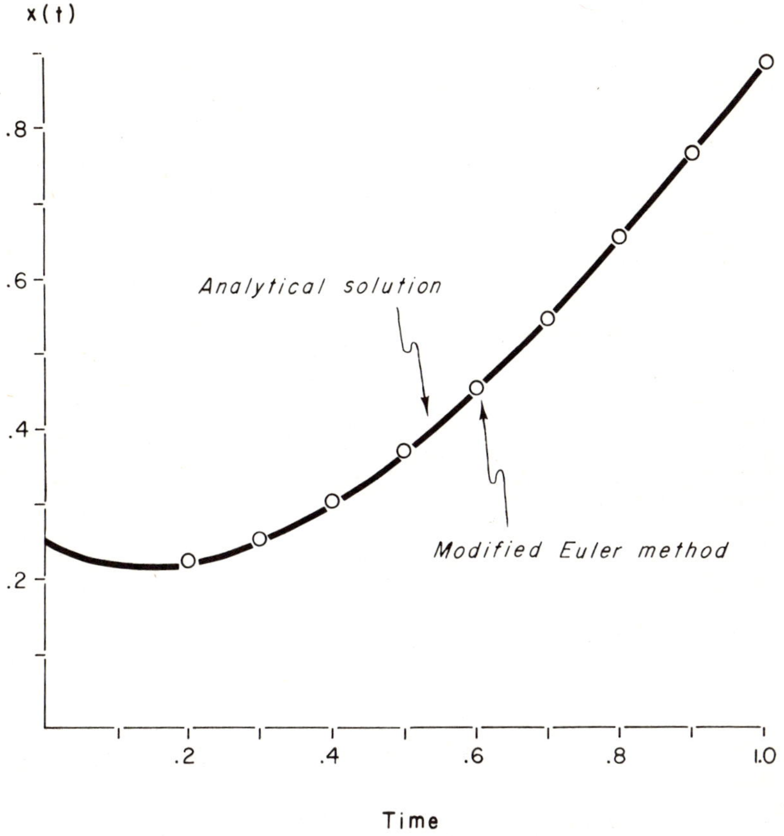

Figure 6.11.1. Predictions by the modified Euler method (points) compared with the true analytical solution (curve).

distinguish the corrected value by an asterisk, the corrector equation is

$$x^*_{n+1} = x_n + \left[\frac{x_n^{(1)} + x_{n+1}^{(1)}}{2}\right] h = x_n + \frac{h}{2}\left\{x_n^{(1)} + x_{n+1}^{(1)}\right\} \qquad 6.11.7.$$

The corrected value is then used in the predictor equation for the next step of the calculation. This refinement serves as a running check on the calculations. If the difference between x_{n+1} and x^*_{n+1} becomes large, e.g. more than 2 or 3 in the third significant figure, the results are likely to be inaccurate, and a smaller value of h should be chosen. If we apply the corrector equation to our example, the series of steps shown in equations 6.11.6 becomes

$$\left.\begin{aligned}
x_0 &= 0.2500 \\
x_1 &= 0.2187 \\
x_2 &= 0.2500 + 0.2\{-2[0.2187] + 3[0.1]\} = 0.2225 \\
x_2^* &= 0.2187 + 0.05\{-0.1374 - 2[0.2225] + 3[0.2]\} = 0.2196 \\
x_3 &= 0.2187 + 0.2\{-2[0.2196] + 3[0.2]\} = 0.2509 \\
x_3^* &= 0.2196 + 0.05\{0.1608 - 2[0.2509] + 3[0.3]\} = 0.2476 \\
x_4 &= 0.2196 + 0.2\{-2[0.2476] + 3[0.3]\} = 0.3006 \\
x_4^* &= 0.2476 + 0.05\{0.4048 - 2[0.3006] + 3[0.4]\} = 0.2978 \\
&\text{etc.}
\end{aligned}\right\} \quad 6.11.8.$$

These calculations are best arranged in the form of a table, similar to the one in Section 6.8, with enough columns so that calculations for both x_n and x_n^* can be carried out in the same row.

While the corrector equation <u>usually</u> improves the accuracy of the estimate it does not always do so. Consider the following example

<u>Example 6.11.1.</u> Given

$$\frac{dx(t)}{dt} = -\frac{t}{x} \qquad 6.11.9.$$

and the two points $x(-0.3) = 0.74162$, $x(0) = 0.80000$, estimate $x(0.3)$ with $h = 0.3$, using the modified Euler method.

The predictor equation gives

$$x(0.3) = 0.74162 + 0.6\left[-\frac{0}{0.80000}\right] = 0.74162 \qquad 6.11.10$$

The corrector equation gives

$$x^*(0.3) = 0.80000 + 0.15\left[0 + \frac{-0.3}{0.74162}\right] = 0.73932 \qquad 6.11.11.$$

The analytical solution gives

$$x_{true}(0.3) = \sqrt{0.64 - [0.3]^2} = 0.74162 \qquad 6.11.12.$$

Hence, in this example, the predicted value is correct while the "corrected" value is off the mark by about 3 parts in 1000. If we plot a curve (Figure 6.11.2) for the analytical solution,

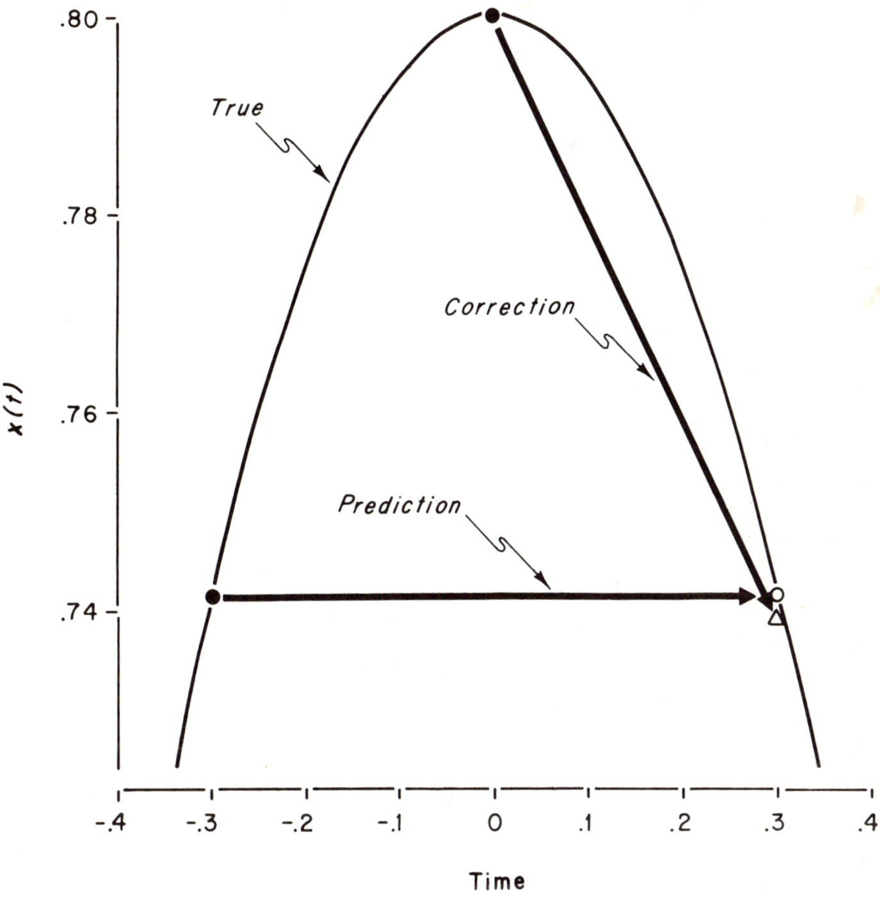

Figure 6.11.2. The "corrector" equation does not necessarily yield a "correcter" estimate! (Example 6.11.1.)

Given $\dot{x}(t) = -t/x(t)$, $x(-0.3) = 0.74162$, and $x(0) = 0.80000$, the predictor equation of the modified Euler method extrapolates from the first point to the third point (open circle) according to the slope of the curve at the second point. The corrector equation extrapolates from the second point to a "corrected" third point (triangle) according to the mean of the slopes at the second point and at the predicted third point. But in this example, the predicted point is correct, while the "corrected" point is slightly off the true curve.

$$x(t) - \sqrt{0.64 - t^2} \qquad 6.11.13.$$

in the region of interest, and show the lines of extrapolation for "prediction" and "correction," the reason for the anomaly becomes clear. But it is equally clear that this kind of situation will not often arise, and that even when it does, no great harm will be done provided h is chosen sufficiently small.

SECTION 6.12. The Runge-Kutta method

The "given" information, namely the differential equation to be solved,

$$\frac{dx(t)}{dt} = f(t, x) \qquad 6.12.1,$$

and the initial condition,

$$x(0) = x_0 \qquad 6.12.2.$$

together with the selected value of h can be used to define a series of constants in terms of which the solution of the equation can be extended from point to point without having to use either a starting formula or any higher derivatives. Methods of this sort are generically referred to as Runge-Kutta methods, though they now include several specific modifications of the original Runge-Kutta algorithms. When stated (as is done here) with neither derivation nor justification a Runge-Kutta method has about it a faintly astrological air. But if we are willing to accept it on faith, it yields astonishingly accurate results. Unfortunately, the magical constants must be reevaluated at each step, which is woefully inconvenient by hand but is nothing to a digital computer. The technique described below is therefore recommended only for those physiologists who either have access to a computer or have been shipwrecked (with slide-rule, pencils, and lots of paper, but without the cartoonist's traditional companion of the opposite sex) on a desert island.

For a 4th-order Runge-Kutta calculation, we define

$$\left.\begin{aligned} k_1 &\equiv hf(t_n, x_n) = hx_n^{(1)} \\ k_2 &\equiv hf(t_n + \frac{h}{2}, x_n + \frac{k_1}{2}) \\ k_3 &\equiv hf(t_n + \frac{h}{2}, x_n + \frac{k_2}{2}) \\ k_4 &\equiv hf(t_n + h, x_n + k_3) \end{aligned}\right\} 6.12.3.$$

Then,

$$x_{n+1} = x_n + \frac{1}{6}\left[k_1 + 2k_2 + 2k_3 + k_4\right] \qquad 6.12.4.$$

As before, suppose we want to solve

$$\left.\begin{aligned} &\frac{dx(t)}{dt} = f(t, x) = -2x(t) + 3t \\ &x(0) = x_0 = 0.2500 \end{aligned}\right\} \quad 6.12.5.$$

With h = 0.1, we evaluate equations 6.12.3 for the <u>first</u> step as

$$\left.\begin{aligned} k_1 &= 0.1\{-2[0.2500] + 0\} = -0.0500 \\ k_2 &= 0.1\{-2\left[0.2500 - \frac{0.05}{2}\right] + 3\left[0 + \frac{0.1}{2}\right]\} = -0.0300 \\ k_3 &= 0.1\{-2\left[0.2500 - \frac{0.03}{2}\right] + 3\left[0 + \frac{0.1}{2}\right]\} = -0.0320 \\ k_4 &= 0.1\{-2[0.2500 - 0.032] + 3[0 + 0.1]\} = -0.0136 \end{aligned}\right\} \quad 6.12.6,$$

whence, by equation 6.12.4,

$$x(0.1) = x_1 = 0.2500 - \frac{1}{6}[0.05 + 0.06 + 0.064 + 0.0136] = 0.218733 \quad 6.12.7,$$

compared with

$$x_{true}(0.1) = 0.218731 \quad 6.12.8.$$

Fantastic! But to carry on with any comfort by hand, even as an example, practically requires us to cast five-significant-figure pearls before three-significant-figure swine. Very well. <u>One</u> more exemplary step, to make sure we know exactly how to program it on the computer,

$$k_1 = 0.1\{-2[0.2187] + 3[0.1]\} = -0.01374$$

$$k_2 = 0.1\left\{-2\left[0.2187 - \frac{0.01374}{2}\right] + 3\left[0.1 + \frac{0.1}{2}\right]\right\} = +0.002634$$

$$k_3 = 0.1\left\{-2\left[0.2187 + \frac{0.002634}{2}\right] + 3\left[0.1 + \frac{0.1}{2}\right]\right\} = +0.000997$$

$$k_4 = 0.1\{-2[0.2187 + 0.00100] + 3[0.1 + 0.1]\} = +0.01606$$

So,

$$x(0.2) = x_2 = 0.2187 + \frac{1}{6}[-0.01374 + 0.00528 + 0.00200 + 0.01606]$$
$$= 0.2203 \quad 6.12.10.$$

$$x_{true}(0.2) = 0.2203 \qquad 6.12.11.$$

Bring on the computer!*

[*There is now available a digital computer language, MIMIC, which makes it quite easy to write programs for solving sets of ordinary linear or nonlinear differential equations simultaneously. The program uses a modified fourth-order Runge-Kutta procedure in which the increment, h, is automatically adjusted between steps so that the accuracy of the approximation remains constant throughout all steps.]

SECTION 6.13. Solution of an equation describing glomerular filtration

As an example of how the solution of a nonlinear differential equation may help us to understand the behavior of a biological system, let us try to find out how the rate of glomerular filtration is influenced by the concentration of protein in plasma. This is a problem of some importance, because, if there were no hormonal feedback regulation of water excretion by the kidneys, one of the few mechanisms available for getting rid of a water load would be the decrease in plasma oncotic pressure (and hence the increase in net glomerular filtration pressure) caused by dilution of the plasma proteins (5).

Consider what happens within a single glomerular capillary. At the beginning of the capillary, where it arises from the afferent arteriole, the intraluminal blood pressure which tends to force water and small solutes out of the capillary into Bowman's capsule considerably exceeds the sum of the opposing hydrostatic pressure in Bowman's capsule and the oncotic pressure of the plasma proteins. The fluid which, as a result, is lost from the plasma becomes part of the glomerular filtrate. The same forces are concerned at every other point in the capillary, but as the plasma flows towards the efferent arteriole there is a gradual decrease in the effective filtration pressure. This decrease is due partly to the fall in intraluminal pressure which occurs because of the resistance to flow through the capillary tube, and partly to the rise in oncotic pressure which occurs as the filtration of fluid increases the concentration of the plasma proteins retained within the capillary.

To describe this process quantitatively, let us consider the model shown in Figure 6.13.1. In the model, all of the millions of glomerular capillaries are lumped together as a single equivalent tube whose characteristics will be so chosen as to represent reasonable values for glomerular filtration in the dog. The symbols, definitions, and exemplary values which we shall use are given in Table 6.13.1. Notice that for this system the independent variable is not time, but is distance, x, measured along the length of the capillary in the direction of blood flow.

The derivation is based upon the following assumptions. (1.) the filtration rate per unit area and per unit of net filtration pressure remains constant. (2.) The capillary radius remains constant. (3.) No plasma protein leaks out of the capillary. (4.) The decrease in blood pressure per unit of capillary length remains constant throughout the capillary. (5.) The hydrostatic pressure in Bowman's capsule remains constant along the capillary.

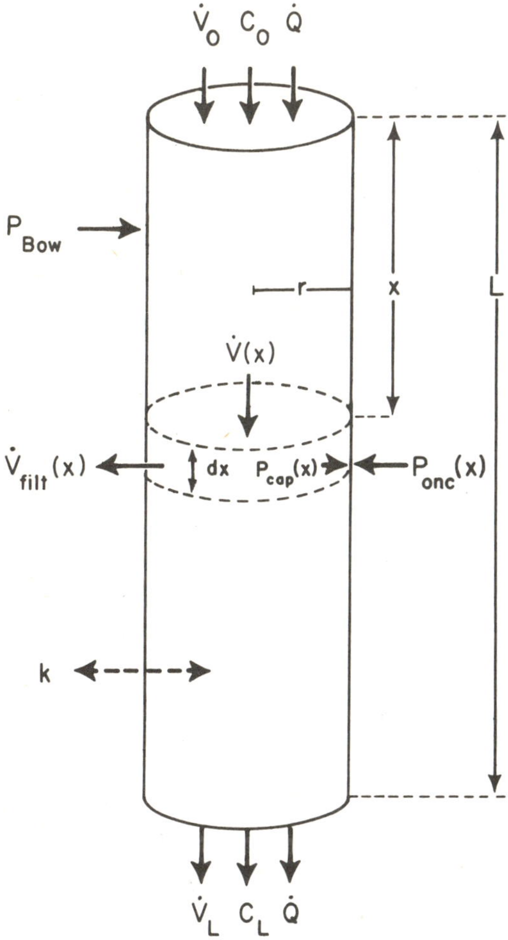

Figure 6.13.1.
The "e pluribus unum" glomerular capillary.

Consider a segment of capillary of length dx lying at a distance x from the beginning of the capillary. The area, A, of the capillary wall of this annular segment is

$$A = 2\pi\ r\ dx \qquad 6.13.1.$$

The net filtration pressure, $P_{filt}(x)$, is

$$P_{filt}(x) = P_{cap}(x) - P_{onc}(x) - P_{Bow} \qquad 6.13.2.$$

Table 6.13.1. Filtration through glomerular capillary. Symbols and definitions.

Type	Symbol	Definition	Exemplary value
Independent coordinate	x	distance along capillary	0 through L
Parameters	L	total length of capillary	?
	r	capillary radius	?
	$\dot{Q}$	grams protein traversing capillary per sec	0.1333 g/sec
	$\dot{V}(0)$	plasma flow into capillary	2.0 ml/sec
	$\dot{V}(L)$	plasma flow out of capillary	1.4 ml/sec
	$P_{cap}(0)$	initial capillary blood pressure	85 mm Hg
	$P_{cap}(L)$	final capillary blood pressure	65 mm Hg
	P_{Bow}	pressure in Bowman's capsule	15 mm Hg
	k	Filtration constant. (Volume filtered per sec, per unit area, and per unit net filtration pressure.)	?
	γ	Empirical constant	2500 mm Hg/[g/ml]^2
	δ	Empirical constant	200 mm Hg/[g/ml]
Dependent variables	$\dot{V}(x)$	plasma flow at x	ml/sec
	$P_{cap}(x)$	capillary blood pressure at x	mm Hg
	$P_{onc}(x)$	plasma oncotic pressure at x	mm Hg
	$C(x)$	plasma protein concentration at x	gm/ml

By assumption 1, $-d\dot{V}(x)$, the decrease in plasma flow caused by the transudation of filtrate into Bowman's capsule through the wall of the annulus is

$$-d\dot{V}(x) = kAP_{filt}(x) \qquad 6.13.3.$$

Combining equations 6.13.1, 6.13.2, and 6.13.3,

$$\frac{-d\dot{V}(x)}{dx} = 2\pi r\, k\,[P_{cap}(x) - P_{onc}(x) - P_{Bow}] \qquad 6.13.4.$$

By assumption 4,

$$P_{cap}(x) = P_{cap}(0) - \frac{x}{L}\,[P_{cap}(0) - P_{cap}(L)] \qquad 6.13.5.$$

Scatchard (6) has studied the relation between the concentration of protein in plasma and its oncotic pressure. His experimental results (with human plasma) within our range of interest (below 0.12 gm protein per ml) can be fitted by the equation,

$$P_{onc}(C) = \gamma C^2 + \delta C \qquad 6.13.6,$$

where γ and δ are parameters of the fitted curve. We shall assume that Scatchard's observations are also valid for dog plasma. By definition,

$$C(x) = \frac{\dot{Q}}{\dot{V}(x)} \qquad 6.13.7,$$

where, by assumption 3, $\dot{Q}$ is a parameter independent of x. Combining equations 6.13.6 and 6.13.7,

$$P_{onc}(C(x)) = P_{onc}(x) = \gamma \left[\frac{\dot{Q}}{\dot{V}(x)}\right]^2 + \delta \left[\frac{\dot{Q}}{\dot{V}(x)}\right] \qquad 6.13.8.$$

Combining equations 6.13.4, 6.13.5, and 6.13.8,

$$-\frac{d\dot{V}(x)}{dx} = 2\pi rk \left\{P_{cap}(0) - \frac{x}{L}\,[P_{cap}(0) - P_{cap}(L)] - \gamma \left[\frac{\dot{Q}}{\dot{V}(x)}\right]^2 - \delta \left[\frac{\dot{Q}}{\dot{V}(x)}\right] - P_{Bow}\right\} \qquad 6.13.9.$$

Equation 6.13.9 now contains only the three variables, x, $\dot{V}(x)$, and $\frac{d\dot{V}(x)}{dx}$, and a number of constants. For most of the constants we can find plausible values obtained from physiological experiments. But how shall we assign numerical values to k, r, x, and L which depend upon the size assumed for the purely imaginary, e pluribus unum capillary in the model of Figure 6.13.1?

First, as far as x and L are concerned, it will be most convenient to replace x by a new independent variable defined as

$$X \equiv x/L \qquad 6.13.10,$$

where X is simply fraction of total length. Since X is a dimensionless ratio we do not have to choose any particular unit of measurement, such as cm, for our model capillary. (In a sense, we measure length in "units of L" so that when x = L, X = 1.) By equation 6.13.10, we can replace x in equation 6.13.9 by LX, thus obtaining

$$-\frac{d\dot{V}(LX)}{LdX} = 2\pi rk \left\{ P_{cap}(0) - X[P_{cap}(0) - P_{cap}(L)] - \gamma \left[\frac{\dot{Q}}{\dot{V}(LX)}\right]^2 - \delta \left[\frac{\dot{Q}}{\dot{V}(LX)}\right] - P_{Bow} \right\} \qquad 6.13.11.$$

Second, let us put all of our remaining difficulties into one package by defining a new constant, B,

$$B \equiv 2\pi rLk \qquad 6.13.12,$$

whose value is to be adjusted so that it will agree with physiological measurements of the actual rate of filtration (see below).

Combining equations 6.13.11, and 6.13.12,

$$-\frac{d\dot{V}(LX)}{dX} = B \left\{ [P_{cap}(0) - P_{Bow}] - [P_{cap}(0) - P_{cap}(L)]\, X - [\gamma \dot{Q}^2] \frac{1}{\dot{V}^2(LX)} - [\delta \dot{Q}] \frac{1}{\dot{V}(LX)} \right\} \qquad 6.13.13.$$

Substituting into equation 6.13.13 the values of the parameters in Table 6.13.1,

$$-\frac{d\dot{V}(LX)}{dX} = B \left\{ 70 - 20X - 2500 \frac{\dot{Q}^2}{\dot{V}^2(LX)} - 200 \frac{\dot{Q}}{\dot{V}(LX)} \right\} \text{ ml/sec} \qquad 6.13.14.$$

Consider what will happen when 2 ml of plasma per sec (120 ml/min) enter the glomerular capillaries. Our initial condition (more specifically called a <u>boundary condition</u> since the independent variable is length, rather than time) will then be

$$\dot{V}(0) = 2 \text{ ml/sec} \qquad 6.13.15.$$

With $\dot{V}(0) = 2$, and C = 1/15 gm protein per ml (= 6.67 gm per 100 ml), $\dot{Q} = 2\ (1/15)$, so 0.1333 gram of protein per sec will enter the capillary, and equation 6.13.14 becomes

$$\frac{d\dot{V}(LX)}{dX} = B\left\{20X + [44.44]\ \frac{1}{\dot{V}^2(LX)} + [26.67]\ \frac{1}{\dot{V}(LX)} - 70\right\}\ \text{ml/sec} \qquad 6.13.16$$

In a normal dog, the filtration fraction, defined as the glomerular filtration rate divided by the "effective plasma flow," is about 0.3. To evaluate the parameter B, let us therefore require that

$$\dot{V}(L) = \dot{V}(0) - 0.3\dot{V}(0) = 1.4\ \text{ml/sec} \qquad 6.13.17.$$

This requirement sets a second boundary condition at the end of the capillary.

Now unfortunately, there is no direct way to evaluate B. We must therefore use a trial-and-error approach. We select some reasonable value of B, and solve equation 6.13.16 for $\dot{V}(L)$, subject to the initial boundary condition $\dot{V}(0) = 2$ ml/sec. We then see how close this calculated value of $\dot{V}(L)$ comes to the 1.4 ml/sec required by the final boundary condition of equation 6.13.17. We next use this preliminary result to revise our estimate of B. Finally, the whole process is repeated as often as is necessary to find a tolerably accurate value for B. Only the most dedicated golfer will enjoy this approach-shot technique; the rest of us will move heaven and earth to get hold of an analog computer so that we can vary B electronically and arrive at a reasonable estimate long before the golfer has finally staggered into the 19th hole.

For purposes of illustration, let us carry on with the golfer.

Even if the rate of filtration remained constant along the capillary, to filter off a total of 0.6 ml/sec from 2 ml/sec of plasma entering would require the value of $d\dot{V}(LX)/dX$ to be -0.6 ml per sec. But we know that the true initial rate of filtration (at LX = 0) must be higher than this to compensate for the progressive decrease in rate of filtration along the capillary. Suppose we begin by guessing that the initial rate is 1.5 times -0.6 ml per sec. Then

$$\left.\frac{d\dot{V}(LX)}{dX}\right|_{X=0} = -\ 0.9\ \text{ml per sec} \qquad 6.13.18.$$

Evaluating equation 6.13.16 for $\dot{V}(0) = 2.0$ ml, and combining the result with equation 6.13.18, we obtain

$$B\,[11.11 + 13.33 - 70] = -\ 0.9\ \text{ml per sec} \qquad 6.13.19,$$

whence,

$$B = 0.9/45.56 = 0.01975\ \text{(ml per sec)/mmHg} \qquad 6.13.19.a.$$

Next, we put this tentative value of B into equation 6.13.16.

$$\frac{d\dot{V}(LX)}{dX} = 0.395X + 0.527\,\dot{V}^{-1}(LX) + 0.878\,\dot{V}^{-2}(LX) - 1.382 \qquad 6.13.20,$$

which we must solve subject to the boundary condition,

$$\dot{V}(0) = 2.00 \text{ ml per sec} \qquad 6.13.15.$$

For our first approach shot, we would be wise to use the simple point-slope method since high accuracy is not required. The slide-rule calculations by this method in Table 6.13.2 using a step increment of $h = \Delta X = 0.1$ give a value of 1.432 for $\dot{V}(L)$, corresponding to a filtration fraction of 0.284 instead of the desired 0.300, a difference of 5.67%. But the true discrepancy must be greater than this, for the point-slope method will certainly over-estimate the volume filtered. We can get a rough idea of how large the overestimate is by calculating $\dot{V}(0.1)$ from $\dot{V}(0)$ using the first three terms of Taylor's series. Equation 6.13.20 is of the form,

$$\frac{d\dot{V}(LX)}{dX} = aX + b\dot{V}^{-1}(LX) + c\dot{V}^{-2}(LX) - k \qquad 6.13.21.$$

Differentiation of equation 6.13.21 to find $d^2\dot{V}(LX)/dX^2$ is most easily done by using the method described in the footnote just under equation 6.9.2. We get

$$\frac{d^2\dot{V}(LX)}{dX^2} = a + \dot{V}^{-2}(LX)\Big\{bk - abX + [2ck - b^2 - 2acX]\dot{V}^{-1}(LX)$$

$$- 3bc\dot{V}^{-2}(LX) - 2c^2\dot{V}^{-3}(LX)\Big\} \qquad 6.13.22.$$

Evaluating equations 6.13.21 and 6.13.22 for $\dot{V}(0) = 2.0$ ml per sec and the numerical coefficients in equation 6.13.20,

$$\dot{V}^{(1)}(LX) = -\,0.899 \text{ ml per sec} \qquad 6.13.23,$$

$$\dot{V}^{(2)}(LX) = +\,0.711 \text{ ml per sec} \qquad 6.13.24.$$

Putting the values from equations 6.13.23 and 6.13.24 into Taylor's series (equation 6.9.1) for $\dot{V}(0.1)$,

$$\dot{V}(0.1) \approx 2.000 - 0.1\,[0.899] + \frac{[0.1]^2}{2}\,[0.711] = 1.9137 \text{ ml per sec} \qquad 6.13.25.$$

a decrease between $X = 0$ and $X = 0.1$ of 0.0863 ml per sec. By the point-slope method (first two terms of equation 6.13.25), the estimated decrease is 0.0899, a difference of about 4%. When added to the 5.67% difference previously

Table 6.13.2. Trial calculation of $\dot{V}(LX)$ from equation 6.13.20 using the point-slope method.

n	X	$\dot{V}_n(LX)$	$\frac{d\dot{V}(LX)}{dX}$	$\dot{V}_{n+1}(LX)$	Δ_1	Δ_2	Δ_3	Δ_4	Δ_5
				2.0000					
					.0899				
0	0.0	2.0000	-.8990	1.9101		.0073			
					.0826		-.0001		
1	0.1	1.9101	-.8257	1.8275		.0074		.0000	
					.0752		-.0001		.0000
2	0.2	1.8275	-.7517	1.7523		.0075		.0000	
					.0677		-.0001		.0003
3	0.3	1.7523	-.6769	1.6846		.0076		-.0003	
					.0601		.0002		-.0003
4	0.4	1.6846	-.6009	1.6245		.0074		.0000	
					.0527		.0002		-.0002
5	0.5	1.6245	-.5273	1.5718		.0072		.0002	
					.0455		.0000		.0006
6	0.6	1.5718	-.4547	1.5263		.0072		-.0004	
					.0383		.0004		-.0004
7	0.7	1.5263	-.3833	1.4880		.0068		.0000	
					.0315		.0004		
8	0.8	1.4880	-.3150	1.4565		.0064		-.0001	
					.0251		.0005		
9	0.9	1.4565	-.2507	1.4314		.0059		-.0001	
					.0192		.0006		
10	1.0	1.4314	-.1906	1.4122		.0053			
					.0139				
11	1.1			1.3983					

(Columns Δ_1 to Δ_5: Successive differences. Boxed values below the step line: values obtained by extrapolation.)

discussed, this suggests that our first guess at the value of B, namely B = 0.01975, was about 9.7% too small. Accordingly, for our next guess we let B = 0.01975 + 0.097 (0.01975) = 0.02167. Equation 6.13.16 now becomes

$$\frac{d\dot{V}(LX)}{dX} = 0.433X + 0.578\dot{V}^{-1}(LX) + 0.963\dot{V}^{-2}(LX) - 1.517 \qquad 6.13.26.$$

subject to the same initial boundary condition, $\dot{V}(0) = 2.00$ as before.

The steps used in deriving equation 6.13.26 have certainly landed us on the green, probably quite close to the cup. If, at this point we were seriously interested in three-figure accuracy, we would want to sink our putt with one of the more precise methods; e.g. the modified Euler method.

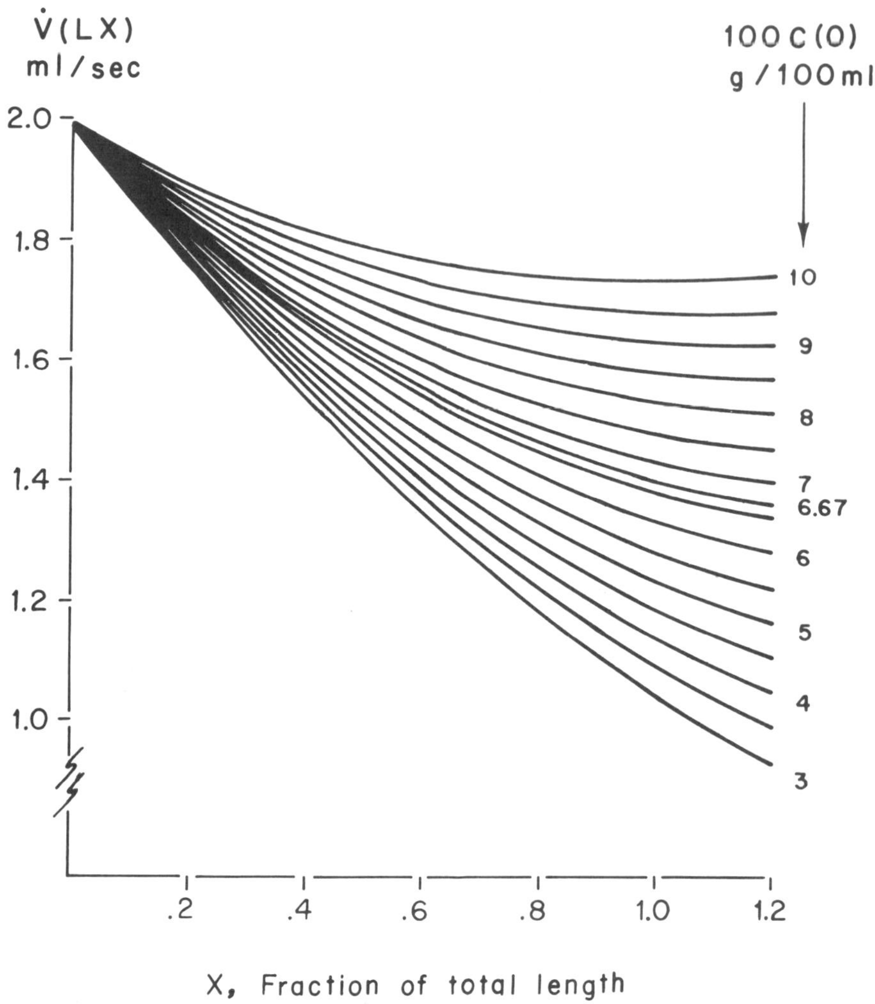

Figure 6.13.2.a. The decrease in capillary flow due to filtration of fluid along the capillary, for various concentrations of plasma protein.

In fact, however, further effort on this particular hole would be a criminal waste of time because (1) the physiological parameters themselves are not known with any precision, and (2) we set out to discover how much a change in plasma protein concentration would alter the glomerular filtration rate. Let us therefore by all means concede the putt and move on to the next hole. For we now realize that the only way to investigate the effect of changing one of the parameters is to solve a new equation for each of several values of that parameter. In doing this, however, we keep all other parameters,

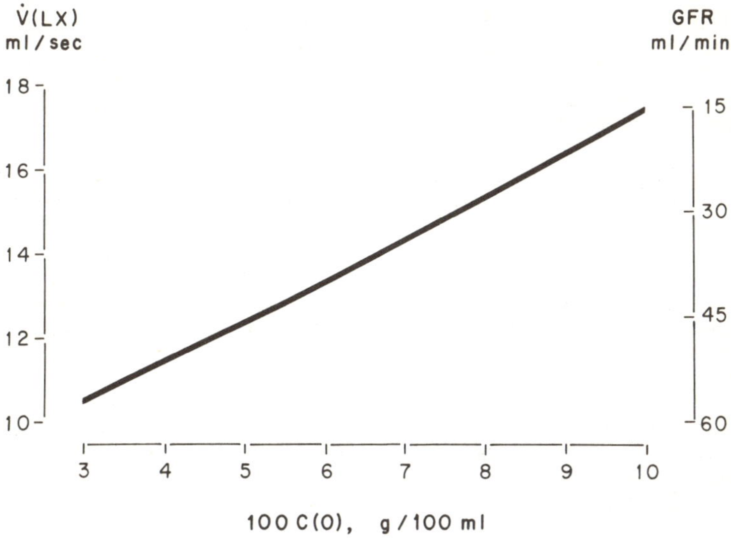

Figure 6.13.2.b. $\dot{V}(L)$, the plasma flow at the end of the glomerular capillary as a function of plasma protein concentration.

The ordinates for this curve are taken from the points where the curves in Figure 6.3.2.a intersect the abscissa for x = 1.0. The glomerular filtration rate (=GFR) in ml/min is 60 $[2.0 - \dot{V}(L)]$

---

including B, constant, so that we do not again have to adopt the trial-and-error approach which was needed initially to find a reasonable value for B. The effect which varying C (and therefore $\dot{Q}$) has upon glomerular filtration is shown in Figure 6.13.2.a (opposite page) and Figure 6.13.2.b (above). But I have to confess that the curves in those figures were obtained by using an analog computer, not by continuing to play a dismal game of arithmetical golf.

SECTION 6.14. Procedures for checking approximate solutions of differential equations

Numerical computations must always be assumed guilty of error until proved innocent.* Suppose that we have solved the equation

$$\frac{dx(t)}{dt} = \dot{x}(t) = f(t, x(t)) \qquad 6.14.1$$

by using some numerical method of approximation so that we have a series of values, x_i, such that $x_0 = x(0)$, $x_1 \approx x(h)$, $x_2 \approx x(2h)$, $x_3 \approx x(3h)$, $\cdots$, $x_n \approx x(nh)$. The x_i's thus estimated cannot be accepted unless they have been carefully checked. Three general methods for checking will be described in this section.

[*This dictum is just as valid for the output of an electronic computer as for the output of a slide rule. I have twice prevented gross academic injustice by looking at computerized examination grades with a jaundiced eye. Once the computer was at fault, once the program. Trust nothing, trust nobody, least of all yourself!]

Graphical checking. Plot the x_i's against t, and try to draw a smooth curve through the points. Contemplate the curve. Does it make physiological sense? If not, a good deal of thought should be devoted both to the derivation of the differential equation and to the method chosen for its solution. This is, by all odds, the most important check. A physiologically preposterous differential equation, carefully solved, will pass mathematical tests as easily as a sensible one. But the numbers it gives will be nonsense. "Garbage in, garbage out!"

Is the curve smooth? A point which deviates from the expected trend should be suspect. It almost always indicates an error in calculation. Mere inspection of the plotted points usually suffices, for the human eye is wonderfully adept at spotting points which are out of line (or, more generally, "out of curve").

Checking by successive differences. This method is illustrated on the right in Table 6.13.2. Each entry in the column headed Δ_1 is the difference between the two values of $\dot{V}$ nearest to it on the left, i.e., $\dot{V}_n - \dot{V}_{n+1}$. Similarly, the Δ_2's are the differences between the successive values of Δ_1; the Δ_3's are differences between successive values of Δ_2 etc. With any smooth, well-behaved function one should eventually reach a column of differences (e.g. the Δ_4 column in Table 6.13.2) whose values are constant (aside from small fluctuations due to rounding errors). The values in the following column (e.g. the Δ_5 column in Table 6.13.2) should therefore fluctuate narrowly about zero as a mean. If, as successively-higher-order differences are calculated, we do not approach a column of substantially constant values because of a jump between two adjacent values, an error is indicated at that point.

Since the method of successive differences is really just a numerical test for smoothness, as a check for computational error it is not much superior to plotting the points carefully on graph paper. However, columns of successive differences may also be used to extend the calculations beyond the point where the more orthodox computation stopped. For example, in Table 6.13.2, the mean of the more-or-less constant Δ_4 differences is about -0.0001. If we assume that this value will stay constant as more rows are added to the table, it is easy enough to extend the columns of differences by working backwards, and thus to estimate additional values of $\dot{V}_n$. This procedure is illustrated at the lower right of Table 6.13.2. While it may be safe to get one or two values this way, we must clearly recognize that the points so cheaply bought have been obtained without any direct reference to the known differential equation. Bargain-basement extrapolations, like bargain-basement clothes, may fit very poorly indeed.

Checking by numerical integration. Let us consider the following definite integral derived from equation 6.14.1

$$\int_{x_a}^{x_b} dx(t) = \int_a^b f(t,\ x(t))dt \qquad 6.14.2$$

or

$$x_b - x_a = \int_a^b f(t,\ x(t))dt \qquad 6.14.3.$$

As the reader undoubtedly knows, if f(t, x(t)) is plotted against t between the limits t = a and t = b, the value of the integral in equation 6.14.3 is the area under the resulting curve. To estimate the area numerically, we divide the interval [a, b] into n equal segments of length Δt, and we erect perpendiculars at the endpoints of the segments namely a, $a + \Delta t$, $a + 2\Delta t$, $a + 3\Delta t$, $\cdots$, $a + (n-1)\Delta t$, b. Let $\dot{x}_a, \dot{x}_1, \dot{x}_2, \dot{x}_3, \cdots, \dot{x}_{n-1}, \dot{x}_b$ be the corresponding values of f(t, x(t)) where the perpendiculars meet the curve. The perpendiculars divide the area into columns, each terminated by a segment of the curve. We approximate these segments by a simpler function of t which makes it easy to calculate the area of each column. The sum of the areas of the columns then approximates the total area under the curve, and hence the integral. Three simple approximating functions for the segments of the curve are illustrated in Figure 6.14.1. These lead to:

A. The rectangular rule. (The columns are rectangular.)

$$\int_a^b f(t,\ x(t))dt \approx \Delta t\,[\dot{x}_a + \dot{x}_1 + \dot{x}_2 + \cdots + \dot{x}_{n-1}] \qquad 6.14.4,$$

or,

$$\int_a^b f(t,\ x(t))dt \approx \Delta t\,[\dot{x}_1 + \dot{x}_2 + \dot{x}_3 + \cdots + \dot{x}_b] \qquad 6.14.5.$$

B. The trapezoidal rule. (The columns are trapezoids.)

$$\int_a^b f(t,\ x(t))dt \approx \frac{\Delta t}{2}\,[\dot{x}_a + 2\dot{x}_1 + 2\dot{x}_2 + \cdots + 2\dot{x}_{n-1} + \dot{x}_b] \qquad 6.14.6.$$

It is obvious by inspection that the value given by equation 6.14.6 is the mean of the two values given by equations 6.14.4 and 6.14.5.

C. Simpson's rule. In this method, n, the number of columns, must be even because two adjacent columns are treated as a single unit. The portion of

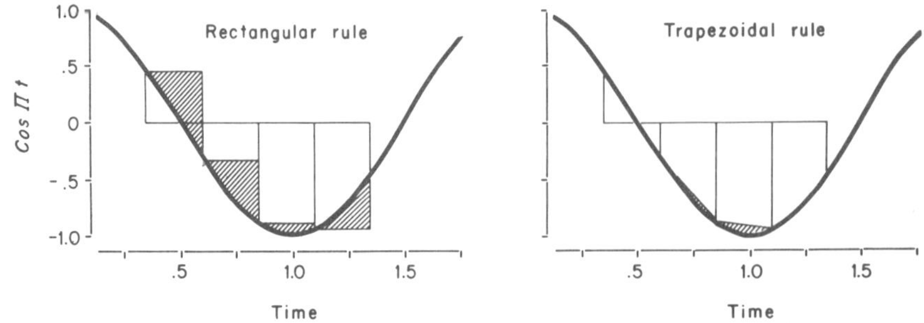

Figure 6.14.1. Numerical approximation of an integral.

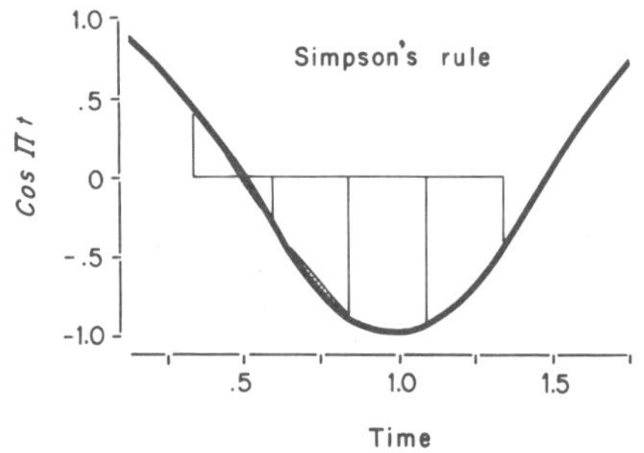

The value of the definite integral $\int_{0.35}^{1.35} \cos \pi t dt$ is here approximated by plotting the integrand, $\cos \pi t$, against t, and estimating the area beneath the resulting curve by dividing the area into columns, each of which is Δt wide. In this example, an unrealistically large value of Δt (Δt = 0.25) has been chosen so as to illustrate clearly the relative accuracy of the three methods discussed in the text.

In each part of the figure, the hatched areas show where the approximation fails to coincide with the true area. The estimate given by the rectangular rule (equation 6.14.4) is in error by about 25% of the true value; the estimate by the trapezoidal rule (equation 6.14.6) is in error by about 5%; the estimate by Simpson's rule (equation 6.14.7) is in error by only about 0.25%.

---

the curve which terminates them is approximated by a segment of a parabola fitted to the three points where the perpendiculars bounding the adjacent columns meet the curve.

$$\int_a^b f(t, x(t))dt \approx \frac{\Delta t}{3}[\dot{x}_a + 4\dot{x}_1 + 2\dot{x}_2 + 4\,\dot{x}_3 + 2\dot{x}_4 + \cdots + 4\dot{x}_{n-1} + \dot{x}_n] \qquad 6.14.7$$

These rules for approximate numerical integration have been stated in general form because their utility is by no means restricted to checking the solutions of non-linear differential equations. The rectangular rule is likely to give quite inaccurate results unless Δt is very small. While the trapezoidal rule is a substantial improvement, the results given by Simpson's rule are usually so nearly correct (if Δt is reasonable) that the small increase in computational work is more than repaid by an increase in accuracy. We shall now use Simpson's rule for checking.

Consider equation 6.14.3 again. Let $a = t_{n-1}$, and let $b = t_{n+1}$. Then we may write the equation as

$$x_{n+1} = x_{n-1} + \int_{t_{n-1}}^{t_{n+1}} \dot{x}(t)dt \qquad 6.14.8.$$

Suppose that we have used the modified Euler method to calculate x_{n+1} from x_{n-1} and x_n. For each of these three values we also have (or can calculate from the given differential equation) the corresponding values of $\dot{x}_{n-1}$, $\dot{x}_n$, and $\dot{x}_{n+1}$. Then, letting $\Delta t = h$, the integral on the right of equation 6.14.8 can be estimated by Simpson's rule as

$$\int_{t_{n-1}}^{t_{n+1}} \dot{x}(t)dt \approx \frac{h}{3}[\dot{x}_{n-1} + 4\dot{x}_n + \dot{x}_{n+1}] \qquad 6.14.9.$$

(Equation 6.14.9 corresponds to plotting $\dot{x}$ against t between the limits $t = t_{n-1}$ and $t = t_{n+1}$, and dividing the area beneath the curve into only <u>two</u> adjacent columns, each of width h.)

Combining equations 6.14.8 and 6.14.9,

$$x_{n+1} = x_{n-1} + \frac{h}{3}[\dot{x}_{n-1} + 4\dot{x}_n + \dot{x}_{n+1}] \qquad 6.14.10$$

For a satisfactory check, the value of x_{n+1} calculated from equation 6.14.10 should agree with the value originally calculated, at least within the desired limits of accuracy.

<u>Example 6.14.1.</u> Check the value $x_3^* = 0.2476$ obtained from equations 6.11.8.

Substituting the appropriate values from those equations into equation 6.14.10,

$$x_3^* = 0.2187 + \frac{0.1}{3}\{-0.1374 + 4[0.1608] + 0.4048\} = 0.2491$$

which differs from the value to be checked by about one part in 165. (The true value is 0.2488).

SECTION 6.15. Approximate solution of differential equations of order greater than one

The methods previously described can also be used to solve higher-order differential equations. We reduce an nth-order equation to n first-order equations by defining phase variables as described in Section 5.2. Then we solve the first-order equations simultaneously by computing the values of all phase variables at t_{n+1} from their known values at t_n before going on to the next step. Since no new principles are involved, it will suffice to outline the steps by an actual example. I gratefully accept Cunningham's example in which the modified Euler method is used to solve a third-order equation. (1). (Since it is a linear equation which can be solved analytically the approximate values obtained numerically can be checked against the true values (Exercise 6.7.)

Example 6.15.1. Solve

$$\frac{d^3x(t)}{dt^3} - \frac{dx(t)}{dt} = t \qquad 6.15.1,$$

subject to the initial conditions,

$$x(0) = x_o = 1.000$$

$$\dot{x}(0) = \dot{x}_o = 0.000 \qquad 6.15.2.$$

$$\ddot{x}(0) = \ddot{x}_o = 1.000$$

Define

$$y(t) \equiv \frac{dx(t)}{dt} = \dot{x}(t) \qquad 6.15.3,$$

$$z(t) \equiv \frac{dy(t)}{dt} = \dot{y}(t) \qquad 6.15.4,$$

so that we have the three simultaneous equations

$$\dot{x}(t) = x^{(1)}(t) = y(t) \qquad 6.15.5,$$

$$\dot{y}(t) = y^{(1)}(t) = z(t) \qquad 6.15.6,$$

$$\dot{z}(t) = z^{(1)}(t) = y(t) + t \qquad 6.15.7,$$

subject to the initial conditions (from equations 6.15.2),

$$\left.\begin{aligned} x_0 &= 1.0000 \\ y_0 &= 0.0000 \\ z_0 &= 1.0000 \end{aligned}\right\} \quad 6.15.8.$$

Let h = 0.1.

Step 1. Calculate x_1, y_1, z_1, from equations 6.15.5, 6.15,6, 6.15.7, and 6.15.8, using the first three terms of Taylor's series.

$$x_1 \approx x_0 + h\dot{x}(0) + \frac{h^2}{2}\ddot{x}(0) = 1.0000 + 0.1\,[0] + \frac{0.01}{2}\,[1.0000] = 1.0050 \qquad 6.15.9,$$

$$y_1 \approx y_0 + h\dot{y}(0) + \frac{h^2}{2}\ddot{y}(0) = 0.0000 + 0.1\,[1.0000] + \frac{0.01}{2}\,[0 + 0]$$

$$= 0.1000 \qquad 6.15.10,$$

$$z_1 \approx z_0 + h\dot{z}(0) + \frac{h^2}{2}\ddot{z}(0) = 1.0000 + 0.1\,[0+0] + \frac{0.01}{2}\,[1+1] = 1.0100 \qquad 6.15.11.$$

Step 2. Calculate x_2 from x_1 and x_0 using the predictor-corrector form of the modified Euler method.

$$x_2 = x_0 + 2h\,[\dot{x}_1] = 1.000 + 0.2\,[y_1] = 1.0200 \qquad 6.15.12,$$

$$x_2^* = x_1 + \frac{h}{2}\,[\dot{x}_1 + \dot{x}_2] = 1.005 + \frac{0.1}{2}\,[0.1000 + y_2] \qquad 6.15.13$$

To evaluate x_2^* by equation 6.15.13, we evidently have first to find y_2.

$$y_2 = y_0 + 2h\,[\dot{y}_1] = 0.0000 + 0.2\,[z_1] = 0.2020 \qquad 6.15.14,$$

hence, by equation 6.15.13,

$$x_2^* = 1.005 + \frac{0.1}{2}\,[0.1000 + 0.2020] = 1.0201 \qquad 6.15.15.$$

Next,

$$y_2^* = y_1 + \frac{h}{2}\,[\dot{y}_1 + \dot{y}_2] = 0.1000 + \frac{0.1}{2}\,[1.0100 + z_2] \qquad 6.15.16,$$

so we must find z_2.

$$z_2 = z_0 + 2h[\dot{z}_1] = 1.000 + 0.2[0.1000 + 0.1000] = 1.0400 \quad 6.15.17,$$

hence, by equation 6.15.16,

$$y_2^* = 0.1000 + \frac{0.1}{2}[1.0100 + 1.0400] = 0.2025 \quad 6.15.18.$$

Finally,

$$z_2^* = z_1 + \frac{h}{2}[\dot{z}_1 + \dot{z}_2] = 1.0100 + \frac{0.1}{2}[0.1000 + 0.1 + 0.2025 + 0.2]$$

$$= 1.0401 \quad 6.15.19,$$

which completes this step. Now that we see the pattern, it is easy to write a series of general equations for the nth step, arranging them in a slightly more orderly way so that we avoid having to hold equations such as 6.15.13 and 6.15.16 in abeyance while we backtrack to find a value for one of their constituents. First, the predictor equations are

$$\left.\begin{aligned} x_{n+1} &= x_{n\;1} + 2h\dot{x}_n = x_{n-1} + 2hy_n \\ y_{n+1} &= y_{n\;1} + 2h\dot{y}_n = y_{n-1} + 2hz_n \\ z_{n+1} &= z_{n\;1} + 2h\dot{z}_n = z_{n-1} + 2h[y_n + t_n] \end{aligned}\right\} \quad 6.15.20.$$

(In using equations 6.15.20 and 6.15.21 the corrected values from previous steps are employed whenever possible.) Second, the corrector equations are

$$\left.\begin{aligned} x_{n+1}^* &= x_n + \frac{h}{2}[\dot{x}_n + \dot{x}_{n+1}] = x_n + \frac{h}{2}[y_n + y_{n+1}] \\ y_{n+1}^* &= y_n + \frac{h}{2}[\dot{y}_n + \dot{y}_{n+1}] = y_n + \frac{h}{2}[z_n + z_{n+1}] \\ z_{n+1}^* &= z_n + \frac{h}{2}[\dot{z}_n + \dot{z}_{n+1}] = z_n + \frac{h}{2}[y_n + y_{n+1} + t_n + t_{n+1}] \end{aligned}\right\} \quad 6.15.21$$

Using these equations, step 3 would be

$$\left.\begin{aligned} x_3 &= 1.0050 + 0.2[0.2025] = 1.0455 \\ y_3 &= 0.1000 + 0.2[1.0401] = 0.3080 \\ z_3 &= 1.0100 + 0.2[0.2025 + 0.2] = 1.0905 \end{aligned}\right\} \quad 6.15.22,$$

$$\left.\begin{aligned} x_3^* &= 1.0200 + \frac{0.1}{2}[0.2025 + 0.3080] = 1.0456 \\ y_3^* &= 0.2025 + \frac{0.1}{2}[1.0401 + 1.0905] = 0.3090 \\ z_3^* &= 1.0401 + \frac{0.1}{2}[0.2025 + 0.3080 + 0.2 + 0.3] = 1.0906 \end{aligned}\right\} \quad 6.15.22,$$

Subsequent steps would follow exactly the same pattern. As usual, the calculations are best arranged in tabular form.

The advantages of using a digital computer for a problem of this kind are too obvious to require emphasis. (See footnote in Section 6.12 about the MIMIC program.)

SECTION 6.16. Stability in nonlinear systems

Testing the stability of a linear system is a simple matter. If the characteristic equation for the system has any roots with non-negative real parts, it is unstable, and that's that. The output of a stable linear system is bounded for any bounded input. Furthermore, for any given constant input, the output will approach a single constant value as the transient response dies away with the passage of time. In nonlinear systems, considerations of stability are much more complicated, and within the scope of the present book it is impossible even to mention all of the various methods which have been (and are still being) developed for testing stability. As the barest introduction to the subject I shall merely try to outline some of the more important problems and concepts; the reader must look elsewhere for a detailed treatment.

Consider a nonlinear system where the output, x(t), is governed by one or more known differential equations. Let us begin by supposing that the system has come into equilibrium with some constant forcing function (which may, in fact, be zero) so that x(t) is no longer changing with time. But if x(t) is no longer changing with time, its equilibrium value, which we shall call x_{eq}, must satisfy the differential equations with all of their time derivatives set to zero, and with time approaching infinity. For a given constant input, a linear system can have but one such equilibrium point, but a nonlinear system may have none, one, or several. If there are several, we must investigate the stability of the system in the neighborhood of each equilibrium point individually.

We shall confine attention to autonomous systems, i.e. to systems with constant coefficients (= stationary systems) which, when displaced from some equilibrium state at time zero, undergo their natural response. Systems with a constant input imposed as a step function at time zero are still considered to be autonomous because a step input at time zero is equivalent to imposing initial conditions with respect to the new equilibrium state required by the constant input. It is, in fact, often convenient to define x(t) as its deviation from whatever x_{eq} is being investigated. When this is done, $x_{eq} = 0$, and there is little point in distinguishing between the instantaneous removal of a previous input at time zero (which is really what initial conditions amount to) and the instantaneous imposition of a constant input at time zero.

Now to study the stability of a given equilibrium state, we displace x(t) from x_{eq} by a small amount at time zero and see whether or not x(t) returns asymptotically to x_{eq} as $t \to \infty$. If it does, that particular equilibrium state is at least locally asymptotically stable. If it does not, that equilibrium state is unstable.* If x(t) eventually returns to x_{eq} regardless of how far we displace it, then x_{eq} is said to be globally asymptotically stable. As an example, let us consider a simple nonlinear equation which will illustrate several of the points discussed above.

Example 6.16.1. Let

$$\dot{x}(t) = x(t)\,[4 - x^2(t)] \qquad 6.16.1.$$

Three equilibrium values of x(t) are found by solving equation 6.16.1 with $\dot{x}(t) = 0$,

$$x_{eq}\,[4 - x_{eq}^2] = 0 \qquad 6.16.2.$$

whence

$$\left.\begin{aligned} x_{eq,1} &= -2 \\ x_{eq,2} &= \;\;0 \\ x_{eq,3} &= +2 \end{aligned}\right\} \qquad 6.16.3.$$

A phase-plane plot of $\dot{x}(t)$ against x(t) is shown in Figure 6.16.1. The equilibrium points are the points where the curve crosses the axis for $\dot{x}(t) = 0$. Consider first stability at $x_{eq,1} = -2$. If x is displaced initially away from -2 to the left, so that $x(0) < -2$, $\dot{x}(t)$ will be positive. Hence x(t) will increase with time, moving along the curve downwards and to the right until it reaches $x_{eq,1} = -2$ again. If x is displaced from -2 towards the right, so that $-2 < x(0) < 0$, $\dot{x}(t)$ will be negative. Hence x(t) will decrease with time, moving along the curve until it reaches $x_{eq,1} = -2$. It is clear, therefore, that for all initial conditions such that $x(0) < 0$, $x_{eq,1} = -2$ is a stable equilibrium point. The same kind of argument shows that for all initial conditions such that $x(0) > 0$, $x_{eq,3} = +2$ is a stable equilibrium point. But $x_{eq,2} = 0$ is clearly an unstable point of equilibrium, because if there is any initial displacement, however small, away from $x_{eq,2} = 0$ on either side, $\dot{x}(t)$ will no longer be zero and will in fact require x to move away from 0 and towards one of the other points of equilibrium. Phase-plane plots of this kind often help to clarify the behavior of first- and second-order systems.

[*Note that in a nonlinear system instability of a given equilibrium point does not imply that the outputs will necessarily increase without bound as $t \to \infty$. The nonlinear system may approach some other equilibrium point which is stable, or it may approach a stable cycle of activity as a limit. (See Section 6.20.)]

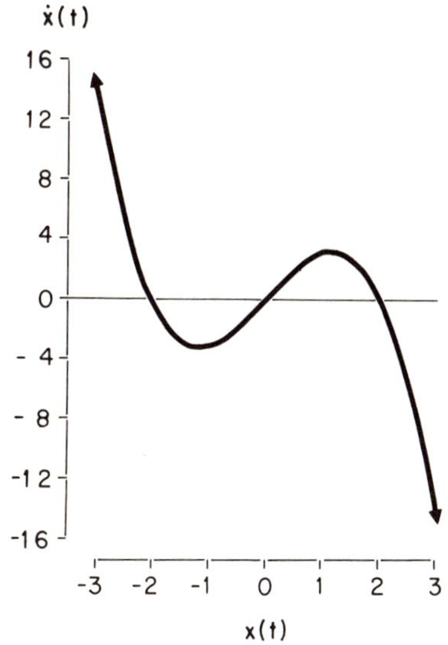

Figure 6.16.1. Phase-plane plot for Example 6.16.1.

For linear systems, we were able to show (Section 4.7) that the output of any stable system remained bounded as long as the input was bounded. If we adopt boundedness as our criterion of stability, we are no longer concerned with particular points of equilibrium but with the over-all behavior of the system. This is important, because, as we shall see, certain nonlinear systems may have no point of stable equilibrium at all but may, nevertheless, have a stable cycle of periodic behavior. (Cf. the thermostat example of Section 6.2.)

Example 6.16.2. Using boundedness as a criterion of stability, is the system described by equation 6.16.1 stable?

Let the system be unforced and at rest (at any equilibrium point) at $t = 0^-$. At $t = 0$, we apply an arbitrarily large step input, K. (K may be regarded as an upper bound for any bounded input.) Then equation 6.16.1 becomes

$$\dot{x}(t) = x(t)\,[4 - x^2(t)] + K \qquad 6.16.4.$$

Since K is large, $\dot{x}(t)$ will be positive. x(t) will therefore increase until $\dot{x}(t) = 0$, when

$$x_{eq}\,[4 - x_{eq}^2] + K = 0 \qquad 6.16.5,$$

or

$$x_{eq}^3 - 4x_{eq} = K \qquad 6.16.5.a.$$

Now as long as the input, K, is finite, equation 6.16.5 can be satisfied only by a finite value of x_{eq}. (Indeed, for large K, x_{eq} will also be large so that $x_{eq}^3 >>> 4x_{eq}$, and $x_{eq} \approx K^{1/3}$.) A similar argument holds if K is negative and of large magnitude. Hence, by the criterion of boundedness the system is stable.

There are no truly general tests for stability in nonlinear systems. But two methods of reasonably wide utility for testing the stability of an equilibrium point will be discussed briefly. These are the perturbation method (Section 6.17) and the second method of Liapunov (Section 6.19).

SECTION 6.17. Local perturbations about an equilibrium point

In the perturbation method, $\underline{x}(t)$ is "perturbed" by moving it a small distance away from $\underline{x}_{eq}$ in state space. The differential equations describing the system are assumed to be linear for this small perturbation, and the familiar tests for stability in linear systems are applied.

Consider a nonlinear system described by the n first-order equations,

$$\left.\begin{aligned} \dot{x}_1(t) &= f_1(x_1(t),\ x_2(t),\ \cdots\ x_n(t)) \\ \dot{x}_2(t) &= f_2(x_1(t),\ x_2(t),\ \cdots\ x_n(t)) \\ &\ \ \vdots \\ \dot{x}_n(t) &= f_n(x_1(t),\ x_2(t),\ \cdots\ x_n(t)) \end{aligned}\right\} \quad 6.17.1.$$

Let $x_{1,eq},\ x_{2,eq},\ \cdots x_{n,eq}$, denote an equilibrium point for the system; i.e., a point where $\dot{x}_1(t) = \dot{x}_2(t) = \cdots = \dot{x}_n(t) = 0$. Then $x_{1,eq} + h_1(0),\ x_{2,eq} + h_2(0),\ \ldots\ x_{n,eq} + h_n(0)$, where all of the initial displacements, $h_i(0)$, are small (much less than unity), represents a point in the immediate neighborhood of the equilibrium point. To apply equations 6.17.1 to this point, we replace $x_1(t)$ by $[x_{1,eq} + h_1(t)]$, $x_2(t)$ by $[x_{2,eq} + h_2(t)]$, etc. thus obtaining

$$\left.\begin{aligned} \dot{h}_1(t) &= f_1(x_{1,eq} + h_1(t),\ x_{2,eq} + h_2(t),\ \cdots,\ x_{n,eq} + h_n(t)) \\ \dot{h}_2(t) &= f_2(x_{1,eq} + h_1(t),\ x_{2,eq} + h_2(t),\ \cdots,\ x_{n,eq} + h_n(t)) \\ &\ \ \vdots \\ h_n(t) &= f_n(x_{1,eq} + h_1(t),\ x_{2,eq} + h_2(t),\ \cdots,\ x_{n,eq} + h_n(t)) \end{aligned}\right\} \quad 6.17.2.$$

Notice that the $\dot{x}_i(t)$ in equation 6.17.1 becomes $\dot{h}_i(t)$ in equations 6.17.2 because, with $x_{i,eq}$ a constant, $\dfrac{d[x_{i,eq} + h(t)]}{dt} = \dfrac{dh(t)}{dt} = \dot{h}(t)$.

In the functions on the right of equations 6.17.2 there may be terms in $h_i^2(t)$, $h_i^3(t)$, etc. as well as linear terms in $h_i(t)$. Since all of the $h_i(t)$'s are taken to be much less than unity, the higher-order terms will be negligible compared with the first-order terms. The higher-order terms can therefore be dropped, provided that the corresponding first-order terms are actually present. (A second-order term is not negligibly small compared with the corresponding first-order term when the first-order term isn't even there!) In the functions on the right of equations 6.17.2 there may also be terms which are not simple powers of h, e.g., $\sin(x_{1,eq} + h)$. These must be approximated by some power series (usually Taylor's) again dropping higher-order terms. (For a specific example, see Exercise 6.10.)

Assuming that all second- or higher-order terms can be neglected, we are left with a set of n linear differential equations for the $\dot{h}_i(t)$'s. If any of the roots of the characteristic equation derived from this set (See Section 4.7) have non-negative real parts, the $h_i(t)$'s will increase with time so that $\underline{x}(t)$ will move away from $\underline{x}_{eq}$ and the equilibrium point is unstable.

Example 6.17.1. Use the perturbation method to test the three equilibrium points of equation 6.16.1,

$$\dot{x}(t) = x(t)\,[4 - x^2(t)] = 4x(t) - x^3(t) \qquad 6.17.3.$$

Substituting $[x_{eq} + h(t)]$ for x(t) in equation 6.17.3,

$$\dot{h}(t) = 4x_{eq} + 4h(t) - x_{eq}^3 - 3h(t)x_{eq}^2 - 3h^2(t)x_{eq} - h^3(t) \qquad 6.17.4.$$

Neglecting terms in h^2 and h^3, equation 6.17.4 becomes

$$\dot{h}(t) + \left[3x_{eq}^2 - 4\right] h(t) = x_{eq}\left[4 - x_{eq}^2\right] \qquad 6.17.5.$$

But by equation 6.16.2, the right side of equation 6.17.5 is zero. Hence,

$$\dot{h}(t) + \left[3x_{eq}^2 - 4\right] h(t) = 0 \qquad 6.17.6.$$

whose characteristic equation,

$$\lambda + \left[3x_{eq}^2 - 4\right] = 0 \qquad 6.17.7.$$

has but the single root,

$$\lambda = -\left[3x_{eq}^2 - 4\right] \qquad 6.17.8.$$

When $x_{eq} = \pm 2$, by equation 6.17.8, $\lambda = -8$, and these equilibrium points are therefore stable. But when $x = 0$, $\lambda = +4$, so the equilibrium point $x_{eq} = 0$ is unstable.

These calculations are hardly needed for a system as simple as this. After all, we reached the same conclusions, and in addition found the exact limits of the regions controlled by the two points of stable equilibrium, by merely looking at Figure 6.16.1. But this might not be possible for more complicated higher-order systems, since at best we can plot only two phase variables on a plane. But for second-order systems in particular, a good deal of information about the behavior of the system as a whole can be deduced from a careful analysis of perturbations about its equilibrium points (See Cunningham (1) Chapter 5, for details.)

As the next section will show, the set of linear differential equations for small perturbations from equilibrium may equally well be obtained by applying Taylor's series for functions of several variables directly to equations 6.17.2.

SECTION 6.18. Using Taylor's series to linearize about an equilibrium point

Over a sufficiently small range, nonlinear functions can usually be treated as if they were linear. Linearization of equations 6.17.2 is achieved by replacing each of the functions $f_1, f_2, \cdots f_n$ by its Taylor Series approximation, retaining only the linear terms of the series expansion. To write a Taylor's series for a function of several variables, we deal with each variable, say x_i, and its corresponding small increment, h_i, separately, expanding in partial derivatives of the function with respect to the ith variable. The total change in the value of the function is simply the sum of all of the partial changes thus obtained. For example, let $f(x_1, x_2, x_3)$ be a function of the three variables, x_1, x_2, and x_3. Let $x_{1,p}$, $x_{2,p}$, $x_{3,p}$ be the specific values of these variables at some point, p, about which we wish to linearize. Let h_1, h_2, h_3 represent small increments of the corresponding variables. Then:

$$f(x_{1,p}+h_1,\ x_{2,p}+h_2,\ x_{3,p}+h_3) = f(x_{1,p}+x_{2,p}+x_{3,p}) +$$

$$\left[h_1\frac{\partial}{\partial x_1} + h_2\frac{\partial}{\partial x_2} + h_3\frac{\partial}{\partial x_3}\right]_{\underline{x}_p} f +$$

$$\frac{1}{2!}\left[h_1\frac{\partial}{\partial x_1} + h_2\frac{\partial}{\partial x_2} + h_3\frac{\partial}{\partial x_3}\right]^2_{\underline{x}_p} f + \cdots +$$

$$\frac{1}{n!}\left[h_1\frac{\partial}{\partial x_1} + h_2\frac{\partial}{\partial x_2} + h_3\frac{\partial}{\partial x_3}\right]^n_{\underline{x}_p} f + \cdots \qquad 6.18.1,$$

where each bracketted sum of operators is to be expanded into mixed and unmixed partial derivatives of the function, f, by formally "multiplying" the sum by itself as often as needed to raise it to the indicated power. To do this, we treat the sum of operators just as if it were an algebraic polynomial. For example,

$$\frac{1}{2!}\left[h_1\frac{\partial}{\partial x_1} + h_2\frac{\partial}{\partial x_2} + h_3\frac{\partial}{\partial x_3}\right]^2_{\underline{x}_p} f \equiv \frac{1}{2!}\left[h_1^2\frac{\partial^2 f}{\partial x_1^2} + h_2^2\frac{\partial^2 f}{\partial x_2^2} + \right.$$

$$h_3^2\frac{\partial^2 f}{\partial x_3^2} + 2h_1h_2\frac{\partial^2 f}{\partial x_1\partial x_2} + 2h_1h_3\frac{\partial^2 f}{\partial x_1\partial x_3} +$$

$$\left. 2h_2h_3\frac{\partial^2 f}{\partial x_2\partial x_3}\right]_{x_{1,p}\ x_{2,p}\ x_{3,p}}$$

As indicated by the subscript $\underline{x}_p$ or, equivalently, $x_{1,p}$, $x_{2,p}$, $x_{3,p}$, all partial derivatives are to be evaluated at the point p.

We shall now apply this method to equations 6.17.2, neglecting all terms of second or higher order, so as to obtain a set of linear differential equations to describe approximately the behavior of the nonlinear system within a small region close to the equilibrium point.

$$\dot{h}_1(t) \approx f_1(x_{1,eq}, x_{2,eq}, \cdots, x_{n,eq}) + \left.\frac{\partial f_1}{\partial x_1}\right|_{\underline{x}_{eq}} h_1(t) + \left.\frac{\partial f_1}{\partial x_2}\right|_{\underline{x}_{eq}} h_2(t) + \cdots + \left.\frac{\partial f_1}{\partial x_n}\right|_{\underline{x}_{eq}} h_n(t)$$

$$\dot{h}_2(t) \approx f_2(x_{1,eq}, x_{2,eq}, \cdots, x_{n,eq}) + \left.\frac{\partial f_2}{\partial x_1}\right|_{\underline{x}_{eq}} h_1(t) + \left.\frac{\partial f_2}{\partial x_2}\right|_{\underline{x}_{eq}} h_2(t) + \cdots + \left.\frac{\partial f_2}{\partial x_n}\right|_{\underline{x}_{eq}} h_n(t)$$

$$\vdots$$

$$\dot{h}_n(t) \approx f_n(x_{1,eq}, x_{2,eq}, \cdots, x_{n,eq}) + \left.\frac{\partial f_n}{\partial x_1}\right|_{\underline{x}_{eq}} h_1(t) + \left.\frac{\partial f_n}{\partial x_2}\right|_{\underline{x}_{eq}} h_2(t) + \cdots + \left.\frac{\partial f_n}{\partial x_n}\right|_{\underline{x}_{eq}} h_n(t) \qquad 6.18.2.$$

But when all of the x's have their equilibrium values, the system must be at rest, with all time derivatives equal to zero. Hence, by equations 6.17.1,

$$f_1(x_{1,eq}, x_{2,eq}, \cdots, x_{n,eq}) = f_2(x_{1,eq}, x_{2,eq}, \cdots, x_{n,eq}) = \cdots$$
$$= f_n(x_{1,eq}, x_{2,eq}, \cdots, x_{n,eq}) = 0 \qquad 6.18.3.$$

These terms may therefore be dropped, and equations 6.18.2 become

$$\left.\begin{aligned}
\dot{h}_1(t) &\approx \left.\frac{\partial f_1}{\partial x_1}\right|_{\underline{x}_{eq}} h_1(t) + \left.\frac{\partial f_1}{\partial x_2}\right|_{\underline{x}_{eq}} h_2(t) + \cdots + \left.\frac{\partial f_1}{\partial x_n}\right|_{\underline{x}_{eq}} h_n(t) \\
\dot{h}_2(t) &\approx \left.\frac{\partial f_2}{\partial x_1}\right|_{\underline{x}_{eq}} h_1(t) + \left.\frac{\partial f_2}{\partial x_2}\right|_{\underline{x}_{eq}} h_2(t) + \cdots + \left.\frac{\partial f_2}{\partial x_n}\right|_{\underline{x}_{eq}} h_n(t) \\
&\vdots \\
\dot{h}_n(t) &\approx \left.\frac{\partial f_n}{\partial x_1}\right|_{\underline{x}_{eq}} h_1(t) + \left.\frac{\partial f_n}{\partial x_2}\right|_{\underline{x}_{eq}} h_2(t) + \cdots + \left.\frac{\partial f_n}{\partial x_n}\right|_{\underline{x}_{eq}} h_n(t)
\end{aligned}\right\} \quad 6.18.4$$

Since all of the partial derivatives in equations 6.18.4 are to be evaluated at some specific equilibrium point, they are simply constant coefficients of the h(t)'s. Hence these equations do indeed represent the desired set of linear differential equations which can be analyzed in place of the original nonlinear set, (equations 6.17.1), provided that we confine our attention to a sufficiently small neighborhood of an equilibrium point, $\underline{x}_{eq}$. So again we could find the characteristic equation for the linearized system, and we could then apply such familiar tests as the Routh-Hurwitz criterion to ascertain whether the system is locally stable in the immediate neighborhood of that equilibrium point.

Notice that even when the original set of equations (equations 6.17.1) is nonhomogeneous because one or more of the equations includes a constant forcing function, the linearized set (equations 6.18.4) is always homogeneous, describing the natural response which occurs when the system has been slightly displaced (e.g., by a small impulse input) away from some point of equilibrium determined by those forcing functions.

Suppose that all of the roots of the characteristic equation for the linear set (equations 6.18.4) have negative real parts. Then in the immediate neighborhood of the equilibrium point about which we have linearized, the system will be stable, and it will show a bounded response to every sufficiently small input. Let $\Delta U_1(t)$, $\Delta U_2(t)$, $\cdots$, $\Delta U_n(t)$ be "sufficiently small" inputs. As the Δ suggests, these inputs actually represent deviations from whatever inputs are already present when the system is at the designated equilibrium point. The behavior of such a forced system will be approximately described by the following nonhomogeneous version of our linearized equations.

$$\dot{h}_1(t) \approx \left.\frac{\partial f_1}{\partial x_1}\right|_{\underline{x}_{eq}} h_1(t) + \left.\frac{\partial f_1}{\partial x_2}\right|_{\underline{x}_{eq}} h_2(t) + \cdots + \left.\frac{\partial f_1}{\partial x_n}\right|_{\underline{x}_{eq}} h_n(t) + \Delta U_1(t)$$

$$\dot{h}_2(t) \approx \left.\frac{\partial f_2}{\partial x_1}\right|_{\underline{x}_{eq}} h_1(t) + \left.\frac{\partial f_2}{\partial x_2}\right|_{\underline{x}_{eq}} h_2(t) + \cdots + \left.\frac{\partial f_2}{\partial x_n}\right|_{\underline{x}_{eq}} h_n(t) + \Delta U_2(t)$$

$$\vdots$$

$$\dot{h}_n(t) \approx \left.\frac{\partial f_n}{\partial x_1}\right|_{\underline{x}_{eq}} h_1(t) + \left.\frac{\partial f_n}{\partial x_2}\right|_{\underline{x}_{eq}} h_2(t) + \cdots + \left.\frac{\partial f_n}{\partial x_n}\right|_{\underline{x}_{eq}} h_n(t) + \Delta U_n(t)$$

6.18.5.

If all of the ΔU's are small step inputs, applied at $t = 0$, the stable system will approach a new equilibrium point as $t \to \infty$. At the new equilibrium, $\dot{h}_1(\infty) = \dot{h}_2(\infty) = \cdots = \dot{h}_n(\infty) = 0$, and equations 6.18.5 become a set of linearized algebraic equations describing the steady-state behavior of the system. Notice, however, that even when the new equilibrium has been reached, the partial derivatives keep the values which were obtained by evaluating them at the old equilibrium. We are, after all, treating them as constants, and we are assuming that the new equilibrium point lies "within linear striking distance" of the old equilibrium point. But having reached the new point, we can then revise the equations so as to linearize about the new equilibrium point.

Much of the analysis of physiological feedback systems in Part II will be based upon the steady-state versions of equations 6.18.5. In the next chapter, these will be derived in a somewhat simpler way directly from the equations which specify the working relationships among the variables and parameters of a system in the steady state. But however different that approach may seem to be, exactly the same general kind of logic is used, and the same end-results are obtained.

SECTION 6.19. Testing stability by the second method of Liapunov

The perturbation method described in Section 6.17 is a valid test of stability only within the immediate neighborhood of an equilibrium point. The second method of Liapunov is a much more general way of proving that a system, starting from some specified initial state, will approach an equilibrium point. Note the wording! Failure of the second method of Liapunov to prove stability does not necessarily imply that the system is unstable. While the mathematical details of the second method must be sought elsewhere (7, 8), its general strategy is reasonably simple. Consider a system with n state variables. Let us choose our coordinates so that the equilibrium point whose stability is at test lies at the origin of the n coordinates of the n-dimensional state space. If we define time as an additional coordinate, the corresponding solution space (See Section 5.2) is (n + 1) - dimensional. Then asymptotic stability about the equilibrium point demands that, as we move from $t = 0$ towards $t = \infty$ we must approach closer and closer to the time axis where all the other coordinates are zero. We therefore try to find a guiding function (called a Liapunov function) whose slope, with reference to the time axis, will surely be negative (or, at worst, zero for only an instant). The Liapunov function must depend upon the state variables in such a way that at time, t, any point $[t, \underline{x}(t)]$ in solution space lies upon the "surface" defined by the Liapunov function, which surface always slopes towards $[\infty, \underline{x}(\infty)] = [\infty, \underline{0}]$, the equilibrium point. It is as if we were trying to show that a skier, starting at some point upon the slope must surely end at the ski lodge in the valley by proving that his route must be downhill, or, at worst, instantaneously level. If we can prove downhillness, we may be sure that (barring non-mathematical accidents due to poor skiing ability or adverse weather conditions) the skier will end up at Equilibrium Lodge. If we cannot prove downhillness, he may still get there via some downhill route that we don't even know about.

Recently Kalman (9) following Popov (10) has shown by a brilliant mathematical argument that we do not actually have to find a specific Liapunov function, it is only necessary to demonstrate that one exists. The general methods developed for demonstrating existence may well represent one of the few highways which have been cut through the nonlinear thickets cluttering the countryside!

SECTION 6.20. Limit cycles and describing functions

The output of a linear feedback system can show sustained oscillations only in response to a periodic input. But certain kinds of nonlinear feedback systems may exhibit sustained oscillations, called limit cycles, even when the input is constant. If the limit cycle is stable, the cyclic behavior characteristic of the system is approached as a limit with increasing time even if the starting point in state space (determined by the initial conditions) is some distance away from the path of the limit cycle in state space. The variation of bath temperature in the example of Section 6.2. (see Figure 6.2.2) is a limit cycle, but limit cycles may also occur in systems where the nonlinearity is of the "no-memory" type. Many cyclic phenomena in biological systems, e.g. variations in the population density of certain species, cyclic alternation of wakefulness and sleep even in a constant environment, the menstrual cycle, "biological clocks," pacemaker activity, are either probably or certainly best explained as limit cycles.

One of the most popular (and most abused) methods for studying limit cycles is the describing function method. Its usefulness is restricted to predicting the approximate amplitude and frequency of the limit cycle in one general class of nonlinear feedback systems. In the following discussion, the restrictions and assumptions are particularly important, since they determine how the final result is to be interpreted.

Consider the generalized, autonomous, nonlinear feedback system in Figure 6.20.1. The block labelled G(s) represents all of the linear components of the system. The block labelled N. L. is a single nonlinearity whose input-output relationship is not a function of s, the complex frequency. The input for the nonlinearity is simply the output of the linear portion with its sign changed, because the scale of measurement has been chosen to make U, the constant input for the whole system, equal to zero. The line of reasoning to be used is as follows: We begin by assuming that a stable limit cycle actually does already exist. This means that we can neglect initial conditions and that we can use phasor analysis. Next we assume that both E(t) and f(E(t)) can be approximate satisfactorily by a Fourier sine series with no constant term and with no cosine terms. This means that we have to limit consideration to nonlinearities where f(E(t)) is an odd function, i.e., where f(E(t)) equals -f(-E(t)) throughout the entire domain of E(t). To ensure this, the coordinates must be chosen so that f(0) = 0. (Cf. the footnote Section 6.2). Finally we assume that E(t) and f(E(t)) can be represented reasonably well by using just the first term of the sine series so that we shall be able to neglect higher harmonics and deal with the fundamental frequency (= first harmonic) only. Now in fact, higher harmonics will always be present in the output from a nonlinearity. This means that we must limit our analysis to systems whose linear portion acts as a low-pass filter so that the open-loop gain for low-frequency signals is substantially greater than for high-frequency signals, and the higher harmonics therefore tend to be suppressed. Fortunately most linear systems do behave in this way.

Now if both E(t) and f(E(t)) can fairly be treated as simple sine waves, a limit cycle can persist only if the sine wave representing E(t) is of exactly the right frequency and magnitude so that, after traversing both the nonlinear and the linear components, it emerges from the linear block as a sine wave of the very same magnitude and frequency but shifted in phase by π radians so that it is -E(t) instead of E(t). (Figure 6.20.1). Then we can use the known functions, G(s), and f(E(t)) to calculate the frequency and magnitude at which this will be true, i.e. at which the input to the nonlinearity is in harmonic balance with the

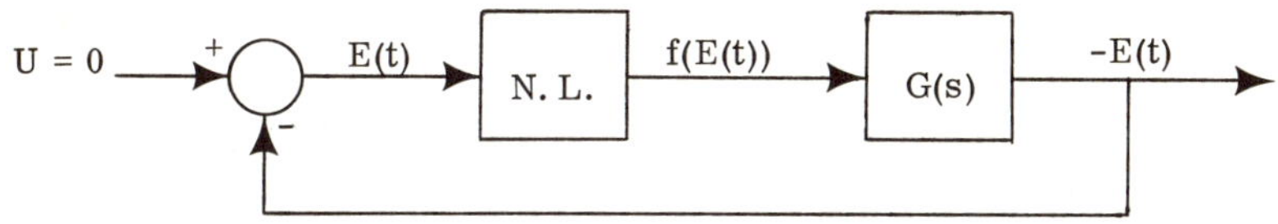

Figure 6.20.1. A generalized block diagram showing the conditions assumed when using the describing function method.

negative of the output from the linear part of the system, so that the signal will continue to circulate unchanged through the loop. If the higher harmonics are indeed negligible, then the magnitude and frequency thus found should be close to the magnitude and frequency of the actual limit cycle. (It would be desirable to require the higher harmonics to balance also, but formidable algebraic difficulties arise except when the nonlinearity is of the strictly on-off, "ideal relay," type.* Then the more accurate Tsypkin-Hammel locus method in which higher harmonics are not neglected, may be used (11)).

[*Sometimes referred to, with more color than clarity, as a "bang-bang" nonlinearity.]

With these considerations in mind, let us analyze the general system in Figure 6.20.1. First consider the nonlinearity. Its true input is

$$E(t) = E_1 \sin(\omega t + \theta_1) + E_2 \sin(\omega_2 t + \theta_2) + \cdots + E_n \sin(\omega_n t + \theta_n) + \cdots \qquad 6.20.1,$$

where

$E_1, E_2, \cdots, E_n, \cdots$ are the magnitudes of the successive harmonics,

$\omega_1, \omega_2, \cdots, \omega_n, \cdots (= \omega_1, 2\omega_1, \cdots, n\omega_1 \cdots)$ are the radian frequencies of the successive harmonics

$\theta_1, \theta_2, \cdots, \theta_n, \cdots$ are the phase angles of the successive harmonics.

But we are assuming that,

$$E(t) \approx E_1 \sin(\omega_1 t + \theta_1) \qquad 6.20.2,$$

and that similarly,

$$f(E(t)) \approx F_1 \sin(\omega_1 t + \delta_1) \qquad 6.20.3,$$

which is a sine wave of the same frequency but not necessarily of the same amplitude or phase angle. Because we are assuming the same frequency throughout the system, we can write equations 6.20.2 and 6.20.3 in phasor form. First, we have to change the sine functions to cosine functions.

$$E(t) = E_1 \cos(\omega t + \theta_1 - \pi/2) \qquad 6.20.4.$$

$$f(E(t)) = F_1 \cos(\omega t + \delta_1 - \pi/2) \qquad 6.20.5.$$

Then,

$$\widetilde{E}(j\omega) = E_1 \, \underline{/\theta_1 - \pi/2} \qquad 6.20.6.$$

$$\widetilde{\mathcal{F}}(E(j\omega)) = F_1 \, \underline{/\delta_1 - \pi/2} \qquad 6.20.7.$$

For the linear part of the system, $\widetilde{\mathcal{F}}(E(j\omega))$ is the phasor input and $-\widetilde{E}(j\omega)$ is the phasor output. Hence,

$$G(j\omega) = \frac{-\widetilde{E}(j\omega)}{\widetilde{\mathcal{F}}(E(j\omega))} = \frac{-E_1 \, \underline{/\theta_1 - \pi/2}}{F_1 \, \underline{/\delta_1 - \pi/2}} = \frac{-E_1}{F_1} \, \underline{/\theta_1 - \delta_1} \qquad 6.20.8.$$

For convenience, define the phase angle, difference, $\underline{/\theta_1 - \delta_1}$, as

$$\underline{/\varphi_1} \equiv \underline{/\theta_1 - \delta_1} \qquad 6.20.9.$$

Then equation 6.20.8 may be written

$$G(j\omega) = \frac{-E_1}{F_1} \, \underline{/\varphi_1} \qquad 6.20.10$$

So far, so good. But how are we to deal with the nonlinearity, for which we cannot even write a transfer function, let alone evaluating it for $s = j\omega$? The answer is that, in fact, we are <u>already</u> assuming that with respect to a sine wave of the fundamental frequency the nonlinearity is behaving exactly like a linear element; for we are assuming that it accepts a sine wave input of frequency ω_1 and puts out a sine wave of the very same frequency. Since we are neglecting all higher frequencies, we can pretend that the nonlinearity has been replaced by a linear element with an <u>equivalent transfer function,</u>* K_{eqv}, defined for our phasor input and output as

$$K_{eqv} \equiv \frac{\widetilde{\mathcal{F}}(E(j\omega))}{\widetilde{E}(j\omega)} \qquad 6.20.11.$$

Combining equations 6.20.8, 6.20.10 and 6.20.11,

$$-\frac{1}{K_{eqv}} = -\frac{E_1}{F_1} \, \underline{/\varphi_1} = G(j\omega) \qquad 6.20.12.$$

For convenience, define D. F., the <u>describing function</u>, as

$$D.F. \equiv -\frac{1}{K_{eqv}} \qquad 6.20.13.$$

Then equation 6.20.12 can be written,

$$D.F. = G(j\omega), \qquad 6.20.14,$$

which (we are assuming) will be valid only if the system is in harmonic balance so that the limit cycle exists.

[*K_{eqv} is usually called "equivalent gain." But the output of the nonlinearity need not have the same physical dimensions as the input, and when the dimensions differ the term "gain" is inapplicable. (See Section 4.11.) It is therefore far better to call K_{eqv} the "equivalent transfer function."]

To find the frequency and magnitude of the limit cycle we evaluate the right side of equation 6.20.14 for various values of ω and plot the results in the manner of a Nyquist plot on the complex plane. We use equation 6.20.13 to evaluate the left side of equation 6.20.14 for various values of E, and plot the results on the same complex plane. If the two curves clearly intersect each other with some margin to spare, ω_{cyc}, the frequency of the limit cycle, is estimated as the value of ω for the Nyquist plot at the point of intersection; E_{cyc}, the magnitude of the quasisinusoidal limit cycle, is estimated as the value of E for the describing function at the point of intersection. If the two curves clearly miss each other with some margin to spare a limit cycle most probably does not exist. But because this method is only approximate, if the two curves barely miss each other, or barely intersect, we cannot be sure whether a limit cycle will, or will not, occur.

Example 6.20.1. Will the system illustrated in Figure 6.20.2 exhibit a limit cycle, and, if so, what will the frequency and magnitude of the cycle be?

We assume that the input to the nonlinearity is

$$E(t) = E_1 \sin\omega_1 t \qquad 6.20.15.$$

Since the nonlinear relationship is

$$f(E(t)) = E^3(t) \qquad 6.20.16.$$

The true output from the nonlinearity for the input of equation 6.20.15 is

$$f(E(t)) = E_1^3 \sin^3 \omega_1 t, \qquad 6.20.17.$$

For convenience, define

$$x \equiv \omega_1 t \qquad 6.20.18,$$

so equation 6.20.17 can be written as

$$f(E(x/\omega_1)) = E_1^3 \sin^3 x \qquad 6.20.19.$$

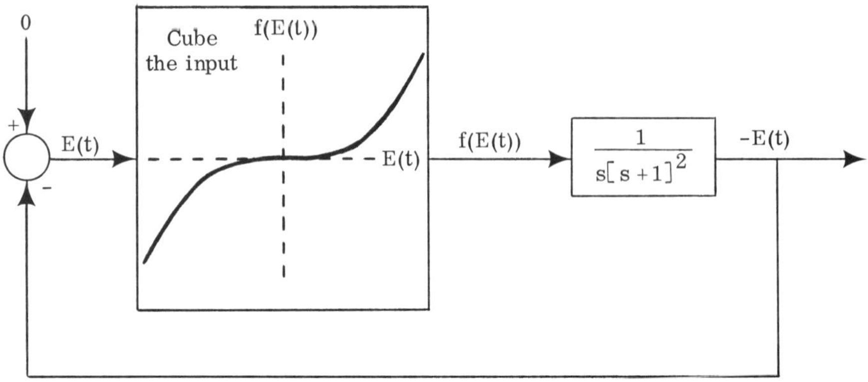

Figure 6.20.2. Block diagram for the nonlinear system of Example 6.20.1.

The Fourier sine series representation of $f(E(x/\omega_1))$ is

$$E_1^3 \sin^3 x = \sum_{n=1}^{\infty} b_n \sin nx \qquad 6.20.20,$$

where

$$b_n = \frac{E_1^3}{\pi} \int_{-\pi}^{\pi} \sin^3 x[\sin nx]dx \qquad 6.20.21.$$

But we are assuming that harmonics higher than the fundamental (i.e., $n > 1$) can be neglected. Hence we let equations 6.20.20 and 6.20.21, be

$$E_1^3 \sin^3 x \approx b_1 \sin x \qquad 6.20.22.$$

$$b_1 = \frac{E_1^3}{\pi} \int_{-\pi}^{\pi} \sin^4 x \, dx \qquad 6.20.23.$$

Evaluating the integral in equation 6.20.23 by consulting a table of integrals,

$$\int_{-\pi}^{\pi} \sin^4 x \, dx = \frac{3}{8}x - \frac{\sin 2x}{4} + \frac{\sin 4x}{32}\Bigg|_{-\pi}^{\pi} = \frac{3}{8}\pi + \frac{3}{8}\pi = \frac{3}{4}\pi \qquad 6.20.24.$$

Combining equations 6.20.22, 6.20.23 and 6.20.24,

$$E_1^3 \sin^3 x \approx \frac{3}{4} E_1^3 \sin x \qquad 6.20.25.$$

Therefore, by equations 6.20.17 and 6.20.18,

$$f(E(t)) \approx \frac{3}{4} E_1^3 \sin \omega_1 t \qquad 6.20.26.$$

In phasor form, the input (equation 6.20.15) to the nonlinearity is

$$\tilde{E}(j\omega) = E_1 \underline{/-\pi/2} \qquad 6.20.27,$$

while the output (equation 6.20.26) is

$$f(E(t)) \approx \frac{3}{4} E_1^3 \underline{/-\pi/2} \qquad 6.20.28.$$

Therefore the equivalent transfer function is

$$K_{eqv} = \frac{3/4\, E_1^3 \underline{/-\pi/2}}{E_1 \underline{/-\pi/2}} = \frac{3}{4} E_1^2 \qquad 6.20.29.$$

(Notice that in this example, as is often true, the nonlinearity causes no phase shift, so that the equivalent transfer function is merely a ratio of magnitudes. When this is so, expressing the sine functions as phasors, while formally correct, is in fact superfluous. We could have obtained equation 6.20.29 directly from equations 6.20.15 and 6.20.26.) The describing function (defined by equation 6.20.13) is

$$D.F. = -\frac{1}{K_{eqv}} = -\frac{4}{3 E_1^2} \qquad 6.20.30.$$

This has been plotted for various values of E as a heavy broken line on the complex plane of Figure 6.20.3. The value of

$$G(j\omega) = \left[\frac{1}{1 + j\omega}\right]^2 \left[\frac{1}{j\omega}\right] \qquad 6.20.31,$$

for various ω has been plotted on the same complex plane. These two plots intersect at the point where D. F. = $G(j\omega)$, i.e., where, by equations 6.20.30 and 6.20.31,

$$-\frac{4}{3 E_1^2} = \frac{1}{j\omega_1 [1 + j\omega_1]^2} \qquad 6.20.32,$$

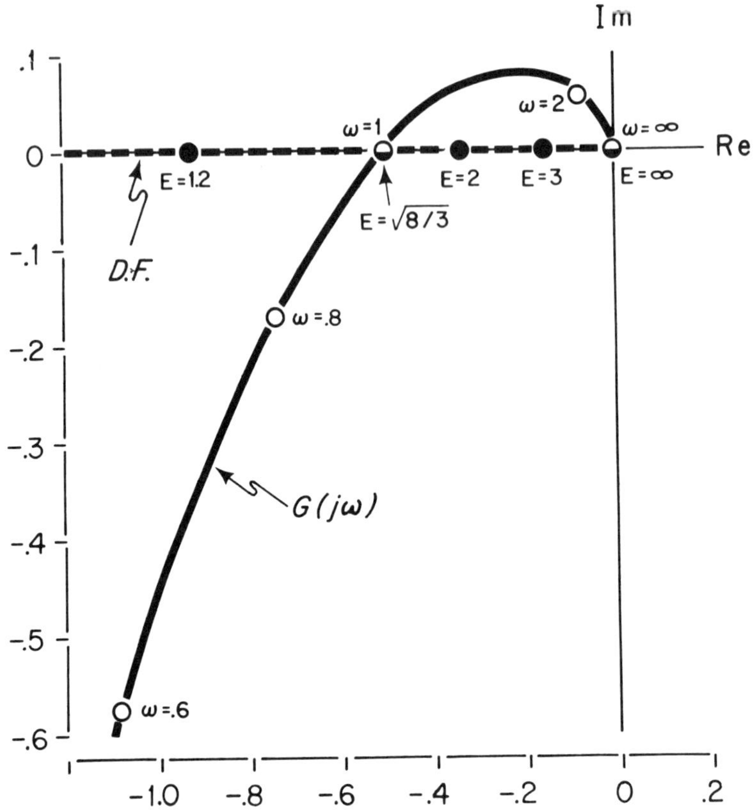

Figure 6.20.3. The describing function, D. F., and G(jω) plotted on the same complex plane for the system of Example 6.20.1.

A limit cycle is clearly predicted because the locus of G(jω) for various ω (solid curve) clearly intersects the locus of the D. F. for various E (broken line). The estimated magnitude of the limit cycle is the value of E for the D. F. at the point of intersection; the estimated frequency of the limit cycle is the value of ω for G(jω) at the point of intersection.

or,

$$E_1^2 + j\,0 = \frac{4}{3}\left\{2\,\omega_1^2 + j\left[\omega_1^3 - \omega_1\right]\right\} \qquad 6.20.32.a,$$

or, since two complex numbers are equal only if their real parts are equal and their imaginary parts are equal,

$$\left.\begin{aligned} E_1^2 &= \frac{8}{3}\,\omega_1^2 \\ 0 &= \omega_1\,[\omega_1^2 - 1] \end{aligned}\right\} \quad 6.20.32.b,$$

whence,

$$\omega_1 = 1 \qquad 6.20.33.$$

$$E_1 = \sqrt{\frac{8}{3}} \qquad 6.20.34.$$

(Check: Treating K_{eqv} (equation 6.20.29) as if it were a genuine transfer function, [OLTF] (jω), the open-loop transfer function (which must include the minus sign at the summing point) for this system, is

$$[OLTF](j\omega) = K_{eqv}\,[G(j\omega)]\,[-1] =$$

$$= -\frac{3}{4}\,E_1^2 \left\{\frac{1}{j\omega_1\,[1 + j\omega_1]^2}\right\} \qquad 6.20.35,$$

which is valid only when E_1 and ω_1 have the values of equations 6.20.33 and 6.20.34, derived on the assumption that for those values the insert signal and the return signal will be identical. Hence, if we substitute those values into equation 6.20.35 we should find that the open-loop gain is unity, and the phase angle is zero;

$$[OLTF](j1) = \left[-\frac{3}{4}\right]\left[\frac{8}{3}\right]\left\{\frac{1}{j1\,[1 + j1]^2}\right\} = [-2]\left[-\frac{1}{2}\right] = 1\underline{/0} \qquad 6.20.36,$$

which is correct.)

Although the values of ω_1 and E_1 may be calculated algebraically, for two reasons it is also important actually to plot the equations for D. F. and G(jω) as has been done in Figure 6.20.3. First, the curve for G(jω) will show whether it is legitimate to assume that the higher harmonics tend to be filtered out by the linear part of the system. If not, the whole procedure is suspect. Second, the plot will indicate whether the two curves barely intersect or barely miss each other. If so, conclusions drawn from the analysis are unreliable. For the present example, Figure 6.20.3 shows that we are safe on both counts. The magnitude of G(jω) steadily and rapidly decreases as ω increases, and the line for the D. F. extends well beyond the point of intersection on both sides.

In Figure 6.20.4, the values of equations 6.20.33 and 6.20.34 have been used to calculate curves showing the assumed input (equation 6.20.15), the corresponding true output (equation 6.20.17) and the assumed output (equation 6.20.26) of the nonlinearity.

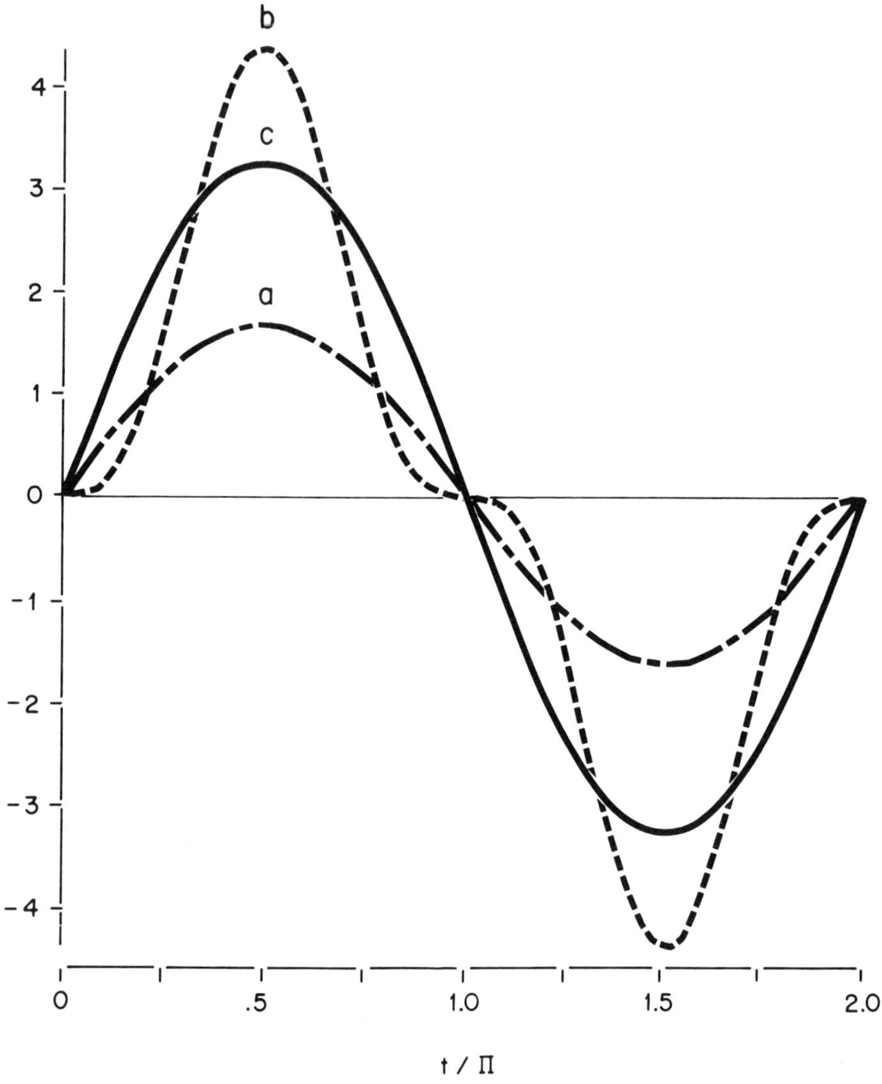

Figure 6.20.4. Curve a. Assumed sine-wave input to the nonlinearity in Figure 6.20.2. Its magnitude, $\sqrt{8/3}$, and its radian frequency, $\omega = 1$, are those predicted by the describing function analysis (Figure 6.20.3). Curve b. The corresponding true output, $[\sqrt{8/3}\ \sin t]^3$, of the nonlinearity, and Curve c. The assumed output, $2\sqrt{8/3}\ \sin t$, which is the first term in the Fourier sine series representation of the true output.

When used circumspectly, the describing function method (or one of its many modifications or extensions (12)) often correctly predicts the existence of a limit cycle, and, what is more, forecasts the frequency and the amplitude of the cycle with astonishing accuracy. But there its talents end. It is not a method for determining the stability of a limit cycle. How could it be, when the whole procedure is founded upon the assumption that, for one particular amplitude and frequency, a stable cycle already exists!* The "equivalent transfer function" is not equivalent to a transfer function save in the exceedingly restricted sense explained previously. Above all, the describing function method is not a valid method for "linearizing" a nonlinear system, even within a restricted range. Under these circumstances, describing function analysis is not likely to play a very important role in the study of physiological systems.

[*A much-quoted criterion for the stability of a limit cycle is whether the system will return to the original cycle after a transient perturbation. As a criterion this is unexceptionable, but the corresponding practical test which has often been proposed (see, for example, Gibson, (12), pp. 350-351) is based upon faulty reasoning: one cannot investigate transient behavior by using a method whose very existence depends on assuming non-transient conditions!]

SECTION 6.21. Steady-state analysis of nonlinear feedback systems

The preceding sections of this chapter have abundantly documented some of the difficulties which beset the analysis of nonlinear systems. Even when the physical components of a nonlinear feedback system, and their input-output relationships, are fully known, it is by no means always possible to predict the behavior of the system for arbitrary inputs at arbitrary instants of time. And when, as is too often true of physiological systems, the characteristics of certain components can only be guessed at, the hope of achieving a complete analysis is indeed illusory. All too often, therefore, the biological scientist must content himself with the more modest goal of studying his feedback systems in the steady state, using some such method as the product-of-first-derivatives method described in Section 4.15 to investigate the open-loop gain of the system. This, in itself, is a worthwhile aim, for by calculating the open-loop gain at several different operating points we can define the range through which a given regulatory feedback system is effective, and we can observe how its effectiveness varies within that range. This approach, which is fully discussed in Chapter 7, will form the backbone of Part II of the present monograph.

Yet it would be pointless to disguise the extent of our loss when we are forced to depend strictly upon steady-state analysis. We then have to confine our quantitative investigations to the ultimate destination of a system, neglecting its transient behavior along the way. We have to restrict our study to autonomous systems which are either unforced or subject to a constant step input; periodic inputs, and hence frequency domain analysis, are beyond our competence to deal with. Worst of all, capacitative and inductive elements may simply vanish into thin air! For example, the steady state approached when a parallel RC circuit is activated by a constant current or voltage is not influenced at all by the capacitance but depends only on the resistance. All this is woefully restrictive. Yet these very restrictions serve as a powerful stimulus to further research aimed at discovering the input-output relationships for every individual component of a given physiological feedback loop. When we have done this, we shall at least be no worse off than our engineering colleagues! In the meantime, "to travel hopefully is a better thing than to arrive."

Exercises

1. Extend each of the numerical approximations discussed in this chapter (point-slope method, Taylor's series approximations, modified Euler method with the predictor-corrector feature, Runge-Kutta method) one or two steps farther than was done in the text so that you are sure you know how to apply them. Check your calculations!

2. Can you find an analytical solution for equation 6.13.26?

3. In Figure 6.13.2.a, the uppermost curve (for 10g of protein per 100 ml plasma) has already reached a minimum before $X = 1.0$ so that the intra-capillary flow actually increases as the end of the capillary is approached. Describe what happens in the model system to account for this behavior.

4. Copy values of ε^{-x} for $x = 0.00, 0.10, 0.20---0.80$ from a published table of exponentials. Apply the method of successive differences to these, and use the results to estimate $\varepsilon^{-0.9}$ and $\varepsilon^{-1.0}$. Compare your estimated values with those in the published table.

5. The curve obtained by plotting $y' \equiv dy/dx$ against x according to the equation

$$y' = a + bx + cx^2$$

passes through the points $[x, y'_1]$, $[x + h, y'_2]$, $[x + 2h, y'_3]$. Show that $y_3 - y_1$ obtained by integration is identical to $y_3 - y_1$ calculated by Simpson's rule. Explain this identity.

6. In example 6.14.1, Simpson's rule was used to check the calculation of x_3^* by the modified Euler method. Could the trapezoidal rule be similarly used to check the calculation of x_3^* by equation 6.11.7?

7. Solve equation 6.15.1, subject to the initial conditions of equation 6.15.2, analytically. Use the solution to evaluate x(0.3), y(0.3) and z(0.3), and compare the results with the approximate values of equations 6.15.22.

8. Given the function $f(x,y) = xy^2$ use equation 6.18.1 to calculate an exact value of $f(a + h, b + k)$ from $f(a,b)$. Check your result numerically by substituting $a = 3$, $b = 2$, $h = 0.2$, $k = 0.1$ into it and showing that it yields $[a + h][b + k]^2 = 3.2[2.1]^2 = 14.112$. Alternatively, check your result algebraically by expanding $[a + h][b + k]^2$.

9. Consider the "Artificial carotid sinus" described in Part II of this book. The sinusoidal fluctuation in arterial blood pressure shown in Figure 3 of that section represents a limit cycle at a frequency of 17 radian/min. Assume that these fluctuations are caused by a dead time (See Section 4.13) between the beginning of norepinephrine infusion and the onset of the blood pressure response, with no other elements of the system contributing to the phase lag. How many seconds of dead time would account for the perpetuation of the limit cycle?

10. A simple pendulum (i.e., one which can be treated as a point mass at the end of a massless rigid rod) hangs motionless in the earth's gravitational field. At time zero, the pendulum is displaced from its resting position by the application of a constant torque, τ. The equation of motion is:

$$m\ell^2 \ddot{\theta}(t) + D\dot{\theta}(t) + mg\ell \sin\theta(t) = \tau$$

where:

τ	= the applied torque	$9.14 \cdot 10^6$ dyne•cm
m	= the mass of the pendulum	300 g
ℓ	= the length of the pendulum	40 cm
D	= the damping constant	10^6 dyne•cm•sec/rad
g	= the acceleration of gravity	980 cm/sec^2
$\theta(t)$	= The angular displacement	radians

A. Using the numerical values listed above, calculate the principal equilibrium values of $\theta(t)$.
B. Test those equilibrium values for stability, using the method of small perturbations.
C. Show that analysis of the linear equation obtained by applying Taylor's series to the nonlinear equation of motion yields the same results as the method of small perturbations.

11. Consider a constant volume, V, of solution into which a substrate, S, is being infused at a constant rate, $\dot{Q}$. S combines (we shall assume irreversibly) with an enzyme, E, to form a substrate-enzyme complex, SE. k_1 is the mass-action proportionality constant for this reaction. SE then decomposes, again irreversibly, into E and a product, P, k_2 being the mass-action constant for this step. P is removed by paths which do not concern us. The total quantity of enzyme remains constant. Diagrammatically:

$$\xrightarrow{\dot{Q}} S + E \xrightarrow{k_1} SE \xrightarrow{k_2} E + P \longrightarrow$$

A. Write a differential equation according to the law of mass action for the concentration of each of the three reactants: S, E, and SE. Also write any relevant algebraic equations describing the system.

B. Are the equations so obtained independent, or can one or more be derived from the others?

C. Calculate the equilibrium values for the concentrations of S, E, and SE, in terms of the parameters of the system.
D. Select the smallest number of differential equations which adequately characterize the system, and linearize them about the equilibrium point, using Taylor's series.
E. Use the characteristic equation for this linearized system to decide whether or not the equilibrium point is stable.

References

1. Cunningham, W.J.: Introduction to Nonlinear Analysis. McGraw-Hill, N.Y., 1958.

2. Sokolnikoff, I.S. and Redheffer, R.M.: Mathematics of Physics and Modern Engineering. McGraw-Hill, N.Y., 1966.

3. Hildebrand, F.B.: Advanced Calculus for Applications. Prentice-Hall, Englewood Cliffs, N.J. 1962.

4. Spiegel, M.R.: Advanced Calculus. Schaum Publishing Co., N.Y., 1963.

5. Millerschoen, N.R. and Riggs, D.S.: Homeostatic control of plasma osmolality in the dog, and the effect of ethanol. Am. J. Physiol., 217, 431-437, 1969.

6. Scatchard, G., Batchelder, A.C., and Brown, A.: Chemical, clinical, and immunological studies on the products of human plasma fractionation. VI. The osmotic pressure of plasma and of serum albumin. J. Clin. Invest., 23, 458-464, 1944.

7. LaSalle, J.P. and Lefschetz, S.: Stability by Liapunov's Direct Method with Applications. Academic Press, N.Y., 1961.

8. Hahn, W.: Theory and Application of Liapunov's Direct Method. Prentice-Hall, Englewood Cliffs, N.J., 1963.

9. Kalman, R.E.: Liapunov functions for the problem of Lur'e in automatic control. Proc. natn. Acad. Sci. U.S.A., 49, 201-205, 1963.

10. Popov, V.M.: Absolute stability of nonlinear systems of automatic control. Avomatika Telemekh. 22, 961-979, 1961.

11. Gille, J.C., Pélegrin, M.J., and Decaulne, P.: Feedback Control Systems: Analysis, Synthesis, and Design. McGraw-Hill, N.Y., 1959.

12. Gibson, J.E.: Nonlinear Automatic Control. McGraw-Hill, N.Y., 1963.

Organisms, composed of material which is characterized by the utmost inconstancy and unsteadyness, have somehow learned the methods of maintaining constancy and keeping steady in the presence of conditions which might reasonably be expected to prove profoundly disturbing.

Walter B. Cannon, THE WISDOM OF THE BODY

7

THE ANALYSIS OF NONLINEAR FEEDBACK SYSTEMS IN THE STEADY STATE

SECTION 7.1. General characteristics of feedback systems in the steady state

Whenever we confine our attention to the steady-state behavior of feedback mechanisms, we automatically accept two important limitations. First, we must restrict our attention to step inputs applied at time zero and maintained constant thereafter, for it is obvious that a forced system cannot reach a steady state unless all of the inputs are themselves constant. Second, we must exclude from consideration systems which exhibit a limit cycle in response to a constant input. However, it is sometimes possible to sidestep this limitation by dealing only with the mean value of each oscillating variable.

Acceptance of these restrictions allows us to broaden our analysis in another direction. When we consider only steady states, there is no longer any real need to distinguish between parameters and inputs. For emphasis, or as a sop to convention, it may still be desirable to designate one or two of the independent variables as the inputs, but this is often a distinction without a difference. We are, after all, confining our attention to the end result of altering any of the independent factors which help to determine the steady-state values of the dependent variables. This end result is as much influenced by what are conventionally called "parameters" as it is by orthodox "inputs." But a caveat must be borne clearly in mind: We must be quite sure that all of the independent variables, --be they "inputs" or "parameters"--really are independent. For example, cardiac output is sometimes regarded as the product of the two parameters, heart rate and stroke volume. But we must not conclude that doubling the heart rate will double cardiac output, because a primary increase in rate causes a secondary decrease in stroke volume.

SECTION 7.2. Steady-state operating points

In the subsequent sections, the concept of an operating point is fundamental. Although we have used this term (Section 4.6) previously, we have used it rather loosely. We must now give it a more precise definition.

Consider any autonomous system containing m output variables, $x_1(t)$, $x_2(t)$, $\cdots$, $x_m(t)$. Let U_1, U_2, $\cdots$, U_n, be n constant step inputs imposed upon the system at time zero. If $x_1(t)$, $x_2(t)$, $\cdots$, $x_m(t)$ approach constant values of $x_1(\infty)$, $x_2(\infty)$, $\cdots$, $x_m(\infty)$ as $t \to \infty$, the point $[U_1, U_2, \cdots, U_n; x_1(\infty), x_2(\infty), \cdots, x_m(\infty)]$ in $[n + m]$-space is an operating point.

From this definition, we see that an operating point comprises both the steady values of the inputs to the system and the steady values of the dependent output variables which are completely determined by these inputs. An operating point is a point of stable equilibrium (See Section 6.16). If, for certain steady inputs, the system is either unstable or exhibits a stable limit cycle, no operating points corresponding to those particular inputs exist.

In the remainder of this monograph, we shall deal almost exclusively with the quasilinear behavior of nonlinear systems in the immediate neighborhood of an operating point. Each such operating point has its own characteristics, and it must therefore be viewed as a separate entity. The reader must learn to think about one operating point at a time!

SECTION 7.3. Extension of the product-of-derivatives method for calculating [OLG](j0) to nonlinear systems

In Section 4.15, we found that for a single linear feedback loop in the steady state, [OLG](j0) could be calculated as follows: Let $x_i(\infty) = f(x_j(\infty))$ and $x_j(\infty) = g(x_i(\infty))$ represent the working equations specifying the relationships between any two variables in the feedback loop. Then by equation 4.15.16,

$$[OLG](j0) = \frac{df(x_j(\infty))}{dx_j(\infty)} \cdot \frac{dg(x_i(\infty))}{dx_i(\infty)} \qquad 7.3.1$$

Although this equation was derived from a specific example where the working equations, 4.15.11, and 4.15.12, were

$$x_2(\infty) = f(x_5(\infty)) = k_1k_2U_1 + k_2U_2 + k_1k_2x_5(\infty) \qquad 7.3.2,$$

and

$$x_5(\infty) = g(x_2(\infty)) = k_3k_4k_5U_3 + k_4k_5U_4 + k_5U_5 + k_3k_4k_5x_2(\infty) \qquad 7.3.3,$$

it was pointed out that equation 7.3.1 could be verified for any pair of loop variables, x_i and x_j, $i \neq j$. Let us accordingly write the two working equations in the general form,

$$x_i(\infty) = f(x_j(\infty)) = C_1 + C_2x_j(\infty) \qquad 7.3.4$$

$$x_j(\infty) = g(x_i(\infty)) = C_3 + C_4x_i(\infty) \qquad 7.3.5$$

where the constants C_1 and C_3 depend upon both the inputs to, and the parameters of the system, while the constants C_2 and C_4 depend only upon the parameters of the open-loop transfer function in the steady state. For example, in equations 7.3.2 and 7.3.3,

$$\left.\begin{aligned} C_1 &= k_1k_2U_1 + k_2U_2 \\ C_3 &= k_3k_4k_5U_3 + k_4k_5U_4 + k_5U_5 \\ C_2 &= k_1k_2 \\ C_4 &= k_3k_4k_5 \end{aligned}\right\} \quad 7.3.6$$

Since equations 7.3.4 and 7.3.5 describe straight lines, their derivatives (the slopes of the lines) are simply the constants, C_2 and C_4. Furthermore, by equation 7.3.1,

$$[OLG](j0) = |C_2\ C_4| \qquad 7.3.7$$

so that the steady-state gain for any linear system is a constant for all values of the inputs, i.e. at all operating points of the system.

Now consider a nonlinear single-loop system in the steady state. In such a system, the feedback equations corresponding to 7.3.4 and 7.3.5 will no longer both be equations for straight lines. Instead, at least one of them will describe a curve of some kind. The slopes (i.e., the derivatives) of such curves will not be everywhere the same, and in general the value of [OLG](j0) will differ from one operating point to another. We must therefore consider each operating point individually, and we must restrict our analysis to the behavior of the system in the immediate neighborhood of the particular operating point under scrutiny. But if we are restricting ourselves to the immediate neighborhood of a point, we might just as well substitute for the nonlinear equations their linear approximations at that point, i.e., the equations for the straight lines tangent to the curves at their point of intersection. This immediately reduces the problem to the simple linear case! Hence we may rewrite equation 7.3.1 for any single-loop nonlinear feedback system as

$$[OLG](j0) = \left| \left.\frac{df(x_j(\infty))}{dx_j(\infty)}\right|_{op} \cdot \left.\frac{dg(x_i(\infty))}{dx_i(\infty)}\right|_{op} \right| \qquad 7.3.8$$

where the subscript "op" is used to remind us that the derivatives are to be evaluated at some particular operating point. But since that point may, in fact, be any operating point, equation 7.3.8 is really quite general.

The preceding description will now be illustrated by a specific example which, though for simplicity artificial, is of the kind likely to be encountered in a real physiological system.

Example 7.3.1. Consider a single-loop system described by the following steady-state working relationships:

$$A(\infty) = f(B(\infty)) = \frac{5B(\infty)}{U + B(\infty)} \qquad 7.3.9,$$

$$B(\infty) = g(A(\infty)) = \frac{2}{A(\infty)} \qquad 7.3.10.$$

$A(\infty)$ and $B(\infty)$ are the steady-state values of the two loop variables. U is the independent variable. 5 and 2 are, in this simple example, fixed parameters, but in a real system they could be represented by appropriate symbols and could be treated as independent variables just as U is being treated. Each step in the analysis will be illustrated by the actual numerical values for the operating point when U = 3.5.

Step 1. Evaluate each of the dependent variables at the operating point by solving the working equations simultaneously. (If an algebraic solution is impossible, the values of the dependent variables for given values of the independent variables can be estimated graphically, for example as the coordinates of the point of intersection of the solid curve and the broken curve in Figure 7.3.1.a) Solving equations 7.3.9 and 7.3.10 simultaneously,

$$A_{op}(\infty) = \frac{20}{2 + \sqrt{4 + 40U}} = 10/7 \qquad \text{when } U = 3.5 \qquad 7.3.11,$$

$$B_{op}(\infty) = \frac{2 + \sqrt{4 + 40U}}{10} = 7/5 \qquad \text{when } U = 3.5 \qquad 7.3.12,$$

where the subscript "op" emphasizes that these values are for the particular operating point determined by the chosen value of U.

Step 2. Calculate the slopes of the tangent lines at the operating point (Figure 7.3.1.b) by evaluating the derivatives of equations 7.3.9 and 7.3.10 at that point. From equation 7.3.9,

$$\left.\frac{df(B(\infty))}{dB(\infty)}\right|_{op} = f'(B) = \frac{5U}{[U + B(\infty)]^2} = 0.729 \quad \text{when} \begin{cases} U = 7/2 \\ B(\infty) = 7/5 \end{cases}$$

7.3.13.

From equation 7.3.10,

$$\left.\frac{dg(A(\infty))}{dA(\infty)}\right|_{op} = g'(A) = \frac{-2}{A^2(\infty)} = -0.980 \qquad \text{when} \quad A(\infty) = \frac{10}{7}$$

7.3.14.

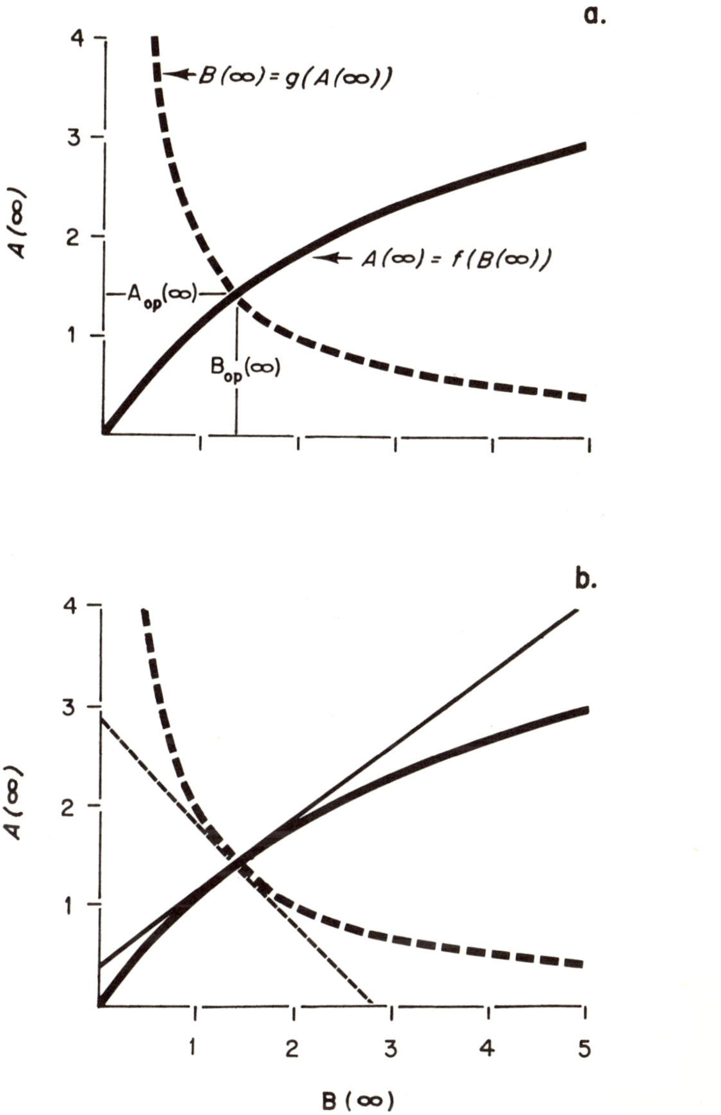

Figure 7.3.1. Panel a. Curves for the working equations 7.3.9 and 7.3.10, and the operating point values determined by their intersection. Panel b. Straight lines tangent to the curves at their intersection represent the linearization of the system at the operating point. These tangent lines show how the feedback system behaves at that point precisely as explained in Figure 4.15.3 for a linear feedback system. See that figure for details.

Step 3. Calculate the value of [OLG](j0) as the absolute value of the product of the derivatives found in step 2.

$$[OLG](j0) = |f'(B)g'(A)| = \left| \frac{10U}{[U + B(\infty)]^2 A^2(\infty)} \right| = 0.714$$

$$\text{when } U = 3.5,\ A(\infty) = 10/7,\ B(\infty) = 7/5 \qquad 7.3.15.$$

In actual practice, it is quite unnecessary to draw the tangent lines. They are shown in Figure 7.3.1.b only to illustrate the underlying rationale of the procedure and to facilitate comparison with the similar figure (Figure 4.15.3) for a linear system. Indeed, the curves themselves need not be drawn, but it is often illuminating to construct them, particularly when several curves, each for a different value of an important independent variable, are plotted on the same graph. Several examples will be found in Part II. Note again that in such graphs, a broken line is used for the function which is plotted unconventionally, i.e. with the independent variable on the vertical axis, the dependent variable on the horizontal axis. The true slope of such a broken line is the reciprocal of the apparent slope, so that the more horizontal the line is, the larger is its true slope.

SECTION 7.4. Calculation of the effect of feedback in systems with more than one feedback loop

The technique of linearizing about an operating point will now be generalized so that it can be used to analyze the steady-state behavior of feedback systems with any number of feedback loops,--positive, negative, or both,--arranged in any pattern. For the sake of simplicity and concreteness, the following discussion will be concerned with a feedback system consisting of only three loop variables, each with a single input. Extension to more complicated systems is straightforward. The objective of the analysis is again to find out how the feedback loops of the system, considered singly or in various combinations, alter the steady-state response of a particular variable to a particular input. But when we use this more general approach, the effect of feedback cannot always be expressed as an [OLG](j0); i.e., as the absolute value of some open-loop transfer function for a linearized system evaluated for $s = j\omega$ when $\omega = 0$. In the following discussion I shall therefore use the more general term gain due to feedback rather than open-loop gain; but I shall postpone a complete discussion of the need for this term until after it has been illuminated by some actual examples. At present, I shall merely define G_{fb}, the gain due to feedback in the steady state, as

$$G_{fb} \equiv \left| \left\{ dy(\infty)/dU \Big|_{\text{open}} \Big/ dy(\infty)/dU \Big|_{\text{closed}} \right\} - 1 \right| \qquad 7.4.1,$$

where $y(\infty)$ is any appropriate dependent variable, U is any independent variable, and the derivatives are to be evaluated at a specified operating point before and after opening one or more specified loops. Note that when $y(\infty)$ is a member of the feedback loop in a single-loop system, G_{fb} is identical to [OLG] (j0). Equation 7.4.1 is, in fact, a generalized version of equation 4.14.9 which was previously derived specifically for a negative feedback loop.

Let A(t), B(t), and C(t) be three time-dependent variables each of which is a member of at least one feedback loop. Let their respective inputs be U_A, U_B and U_C. (Since these independent variables must be step inputs which are constant for all $t > 0$, I shall not bother to designate them $U_A(t)$ etc.). Let

$$dA(t)/dt = \varphi_A(U_A, A(t), B(t), C(t)) \qquad 7.4.2,$$

$$dB(t)/dt = \varphi_B(U_B, A(t), B(t), C(t)) \qquad 7.4.3,$$

$$dC(t)/dt = \varphi_C(U_C, A(t), B(t), C(t)) \qquad 7.4.4,$$

represent the differential equations which fully describe the behavior of the system at every instant of time following imposition of one or more of the step inputs at time zero. Equations 7.4.2, 7.4.3 and 7.4.4 are, in general, nonlinear. Moreover, for most physiological systems, the differential equations themselves are unknown. Hence, much as we would like to obtain an explicit solution of these equations, it is not usually possible, and we must turn to the more tractable steady-state analysis.

As usual, let A(∞), B(∞), and C(∞) be the steady-state values approached by the variables as $t \to \infty$. At the steady state the first derivatives in equations 7.4.2, 7.4.3 and 7.4.4 vanish and we obtain a set of algebraic equations which again are, in general, nonlinear.

$$\varphi_A(U_A, A(\infty), B(\infty), C(\infty)) = 0 \qquad 7.4.5,$$

$$\varphi_B(U_B, A(\infty), B(\infty), C(\infty)) = 0 \qquad 7.4.6,$$

$$\varphi_C(U_C, A(\infty), B(\infty), C(\infty)) = 0 \qquad 7.4.7.$$

Solving equation 7.4.5 for A(∞), 7.4.6 for B(∞) and 7.4.7 for C(∞),

$$A(\infty) = f_A(U_A, B(\infty), C(\infty)) \qquad 7.4.8$$

$$B(\infty) = f_B(U_B, A(\infty), C(\infty)) \qquad 7.4.9$$

$$C(\infty) = f_C(U_C, A(\infty), B(\infty)) \qquad 7.4.10$$

The specific algebraic equations corresponding to equations 7.4.8, 7.4.9 and 7.4.10 for any particular system show the working relationships between the independent input variables and the dependent loop variables of the system in the steady state. In the present derivation this is obvious because, for example, equation 7.4.8, the working equation for variable A, was obtained from an equation (equation 7.4.2) written specifically for the rate of change of A(t) with time. But if equations such as 7.4.8, 7.4.9 and 7.4.10 have been obtained by considering the steady-state situation directly and not by evaluating a known differential equation at the limit as $t \to \infty$, it may be necessary to apply the method of concomitant variations (Section 1.4) to discover the proper working relationships.

Before going further, it is important to emphasize that in equations 7.4.8, 7.4.9 and 7.4.10, time is no longer the independent variable. From now on, we inquire not how a particular variable changes with time, but how its steady-state value is influenced by the other variables of the system, both independent and dependent. While the specific working equations corresponding to 7.4.8, 7.4.9 and 7.4.10 describe these influences precisely, it is often useful to summarize their pattern and direction in the form of a symbol-and-arrow diagram, (Sections 1.5 thru 1.8). The utility of such a diagram may be illustrated even with the present very general equations, 7.4.8, 7.4.9 and 7.4.10. By simply inspecting those equations, it is not easy to see how many different feedback loops might be present in such a system. But a symbol-and-arrow diagram (Figure 7.4.1) immediately shows that the maximum number of different loops (designated by Roman numerals) is five. (For any particular system, any arrow in this generalized diagram might be solid or broken, present or missing, so that even with but three loop variables the number of possible patterns is large.)

Since equations 7.4.8, 7.4.9 and 7.4.10 are, in general, nonlinear, we cannot apply transfer function analysis directly to calculate quantities such as open-loop gain, or, more generally, gain due to feedback. Furthermore, as we have already noted, in nonlinear systems such quantities differ from operating point to operating point, so that again we are obliged to consider each operating point individually. But just as with the single-loop system, if we have to consider each operating point individually, we might as well replace the equations 7.4.8, 7.4.9 and 7.4.10 by their linear approximations, i.e. by the equations for the tangent (which, in general, will be a tangent hyperplane*) at that particular point. This is the key to the whole analysis!

[*The general equation for a plane in 3-space is

$$x_1(x_2, x_3) = a_1 + a_2x_2 + a_3x_3$$

where a_1 is the value of $x_1(0, 0)$, a_2 is the slope of the plane in the x_2 direction, and a_3 is the slope of the plane in the x_3 direction. For a plane tangent at a given point, p, to the curved surface defined by $x_1 = f(x_2, x_3)$, a_2 is $\partial f(x_2, x_3)/\partial x_2\big|_p$, i.e., the slope of the curved surface in the x_2 direction at the point, and a_3 is $\partial f(x_2, x_3)/\partial x_3\big|_p$, the slope of the curved surface in the x_3 direction at the point. The intercept, a_1 may be evaluated by requiring a plane with these slopes to pass through the given point.

These equations are easily generalized to n-dimensional space where the coordinates are $x_1, x_2, x_3, \cdots, x_n$. The general equation for a hyperplane in n-space is

$$x_1(x_2, x_3, \cdots, x_n) = a_1 + a_2x_2 + a_3x_3 + \cdots + a_nx_n,$$

and the general equation for a hyperplane tangent to the curved hypersurface $x_1 = f(x_2, x_3, \cdots, x_n)$ at a point p specified by a given set of values of $x_1, x_2, x_3, \cdots, x_n$ is

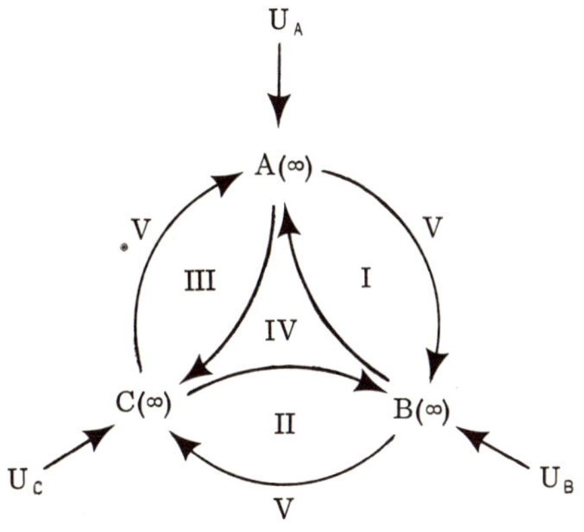

Figure 7.4.1. Symbol-and-arrow diagram for a multiple-loop system

---

$$x_{1,tan} = A + \frac{\partial f}{\partial x_2}\bigg|_p x_2 + \frac{\partial f}{\partial x_3}\bigg|_p x_3 + \cdots + \frac{\partial f}{\partial x_n}\bigg|_p x_n$$

In these general equations, the choice of x_1, as the dependent variable was purely arbitrary. Mathematically, it might have been any of the n variables. But in real systems the dependent variable is determined by the working relationships governing the behavior of the system.]

We have altogether six variables: the three inputs, U_A, U_B and U_C and the three loop variables $A(\infty)$, $B(\infty)$ and $C(\infty)$. A particular operating point is therefore a point in 6-space.* Let this point be [$U_{A,op}$, $U_{B,op}$, $U_{C,op}$; $A_{op}(\infty)$, $B_{op}(\infty)$, $C_{op}(\infty)$]. It will be convenient to place the origin of coordinates at the particular operating point of current interest. (Merely shifting the origin of coordinates will in no way alter the slopes of any of the curved surfaces in 6-space defined by equations 7.4.8, 7.4.9, and 7.4.10, but will enable us to get rid of their troublesome intercepts which have nuisance value only.) Then the linear approximations of equations 7.4.8, 7.4.9 and 7.4.10 at the operating point are

$$\Delta A = \frac{\partial f_A}{\partial U_A}\bigg|_{op} \Delta U_A + \frac{\partial f_A}{\partial B(\infty)}\bigg|_{op} \Delta B + \frac{\partial f_A}{\partial C(\infty)}\bigg|_{op} \Delta C \qquad 7.4.11,$$

$$\Delta B = \frac{\partial f_B}{\partial U_B}\bigg|_{op} \Delta U_B + \frac{\partial f_B}{\partial A(\infty)}\bigg|_{op} \Delta A + \frac{\partial f_B}{\partial C(\infty)}\bigg|_{op} \Delta C \qquad 7.4.12,$$

$$\Delta C = \frac{\partial f_C}{\partial U_C}\bigg|_{op} \Delta U_C + \frac{\partial f_C}{\partial A(\infty)}\bigg|_{op} \Delta A + \frac{\partial f_C}{\partial B(\infty)}\bigg|_{op} \Delta B \qquad 7.4.13.$$

where ΔU_A, ΔU_B, etc. are each defined as the deviation of the indicated variable from its operating point value; for example, $\Delta U_A \equiv U_A - U_{A,op}$; $\Delta U_B \equiv U_B - U_{B,op}$; etc. Each of the partial derivatives gives the slope of the hyperplane tangent at the operating point along the designated coordinate. This slope is, of course, obtained by partially differentiating the appropriate equation, 7.4.8, 7.4.9 or 7.4.10, and evaluating the partial derivative at the operating point.

[*This is not the state space of the system. If we know the working equations, the state (now the steady state) of the system is completely determined by the values of the independent variables. In the present example, therefore, state space would be three-dimensional.]

In the linear approximations (the tangent hyperplanes) of equations 7.4.11, 7.4.12 and 7.4.13, each of these partial-derivative-at-an-operating-point slopes is a constant, just as the slope of a straight line is a constant. For convenience, therefore, let us define

$$\left.\begin{aligned} \beta_1 &\equiv \left.\frac{\partial f_A}{\partial U_A}\right|_{op} ; \quad \beta_2 \equiv \left.\frac{\partial f_A}{\partial B(\infty)}\right|_{op} ; \quad \beta_3 \equiv \left.\frac{\partial f_A}{\partial C(\infty)}\right|_{op} \\ \beta_4 &\equiv \left.\frac{\partial f_B}{\partial U_B}\right|_{op} ; \quad \beta_5 \equiv \left.\frac{\partial f_B}{\partial A(\infty)}\right|_{op} ; \quad \beta_6 \equiv \left.\frac{\partial f_B}{\partial C(\infty)}\right|_{op} \\ \beta_7 &\equiv \left.\frac{\partial f_C}{\partial U_C}\right|_{op} ; \quad \beta_8 \equiv \left.\frac{\partial f_C}{\partial A(\infty)}\right|_{op} ; \quad \beta_9 \equiv \left.\frac{\partial f_C}{\partial B(\infty)}\right|_{op} \end{aligned}\right\} \quad 7.4.14.$$

Substituting the values from equations 7.4.14 into equations 7.4.11, 7.4.12 and 7.4.13:

$$\Delta A = \beta_1 \Delta U_A + \beta_2 \Delta B + \beta_3 \Delta C \qquad 7.4.15$$

$$\Delta B = \beta_4 \Delta U_B + \beta_5 \Delta A + \beta_6 \Delta C \qquad 7.4.16$$

$$\Delta C = \beta_7 \Delta U_C + \beta_8 \Delta A + \beta_9 \Delta B \qquad 7.4.17$$

Equations 7.4.15, 7.4.16 and 7.4.17 constitute a set of linear algebraic equations which, at, and in the immediate neighborhood of the operating point being considered, can be analyzed instead of the nonlinear equations 7.4.8, 7.4.9 and 7.4.10. If we wish, we can even draw a conventional block diagram to represent this "linear-at-a-particular-operating-point" system. (Figure 7.4.2).

Now suppose we want to find out how feedback alters the change in $A(\infty)$ caused by a (very small) step change in U_A. To do this we let $\Delta U_A > 0$, but keep the other inputs of Figure 7.4.2 unchanged at zero. Equations 7.4.15, 7.4.16 and 7.4.17 then become,

$$\Delta A = \beta_1 \Delta U_A + \beta_2 \Delta B + \beta_3 \Delta C \qquad 7.4.15$$

$$\Delta B = \beta_5 \Delta A + \beta_6 \Delta C \qquad 7.4.18$$

$$\Delta C = \beta_8 \Delta A + \beta_9 \Delta B \qquad 7.4.19$$

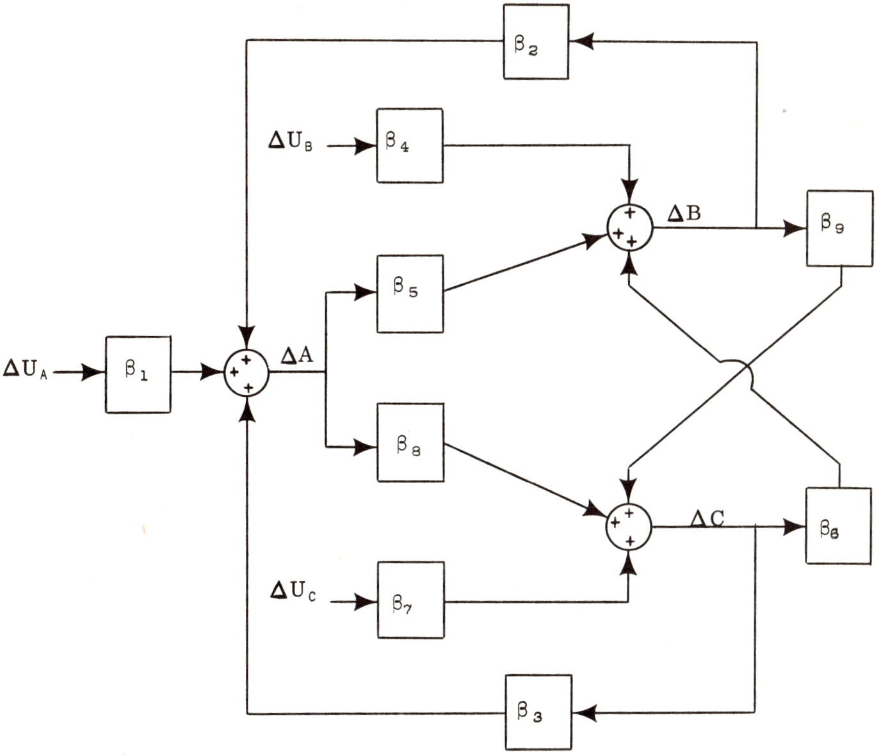

Figure 7.4.2. Conventional block diagram for the multiple-loop steady-state system of Figure 7.4.1 after it has been linearized about an operating point.

The Δ's are deviations from the operating-point values. The β's are the partial derivatives defined by equations 7.4.14.

Combining equations 7.4.15, 7.4.18 and 7.4.19 by eliminating ΔB and ΔC and solving for ΔA,

$$\Delta A = \left[\frac{\beta_1}{1 - \beta_3\beta_8 - \left\{ \dfrac{[\beta_5 + \beta_6\beta_8][\beta_2 + \beta_3\beta_9]}{1 - \beta_6\beta_9} \right\}} \right] \Delta U_A \qquad 7.4.20$$

which gives ΔA as a function of ΔU_A with all of the feedback loops closed. From equation 7.4.20, the change in $A(\infty)$ caused by a step change in U_A at a particular operating point is

$$\left.\frac{dA}{dU_A}\right|_{op,\,closed} = \frac{\beta_1}{\dfrac{\beta_2\beta_5 + \beta_3\beta_8 + \beta_2\beta_6\beta_8 + \beta_3\beta_5\beta_9}{1 - \beta_8\beta_9}} \qquad 7.4.21.$$

Now suppose that we want to compare this closed-loop change with the change which would occur when all of the loops influencing the response of ΔA to ΔU_A have been opened, for example by letting $\beta_5 = \beta_8 = 0$*. Then simultaneous solution of equations 7.4.15, 7.4.18 and 7.4.19 gives

$$\Delta A = \beta_1 \Delta U_A + \beta_2\beta_6 \Delta C + \beta_3\beta_9 \Delta B \qquad 7.4.22,$$

whence

$$\left.\frac{dA}{dU_A}\right|_{op,\,open} = \beta_1 \qquad 7.4.23.$$

But by equation 7.4.1, the gain due to feedback with all loops acting together is

$$G_{fb,\,all} = \left| \left\{ \left. dA/dU_A \right|_{open} \Big/ \left. dA/dU_A \right|_{closed} \right\} - 1 \right| \qquad 7.4.24.$$

Substituting the values from equations 7.4.21 and 7.4.23 into equation 7.4.24

$$G_{fb,\,all} = \left| \frac{\beta_2\beta_5 + \beta_3\beta_8 + \beta_2\beta_6\beta_8 + \beta_3\beta_5\beta_9}{1 - \beta_8\beta_9} \right| \qquad 7.4.25.$$

Note that this value of G_{fb} is valid only for the joint effect of <u>all</u> of the relevant feedback loops upon the response of <u>A(∞)</u> to an infinitesimal change in <u>U_A</u>. If we choose to open some loops but not others, or if we want to know the value of G_{fb} for all loops jointly but for a different output variable or for a different input, we shall, in general, obtain a different value for G_{fb}.

[*In this particular example, it is not necessary to open Loop II by letting β_8 or $\beta_9 = 0$, because Loop II cannot influence the response of A to U_A when all other loops have been opened.]

Since the preceding method for calculating the gain due to feedback depends upon replacing nonlinear equations by their linear approximations at an operating point, it might be supposed that the gains thus calculated would themselves be approximate. Not so! For in calculating the gain, we have considered only infinitesimal deviations from the operating point, and for infinitesimal deviations the linear approximations are just as accurate as the original nonlinear equations.

We shall now apply these analytical steps to a specific example.

<u>Example 7.4.1.</u> Consider a system whose steady-state behavior is described by the following working equations,

$$A(\infty) = f_A(B, C, W, U) = \frac{WB(\infty)}{U + B(\infty)} + C(\infty) \qquad 7.4.26$$

$$B(\infty) = f_B(A, M) = M/A(\infty) \qquad 7.4.27$$

$$C(\infty) = f_C(B, C_O, K) = C_O\varepsilon^{-KB(\infty)} \qquad 7.4.28$$

The corresponding symbol-and-arrow diagram,

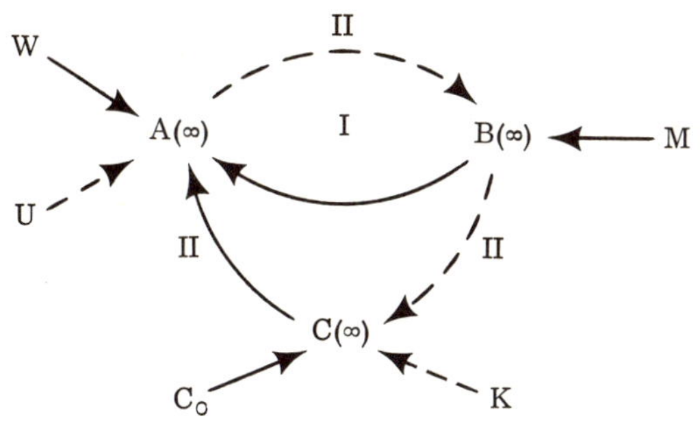

shows that in this system there are two feedback loops, one negative (Loop I), one positive (Loop II). To illustrate the method outlined above, let us find out how the effect of U upon $C(\infty)$ is influenced by opening Loop I, or Loop II, or both loops together. Using capital deltas as before to designate the deviations of the inputs and variables from their operating point values in 8-space, we write the equations for the tangent hyperplanes at that point.

$$\Delta A = \frac{\partial f_A}{\partial B(\infty)}\bigg|_{op} \Delta B + \frac{\partial f_A}{\partial C(\infty)}\bigg|_{op} \Delta C + \frac{\partial f_A}{\partial W}\bigg|_{op} \Delta W + \frac{\partial f_A}{\partial U}\bigg|_{op} \Delta U \qquad 7.4.29$$

$$\Delta B = \frac{\partial f_B}{\partial A(\infty)}\bigg|_{op} \Delta A + \frac{\partial f_B}{\partial M}\bigg|_{op} \Delta M \qquad 7.4.30$$

$$\Delta C = \frac{\partial f_C}{\partial B(\infty)}\bigg|_{op} \Delta B + \frac{\partial f_C}{\partial C_O}\bigg|_{op} \Delta C_O + \frac{\partial f_C}{\partial K}\bigg|_{op} \Delta K \qquad 7.4.31$$

Evaluating the partial derivatives by differentiating equations 7.4.26, 7.4.27 and 7.4.28, we obtain

$$\left.\begin{aligned} \frac{\partial f_A}{\partial B(\infty)}\bigg|_{op} &= \frac{WU}{[U + B(\infty)]^2} \equiv \beta_1 \\ \frac{\partial f_A}{\partial C(\infty)}\bigg|_{op} &= 1 \equiv \beta_2 \end{aligned}\right)$$

$$\left.\frac{\partial f_A}{\partial W}\right|_{op} = \frac{B(\infty)}{U + B(\infty)} = \beta_3$$

$$\left.\frac{\partial f_A}{\partial U}\right|_{op} = - \frac{WB(\infty)}{[U + B(\infty)]^2} \equiv - \beta_4$$

$$\left.\frac{\partial f_B}{\partial A(\infty)}\right|_{op} = - \frac{M}{A^2(\infty)} \equiv - \beta_5$$

$$\left.\frac{\partial f_B}{\partial M}\right|_{op} = \frac{1}{A(\infty)} \equiv \beta_6$$

$$\left.\frac{\partial f_C}{\partial B(\infty)}\right|_{op} = - KC_0 \epsilon^{-KB(\infty)} \equiv - \beta_7 \qquad 7.4.32$$

$$\left.\frac{\partial f_C}{\partial C_0}\right|_{op} = \epsilon^{-KB(\infty)} \equiv \beta_8$$

$$\left.\frac{\partial f_C}{\partial K}\right|_{op} = - B(\infty)\, C_0 \epsilon^{-KB(\infty)} \equiv - \beta_9$$

In equations 7.4.32, the independent and dependent variables must have the values appropriate for any one particular operating point,* though for economy of notation the subscript "op" has been omitted from them. Notice that the β's have all been defined as positive quantities.

[*Obtained, as mentioned in Section 7.3, by solving the working equations (7.4.26, 7.4.27 and 7.4.28) simultaneously. In the present example, this cannot be done algebraically, because, in equation 7.4.28, $B(\infty)$ appears in the exponent. A graphical solution (see Appendix) must therefore be used.]

For convenience, we now rewrite equations 7.4.29, 7.4.30 and 7.4.31 in terms of the β's defined by equations 7.4.32.

$$\Delta A = \beta_1 \Delta B + \beta_2 \Delta C + \beta_3 \Delta W - \beta_4 \Delta U \qquad 7.4.33$$

$$\Delta B = - \beta_5 \Delta A + \beta_6 \Delta M \qquad 7.4.34$$

$$\Delta C = - \beta_7 \Delta B + \beta_8 \Delta C_0 - \beta_9 \Delta K \qquad 7.4.35$$

The corresponding block diagram is shown in Figure 7.4.3. To find the effect of U upon $C(\infty)$ with both loops closed, we let all the other independent variables remain at the operating point so that $\Delta W = \Delta M = \Delta K = \Delta C_0 = 0$. We then solve equations 7.4.33, 7.4.34 and 7.4.35 simultaneously for ΔC in terms of ΔU.

$$\Delta C_{closed} = - \frac{\beta_4 \beta_5 \beta_7}{1 + \beta_1 \beta_5 - \beta_2 \beta_5 \beta_7} \Delta U \qquad 7.4.36$$

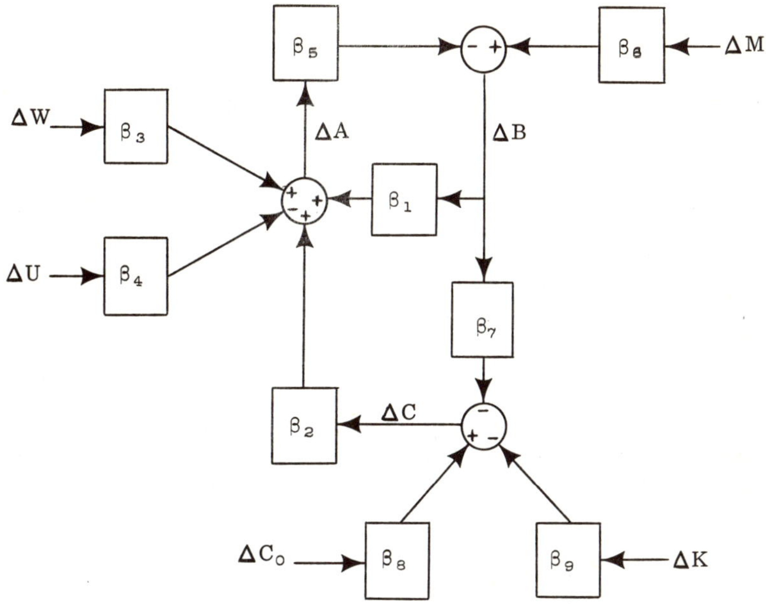

Figure 7.4.3. Conventional block diagram for the system of example 7.4.1 after it has been linearized about an operating point.

The Δ's are deviations from the operating-point values. The β's are the partial derivatives defined by equations 7.4.32.

---

whence,

$$\left.\frac{dC}{dU}\right|_{closed} = -\frac{\beta_4\beta_5\beta_7}{1 + \beta_1\beta_5 - \beta_2\beta_5\beta_7} \qquad 7.4.37$$

To open Loop I only, let $\beta_1 = 0$. Then equation 7.4.36 becomes

$$\Delta C_{I,\,open} = -\frac{\beta_4\beta_5\beta_7}{1 - \beta_2\beta_5\beta_7}\,\Delta U \qquad 7.4.38$$

whence,

$$\left.\frac{dC}{dU}\right|_{I,\,open} = -\frac{\beta_4\beta_5\beta_7}{1 - \beta_2\beta_5\beta_7} \qquad 7.4.39$$

Now let $G_{fb,I}$ be the gain due to feedback* via Loop I with regard to the influence of U on C(∞). By equation 7.4.1,

$$G_{fb,I} = \left| \frac{dC/dU|_{I,\, open}}{dC/dU|_{closed}} - 1 \right| = \left| \frac{\beta_1 \beta_5}{1 - \beta_2 \beta_5 \beta_7} \right| \qquad 7.4.40$$

Similarly, by letting $\beta_2 = 0$, we find that the gain due to feedback for Loop II alone is

$$G_{fb,II} = \left| - \frac{\beta_2 \beta_5 \beta_7}{1 + \beta_1 \beta_5} \right| = \frac{\beta_2 \beta_5 \beta_7}{1 + \beta_1 \beta_5} \qquad 7.4.41$$

Finally, by letting both β_1 and β_2 equal zero we find that the gain due to feedback for the joint effect of both loops is

$$G_{fb,\,both} = \beta_5 \left| \beta_1 - \beta_2 \beta_7 \right| \qquad 7.4.42$$

[*Notice once more that "gain due to feedback" is not necessarily the same thing as [OLG] (j0). We have to abide by the standard definition of "open-loop gain," namely the magnitude of the open-loop transfer function around a feedback loop when $s = j\omega$. For Loop I by itself, [OLG] (j0) would be $|\beta_1 \beta_5|$, a quantity which has no direct bearing upon our present inquiry.]

Some of the attributes of this system are illustrated by the numerical examples**in Table 7.4.1.

[**It is worth constructing such numerical examples, even at the cost of considerable labor, because if we concentrate only upon the algebraic results derived from the original working equations we may forget other important constraints upon the system, constraints which are not embodied in those equations. In particular we must limit our analysis to values which are physically or physiologically possible. Calculating how the cardiovascular system behaves when the heart rate is negative would be a pretty sterile exercise!]

Consider first the joint effect of both loops upon the response of $C(\infty)$ to a change in U. Does the system, taken as a whole, act as a positive or a negative feedback system? To answer this question it is useless to look merely at the gain specified by equation 7.4.42, because gains between zero and unity are ambiguous; they may arise either from positive or from negative feedback (see Figure 7.5.4). Instead, let us ask how much the effect of U upon $C(\infty)$ is magnified by feedback. Like gain, the magnification, previously defined for linear systems in equation 4.16.1, is an intrinsically positive quantity, but it is not ambiguous. We shall now define magnification more generally as

$$[\text{Magnification}] = \frac{dC/dU|_{closed}}{dC/dU|_{open}} \qquad 7.4.43.$$

With positive feedback, the magnification is always greater than unity, while with negative feedback, it is always less than unity. In the present system, where

$$\frac{dC/dU|_{closed}}{dC/dU|_{both\ open}} = \frac{1}{1 + \beta_5 [\beta_1 - \beta_2 \beta_7]}, \qquad 7.4.44,$$

Table 7.4.1. Effect of feedback on the response of C(∞) to U in Example 7.4.1.

Line No	Values of variables at operating point* M	C_o	A(∞)	B(∞)	C(∞)	Loops concerned	Kind of effect	Magni-fication	Gain due to Feedback
1	0.275	0.1597	0.1250	2.200	0.0177	I & II	-	0.759	0.317
2	0.275	0.2770	0.1375	2.000	0.0375	I & II	0	1	0
3	0.275	0.3660	0.1528	1.800	0.0605	I & II	+	1.330	0.248
4	1.000	2.548	0.9346	1.070	0.874	I	-	0	∞
5	1.000	2.563	1.0000	1.000	0.9428	I	-	0.538	0.857
6	1.000	2.956	0.5000	2.000	0.400	I	+	1.333	0.250

[*W = 0.4, U = 6.0, K = 1 for all of the operating points in this table.]

the joint effect of both loops upon the response of C(∞) to U will depend upon what operating point is being investigated. If the independent variables are such that $\beta_1 - \beta_2\beta_7$ is a positive number (e.g. line 1, Table 7.4.1) the magnification will be less than unity and the joint effect of the feedback loops for the variables U and C(∞) which are under scrutiny will be negative. If the independent variables are such that $\beta_1 - \beta_2\beta_7$ is zero, the joint effect of the loops upon the response of C(∞) to U is nil (e.g. line 2, Table 7.4.1). If the independent variables are such that $-1 < \beta_5[\beta_1 - \beta_2\beta_7] < 0$, (e.g. line 3, Table 7.4.1) the joint effect of the feedback loops is positive. (I shall defer to the next section any discussion of the kind of situation which would arise if $\beta_5[\beta_1 - \beta_2\beta_7] \leq -1$.)

Since both a positive and a negative loop are present in this system, these results are understandable enough. Far more surprising is the apparent result of opening only Loop I. Loop I is unquestionably a negative feedback loop. A priori, one might therefore suppose that, whatever our choice of independent variables, the effect of Loop I would be to decrease the response of any of the dependent variables of the system to any of the independent variables. But if we calculate the magnification as before,

$$\frac{dC/dU\,|_{\text{closed}}}{dC/dU\,|_{\text{I, open}}} = \frac{1 - \beta_2\beta_5\beta_7}{1 - \beta_2\beta_5\beta_7 + \beta_1\beta_5} \qquad 7.4.45,$$

we observe that the kind of influence exerted by Loop I alone seems to be ambiguous. For certain ranges of values of the independent variables, Loop I can indeed decrease the change in C(∞) caused by a change in U (line 5, Table 7.4.1); in fact, at certain critical operating points, the gain due to feedback may be infinite (line 4, Table 7.4.1). Yet for other ranges of values of the independent variable, the magnification of the response of C(∞) to a change in U appears to be greater than unity (line 6, Table 7.4.1) just as if Loop I were acting as a positive feedback loop! In contrast, the magnification of the effect of U on C(∞) due to the positive feedback Loop II,

$$\frac{dC/dU\,|_{closed}}{dC/dU\,|_{II,\,open}} = \frac{[1+\beta_1\beta_5]}{[1+\beta_1\beta_5]-\beta_2\beta_5\beta_7}\ ;$$

$$[1+\beta_1\beta_5] > \beta_2\beta_5\beta_7 \qquad 7.4.46,$$

can never be less than unity. With respect to the variables under scrutiny, therefore, Loop II always behaves as a positive feedback loop.

Let us analyze the apparently contradictory performance of Loop I more closely by writing our linear equations, 7.4.33, 7.4.34 and 7.4.35 in terms of the actual numerical values for the β's calculated from line 6 in Table 7.4.1 (i.e. the line which shows Loop I apparently increasing the effect of U on C(∞)). With both loops closed, we have

$$\Delta A = 0.0375\,\Delta B + \Delta C - 0.0125\,\Delta U \qquad 7.4.47,$$

$$\Delta B = -\,4.0\,\Delta A \qquad 7.4.48,$$

$$\Delta C = -0.4\,\Delta B \qquad 7.4.49.$$

Solving these equations simultaneously for ΔC

$$\Delta C = 0.0444\,\Delta U \qquad 7.4.50,$$

whence,

$$dC/dU|_{closed} = 0.0444 \qquad 7.4.51.$$

After opening Loop I by letting $\beta_1 = 0$, we have

$$\Delta A = \Delta C - 0.0125\Delta U \qquad 7.4.52,$$

plus equations 7.4.48 and 7.4.49 as before. Solving simultaneously,

$$\Delta C = 0.0333\,\Delta U \qquad 7.4.53,$$

whence,

$$dC/dU_{I,\,open} = 0.0333 \qquad 7.4.54$$

Combining equations 7.4.51 and 7.4.54, the magnification,

$$\frac{dC/dU\,|_{closed}}{dC/dU\,|_{I,\,open}} = \frac{0.0444}{0.0333} = 1.33 \qquad 7.4.55,$$

seems to confirm the puzzling result we are trying to explain. But if we draw a block diagram according to equations 7.4.48, 7.4.49 and 7.4.52,

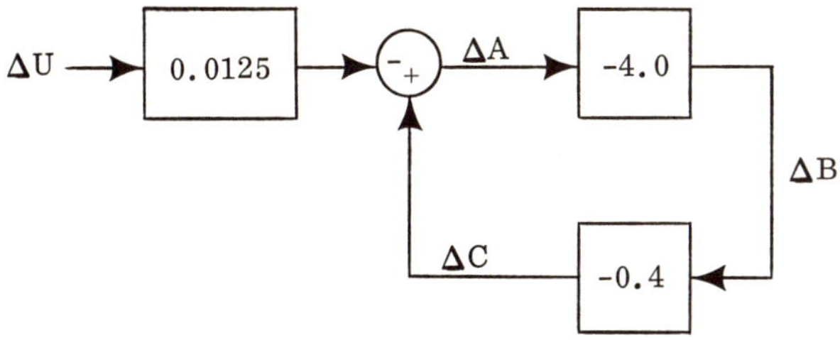

we see at once that the open-loop gain for the positive feedback loop which remains after we have opened Loop I is 1.6, a value which is incompatible with the assumption of a stable steady state. But the assumption of a stable steady state underlies the whole analysis! Bound as we are by this steady-state restriction, it is sheer nonsense to try to compare a closed-loop steady state with an open-loop instability. In calculating the effect of opening Loop I, we must therefore confine ourselves to such values of the variables as will leave us with a stable Loop II. This means that $\beta_2\beta_5\beta_7 < 1$, a restriction which completely rules out of consideration all values of the variables which would make Loop I behave like a positive feedback Loop. To emphasize this restriction, we should include it explicitly as part of equations 7.4.38, 7.4.39, 7.4.40 and 7.4.45. For example, equation 7.4.45 should be rewritten

$$\frac{dC/dU\big|_{closed}}{dC/dU\big|_{I,\ open}} = \frac{1 - \beta_2\beta_5\beta_7}{1 - \beta_2\beta_5\beta_7 + \beta_1\beta_5}\ ;\ 0 < \beta_2\beta_5\beta_7 < 1 \qquad 7.4.45.a$$

SECTION 7.5. The effect of a feedback loop upon a variable which is not itself a member of the loop

In analyzing the two-loop example of the previous section, we turned a blind eye towards a problem of very general importance simply because it did not force itself upon our attention. Consider the working relationships among the variables of that system as displayed in the symbol and arrow diagram just after equation 7.4.28. $C(\infty)$, the variable whose behavior we analyzed, is a member of Loop II but not of Loop I. We shall now see that the concepts of "open-loop gain" and "magnification" are not, in general, applicable to the effect of a feedback loop upon the behavior of a variable which is not itself a member of that loop.

Example 7.5.1. Consider a system whose steady-state behavior is governed by the following working equations,

$$A(\infty) = W - B(\infty) \qquad 7.5.1.$$

$$B(\infty) = U + 3A(\infty) \qquad 7.5.2.$$

$$C(\infty) = U\ A(\infty) \qquad 7.5.3.$$

Simultaneous solution gives $A(\infty) = [W-U]/4$, $B(\infty) = [3W+U]/4$, $C(\infty) = [WU-U^2]/4$ for the values of the dependent variables at an operating point. A

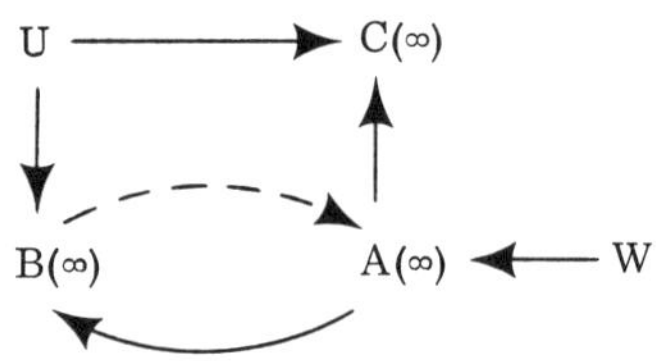

symbol-and-arrow diagram shows that A(ω) and B(ω) are members of the negative feedback loop, but that C(∞) is not. The equations for linearization about an operating point in 5-space are

$$\Delta A = \frac{\partial A(\infty)}{\partial W}\bigg|_{op} \Delta W + \frac{\partial A(\infty)}{\partial B(\infty)}\bigg|_{op} \Delta B \qquad 7.5.4.$$

$$\Delta B = \frac{\partial B(\infty)}{\partial U}\bigg|_{op} \Delta U + \frac{\partial B(\infty)}{\partial A(\infty)}\bigg|_{op} \Delta A \qquad 7.5.5.$$

$$\Delta C = \frac{\partial C(\infty)}{\partial U}\bigg|_{op} \Delta U + \frac{\partial C(\infty)}{\partial A(\infty)}\bigg|_{op} \Delta A \qquad 7.5.6.$$

By calculating the partial derivatives from equations 7.5.1, 7.5.2, and 7.5.3 and evaluating them at the operating point we obtain

$$\Delta A = \Delta W - \Delta B \qquad 7.5.7.$$

$$\Delta B = \Delta U + 3\Delta A \qquad 7.5.8.$$

$$\Delta C = \left[\frac{W - U}{4}\right] \Delta U + U\Delta A \qquad 7.5.9.$$

We can now analyze the linear equations 7.5.7, 7.5.8 and 7.5.9 instead of the original working equations.

Let us inquire how the feedback loop influences the response of C to a change in U. Keeping all other independent variables (here, only W) constant, we have, from equations 7.5.7, 7.5.8 and 7.5.9, the following closed-loop situation,

$$\Delta A = -\Delta B$$

$$\Delta B = \Delta U + 3\Delta A$$

$$\underbrace{\Delta C = \left[\frac{W - U}{4}\right] \Delta U + U\Delta A}$$

$$\Delta C = \left[\frac{W - 2U}{4}\right] \Delta U \qquad 7.5.10$$

$$\frac{dC}{dU}\bigg|_{closed} = \frac{W - 2U}{4} \qquad 7.5.11.$$

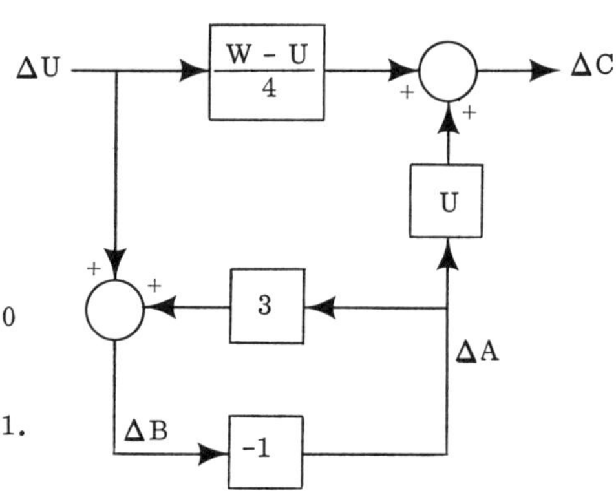

The open-loop situation is,

$\Delta A = -\Delta B$

$\Delta B = \Delta U$

$$\Delta C = \left[\frac{W - U}{4}\right] \Delta U + U \Delta A$$

$$\Delta C = \left[\frac{W - 5U}{4}\right] \Delta U \qquad 7.5.12$$

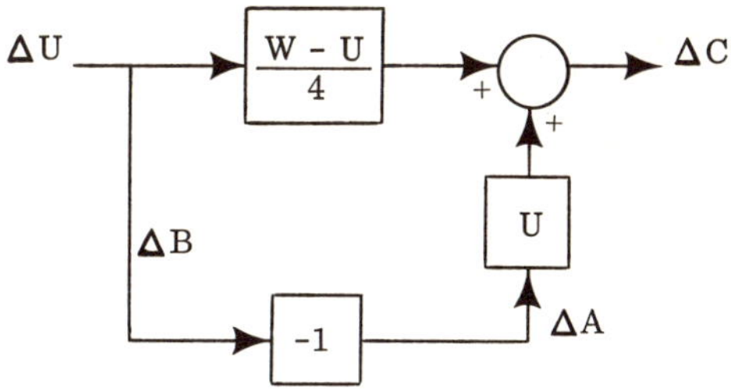

$$\left.\frac{dC}{dU}\right|_{open} = \frac{W - 5U}{4} \qquad 7.5.13$$

As a concrete example, let W = 5. Consider first the operating points determined by $0 \leq U < 1$. Within this range, $\left.\frac{dC}{dU}\right|_{closed}$ and $\left.\frac{dC}{dU}\right|_{open}$ will both be positive, so that an increase in U will cause an increase in $C(\infty)$ whether the loop is open or closed. It is therefore perfectly legitimate to calculate, as we did in the previous section, both the gain due to feedback,*

$$G_{fb} = \left| \frac{dC/dU\,|_{open}}{dC/dU\,|_{closed}} - 1 \right| = \left| \frac{5 - 5U}{5 - 2U} - 1 \right| = \frac{3U}{5 - 2U} \qquad 7.5.14.$$

and the magnification of the effect of U on C,

$$\text{Magnification} = \frac{dC/dU\,|_{closed}}{dC/dU\,|_{open}} = \frac{5 - 2U}{5 - 5U} \qquad 7.5.15$$

When U = 0, the magnification is unity; the gain, zero. As $U \to 1$ from below, the magnification $\to \infty$ and the gain $\to 1$.

[*Note again that this is not the [OLG] (j0). By inspection of the block diagram just preceding equation 7.5.10, [OLG] (j0) = 3 regardless of what values are assigned to U and W.]

Consider next the operating points determined by $2.5 < U < \infty$. Within this range $\left.\frac{dC}{dU}\right|_{closed}$ and $\left.\frac{dC}{dU}\right|_{open}$ will both be negative, so that an increase in U will cause a decrease in C, whether the loop is open or closed. Again we can calculate the gain due to feedback, using equation 7.5.14, and the magnification, using equation 7.5.15. Within this range as $U \to 2.5$ from above, the magnification $\to 0$ and the gain $\to \infty$. As $U \to \infty$, the magnification $\to 0.4$; the gain $\to 1.5$.

Finally, consider the operating points determined by $1 < U < 2.5$. Within this range, $dC/dU|_{closed}$ is positive so that an increase in U will cause an increase in C when the loop is closed; but $dC/dU|_{open}$ is negative so that an increase in U will cause a decrease in C when the loop is open. Under these circumstances, if we try to use equation 7.5.15 we get negative values for the magnification. But negative magnifications are meaningless. (How does one magnify a picture minus four-fold? !) We simply cannot use the term "magnification" when the feedback loop actually reverses the direction of the response. The term "gain" is equally meaningless in this situation. To avoid confusion,* it would seem best not to use the terms "gain" or "magnification" at all for any response whose direction may be altered by opening the loop or loops under consideration, even when the terms might be justified within certain ranges of the variables concerned. In contrast, the ratio of derivatives itself, with its appropriate sign, can be used under all circumstances to describe the effect of any or all of the feedback loops on the steady-state response of any dependent variable to any independent variable. For convenience, I shall define RD, the ratio of derivatives,

$$\mathrm{RD} \equiv \frac{y'_{\text{closed}}}{y'_{\text{open}}} \equiv \frac{dy/dx\,\big|_{\text{closed}}}{dy/dx\,\big|_{\text{open}}} \qquad 7.5.16$$

as a general symbol for the effect of one or more specified feedback loops upon the steady-state response of a specified dependent variable, y, to a step change in a specified independent variable, x, at some particular operating point, but I shall not further muddy the terminological waters by giving this most general measure of the steady-state effect of feedback a name. The relationships between this ratio of derivatives, magnification, and gain are depicted in Figure 7.5.1, which also shows their ranges of applicability. I suggest that when a feedback loop actually reverses the direction of response we should call the effect of feedback "reversive."

[*a physiological example of which I have discussed elsewhere (1).]

SECTION 7.6. A comment upon the relationship between the original working equations and their linear approximations

In the preceding two sections, our analysis of the steady-state behavior of feedback loops was based entirely upon a set of linear equations for the hyperplanes tangent to the curved surfaces defined by the original working equations. Although a mathematician takes this sort of vicarious analysis in stride, those of us who are biological scientists may well feel happier if we can see exactly what is going on in terms of the original equations. The following discussion is therefore intended only for those readers who are as ill at ease with planar and curved surfaces meandering about in hyperspace as I am. I apologize to the more sophisticated.

I shall begin by illustrating the wrong way to work with the original equations, using the simple example of the previous section. As a concrete illustration, let W = 5, U = 2. Equation 7.5.11 says that at the closed-loop operating point determined by these independent variables, $dC/dU\,\big|_{\text{closed}} = 0.25$. At the operating point, this is the precise rate-of-increase of C(∞) with an increase in U; but it should be almost correct for finite increments of U small enough so that U stays within the immediate neighborhood of the operating point.

Very well. Let us confirm this statement by using the original working equations, 7.5.1, 7.5.2 and 7.5.3 for the closed-loop situation. We shall solve these simultaneously for $C(\infty)$, both when $W = 5$, $U = 2$, and when $W = 5$, $U = 2 + h$, h being our "small enough" increment. We find that

when $U = 2 + h$, $C(\infty) = 1.5 + 0.25h - 0.25h^2$

when $\underline{U = 2}$, $C(\infty) = \underline{1.5}$

Change (difference) $\Delta U = h$, $\Delta C(\infty) = 0.25h - 0.25h^2$

whence,

$$\frac{\Delta C(\infty)}{\Delta U} = \frac{0.25h - 0.25h^2}{h} = 0.25 - 0.25h \approx 0.25 = \left.\frac{dC(\infty)}{dU}\right|_{\text{closed}} \qquad 7.6.1.$$

just as predicted from the linearized model by equation 7.5.11. (If the reader accuses me of going back to Chapter I, Section 1, Paragraph 1 of his elementary calculus textbook, I can only reply that that is exactly what I am doing. But I myself find it comforting to revert to first principles once in a while.)

We shall now apply the same line of reasoning to the open-loop situation. When $W = 5$, $U = 2$, equation 7.5.13 states that $dC/dU\,|_{\text{open}} = -1.25$. So we should get this same value by using the appropriate open-loop versions of our original equations to calculate the response of $C(\infty)$ to a sufficiently small increment of U. When the loop has been opened as shown in the block diagram just above equation 7.5.12, $A(\infty)$ can no longer influence $B(\infty)$. Hence, the working equations for the open-loop situation must be

$$A(\infty) = W - B(\infty) \qquad 7.6.2.$$

$$B(\infty) = U \qquad 7.6.3.$$

$$C(\infty) = U\,A(\infty) \qquad 7.6.4.$$

These equations are the same as the original equations 7.5.1, 7.5.2 and 7.5.3 except that, to open the loop, we have deleted from equation 7.5.2 the term which made $B(\infty)$ dependent on $A(\infty)$. Solving these equations simultaneously for $C(\infty)$, both when $W = 5$, $U = 2$, and when $W = 5$, $U = 2 + h$ as before,

when $U = 2 + h$, $C(\infty) = 6 + h - h^2$

when $U = \underline{2}$, $C(\infty) = \underline{6}$

Change (difference) $\Delta U = h$, $\Delta C(\infty) = h - h^2$

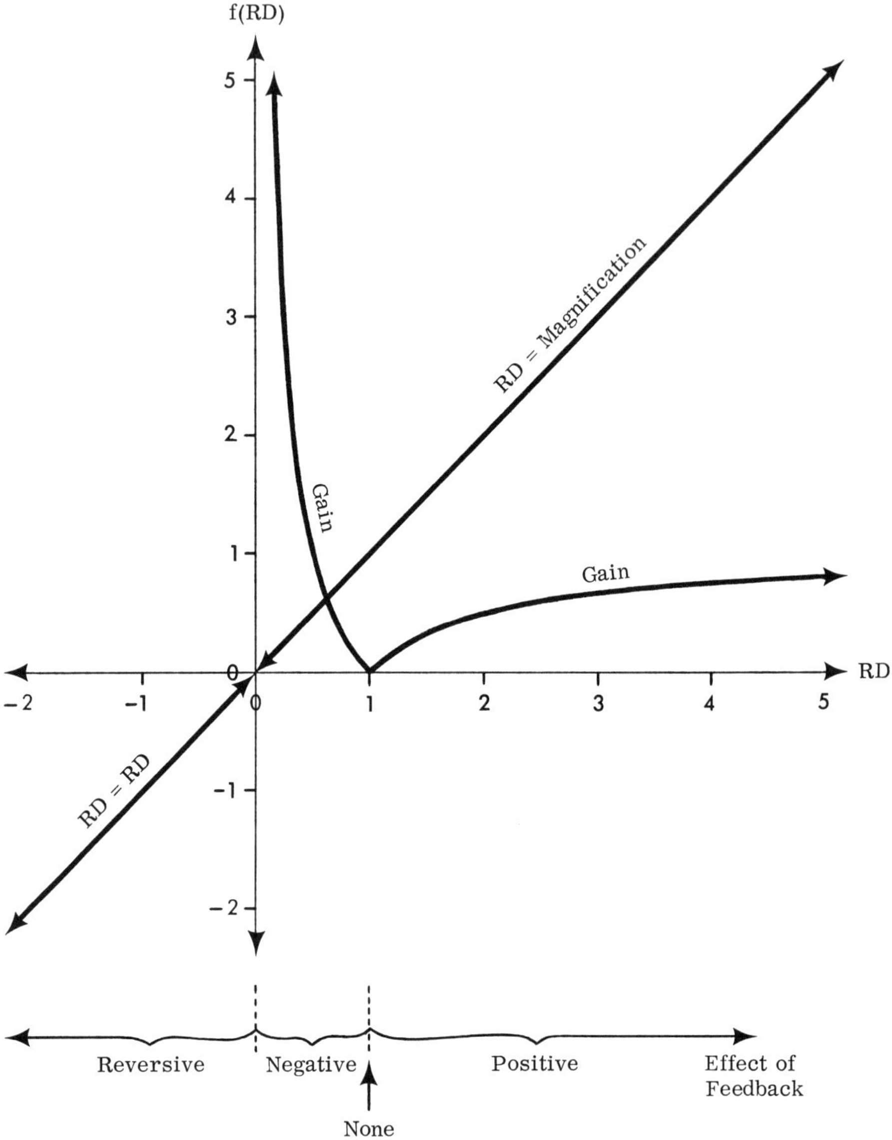

<u>Notice</u> 1.) that gains and magnifications are intrinsically positive, and
2.) that steady-state gains between zero and unity may be encountered both with negative and with positive feedback.

Figure 7.5.1. Relationships among various measures of the effect of feedback in the steady state. (See figure on the opposite page.)

Let $y(\infty)$ be any dependent variable (= output). Let U be any independent variable (= input). Then the quantity $dy(\infty)/dU$, evaluated at a given operating point, is a measure of the response of $y(\infty)$ to an infinitesimal step increment of U.

In the steady state, the most general measure of the effect of feedback is RD, the ratio of derivatives, defined as:

$$RD \equiv dy(\infty)/dU\big|_{closed} / dy(\infty)/dU\big|_{open}$$

where both derivatives are to be evaluated at the same operating point.

If the numerator and the denominator of RD have the same sign, Magnif., the magnification of the response by one or more particular feedback loops, is identically equal to RD.

For simplicity, the following table deals with the effect of but one feedback loop at a time, whether it be the only loop in the system, or one of several.

<table>
<tr><th></th><th>SIGNS of numer. & denom. in RD are</th><th>Effect of the feedback loop on $dy(\infty)/dU$</th><th colspan="2">Relationships</th></tr>
<tr><td rowspan="2">$y(\infty)$ is a member of the loop being studied</td><td>SAME</td><td>$|dy(\infty)/dU|$ is decreased by the loop, which must therefore be negative</td><td>$G_{fb} = \frac{1}{\text{Magnif.}} - 1$</td><td rowspan="2">With a single-loop system, $G_{fb} = [OLG](j0) = |OLTF|$ in the steady state.</td></tr>
<tr><td>SAME</td><td>$|dy(\infty)/dU|$ is increased by the loop, which must therefore be positive.</td><td>$G_{fb} = 1 - \frac{1}{\text{Magnif.}}$</td></tr>
<tr><td rowspan="3">$y(\infty)$ is not a member of the loop being studied</td><td>SAME</td><td>$|dy(\infty)/dU|$ is decreased by the loop.</td><td>$G_{fb} = \frac{1}{\text{Magnif.}} - 1$</td><td rowspan="2">Neither the $[OLG](j0)$, nor the nature of the loop can, in general, be deduced from the magnification.</td></tr>
<tr><td>SAME</td><td>$|dy(\infty)/dU|$ is increased by the loop.</td><td>$G_{fb} = 1 - \frac{1}{\text{Magnif.}}$</td></tr>
<tr><td>OPPOSITE</td><td>loop reverses sign of $dy(\infty)/dU$</td><td colspan="2">Gains and magnifications do not even exist. RD is negative.</td></tr>
</table>

whence,

$$\frac{\Delta C(\infty)}{\Delta U} = \frac{h - h^2}{h} = 1 - h \approx +1 \neq \left.\frac{dC(\infty)}{dU}\right|_{open} = -1.25 \qquad 7.6.5.$$

Not just "not equal to"; it doesn't even have the right sign! What can account for this disaster?

The explanation is simple. In our eagerness to show how well the predictions of the linearized equations are borne out by the actual working behavior of the system, we disregarded the fundamental restriction which underlies the whole analysis: We have to examine the behavior of the system at a single operating point! Instead of abiding by that requirement, we tried to compare closed-loop behavior at one operating point with open-loop behavior at quite a different operating point. For although we assumed the same values for the independent variables, the corresponding values of the dependent variables calculated from equations 7.5.1, 7.5.2, and 7.5.3, namely, $A(\infty) = 0.75$, $B(\infty) = 4.25$, $C(\infty) = 1.5$, are not at all the same as the ones calculated from equations 7.6.2, 7.6.3 and 7.6.4, namely $A(\infty) = 3$, $B(\infty) = 2$, $C(\infty) = 6.0$.

Correction of this wrong procedure is easy. When rewriting the original working equations for the open-loop situation, we must never simply delete the term corresponding to the working relationship which has been interrupted by opening the loop. Instead, we must replace the variable in that term by a constant, namely the value which the variable actually has at the particular operating point we have chosen to investigate. By thus substituting a constant for a variable we can both keep the operating point unchanged, and interrupt the feedback loop. (Notice that the partial derivative of any function with respect to a constant does not exist, so that substituting a constant for a variable effectively deletes the corresponding pathway from our block diagram, and opens the loop.)

Let us apply this correct procedure to the present problem. Instead of just dropping the $3A(\infty)$ term from equation 7.5.2, we must replace $A(\infty)$ by its value at the operating point, namely 0.75. For the open-loop situation, we then have,

$$A(\infty) = W - B(\infty) \qquad 7.6.2,$$

$$B(\infty) = U + 2.25 \qquad 7.6.6,$$

$$C(\infty) = U\,A(\infty) \qquad 7.6.4,$$

as the true open-loop working equations at the operating point determined by $W = 5$ and $U = 2$. Solving these equations simultaneously for $C(\infty)$, both when $W = 5$, $U = 2$, and when $W = 5$, $U = 2 + h$ as before, when $U = 2 + h, \quad C(\infty) = 1.5 - 1.25h - h^2$

when $U = 2 \quad , \quad C(\infty) = 1.5$

Change (difference) $\Delta U = h, \quad \Delta C(\infty) = - 1.25h - h^2$,

whence,

$$\frac{\Delta C(\infty)}{\Delta U} = \frac{-1.25h - h^2}{h} = -1.25 - h \approx -1.25 = \left.\frac{dC}{dU}\right|_{open} \qquad 7.6.7,$$

precisely as predicted by equation 7.5.13.

The requirement that open-loop and closed-loop changes must be compared at exactly the same operating point has an important bearing upon the design of experiments in which feedback is studied by actually opening a loop. (See Example 7.13.1.)

SECTION 7.7. General kinds of working equations

In Part II which deals with specific physiological feedback systems three general types of equations will be used to describe working relationships in the steady state.

When part or all of a physiological system can be represented by a physical or physicochemical model whose component parts are known, the laws governing the components may often be used to derive one or more theoretical equations. Theoretical equations are far more valuable than any other type. "Because of their generality, and because of their power to aid the logical development of fundamental knowledge, theoretical equations have an elegance and an intellectual appeal quite lacking in empirical equations." (2). Moreover, if the entire system can be analyzed using theoretical equations, the results of the analysis can often be tested experimentally, leading to confirmation, modification, or rejection of the physical model being used (see Section 7.14). Unfortunately, though some physiological feedback systems are simple enough, and well enough understood, to be described entirely by theoretical equations (see Example 7.8.1) they are few and far between.

When theoretical equations are not available, it is often possible to make experimental observations of how one variable affects another, and to fit the observations by an empirical equation. Since an empirical equation merely describes experimental results and is not derived from any underlying theoretical model, the form of the equation may be chosen arbitrarily. One should try to select the simplest equation (i.e. the one with the fewest parameters) which will describe the observations adequately over their entire range, paying particular attention to upper and lower limits which may be known from theory, or from previous work, even if they have not been approached in the particular experiments being analyzed. If two or more equally simple forms will fit the data equally well, choose the one which is easiest to manipulate mathematically, e.g., easiest to differentiate, or to combine algebraically with other equations. I have included a brief section on empirical equations in the Appendix of the present book, and I have tried to give additional hints about curve-fitting elsewhere (2), but I am afraid it will always be more of an art than a science.

In a few of the descriptions of individual systems I have used equations which can most charitably be called suppositious. These describe fancied relationships which have been obtained neither from theory, nor from experimental observation, but from a desperate desire to close the loop with something quantitative. The reader may form his own opinion about how much such an equation is worth!

SECTION 7.8. Derivation of working equations from a physical model

If the differential equations describing the time-dependent behavior of the variables are known, the steady-state equations may be obtained by setting each time derivative equal to zero and then solving for the dependent variable. Alternatively, it may be possible to derive theoretical equations directly from a consideration of steady-state relationships in the model being used. The following detailed example illustrates the steps involved.

Example 7.8.1. Activation of an enzyme by a product.

Chance and his collaborators (3 , 4) have proposed the following biochemical system as a model which can display limit-cycle activity of the sort which might account for certain rhythmic activities, e.g. "biological clocks," in living systems.

Consider a fixed volume, V, of a solution containing fixed total amounts of two enzymes, E_{tot} and F_{tot}. A substrate, S, is fed into the system at a constant rate, $\dot{Q}$. By irreversible steps, E converts S to a product P, and F converts P to some third substance, X, which need not concern us. Finally, E, the active form of the enzyme, is the complex formed when P combines reversibly with an inactive enzyme precursor, e. The reactions involved are summarized in Figure 7.8.1 which also serves to define the mass-action rate constants. In the following derivation, the letter-symbols for the various substances will also be used to designate their molar concentrations except that the concentrations of the enzyme-substrate complexes will be designated [SE](t) and [PF](t) to distinguish them from S(t)E(t), the product of the two concentrations S(t) and E(t), and P(t)F(t), the product of the two concentrations P(t) and F(t).*

[*The symbols used here differ from those used by Chance et al. (3) for their illustrative biochemical example. The corresponding symbols are: $\dot{Q}$ = rate of supply of F-6-P (= fructose-6-phosphate) in the form of its precursor, glucose. S = F-6-P; E = E_1; SE = $E_1 \cdot$ F-6-P; P = FDP (= fructose diphosphate) e = E; F = E_2; PF = FDP$\cdot E_2$; X = GAP (= glyceraldehyde phosphate). Higgins (4) uses somewhat different symbols for the same system of reactions.]

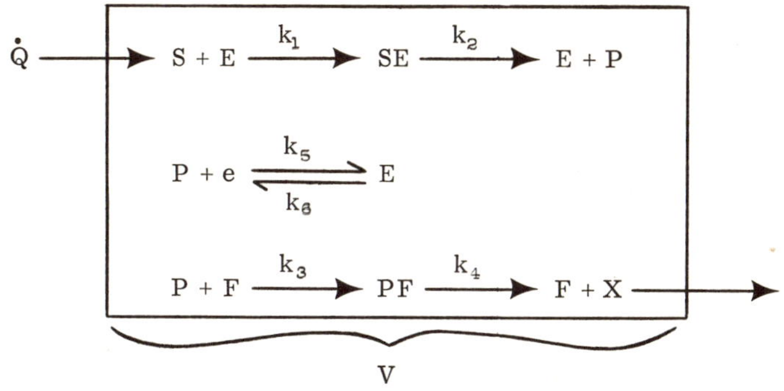

Figure 7.8.1. Model of the enzyme system of Example 7.8.1.

The volume, V, in which the reactions occur is constant and well-stirred. S is added to V at the constant rate, $\dot{Q}$. Each of the k's is the proportionality constant (= rate constant) for the law of mass action applied to the reaction indicated by the adjacent arrow. As usual, we assume that the law of mass action can be written in terms of molar concentrations throughout the range of interest. For example, the rate at which E(t), the concentration of E, is being increased by the combination of P and e is assumed to be proportional to the product of P(t) and e(t), the concentrations of the two reactants, k_5 being the proportionality constant. The total rate of change of E(t) as stated by equation 7.8.5 is the algebraic sum of the individual rates of change in all of the reactions where E is either a reactant or a product.

From the preceding qualitative description, we surmise that [SE](t), E(t), and P(t) belong to a positive feedback loop. Consider, for example, the response of the system to an increase in $\dot{Q}$. An increase in $\dot{Q}$ will cause an increase in S(t). An increase in S(t) will cause an increase in [SE](t); the increase in [SE](t) will increase P(t); the increase in P(t) will increase E(t); and the increase in E(t) will still further increase [SE](t). It is this loop which we wish to investigate quantitatively.

We begin by writing the differential equations, based on the law of mass action, for the rate at which the concentration of each substance changes with time (see legend of Figure 7.8.1). (If we multiply both sides of any such equation by V, we see that it is merely an equation of conservation, stating that, at any instant, the rate at which the given substance is being added to V must equal the sum of the rate at which it is accumulating in V and the rate at which it is being removed from V).

$$\frac{dS(t)}{dt} = \frac{\dot{Q}}{V} - k_1 S(t)E(t) \qquad 7.8.1.$$

$$\frac{d[SE](t)}{dt} = k_1 S(t)E(t) - k_2 [SE](t) \qquad 7.8.2.$$

$$\frac{dP(t)}{dt} = k_2 [SE](t) + k_6 E(t) - [k_3 F(t) + k_5 e(t)] P(t) \qquad 7.8.3.$$

$$\frac{d[PF](t)}{dt} = k_3 P(t) F(t) - k_4 [PF](t) \qquad 7.8.4.$$

$$\frac{dE(t)}{dt} = k_2 [SE](t) + k_5 P(t)\, e(t) - [k_1 S(t) + k_6] E(t) \qquad 7.8.5.$$

$$\frac{de(t)}{dt} = k_6 E(t) - k_5 P(t) e(t) \qquad 7.8.6.$$

$$\frac{dF(t)}{dt} = k_4 [PF](t) - k_3 P(t) F(t) \qquad 7.8.7.$$

In addition, we have the restrictions that

$$E(t) + [SE](t) + e(t) = E_{tot} \qquad 7.8.8,$$

$$F(t) + [PF](t) = F_{tot} \qquad 7.8.9.$$

Equations 7.8.1 through 7.8.9 constitute a set of nine equations in seven variables. Two equations must therefore be dependent on (i.e., derivable from) others. And indeed we now see that by differentiating equation 7.8.8 with respect to time,

$$\frac{dE(t)}{dt} + \frac{d[SE](t)}{dt} + \frac{de(t)}{dt} = 0 \qquad 7.8.10.$$

we obtain the same result as we would by adding equations 7.8.2, 7.8.5 and 7.8.6. By using equation 7.8.8 for suitable substitution, we could therefore eliminate either E(t), [SE](t) or e(t) together with the equation for the rate of change of the eliminated variable. Similarly, we could use equation 7.8.9 to eliminate F(t) or [PF](t) and thus either equation 7.8.4 or 7.8.7. If we were planning to solve this system of equations on a computer, we could thus reduce it to a set of five differential equations in five variables, so that the state space would be five-dimensional. (See Section 5.2.)

But at present we are interested in getting a set of steady-state working equations describing the relationships among <u>all</u> of the variables of the system. Let us therefore delay any such substitutions until after we have set each derivative in equations 7.8.1 through 7.8.7 equal to zero and solved each of the resulting steady-state equations for the variable whose rate of change with time it expressed. Taking them in order, we obtain,

$$S(\infty) = \frac{\dot{Q}/V}{k_1 E(\infty)} \qquad 7.8.11,$$

$$[SE](\infty) = \frac{k_1}{k_2} S(\infty)\, E(\infty) \qquad 7.8.12,$$

$$P(\infty) = \frac{k_2[SE](\infty) + k_6E(\infty)}{k_3F(\infty) + k_5e(\infty)} \qquad 7.8.13,$$

$$[PF](\infty) = \frac{k_3}{k_4} P(\infty)\ F(\infty) \qquad 7.8.14,$$

$$E(\infty) = \frac{k_2[SE](\infty) + k_5P(\infty)e(\infty)}{k_1S(\infty) + k_6} \qquad 7.8.15,$$

$$e(\infty) = \frac{k_6E(\infty)}{k_5P(\infty)} \qquad 7.8.16,$$

$$F(\infty) = \frac{k_4[PF](\infty)}{k_3P(\infty)} \qquad 7.8.14.a.$$

These seven steady-state relationships derived from the original differential equations, plus equations 7.8.8 and 7.8.9, again give us a set of nine equations in seven variables. Two equations must be dependent and may therefore be deleted with no loss of information. It is clear by inspection that equations 7.8.14 and 7.8.14.a are mere variants of the same equation. It does not matter which of these we discard, because we know that the remaining equation will be a valid working equation whether we solve it for $F(\infty)$ or $[PF](\infty)$. Furthermore, any one of the three equations 7.8.12, 7.8.15 and 7.8.16 may be derived from the other two. If we clear each of these of fractions, we see that equation 7.8.15 is really just the sum of equations 7.8.12 and 7.8.16. So if we were to discard equation 7.8.16 for example, we would really be retaining the terms of equation 7.8.12 in two equations; 7.8.12 itself and 7.8.15. It is almost always best to keep the simplest equations. Accordingly we shall discard 7.8.15, again remembering that we can recover a valid working equation for $E(\infty)$ by combining 7.8.12 and 7.8.16 and solving for $E(\infty)$. (Note that equation 7.8.16 expresses an equilibrium, and thus remains as valid a working equation when solved for $E(\infty)$ as it is in its present form.)

We now have a set of 7 equations (7.8.8, 7.8.9, 7.8.11, 7.8.12, 7.8.13, 7.8.14 and 7.8.16) in 7 variables. Before going further it is worth noting that essentially this same set can be obtained by considering the steady-state situation directly. For in the steady state, the rate of each successive irreversible reaction is determined by the rate at which the preceding step supplies reactant. This fact, plus the information in Figure 7.8.1 allows us to write

$$k_1S(\infty)\ E(\infty) = \dot{Q}/V \qquad 7.8.11.a,$$

$$k_2[SE](\infty) = k_1S(\infty)\ E(\infty) \qquad 7.8.12.a,$$

$$k_3P(\infty)\ F(\infty) = k_2[SE](\infty) \qquad 7.8.13 \text{ and } 7.8.16,$$

$$k_4[PF](\infty) = k_3P(\infty)F(\infty) \qquad 7.8.14.b,$$

which, together with equations 7.8.8 and 7.8.9 for total enzyme, and equation 7.8.16 for the reversible equilibrium involving $e(\infty)$, constitute the 7 equations which are needed. Equation 7.8.13 emerges in a somewhat simpler form (equivalent to eliminating e(∞) between 7.8.13 and 7.8.16), but that is a trivial difference. More important is the difficulty we may sometimes have in identifying the dependent variable or variables in each equation. We may need to use the method of concomitant variation (Section 1.4) to decide what the true working relationship is.

The proposed feedback loop involves P(∞), E(∞) and [SE](∞). Let us solve for these three variables, eliminating the other four. Combine equations 7.8.11 and 7.8.12.

$$[SE](\infty) = \dot{Q}/k_2V \tag{7.8.17.}$$

Combine equations 7.8.9 and 7.8.14 to eliminate [PF](∞). Use the result to eliminate F(∞), and use equation 7.8.16 to eliminate e(∞) from equation 7.8.13.

$$P(\infty) = \frac{1}{k_3\left\{\dfrac{F_{tot}}{k_2[SE](\infty)} - \dfrac{1}{k_4}\right\}} \tag{7.8.18.}$$

Solve equation 7.8.16 for E(∞), and use equation 7.8.8 to eliminate e(∞) from the result.

$$E(\infty) = \frac{E_{tot} - [SE](\infty)}{1 + \dfrac{k_6}{k_5P}} \tag{7.8.19.}$$

Equation 7.8.17 shows that [SE](∞) depends wholly upon independent variables. Therefore we might just as well use equation 7.8.17 to eliminate [SE](∞) from equations 7.8.18 and 7.8.19.

$$P(\infty) = \frac{1}{k_3\left\{\dfrac{F_{tot}}{\dot{Q}/V} - \dfrac{1}{k_4}\right\}} \tag{7.8.20}$$

$$E(\infty) = \frac{E_{tot} - \dfrac{\dot{Q}}{k_2V}}{\dfrac{k_6}{k_5P(\infty)} + 1} \tag{7.8.21}$$

Equations 7.8.17, 7.8.20 and 7.8.21 together summarize the relevant steady-state behavior of the system. But what has happened to our feedback loop? As the symbol-and-arrow diagram of Figure 7.8.2 shows, it has vanished! <u>There is no feedback among the variables of this system in the steady state.</u> Does this mean that the whole concept of positive feedback was illusory? Of course not! For we have analyzed <u>only</u> the steady-state situation in which the major variables must slavishly conform to the rates of reaction imposed upon the system by $\dot{Q}$. In fact, however, the present system is of interest chiefly because of the <u>non-steady-state</u> behavior which it may exhibit; namely, limit cycle activity. Merely solving the set of algebraic equations obtained by putting

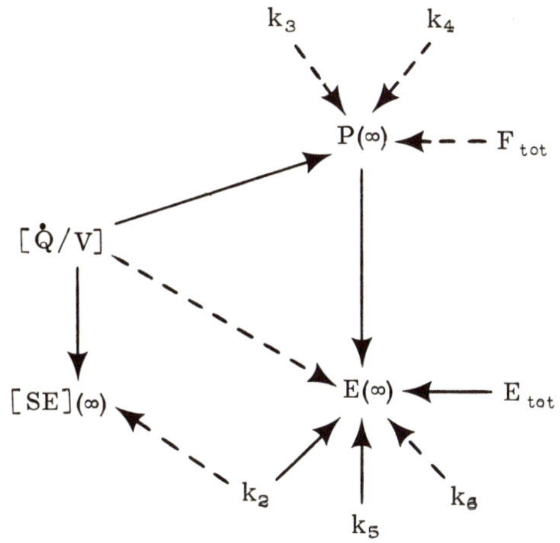

Figure 7.8.2. Symbol-and-arrow diagram corresponding to equations 7.8.17, 7.8.20 and 7.8.21.

all time derivatives equal to zero can do no more than identify the points of equilibrium which are determined by given values of the inputs and parameters. It cannot tell us whether such an equilibrium point is stable, or, if unstable, whether a stable limit cycle exists. In theory, the methods of Chapter 6 should at least allow us to test the stability of a given point of equilibrium by first linearizing the original differential equations (equations 7.8.1 through 7.8.7) about the point and then applying such stability criteria as the Routh-Hurwitz test. In the present example, this would require finding the characteristic equation for a fifth-order linearized system; a prospect which, dismal enough in itself, is made all the more dismal when we realize that proving instability is by no means equivalent to proving the existence of a stable limit cycle. We can fully appreciate why Higgins chose to investigate limit cycle activity in this system by "fishing" for oscillatory solutions in multidimensional space with an analog computer. Some of the various limit-cycle fish thus caught are on display in his paper (4). (See Exercise 7.2).

In the present example, a complete physicochemical model of the system under scrutiny was available. Moreover, we were able to write a complete set of nonlinear differential equations describing the behavior of the model. Though the nonlinear characteristics of such a system usually preclude a full analytical solution, modern computers do enable us to find the values of all output variables at any instant of time for any specified set of inputs and parameters. In principle, then, a complete analysis of both the dynamic behavior and the steady-state behavior of so well-characterized a model can be achieved. But this happy situation is not often encountered when we study biological systems. In the next several sections, therefore, we shall consider how to make the best use of more limited information about biological feedback systems.

SECTION 7.9. Derivation of working equations from experimental observations

Consider a system which contains only a single feedback loop. Each variable is either independent, dependent and a member of the loop, or dependent and not a member of the loop. The dependent variables which are not members of the loop can either be eliminated by substitution, or neglected, so as to reduce the system to one containing independent variables and loop variables only.

Let the members of the loop be $L_1(\infty)$, $L_2(\infty)$, $\cdots$ $L_n(\infty)$. Let the independent variables be designated U, with subscripts to indicate which of the loop variables they directly influence. For example, $U_{2,3}$ would be an independent variable which has a direct influence on both $L_2(\infty)$ and $L_3(\infty)$, but which influences the other $L(\infty)$'s only via part of the loop. The working equations for the entire system are of the form,

$$\left.\begin{aligned} L_1(\infty) &= f_1(L_n(\infty);\ U_{1,\ldots}) \\ L_2(\infty) &= f_2(L_1(\infty);\ U_{\ldots,2,\ldots}) \\ &\ \ \vdots \\ L_n(\infty) &= f_n(L_{n-1}(\infty);\ U_{\ldots,n}) \end{aligned}\right\} \quad 7.9.1,$$

where $U_{\ldots,i,\ldots}$ designates the set of all U's with i in the subscript. Now suppose that we are able to measure any two members of the feedback loop, say $L_i(\infty)$ and $L_j(\infty)$, experimentally. By suitably combining the several equations in the 7.9.1 set, we can eliminate all of the other variables, reducing the working equations to

$$L_i(\infty) = g(L_j(\infty);\ U_i,\ U_{i,j}) \qquad 7.9.2,$$

$$L_j(\infty) = h(L_i(\infty);\ U_j,\ U_{i,j}) \qquad 7.9.3,$$

where U_i, U_j, and $U_{i,j}$ represent sets of independent variables with their original subscripts altered to conform with the simplified system of equations 7.9.2 and 7.9.3. (For example, if L_3 has been eliminated, U_3 would now belong either to the U_i set or to the U_j set.)

Finally, let us suppose that, keeping all other independent variables constant we can change one of the U_i's (not one of the $U_{i,j}$'s) stepwise over a considerable range. At each step we measure $L_i(\infty)$ and $L_j(\infty)$ after a new steady state has been reached. It should be clear by inspection of equations 7.9.2 and 7.9.3 that by manipulating one of the U_i's we change $L_i(\infty)$ directly, but that the changes which we observe in $L_j(\infty)$ must be due to the changes in $L_i(\infty)$. Hence, if we plot $L_j(\infty)$ as a function of $L_i(\infty)$ and fit the resulting points by an empirical equation we should obtain an approximation of equation 7.9.3 for the particular values of the U_j's and $U_{i,j}$'s which are being kept constant. If, in a different series of experiments, we can similarly manipulate one of the U_j's experimentally we can also find an empirical approximation of equation 7.9.2. We thus get two working equations for the two loop variables, without ever having had to open the feedback loop. The advantages of being able to study a physiological feedback loop without interfering with its normal operation can hardly be overestimated.

Example 7.9.1. Consider a physiological system whose steady-state behavior is fully described by the following five working equations.

$$A(\infty) = D(\infty)/C(\infty) \qquad 7.9.4,$$

$$B(\infty) = A(\infty)[U_{B,E} + U_{B,C}] \qquad 7.9.5,$$

$$C(\infty) = [B(\infty)/U_{B,C}] + U_C \qquad 7.9.6,$$

$$D(\infty) = U_{D,1} - U_{D,2} \qquad 7.9.7,$$

$$E(\infty) = B(\infty)/U_{B,E} \qquad 7.9.8.$$

The corresponding symbol-and-arrow diagram,

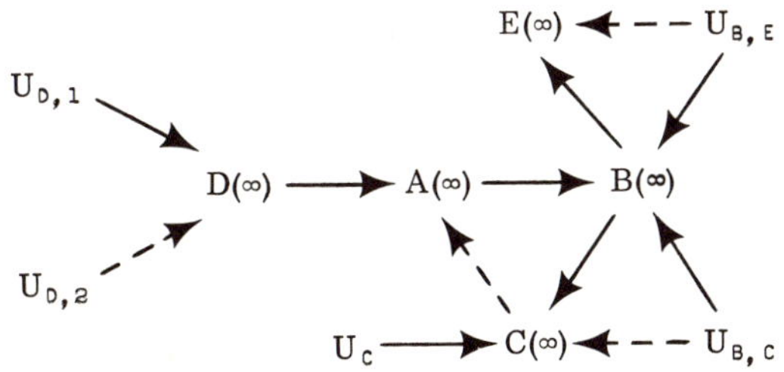

Figure 7.9.1.

shows that $A(\infty)$, $B(\infty)$ and $C(\infty)$ are members of a single feedback loop, but that $D(\infty)$ and $E(\infty)$ are not loop variables. $E(\infty)$ is completely dependent. Both it and its working equation, 7.9.8, may therefore be neglected, for they can have no influence upon the operation of the feedback loop. Furthermore, we may eliminate $D(\infty)$, the other non-loop variable, by combining equations 7.9.4 and 7.9.7.

$$A(\infty) = [U_{D,1} - U_{D,2}]/ C(\infty) \qquad 7.9.9.$$

Finally, suppose that we can measure $B(\infty)$ and $C(\infty)$, but not $A(\infty)$, experimentally. By combining equations 7.9.5 and 7.9.9,

$$B(\infty) = [U_{D,1} - U_{D,2}][U_{B,E} + U_{B,C}]/C(\infty) \qquad 7.9.10$$

we can eliminate $A(\infty)$. We now have two working equations 7.9.6 and 7.9.10, which describe the feedback relationships between $B(\infty)$ and $C(\infty)$ in terms of independent variables only. The value of [OLG](j0) for this single-loop system is therefore the absolute value of the product of the derivatives of those two equations.

$$[OLG](j0) = \left| \frac{-[U_{D,1} - U_{D,2}][U_{B,E} + U_{B,C}]}{U_{B,C}\, C^2(\infty)} \right| \qquad 7.9.11,$$

where, (by simultaneous solution of equations 7.9.6 and 7.9.10),

$$C(\infty) = \frac{U_C + \left\{ U_C^2 + \dfrac{4[U_{D,1} - U_{D,2}][U_{B,E} + U_{B,C}]}{U_{B,C}} \right\}^{1/2}}{2} \qquad 7.9.12$$

For example, suppose that under normal conditions the values of the five independent variables are

$U_{D,1}$	$U_{D,2}$	$U_{B,C}$	$U_{B,E}$	U_C
7	5	3	4	0

Then the values of $B(\infty)$, $C(\infty)$ and $[OLG](j0)$ at the normal operating point will be

$B(\infty)$	$C(\infty)$	$[OLG](j0)$
6.48	2.16	1.00

Thus far we have analyzed the system as if we actually knew the theoretical equations governing its behavior. Let us now forget all about these theoretical equations and the results obtained by analyzing them, and let us see how much we could learn by a purely experimental approach. As a minimum, we must know the general pattern of relationships among the variables of the system (Figure 7.9.1). In particular, we must know that $B(\infty)$ and $C(\infty)$ are measurable members of the loop. We must know how to manipulate one independent variable which alters $B(\infty)$ directly but affects $C(\infty)$ only via $B(\infty)$, and another which alters $C(\infty)$ directly, but affects $B(\infty)$ only via $C(\infty)$. Finally, we must know what the other independent variables of the system are, and we must design our experiments so that they will stay reasonably constant. (Notice that all the above requirements are qualitative. At no stage of our investigation do we need to know the actual numerical values of any independent variables, -- not even the ones which we are going to manipulate experimentally!)

In a first series of experiments, let us alter U_C stepwise and measure the corresponding steady-state values of $B(\infty)$ and $C(\infty)$ (left side of Table 7.9.1). (In constructing the hypothetical data of Table 7.9.1, I have assumed, as will in fact often be true, that U_C can only be increased above, and not decreased below the value which it has at the normal operating point.) In Figure 7.9.2.a, the observed values of $B(\infty)$ have been plotted against the observed values of $C(\infty)$. The relationship seems to be curvilinear and this suggests that we may need to fit the points by an equation with at least three parameters. The trend of the points suggests that $B(\infty)$ may be decreasing exponentially towards some asymptote as $C(\infty)$ increases. Indeed, it is easy to find an exponential equation,

Table 7.9.1.

Step Input U_C	Measured values $C(\infty)$	$B(\infty)$	Step Input $U_{B,E}$	Measured values $B(\infty)$	$C(\infty)$
0	2.1	6.5	4	6.5	1.8
0.5	2.4	5.7	8	8.2	2.5
1.0	2.6	4.9	12	9.1	3.2
1.5	3.1	4.7	16	10.4	3.6
2.0	3.5	4.3	20	11.8	3.7
2.5	3.6	3.5	24	12.5	4.3
3.0	4.2	3.5	28	13.6	4.7
3.5	4.8	3.2	32	14.5	5.3
4.0	5.0	2.8	36	15.4	5.3
4.5	5.4	2.5			
5.0	5.9	2.5			

$$B(\infty) = 1.5 + 11.9\epsilon^{-0.439C(\infty)} \qquad 7.9.13,$$

which adequately represents the trend of the data.*

[*For reasons which I have set forth ad nauseam elsewhere, (2), there is no use quibbling about whether this is the "best" three-parameter curve, or even the "best" exponential curve for these particular points. It does the job, and that is all we can ask of a purely empirical equation.]

The right side of Table 7.9.1, and Figure 7.9.2.b give the values of $B(\infty)$ and $C(\infty)$ obtained in a second series of experiments in which $U_{B,E}$ was increased stepwise, keeping all other independent variables unchanged at their normal operating point values. This time the points seem to follow a straight-line trend which is adequately described by

$$C(\infty) = 0.405\ B(\infty) - 0.76 \qquad 7.9.14$$

In both of these empirical equations, the parameters are subject to the statistical uncertainties caused by the random scatter of the observed points. Nevertheless, within the range of the observations, and for the particular values of the independent variables maintained during the experiments, equation 7.9.13 is an empirical approximation of the "true" working equation, 7.9.10.

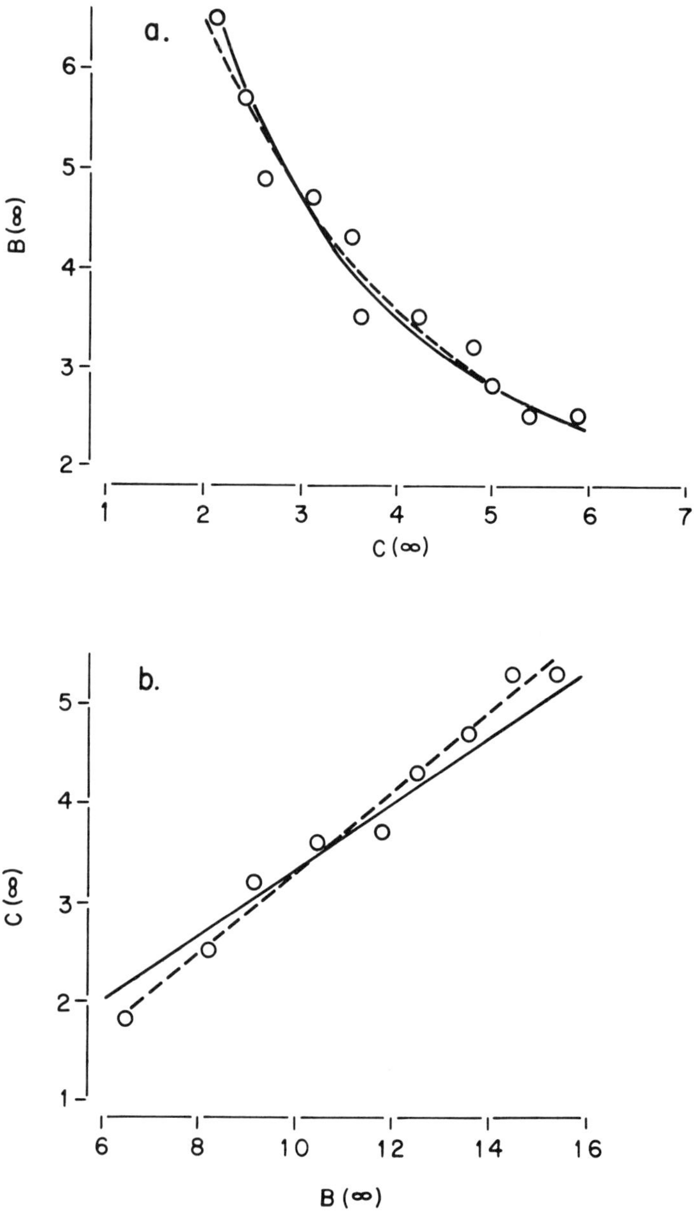

Figure 7.9.2. A comparison of the empirical curves (broken lines), fitted to the experimental points of Table 7.9.1, with their theoretical "true" counterparts (solid lines).

Similarly equation 7.9.14 is an empirical approximation of the "true" working equation, 7.9.6 (Figure 7.9.2). Hence, simultaneous solution of equations 7.9.13 and 7.9.14 provides an estimate of the operating point values, $B(\infty) = 6.62$, $C(\infty) = 1.92$. Furthermore, from the first derivative of equation 7.9.13 evaluated at that operating point,

$$\frac{dB(\infty)}{dC(\infty)} = -5.22\epsilon^{-0.439[1.92]} = -2.25, \qquad 7.9.15,$$

and from the first derivative of equation 7.9.14,

$$dC(\infty)/dB(\infty) = 0.405, \qquad 7.9.16,$$

we can estimate

$$[OLG](j0) = \left|[-2.25]\,[0.405]\right| = 0.906 \qquad 7.9.17,$$

for the value of the open-loop gain at the operating point.

Since this is an artificial example, we can compare these "experimental" estimates with their "true" values,--a comparison which would be quite impossible in a real physiological system,

Variable:	B(∞)	C(∞)	B′(C(∞))	C′(B(∞))	[OLG](j0)
"True" value:	6.48	2.16	- 3.00	0.333	1.000
Experimental Estimate:	6.62	1.92	- 2.25	0.405	0.906

This comparison makes it obvious that when we rely upon empirical equations fitted to experimental points the true values of the derivatives may be overestimated or underestimated by a substantial amount, even when, as was true in this example (Figure 7.9.2), the observed points cluster fairly closely about the "true" line.

Whenever we have to base our analysis <u>entirely</u> upon empirical equations, as we did in the above example, two series of experiments must be performed before we can estimate the open-loop gain at a <u>single</u> operating point. To find out how [OLG](j0) changes as we go from one operating point to another may therefore require us to complete a formidable experimental program. We should grasp all the more gratefully at whatever theoretical straws float towards us on the turbulent empirical waters. For example, if, in the above problem, we knew on theoretical grounds that the equation for C(∞) as a function of B(∞) would always be an equation for a straight line of constant slope, we could immediately estimate [OLG](j0) at <u>any</u> operating point along the observed curve for B(∞) as a function of C(∞).

SECTION 7.10. Local manipulation of a loop variable at a sensing element

In many homeostatic systems, a localized sensing mechanism is an essential part of the feedback loop. This special mechanism acts as a transducer, responding to changes in the local value of the regulated variable by generating a signal which initiates an opposite change in the general value of the regulated variable. The local variable, say $C_{loc}(\infty)$, and the general variable, say $C_{gen}(\infty)$, may then be treated as two different members of the feedback loop whose influence upon each other may be studied experimentally, using the method of Section 7.9. The investigator alters the intensity of the regulated variable at the sensing device only, and measures both $C_{loc}(\infty)$ and $C_{gen}(\infty)$. These observations can then be described by an empirical working equation for $C_{gen}(\infty)$ as a function of $C_{loc}(\infty)$. But the second working equation, i.e. for $C_{loc}(\infty)$ as a function of $C_{gen}(\infty)$, is usually an obvious theoretical equation which may be nothing more than $C_{loc}(\infty) = C_{gen}(\infty)$. Obtaining one of the two equations,--and a theoretical one at that!--with such effortless ease, is a tremendous advantage.

Although the experimental procedure described in this section does not require interrupting the feedback loop, in actual practice it is often convenient to isolate the localized sensing mechanism so that $C_{gen}(\infty)$ cannot influence $C_{loc}(\infty)$. The investigator can then more easily control the intensity of $C_{loc}(\infty)$ when studying its effect on $C_{gen}(\infty)$.

Example 7.10.1. For a specific example, see the description of the arterial blood pressure--carotid sinus loop in Part II.

SECTION 7.11. Derivation of an empirical working equation from a known theoretical equation and experimental observation of only one loop variable

In Section 7.9 it was pointed out that if we have an empirical equation for $B(\infty) = g(A(\infty))$ and a theoretical equation for $A(\infty) = f(B(\infty); U_A)$, we can calculate the values of $A(\infty)$ and $B(\infty)$, and the corresponding values of $[OLG](j0)$, at a series of different operating points determined by different values of U_A. The reverse process (obtaining an empirical equation from a theoretical equation and a series of values observed at different operating points) is important when either one of the two loop variables, but not the other, can be measured. For concreteness, let us suppose that $A(\infty)$ is measurable but $B(\infty)$ is not. We vary U_A stepwise and observe the value of $A(\infty)$ at each step. Suppose that the theoretical equation corresponding to $A(\infty) = f(B(\infty); U_A)$ is known. Then for each value of U_A, say $U_{A,1}$, which was employed experimentally, we can plot the corresponding theoretical curve for $f(B(\infty); U_{A,1})$ on the $A(\infty)$, $B(\infty)$ plane. Now $A_1(\infty)$, the value of $A(\infty)$ which was actually observed when $U_A = U_{A,1}$, must conform to the theoretical equation. Accordingly we plot $A_1(\infty)$ on the $A(\infty) = f(B(\infty); U_{A,1})$ curve; and we do likewise for each of the observed values of $A(\infty)$. But since the points thus generated are steady-state points, they must obey both working equations. Therefore they must also lie on, or (to give the devil of random error his due) near the unknown curve for $B(\infty) = g(A(\infty))$. Hence, an estimate of $B(\infty) = g(A(\infty))$ can be obtained by finding an empirical equation which fits the plotted points.

Example 7.11.1. For a specific example, see the description of the parathyroid hormone--plasma calcium feedback loop in Part II.

SECTION 7.12. Can theoretical silk purses be made from empirical sow's ears?

We have repeatedly noted the feasibility of reducing a many-variable feedback loop to one which contains only two variables, so that only two working equations are needed to describe its behavior. In each step of this process of reduction, we combine the relatively simple working equations for some pair of adjacent loop variables, so as to obtain a single more complicated working equation from which the "upstream"* member of the pair has been eliminated. Now in theory this process is reversible, so that, given the more complicated equation and one of the two equations from which it was derived, we should be able to recover the second equation used for the derivation. But if one of the given equations is empirical, the other theoretical, combining them may yield an equation,which, though mathematically correct, is not at all the working equation we hoped to obtain. This puzzling result is best explained by means of an actual example.

[*"upstream" relative to signal flow around the loop]

Example 7.12.1. Consider the model of the carotid sinus-arterial blood pressure loop described in Part II. Suppose that the true working equations for this model were

$$P_{sinus}(\infty) = P_{art}(\infty) \qquad 7.12.1,$$

$$\dot{N}_{sinus}(\infty) = -a + bP_{sinus}(\infty) \qquad 7.12.2,$$

$$\dot{N}_{vaso}(\infty) = \frac{C}{1 + e^{-A + B\dot{N}_{sinus}(\infty)}} \qquad 7.12.3,$$

$$R(\infty) = \alpha + \beta \dot{N}_{vaso}(\infty) \qquad 7.12.4,$$

$$P_{art}(\infty) = P_{ven} + R(\infty)\,\dot{V}_{out} \qquad 7.12.5,$$

where

$P_{sinus}(\infty)$ = mean blood pressure within the carotid sinus

$P_{art}(\infty)$ = mean arterial blood pressure

$\dot{N}_{sinus}(\infty)$ = mean traffic of afferent impulses in the sinus nerves

$\dot{N}_{vaso}(\infty)$ = mean traffic of efferent vasoconstrictor impulses

$R(\infty)$ = peripheral resistance to blood flow

$\dot{V}_{out}$ = cardiac output

P_{ven} = venous blood pressure

$\left.\begin{array}{l} a,\ b \\ A,\ B,\ C \\ \alpha,\ \beta \end{array}\right\}$ = various parameters of the system.

The symbol-and-arrow diagram corresponding to equations 7.12.1 through 7.12.5 is shown in Figure 7.12.1.a. Now suppose that of the five working equations given above, only 7.12.1 and 7.12.5 are known to us. (Sound theoretical equations, both of them, and not to be spurned!) But suppose also that we have investigated experimentally the dependence of $P_{art}(\infty)$ upon $P_{sinus}(\infty)$, and that we have fitted the observed points by the empirical equation,

$$P_{art}(\infty) = P_{min} + \frac{P_{max} - P_{min}}{1 + \epsilon^{-k + mP_{sinus}(\infty)}} \qquad 7.12.6,$$

where P_{min}, P_{max}, k, and m are empirical parameters. Furthermore, we know that the working relationships among the three loop variables of equations 7.12.1, 7.12.5 and 7.12.6 can be diagrammed as in Figure 7.12.1.b. Now it would be of considerable interest to find equation 7.12.7, the working equation which shows how $R(\infty)$ depends upon $P_{sinus}(\infty)$. But surely that ought to be easy, because our empirical equation, 7.12.6 for the dependence of $P_{art}(\infty)$ on $P_{sinus}(\infty)$ must represent the result of combining the known equation 7.12.5 and the unknown equation 7.12.7, with the elimination of $R(\infty)$. Hence we should be able to recover equation 7.12.7 by the reverse procedure; i.e., by combining the known equations 7.12.5 and 7.12.6, eliminating $P_{art}(\infty)$ and solving for $R(\infty)$.

$$R(\infty) = \frac{1}{\dot{V}_{out}} \left\{ P_{min} - P_{ven} + \frac{P_{max} - P_{min}}{1 + \epsilon^{-k + mP_{sinus}(\infty)}} \right\} \qquad "7.12.7"$$

Now mathematically there is nothing wrong with equation "7.12.7." But it does <u>not</u> express a true working relationship and is <u>not</u> the equation 7.12.7 of Figure 7.12.1.b which we seek. For, viewed as a <u>working</u> equation, it states that an increase in venous pressure or an increase in cardiac output would cause a decrease in resistance to blood flow, which is nonsense!

The <u>true</u> working relationship for $R(\infty)$ as a function of $P_{sinus}(\infty)$ can be obtained by combining equations 7.12.2, 7.12.3 and 7.12.4, eliminating the intermediate variables, $\dot{N}_{sinus}(\infty)$ and $\dot{N}_{vaso}(\infty)$.

$$R(\infty) = \alpha + \left[\frac{\beta C}{1 + \epsilon^{-A - aB + bBP_{sinus}(\infty)}} \right] \qquad 7.12.7.$$

Notice that equation 7.12.7 does not contain either P_{ven} or $\dot{V}_{out}$, the two quantities whose influence upon $R(\infty)$ was so flagrantly misstated by equation "7.12.7."

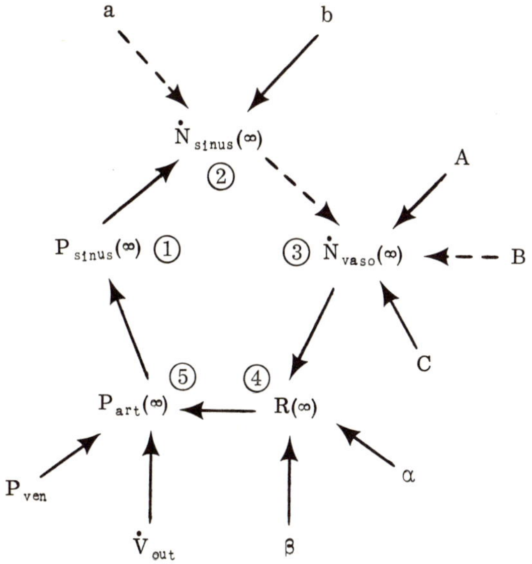

Figure 7.12.1.a.

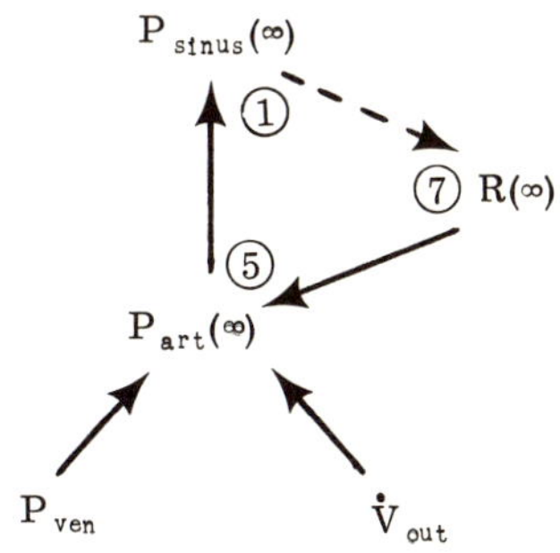

Figure 7.12.1.b.

In this artificial example, the reason for the "7.12.7" fiasco is easy to find because we know all of the underlying theoretical equations. (This would not be true in any real-life situation; if we already know all of the working equations we shall hardly be tempted to derive one of them from an empirical equation.) Let us again go through the steps leading to equation "7.12.7", but instead of using the empirical equation 7.12.6 let us use its theoretical counterpart obtained by combining equations 7.12.2 through 7.12.5.

$$P_{art}(\infty) = P_{ven} + \dot{V}_{out}\left\{\alpha + \beta\left[\frac{C}{1 + \epsilon^{-[A + aB] + bBP_{sinus}(\infty)}}\right]\right\} \qquad 7.12.8.$$

To get "7.12.7," we must combine equations 7.12.8 and 7.12.5 by eliminating $P_{art}(\infty)$ and solve the result for $R(\infty)$.

$$R(\infty) = \frac{1}{\dot{V}_{out}}\left\{[P_{ven} + \dot{V}_{out}\,\alpha] - P_{ven} + \left[\frac{\dot{V}_{out}\beta C}{1 + \epsilon^{-[A + aB] + bBP_{sinus}(\infty)}}\right]\right\}$$

7.12.7.a.

which immediately reduces to equation 7.12.7 because P_{ven} gets subtracted from itself and the $\dot{V}_{out}$'s all cancel! But when written in the form of 7.12.7.a, our theoretical equation can be compared, item by item, with equation "7.12.7"; a comparison which at once reveals the source of our difficulty. For we now see that the terms needed to eliminate P_{ven} and $\dot{V}_{out}$ from equation "7.12.7" are hidden away, unknown and unknowable, in the empirical parameters of equation 7.12.6. In real life we could observe these parameters, P_{max}, P_{min}, k, and m, only as numerical constants, and we would never be able to see, as we do here, that in fact,

$$P_{min} = P_{ven} + \dot{V}_{out}\,\alpha \qquad 7.12.9,$$

$$P_{max} = P_{ven} + \dot{V}_{out}\,[\alpha + \beta C] \qquad 7.12.10,$$

$$k = A + aB \qquad 7.12.11,$$

$$m = bB \qquad 7.12.12.$$

Moral: Theoretical relationships can be derived only from a physicochemical model of the system being studied. Empirical equations merely describe the observed trend of a limited number of experimental observations. We can test theory by experimental observations (see Section 7.14), but we cannot construct theory from experimental observations. Hence, do not try to derive theoretical silk-purse relationships from empirical sow's-ear equations.

SECTION 7.13. Calculation of open-loop gain from direct observation of open-loop and closed-loop responses

In the preceding sections we have described various ways of finding two working equations from which the gain due to feedback can be calculated. But suppose we have a system in which only one of the loop variables can be measured. Suppose further that we do not have even a partial physicochemical model of the system from which a theoretical working equation could be derived. Then it is impossible to obtain even a single working equation to describe the dependence of one loop variable upon another. Indeed, under these circumstances the only way in which the operation of the feedback loop can be studied quantitatively is to observe how the measurable loop variable responds under both closed-loop and open-loop conditions to step changes in one of the independent variables upon which it depends. For example, suppose that we can vary U_A stepwise and measure $A(\infty)$ at each step with the feedback loop intact. We can then plot $A(\infty)$ against U_A and fit the plotted data by an empirical equation whose derivative at any point provides a direct estimate of $dA(\infty)/dU_A|_{closed}$ at that point. If these maneuvers can be repeated after the feedback loop has been interrupted, then we can also obtain a direct estimate of $dA(\infty)/dU_A|_{open}$. From these two derivatives we can calculate

$$\text{Magnification} = \frac{dA(\infty)/dU_A|_{closed}}{dA(\infty)/dU_A|_{open}} \qquad 7.13.1,$$

and

$$[OLG](j0) = \left| \frac{1}{\text{Magnification}} - 1 \right| \qquad 7.13.2$$

where both derivatives must be evaluated at the same operating point.

The requirement that $dA(\infty)/dU_A|_{closed}$ and $dA(\infty)/dU_A|_{open}$ must be compared at the same operating point has already been stressed in Section 7.6. It is a requirement by no means always easy to satisfy in experimental practice. Under particularly favorable circumstances, it may be possible to open a physiological feedback loop without committing surgical or pharmacological mayhem. For example, Stark and his collaborators (5 , ,6) have brilliantly exploited the technique of opening the loop responsible for the pupillary reaction to light by illuminating the retina with a beam of light too narrow to be occluded by pupillary constriction. Unfortunately most feedback mechanisms do not have any such conveniently placed hole through which an un-feed-backable signal can be inserted, and the investigator may be forced to open the loop with drug or scalpel. After such heavy-handed methods, restoration of the system to its original operating point may not be easy.

Example 7.13.1. Hausmann and Riggs (7) attempted to study the hormonal regulation of plasma calcium by measuring the steady-state concentration of calcium in plasma during constant intravenous infusions of calcium gluconate or of the calcium-chelating drug, disodium ethylenediaminetetra-acetic acid. One series of experiments was performed under closed-loop conditions, using normal dogs; a second under open-loop conditions, using dogs from whom the parathyroid glands had been removed surgically. Now the only practical way to excise all parathyroid tissue in the dog is to perform a complete thyro-parathyroidectomy. This removes the source not just of parathyroid hormone but of both thyrocalcitonin and thyroid hormone as well. Ideally, then, to restore the system to one of its original operating points: 1.) We should have repaired the metabolic defects caused by hypothyroidism. This we attempted to do by administering exogenous thyroid hormone. 2.) We should have infused thyrocalcitonin at a variable rate equal to its normal rate of secretion at the particular plasma calcium concentration present in each experiment. But even if thyrocalcitonin had been available to us, we could not have done this because too little is known about the rate of secretion of thyrocalcitonin. 3.) We should have infused parathyroid hormone at a constant rate just sufficient to maintain the plasma calcium concentration at the desired operating point with no calcium infusion but with the dogs maintained on a diet containing a normal quantity of calcium. For several reasons this was not technically feasible. Instead, in order to achieve the same plasma calcium concentration in the thyroparathyroidectomized dogs as in the normal dogs, substantially higher rates of calcium infusion were used during an experiment; and to avoid frank tetany large supplements of calcium were added to the diet. It is all too obvious that the actual procedure which we used fell far short of the ideal, and that conclusions drawn from such crude experiments must be regarded as tentative. But one has to start somewhere.

SECTION 7.14. A general strategy for the experimental study of biological feedback systems in the steady state

In the immediately preceding sections, several methods for studying feedback systems in the steady state have been discussed in detail and have been illustrated by specific examples. If, for simplicity, we limit consideration to single feedback loops which have been reduced to only two variables, all such methods can be displayed in a single table (Table 7.14.1) which not only summarizes their salient features but also suggests the general strategy to be followed when investigating the steady-state behavior of a physiological feedback system.

The table presents a two-way classification: by rows, according to what sort of physicochemical model of the system is available; by columns, according to which loop variables can be measured in the laboratory. (I am assuming that experimental manipulation of the appropriate independent variables is no problem). The table is so arranged that the completeness of our theoretical model decreases from above downwards, the versatility of our experimental approach decreases from left to right. The weakest possible situation is therefore represented by the box in the lower right corner, a situation so weak, in fact, that it precludes any quantitative description of the behavior of the system. The strongest possible situation is represented by the box diagonally opposite, at the upper left. Here, both equations can be derived from theory, and both can also be derived empirically. Under these circumstances, the adequacy of our physicochemical model (which is really nothing more than a working hypothesis about the behavior of the system) can be completely tested by seeing whether each of the two empirical equations, derived solely from experimental observations, agrees satisfactorily with its theoretical counterpart. Notice that in addition to this box, only the middle box in the upper row and the middle box in the left column represent situations which allow us to test theory by experiment. However, all but three of the nine situations allow us, in one way or another, to get our two working equations.

Our strategy is now plain enough. We should start as far away from the lower right corner as our current working hypothesis and analytical techniques allow, and we should try to develop both the theoretical insight and the experimental prowess needed to move closer and closer to the upper left corner. Whenever we reach one of the three boxes which allow experimental testing, test we must, even though the result may force us to scrap part or all of our current theory and to retreat to a box lower in the hierarchy of Table 7.14.1. When we have finally reached the upper left corner, and both of our theoretical equations have survived experimental scrutiny, we can have some confidence in the adequacy of our model. But this game is never really won. For there will always be other, often closely related, feedback systems waiting for us in the lower boxes!

SECTION 7.15. Nonlinear analysis at a glance

In Figure 7.15.1, I have attempted to summarize the sequences of steps which may be used in the analysis of nonlinear systems. While the title, "Nonlinear analysis at a glance" may be a trifle optimistic, I believe that the figure does help to clarify the relationships between the solution of nonlinear differential equations which was discussed in Chapter 6, and the analysis of steady-state working equations which has been the main theme of the present chapter.

When the nonlinear differential equations describing the system are known, we can start at the upper left circle and follow any suitable path from that point.

Table 7.14.1.

		Which of the two loop variables can be measured?		
		Both	One	Neither
What kind of simple* physico-chemical model of the system is available?	COMPLETE	Number of equations derivable 2 theoretical 2 empirical Complete model can be completely tested.	Number of equations derivable 2 theoretical 1 empirical Complete model can be partially tested.	Number of equations derivable 2 theoretical 0 empirical Model cannot be tested.
	PARTIAL**	Number of equations derivable 1 theoretical 2 empirical Partial model can be completely tested.	Number of equations derivable 1 theoretical 1 empirical Model cannot be tested.	Number of equations derivable 1 theoretical 0 empirical Model cannot be tested.
	FRAGMENTARY or NONE	Number of equations derivable 0 theoretical 2 empirical There is no model to test.	Number of equations derivable 0 theoretical 0 empirical Only a direct comparison of open-loop and closed- loop responses can be made.	Number of equations derivable 0 theoretical 0 empirical System can only be described qualitatively.

*Simple enough to be described by specific theoretical equations.
**But sufficient for the derivation of <u>one</u> working equation.

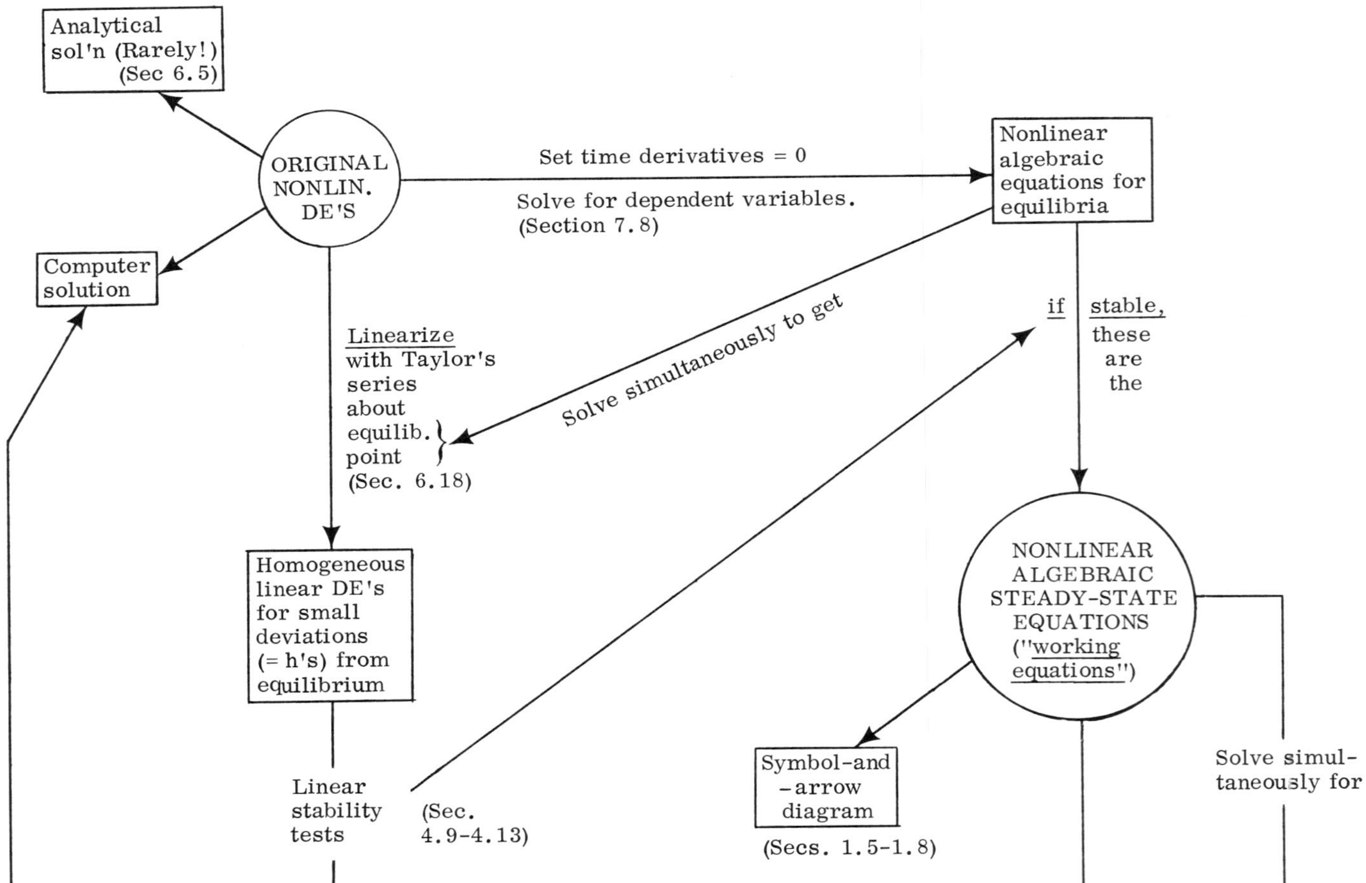
Analytical
sol'n (Rarely!)
(Sec 6.5)
ORIGINAL
NONLIN.
DE'S
Computer
solution
Set time derivatives = 0
Solve for dependent variables.
(Section 7.8)
Nonlinear
algebraic
equations for
equilibria
Linearize
with Taylor's
series
about
equilib.
point
(Sec. 6.18)
Solve simultaneously to get
if stable,
these
are
the
Homogeneous
linear DE's
for small
deviations
(= h's) from
equilibrium
NONLINEAR
ALGEBRAIC
STEADY-STATE
EQUATIONS
("working
equations")
Linear
stability
tests
(Sec.
4.9-4.13)
Symbol-and
-arrow
diagram
(Secs. 1.5-1.8)
Solve simul-
taneously for

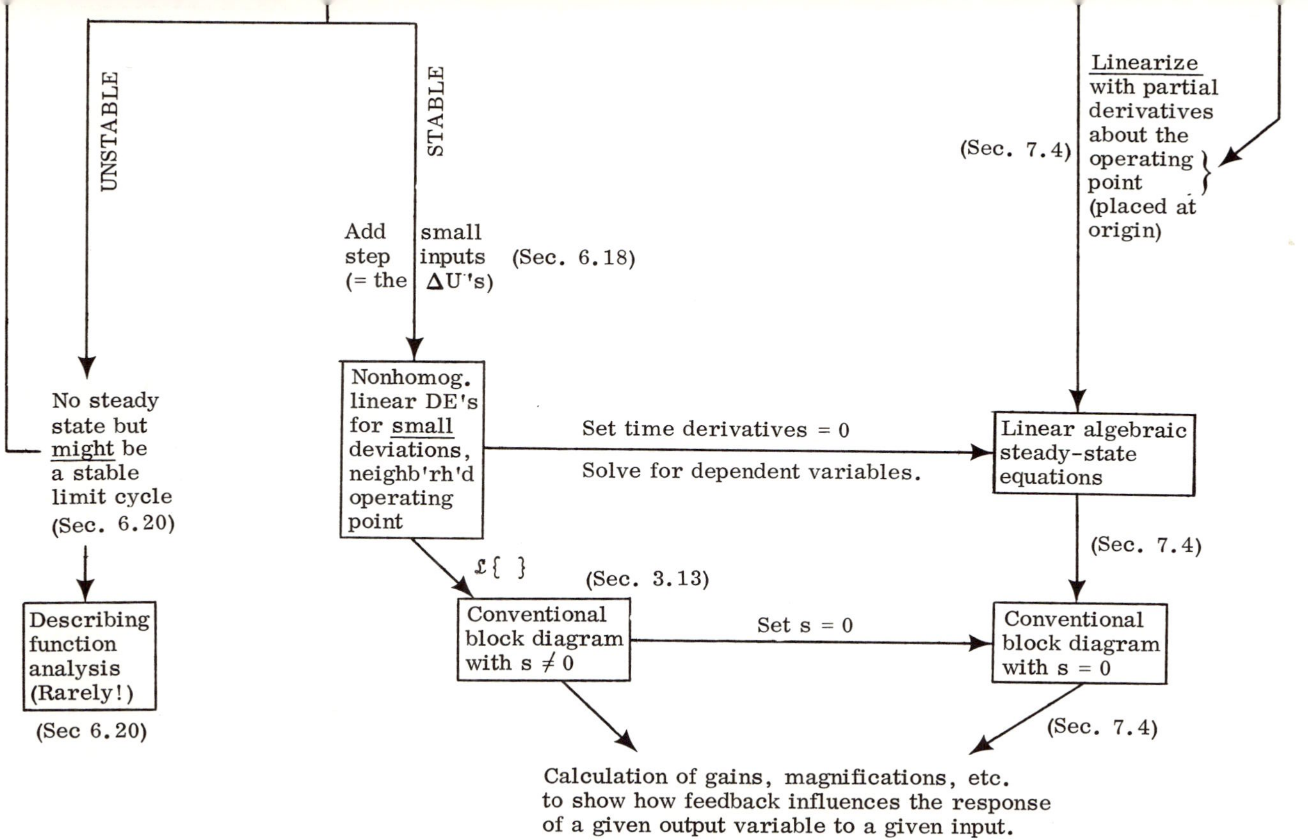

Figure 7.15.1. Nonlinear analysis at a glance

ORIG. NONLIN. DIFF. EQNS.

$$\frac{dX(t)}{dt} = \frac{1}{V}[-F_1(X) + F_2(Y) + \dot{I}]$$

$$\frac{dY(t)}{dt} = \frac{1}{V}[-F_3(X, Y) - F_4(X, Y) + \dot{G}]$$

$t \to \infty$

$$-F_1(X) + F_2(Y) + \dot{I} = 0$$

$$-F_3(X, Y) - F_4(X, Y) + \dot{G} = 0$$

Cannot solve for $X(\infty)$ and $Y(\infty)$ to get working equations, because the F's are unknown.

Can write an incomplete S-and-A diagram based on known physiology

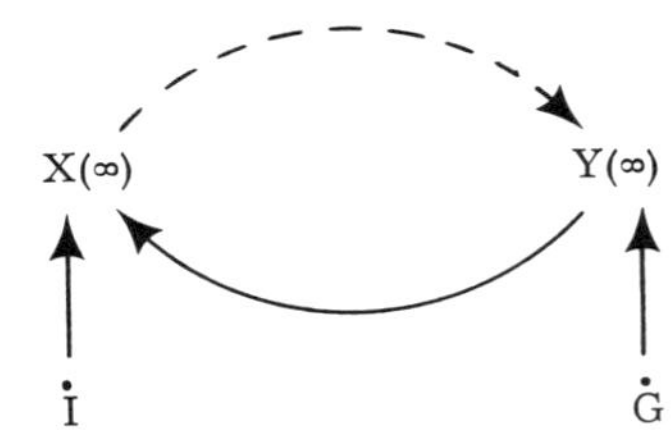

Linearize with Taylor's series for $F_i(X_o + x(t),\ Y_o + y(t))$.

$i = 1, 2, 3, 4$, around $[X_o,\ Y_o;\ \dot{I} = \dot{G} = 0]$

$$\frac{dx(t)}{dt} = \underbrace{\frac{-\partial F_1(X)}{V \partial X(t)}\Big|_{X_o}}\, x(t) + \underbrace{\frac{\partial F_2(Y)}{V \partial Y(t)}\Big|_{Y_o}}\, y(t)$$

$$\frac{dx(t)}{dt} = -\alpha \quad x(t) \quad +\beta \quad y(t)$$

$$\frac{dy(t)}{dt} = -\underbrace{\frac{\partial F_3(X, Y)}{V \partial X(t)}\Big|_{X_o, Y_o}}\, x(t) - \underbrace{\frac{\partial F_4(X, Y)}{V \partial Y(t)}\Big|_{X_o, Y_o}}\, y(t)$$

$$\frac{dy(t)}{dt} \quad -\gamma \quad x(t) \quad -\delta \quad y(t)$$

Stability test not needed (2nd-order syst. is stable if all terms of char. eqn, have same sign)

Add small step inputs

$$\frac{dx(t)}{dt} = -\alpha x(t) + \beta y(t) + \frac{\Delta \dot{I}}{V}$$

$$\frac{dy(t)}{dt} = -\gamma x(t) - \delta y(t) + \frac{\Delta \dot{G}}{V}$$

$t \to \infty$

$$x(y(\infty)) = \frac{\beta}{\alpha} y(\infty) + \frac{1}{\alpha V} \Delta \dot{I}$$

$$y(x(\infty)) = -\frac{\gamma}{\delta} x(\infty) + \frac{1}{\delta V} \Delta \dot{G}$$

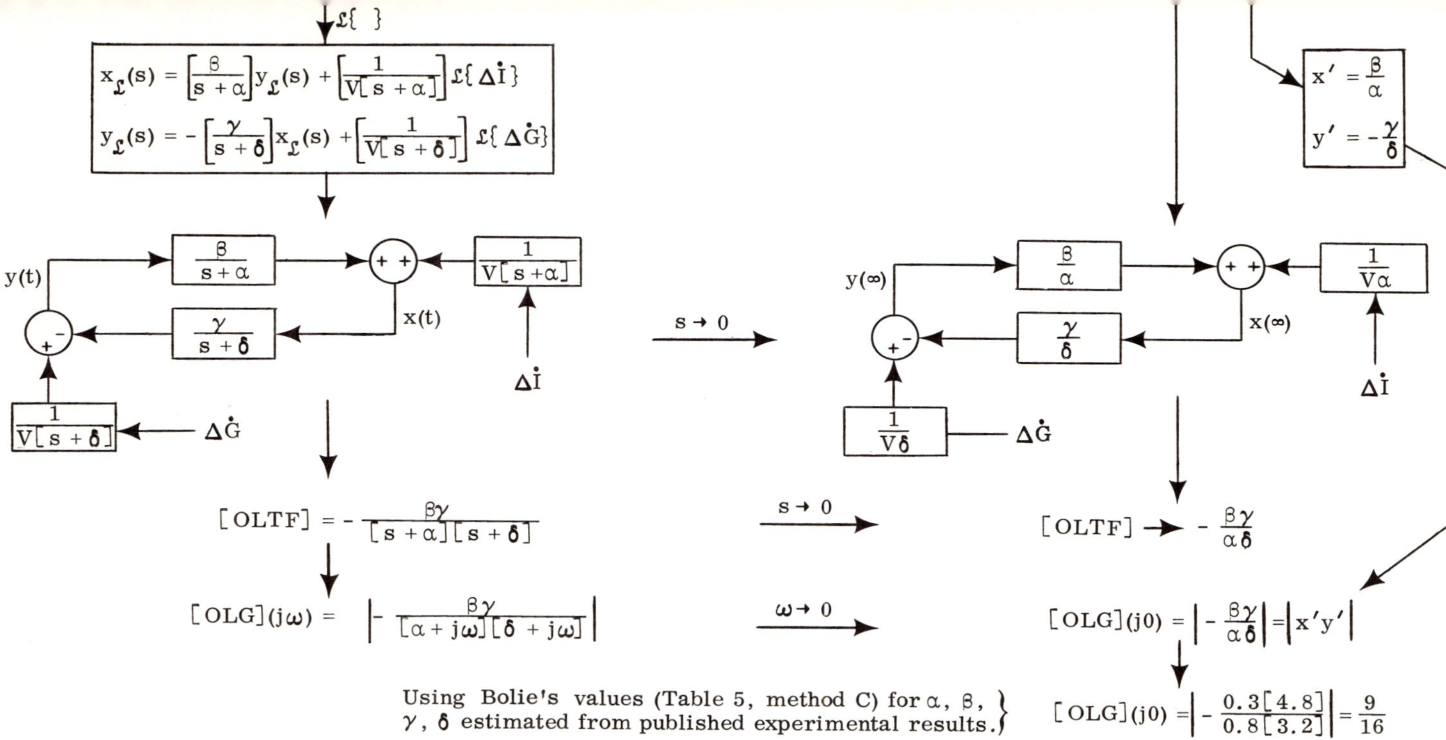

Figure 7.15.2. Blood glucose-insulin feedback loop (Example 7.15.1)

Note, in particular, that the same set of linear, algebraic steady state equations (lower right) will be obtained whether we linearize first and then set the derivatives equal to zero (left portion of diagram), or set the derivatives to zero first and then linearize (right portion of diagram). But if we lack the original differential equations, we have to start at the "working equations" circle whence only one analytical path is open to us.

Notice also that while we can construct a conventional block diagram according to the linearized versions of the original differential equations, the input forcing functions used in the corresponding transfer function analysis must be kept small so that the system will stay in the neighborhood of the operating point about which we have linearized. ("Small-signal analysis"). In particular, this means that if we want to explore the frequency domain by applying sinusoidal forcing to biological systems, we shall have to steer a precarious course between the Scylla of inputs so large that the system is forced away from its linear neighborhood, and the Charibdis of inputs so small that we cannot measure the corresponding output oscillations against the background of random "noise."

Figure 7.15.2 on pages 360-361 follows the pattern of Figure 7.15.1, and illustrates many of the general steps therein by the following specific example.

Example 7.15.1. The blood glucose-insulin feedback loop.

Bolie (8) has proposed the following differential equations to describe a much-simplified model of the feedback system by which insulin helps to regulate the concentration of glucose in blood.

$$V[dX/dt] = \dot{I} - F_1(X) + F_2(Y) \qquad 7.15.1$$

$$V[dY/dt] = \dot{G} - F_3(X, Y) - F_4(X, Y) \qquad 7.15.2$$

where

t = time (hr)

V = extracellular fluid volume, assumed to be the volume of distribution for both glucose and insulin. (liters)

$\dot{I}$ = rate of insulin infusion (units/hr).

$\dot{G}$ = rate of glucose infusion (g/hr)

$X(t)$ = concentration of insulin in V (units/liter)

$Y(t)$ = concentration of glucose in V (g/liter)

$F_1(X)$ = rate of insulin destruction (units/hr).

$F_2(Y)$ = rate of insulin secretion (units/hr)

$F_3(X, Y)$ = rate of accumulation of glucose in liver (g/hr)

$F_4(X, Y)$ = rate of utilization of glucose by tissues (g/hr)

The functions F are unknown. But by assuming that they are linear within the immediate neighborhood of the "normal" operating point, $[X_0, Y_0]$, with $\dot{I} = \dot{G} = 0$, Bolie was able to carry the analysis forward in terms of $x(t) = X(t) - X_0$; $y(t) = Y(t) - Y_0$, the deviations of the variables from their operating point values. Figure 7.15.2 shows the further steps which can be taken and which ultimately lead to a numerical value for the [OLG](j0) calculated from experimentally-derived estimates of the parameters.

It is appropriate for Part I to end with a biological example such as this, where, to achieve even a partial analysis, we have had to assume that a grossly oversimplified model behaves in a linear fashion throughout some unspecified range in the vicinity of the normal operating point. Yet in spite of its inadequacies, the analysis has given us at least a crude quantitative estimate of how well the insulin-glucose feedback loop minimizes the changes of blood sugar caused by imposition of a constant load. Furthermore, the small magnitude of the [OLG](j0) thus estimated at once suggests that this feedback loop by itself is a rather poor regulator of blood sugar at least in the steady state. But no final conclusions about the overall effectiveness of blood glucose regulation are warranted until we have much more information about the transient and steady-state behavior not only of the insulin loop but also of the numerous other loops,--such as those involving glucagon, epinephrine, glucocorticoids, and growth hormone,--which influence carbohydrate metabolism.

Look at the end of the work, contrast
The petty done, the undone vast--

Robert Browning,
THE LAST RIDE TOGETHER

Exercises

1. In Chapter 4 it was stated that only an integral controller (requiring a physically-real set point) could eliminate steady-state error. But in the multiple-loop system responsible for the regulation of body temperature (See Part II), the value of [OLG](j0) at one particular point seems to be infinite, implying that at that point negative feedback would hold the body temperature perfectly constant in the face of any sufficiently small change in ambient temperature. What accounts for this perfect stabilization in the absence of any physically-real set point?

2. In the text, it was stated that the system of example 7.8.1 could be described by a set of five first-order differential equations in five variables (plus two algebraic equations which, for present purposes, we can neglect.)

   A. Write such a set of five equations.
   B. Use Taylor's Series (see Section 6-18) to linearize your equations about an equilibrium point.
   C. Write the matrix, $[sI-A]$ (see Section 5.4) for your linearized equations.
   D. Describe what further steps you might take if you wanted to find out about the stability of the equilibrium points determined by various values of the parameters.

3. A feedback system for which U_x is the input is described by the following working equations:

$$x(\infty) = U_x - \frac{3y(\infty)}{1 + y(\infty)}$$

$$y(\infty) = 1 + [x^2(\infty)/4]$$

   A. Assuming that $x(\infty) \geq 0$, what is the minimum value of U_x? What is the value of [OLG](j0) at that point?
   B. Calculate [OLG](j0) for $U_x = 2, 3, 4, 5, 6, 7$ and 8.
   C. Use your results from A and B to estimate the value of U_x at which [OLG](j0) is a maximum.

4. Many an important biological feedback system has not been described in Part II of this book. Choose one from the following list and try to analyze its behavior. A key reference or two is given for each. Use these references critically!

   A. Feedback control of genetic expression. (9, 10, 11).
   B. Control of glucocorticoid secretion (12).
   C. The regulation of cerebral blood flow by carbon dioxide. Start with figure 8, page 955, in Reference 13.
   D. Blood clotting. (14, 15).
   E. Sex cycles in the rat. (16).
   F. Erythropoietin--red cell production. (17).

GENERAL EXERCISES

As a general review of some of the topics discussed in this book, comment briefly on each of the following:

1. "Every feedback mechanism or cybernetic system, in addition to a regulator such as the kidney and the substance being regulated, includes a storage reservoir." For the regulation of plasma calcium concentration, "The bone mineral serves this purpose; the bone blood continuum acts as a closed cycle or servosystem." pp. 108-109 in Reference 18.

2. "The effectiveness of any reflex control depends in major part on the sensitivity or "gain." This is, in general, the ratio of a change in output to a change in input---" p. 672 in Reference 19.

3. Figure 1, p. 307 in Reference 20, which is a block diagram for a "regenerative" (i.e. positive) feedback amplifier. Is the statement in the legend about open-loop gain correct?

4. Figure 2-10, p. 23, in Reference 21. In that figure, the author treats simple exponential growth (or decay) as a feedback system. Is it? In his block diagram, the input for one of the two blocks in the feedback loop is dP(t)/dt and the output is P(t). In the block, integration of the input with respect to time is shown. For the second block, P(t) is the input, dP(t)/dt is the output. In the block is the rate constant, k, in the expression dP(t)/dt = kP(t). If we accept the inputs and the outputs as defined by the author, what <u>should</u> the transfer functions for the two blocks be? If k is negative, will the feedback loop be negative as the author claims? Calculate [OLG](j0) from <u>your</u> two transfer functions, and explain its value. How would <u>you</u> draw a block diagram for simple exponential growth?

5. "---the fact that arterial P_{CO_2} remains remarkably constant and varies little with small changes in inspired P_{CO_2}---raises doubts whether it can be effective in regulating its own level." p. 811 in Reference 22.

6. "At normal arterial pressures, the gain of the pressoreceptor control system is approximately -2, whereas the positive feedback gain that results from depressed cardiac nutrition when the cardiac output falls is somewhere in the order of 0.1. Thus the net gain is -1.9. Under these conditions, the circulation will be very stable." p. 6 in Reference 23.

7. "W.M. Bayliss (1902) originally suggested that increasing the tension on the smooth muscle in the arteriolar walls caused it to contract. That is, a rise in pressure would cause a vasoconstriction.---this mechanism would appear to act as a positive feedback loop and so be completely unstable. Thus a rise in blood pressure would cause a vasoconstriction which would cause a further rise in blood pressure." p. 141 in Reference 24. This criticism has more recently been quoted by (and accepted by) Rushmer, pp. 111-112 in Reference 25.

8. In Reference 26,

   A. "Reversal of the feedback from negative to positive results in a run-away, either to maximum or to zero." p. 3.
   B. "Oscillation invariably betokens the presence of stabilizing feedback." p. 7.
   C. "Stability (including oscillation) invariably demands negative feedback." p. 14.

   "Any system which maintains its stability demands negative refection." p. 16. (The authors use "refection" as a synonym for "feedback.")

9. <u>You can't win!</u>

   A. "If, as McLean postulates, the bone were the only means of regulating" [calcium activity] "of plasma in conjunction with the parathyroids, the resulting feedback system would lead to wide oscillations in the level of" [calcium activity] "in the plasma." p. 124 in Reference 27.

      also,

      "Parathormone is slow-acting and if it were the sole means of calcium regulation, this would lead to oscillations in the plasma-calcium level." p. 1305 in Reference 28.
   B. "---when antogonistic activities are present" [such as, for example, insulin and glucagon, or parathyroid hormone and calcitonin] "there is a danger of instability ('vicious cycles')." p. 91 in Reference 29.

References

1. Riggs, D.S.: Feedback: Fundamental relationship or frame of mind? Adv. Enz. Regul., <u>5</u>, 357-382, 1967.

2. Riggs, D.S.: <u>The Mathematical Approach to Physiological Problems.</u> Williams and Wilkins, Baltimore, 1963.

3. Chance, B., Ghosh, A., Higgins, J.J., and Maitra, P.K.: Cyclic and oscillatory responses of metabolic pathways involving chemical feedback and their computer representations. Ann. N.Y. Acad. Sci., <u>115</u>, 1010-1024, 1964.

4. Higgins, J.: The theory of oscillating reactions. Indust. engng. Chem., <u>59</u>, 19-62, 1967.

5. Stark, L. Stability, oscillations and noise in the human pupil servomechanism. Proc. Inst. Radio Engrs., <u>47</u>, 1925-1939, 1959.

6. Stark, L.: <u>Neurological Control Systems: Studies in Bioengineering</u>. Plenum Press, N.Y., 1968.

7. Hausmann, E., and Riggs, D.S.: The effectiveness of negative feedback in regulating plasma calcium in the dog. J. Theoret. Biol., 12, 350-363, 1966.

8. Bolie, V.W.: Coefficients of normal blood glucose regulation. J. appl. Physiol., 16, 783-788, 1961.

9. Changeux, J-P.: The control of biochemical reactions. Sci. Amer., 212, 36-45, 1965.

10. Jacob, F.: Genetics of the bacterial cell. Science, 152, 1470-1478, 1966.

11. Bretscher, M.S.: How repressor molecules function. Nature, 217, 509-511, 1968.

12. Yates, F.E., Brennan, R.D., and Urquhart, J.: Adrenal glucocorticoid control system. Federation Proc., 28, 71-83, 1969.

13. Patterson, J.L.: Circulation through the brain. Pp. 950-958, in Ruch, T.C., and Patton, H.D.: Physiology and Biophysics. 19th edition, W.B. Saunders, Philadelphia, 1965.

14. MacFarlane, R.G.: An enzyme cascade in the blood clotting mechanism, and its function as a biochemical amplifier. Nature, 202, 498-499, 1964.

15. Biggs, R. and MacFarlane, R.G., editors: Treatment of Haemophilia and other Coagulation Disorders. F.A. Davis Co. Philadelphia. 1966. Chapter 1.

16. Schwartz, N.B.: A model for the regulation of ovulation in the rat. Rec. Prog.Horm. Res., 25, 1-55, 1969.

17. (Various authors) Erythropoietin. Ann. N.Y. Acad. Sci., 149, 1-583, 1968.

18. Urist, M.R.: Bone-body fluid continuum. Perspect. Biol. Med., 6, 75-115, 1962.

19. Scher, A.M.: Control of arterial blood pressure: Measurement of pressure and flow. Chapter 34 in Ruch, T.C., and Patton, H.D.: Physiology and Biophysics. 19th edition, W.B. Saunders, Philadelphia, 1965.

20. Hirsch, H.R., and Engelberg, J.: An analog of cell-culture growth. J. Theoret. Biol., 9, 303-312, 1965.

21. Milsum, J.H.: Biological Control Systems. McGraw-Hill, N.Y., 1966.

22. Hornbein, T.F.: The chemical regulation of ventilation. Chapter 42 in Ruch, T.C., and Patton, H.D.: Physiology and Biophysics. 19th edition, W.B. Saunders, Philadelphia, 1965.

23. Guyton, A. C. and Crowell, J. W.: Cardiac deterioration in shock: I. Its progressive nature. Pp. 1-12, in Hershey, S. G., editor, Shock. Little, Brown, and Co., Boston, 1964.

24. McDonald, D. A., and Taylor, M. G.: The hydrodynamics of the arterial circulation. Prog. Biophys. biophys. Chem., 9, 105-173, 1959.

25. Rushmer, R. F.: Cardiovascular Dynamics. 2nd edition, W. B. Saunders, Philadelphia, 1961.

26. Stanley-Jones, D., and Stanley-Jones, K.: The Kybernetics of Natural Systems. A Study in Patterns of Control. Pergamon Press, Oxford, 1960.

27. Rasmussen, H.: Parathyroid hormone. Nature and mechanism of action. Am. J. Med., 30, 112-128, 1961.

28. Foster, G. V., Baghdiantz, A., Kumar, M. A., Slack, E., Soliman, H. A., and MacIntyre, I.: Thyroid origin of calcitonin. Nature, 202, 1303-1305, 1964.

29. Goldman, S.: Cybernetic aspects of homeostasis. Chapter 3 in Comar, C. L. and Bronner, F.: Mineral Metabolism. Vol. 1. Part A. Academic Press, N. Y., 1960.

PART II

PHYSIOLOGICAL FEEDBACK MECHANISMS

The coordinated physiological processes which maintain most of the steady states in the organism are so complex and so peculiar to living beings--involving, as they may, the brain and nerves, the heart, lungs, kidneys, and spleen, all working cooperatively--that I have suggested a special designation for these states, homeostasis.

Walter B. Cannon, THE WISDOM OF THE BODY

INTRODUCTION

Part II is an incomplete and deliberately superficial survey of a number of physiological feedback systems. For simplicity and uniformity, I have concentrated upon steady-state behavior, mentioning transient behavior only occasionally and in passing. In particular, I have slighted the wealth of information which is available about the transient behavior of such thoroughly-investigated mechanisms as the ocular reflexes, and the chemical control of breathing. To have given these systems their due would have overburdened the book and would have over-reached my personal competence.

My primary aim in this part of the book has been to gather together a series of more-or-less plausible models, and to present them in a reasonably uniform manner. Many of the models are descriptive, not mathematical. Many are grossly oversimplified. Some are highly speculative; others are biased. Indeed, the expert in any of the fields I touch upon is likely to find my presentations at best naive, at worst misleading. In spite of these faults, I hope the models will serve their primary purpose which is to stimulate further work designed to better them. If nothing else, they abundantly illustrate how difficult it is to obtain really quantitative information about the behavior,--even the simple steady-state behavior--of physiological feedback systems.

In preparing Part II, I have been assisted continuously by two small Devils. The one whose name is Caution sat upon my left shoulder. "It will not matter" he kept saying to me "what you leave out. You will be judged only by what you put in." Laziness sat upon my right, whispering "Who cares? Come have a drink!" The reader can blame these two for my disregard of such important matters as physiological limit cycles and "biological clocks" (1,2,3), unidirectional rate sensitivity (4), neuromuscular (5,6,7,8) and behavioral (9,10,11) feedback systems, and those fascinating, but still poorly understood regulatory mechanisms which account for long-term adjustments such as work hypertrophy, regeneration of partially excised organs (12), acclimatization and the like. Furthermore, even for those topics which are included, I have not undertaken anything like a comprehensive survey of the relevant literature. This is, after all, a textbook, not an exhaustive survey of published work. In fact, Part II is best regarded as an extension of Chapter 7, an extension which I hope will clearly illustrate the value of applying steady-state analysis to biological feedback mechanisms.

Part II depends largely upon the technique of linearization about an operating point. This technique is relatively easy to understand and to use. It can be applied directly to the equations,--be they theoretical or empirical--which describe how the living system, or a model thereof, actually works. It requires only modest competence in mathematics. The physiologist can use it profitably even when he knows little about the concepts and the specialized methods of control theory. Best of all, it is rigorously applicable to nonlinear systems. For these reasons, I shall be happy if Part II helps to popularize steady-state analysis among life scientists. But we must not forget that steady-state analysis also has serious limitations (Section 6.21). The zealous investigator will not long be satisfied to study just the steady-state behavior of a biological feedback system; he will want to explore its transient behavior as well. I trust that the paucity of references in Part II to time-dependent behavior will not delude the beginner into thinking that the mathematical analysis of transient states is a neglected art. Quite to the contrary, in many fields there is a rich and rapidly-increasing literature which deals with various aspects of transient behavior. Yet even these highly sophisticated studies are often easier to understand and to interpret when we are thoroughly familiar with the steady-state situation.

In general, each section in Part II starts with a list of symbols for, and definitions of the variables involved. The variables are classified according to whether they are independent inputs, ("Ind"), parameters, ("Par"), or dependent variables, ("Dep"). Dependent variables are distinguished either as functions of time, for example $\dot{V}(t)$, or (by replacing t by ∞) as time-dependent variables in the steady state, for example, $\dot{V}(\infty)$. For the sake of concreteness, a specific numerical example is usually given. I have tried to choose plausible exemplary values, but they should not be regarded as "standard" or "normal" values. Indeed, their chief merit is that they constitute an internally-consistent set which can be used in checking the equations. This is the only excuse for listing them with three, or occasionally even four significant figures.

The symbols, definitions, and exemplary values are followed by a description of a physico-chemical model of the system and of the simplifying assumptions upon which the derivation will be based. In the derivation itself, the number of each primary "building-block" equation is followed by a letter, indicating whether it is a theoretical ("T") equation based upon the known characteristics of the model, an empirical ("E") equation, describing experimental observations, or, rarely, a suppositious ("S") equation, describing a fancied relationship (see Section 7.7). These primary equations are then combined to derive a smaller set of working equations which can be solved simultaneously to obtain operating-point values, and which can also be linearized by using the appropriate partial derivatives (the β's). The linearized equations are used to calculate ratios of derivatives, magnifications, or gains under a variety of circumstances. The concluding portions of the section include general comments, remarks on transient behavior, and references.

I do not know what the useful half-life of the material in Part II is. Judging from present interest in quantitative aspects of biological regulation, and from the current rate of progress in understanding physiological feedback mechanisms, its half-life can hardly be more than three years. This means that the continued value of Part II will depend upon its being revised at frequent intervals. I shall welcome any suggestions for additions, expansions, and revisions which fall within the scope of steady-state analysis.

References

1. Various authors: Biological clocks. Cold Spring Harbor Symposia on Quantitative Biology. 25, 1960.

2. Various authors: Rhythmic functions in the living system. Ann. N.Y. Acad. Sci., 98, 755-1326, 1962.

3. Sollberger, A.: Biological Rnythm Research. American Elsevier Publication Co., N.Y., 1965.

4. Various authors: Rein control, or unidirectional rate sensitivity, a fundamental dynamic and organizing function in biology. Ann. N.Y. Acad. Sci., 156, 627-968, 1969.

5. Matthews, P.B.C.: Muscle spindles and their motor control. Physiol. Rev., 44, 219-288, 1964.

6. Granit, R., editor,: Muscle Afferents and Motor Control. First Nobel Symposium. John Wiley and Sons, N.Y., 1966.

7. Yahr, M.D., and Purpura, D.P., editors,: Neurophysiological Basis of Normal and Abnormal Motor Activities, Raven Press, Hewlett, N.Y., 1967.

8. Stark, L.: Neurological Control Systems. Plenum Press, N.Y., 1968.

9. Bartholomew, G.A.: The roles of physiology and behavior in the maintenance of homeostasis in the desert environment. Sympos. Soc. exper. Biol., 18, 7-29, 1964.

10. Epstein, A.N., and Milestone, R.: Showering as a coolant for rats exposed to heat. Science, 160, 895-896, 1968.

11. Corbit, J.D.: Behavioral regulation of hypothalamic temperature. Science, 166, 256-258, 1969.

12. Karnofsky, D.A., and Clarkson, B.D.: Cellular effects of anticancer drugs. Annual Rev. Pharmacol., 3, 357-428, 1963. See p. 372 for references to the control of liver regeneration, etc.

General References

The following recent textbooks and symposia are useful not only for their own content but also for their references to published work on various topics.

Bayliss, L.E.: Living Control Systems. W.H. Freeman and Co., San Francisco, 1966.

Grodins, F.S.: Control Theory and Biological Systems. Columbia University Press, N.Y., 1963.

Kalmus, H., editor,: Regulation and Control in Living Systems. John Wiley and Sons, N.Y., 1966.

Milhorn, H.T. Jr.: The Application of Control Theory to Physiological Systems. W.B. Saunders Co., Philadelphia, 1966.

Milsum, J.H.: Biological Control Systems Analysis. McGraw-Hill Book Co., N.Y., 1966.

Stark, L., Young, L.R., Taub, R., Taub, A., and Katona, P.G.: Biological Control Systems. National Aeronautics and Space Administration, Report CR-577, 1966.

Yamamoto, W.S., and Brobeck, J.R., editors: Physiological Controls and Regulations. W.B. Saunders Co., Philadelphia, 1965.

Homeostasis and Feedback Mechanisms. Sympos. Soc. exper. Biol., 18, 1964.

PRODUCT INHIBITION OF A SYNTHETIC PATHWAY: REVERSIBLE COMPETETIVE INHIBITION

NEGATIVE

Type	Symbol	Definition	Example
Ind.	$\dot{Q}_S$	Rate at which S is supplied to the system	8 μ mole/min
	$\dot{Q}_{I,ex}$	Rate at which exogenous I is supplied to the system	0 μ mole/min
Par.	k_P	Michaelis constant for conversion of S to P	50 μ mole/L
	k_I	Michaelis constant for conversion of S to I	45 μ mole/L
	k_R	Michaelis constant for conversion of I to R	80 μ mole/L
	k_{eq}	Equilibrium dissociation constant for IF complex	1 μ mole/L
	$\dot{Q}_{P,max}$	Maximum rate of production of P	68 μ mole/min
	$\dot{Q}_{I,max}$	Maximum rate of production of I	40 μ mole/min
	$\dot{Q}_{R,max}$	Maximum rate of production of R	60 μ mole/min
	V	Volume in which the reactions occur	L
Dep.	$\dot{Q}_P(t)$	Rate of production of P from S	6.509 μ mole/min
	$\dot{Q}_I(t)$	Rate of production of endogenous I from S	1.491 μ mole/min
	$\dot{Q}_R(t)$	Rate of production of R from I	1.491 μ mole/min
	$C_S(t)$	Concentration of S	5.297 μ mole/L
	$C_I(t)$	Concentration of I	2.039 μ mole/L

Steady-state values (Dep. examples)

Model

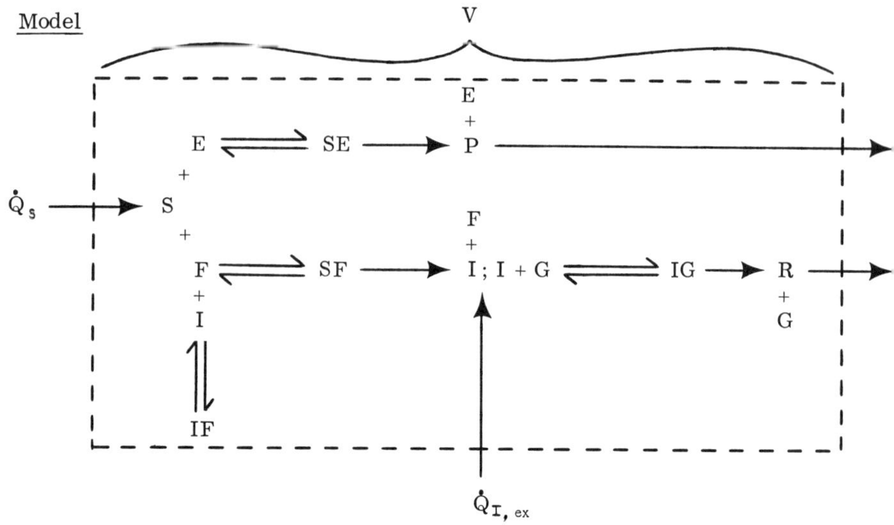

Figure 1.

The system comprises three enzymes, E, F and G which catalyze the three irreversible reactions depicted above. The pathway via F and G synthesizes a final product, R, from the substrate S. I, the last intermediate before R is synthesized, competes reversibly with S for occupation of the enzyme F at the "branch point" where a given molecule of S can enter either the synthetic path leading to R or the alternate path leading to the product P. S is supplied to the system at a constant rate of $\dot{Q}_S$. Besides the I formed from S endogenously, exogenous I may be supplied at a constant rate, $\dot{Q}_{I,\,ex}$.

Simplifying Assumptions:

1. The formation of product from each enzyme-substrate complex is irreversibl

2. When using the Michaelis-Menten formulation (as we shall in this section) it is necessary to assume that the total quantity of each enzyme remains constant. Usually it is also necessary to assume that the concentration of enzyme is small compared with the concentration of substrate. But when we confine attention to steady states, this restriction is no longer necessary.

Derivation

The Michaelis-Menten equations (1.) for the reactions in this system are

$$\dot{Q}_P(t) = \frac{\dot{Q}_{P,\,max}\, C_S(t)}{k_P + C_S(t)} \qquad 1.T,$$

$$\dot{Q}_I(t) = \frac{\dot{Q}_{I,max}\, C_S(t)}{k_I + C_S(t) + \frac{k_I}{k_{eq}} C_I(t)} = \frac{\dot{Q}_{I,max}}{\frac{k_I}{C_S(t)}\left\{1 + \frac{C_I(t)}{k_{eq}}\right\} + 1} \qquad 2.T,$$

$$\dot{Q}_R(t) = \frac{\dot{Q}_{R,max}\, C_I(t)}{k_R + C_I(t)} \qquad 3.T.$$

Conservation of S requires that

$$V \frac{dC_S(t)}{dt} = \dot{Q}_S - \dot{Q}_P(t) - \dot{Q}_I(t) \qquad 4.T.$$

Similarly, conservation of I requires that

$$V \frac{dC_I(t)}{dt} = \dot{Q}_{I,ex} + \dot{Q}_I(t) - \dot{Q}_R(t) \qquad 5.T.$$

Combine equations 1 and 4, eliminating $\dot{Q}_P(t)$

$$V \frac{dC_S(t)}{dt} = \dot{Q}_S - \dot{Q}_I(t) - \frac{\dot{Q}_{P,max}\, C_S(t)}{k_P + C_S(t)} \qquad 6.$$

In the steady state, equation 6 becomes:

$$C_S(\infty) = \frac{k_P[\dot{Q}_S - \dot{Q}_I(\infty)]}{\dot{Q}_{P,max} - [\dot{Q}_S - \dot{Q}_I(\infty)]} = \frac{k_P}{\left\{\frac{\dot{Q}_{P,max}}{\dot{Q}_S - \dot{Q}_I(\infty)} - 1\right\}} \qquad 7.$$

Combine equations 3 and 5, eliminating $\dot{Q}_R(t)$:

$$V \frac{dC_I(t)}{dt} = \dot{Q}_{I,ex} + \dot{Q}_I(t) - \frac{\dot{Q}_{R,max}\, C_I(t)}{k_R + C_I(t)} \qquad 8.$$

In the steady state, equation 8 becomes:

$$C_I(\infty) = \frac{k_R[\dot{Q}_{I,ex} + \dot{Q}_I(\infty)]}{\dot{Q}_{R,max} - [\dot{Q}_{I,ex} + \dot{Q}_I(\infty)]} = \frac{k_R}{\frac{\dot{Q}_{R,max}}{\dot{Q}_{I,ex} + \dot{Q}_I(\infty)} - 1} \qquad 9.$$

We now have three working equations, 2, 7 and 9, for which the symbol-and-arrow diagram is shown in Figure 2.

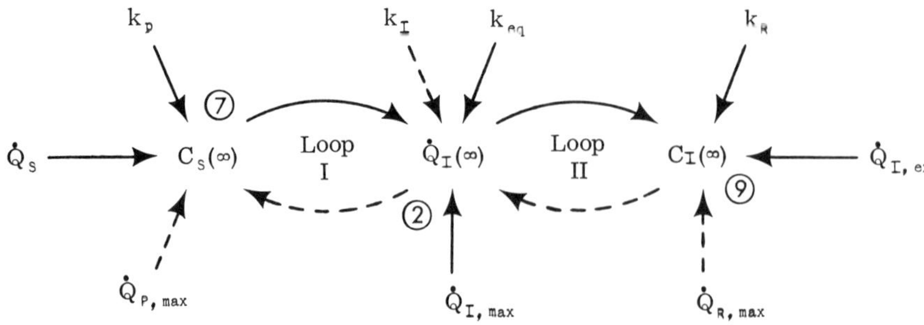

Figure 2.

In this system we are really interested only in Loop II, since only that loop is opened when feedback inhibition of F by I cannot occur (i.e., when $k_{eq} \to \infty$). (See comment 2).

To simplify further analysis, rewrite the working equations using the listed numerical values for all parameters except $\dot{Q}_s$:

$$C_s(\infty) = \frac{50}{\dfrac{68}{\dot{Q}_s - \dot{Q}_I(\infty)} - 1} \qquad 10.$$

$$\dot{Q}_I(\infty) = \frac{40}{\dfrac{45}{C_s(\infty)}\left\{1 + \dfrac{C_I(\infty)}{1}\right\} + 1} \qquad 11.$$

$$C_I(\infty) = \frac{80}{\dfrac{60}{\dot{Q}_I(\infty)} - 1} \qquad 12.$$

Equations 10, 11 and 12 describe the system in 4-space. Their linear approximations at any given operating point are

$$\Delta C_s = \beta_1 \Delta \dot{Q}_s - \beta_2 \Delta \dot{Q}_I \qquad 13.$$

$$\Delta \dot{Q}_I = \beta_3 \Delta C_s - \beta_4 \Delta C_I \qquad 14.$$

$$\Delta C_I = \beta_5 \Delta \dot{Q}_I \qquad 15.$$

where

$$\beta_1 = \left.\frac{\partial C_s(\infty)}{\partial \dot{Q}_s}\right|_{op} = \frac{3400}{[68 - \dot{Q}_s + \dot{Q}_I(\infty)]^2} \qquad 16.$$

$$\beta_2 = -\left.\frac{\partial C_s(\infty)}{\partial \dot{Q}_I(\infty)}\right|_{op} = \frac{3400}{[68 - \dot{Q}_s + \dot{Q}_I(\infty)]^2} = \beta_1 \qquad 17.$$

$$\beta_3 = \left.\frac{\partial \dot{Q}_I(\infty)}{\partial C_s(\infty)}\right|_{op} = \frac{1800\,[C_I(\infty) + 1]}{[C_s(\infty) + 45C_I(\infty) + 45]^2} \qquad 18.$$

$$\beta_4 = -\left.\frac{\partial Q_I(\infty)}{\partial C_I(\infty)}\right|_{op} = \frac{1800\, C_s(\infty)}{[C_s(\infty) + 45C_I(\infty) + 45]^2} \qquad 19.$$

$$\beta_5 = \left.\frac{\partial C_I(\infty)}{\partial \dot{Q}_I(\infty)}\right|_{op} = \frac{4800}{[60 - \dot{Q}_I(\infty)]^2} \qquad 20.$$

The block diagram corresponding to equations 13, 14 and 15 is

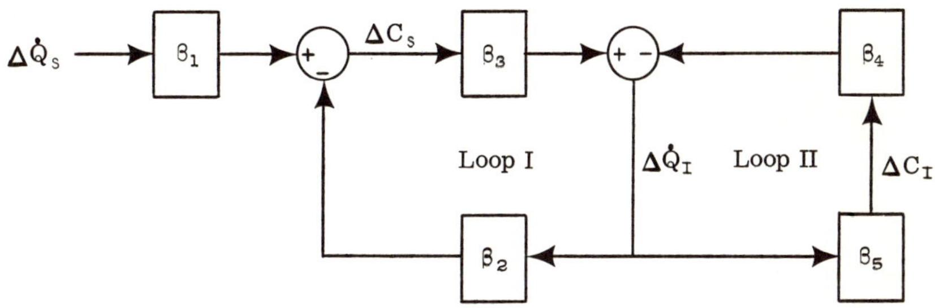

Figure 3.

We want to find the gain due to feedback loop II with respect to the influence of $\Delta\dot{Q}_s$ upon $\Delta\dot{Q}_I$. In the closed-loop condition, simultaneous solution of equations 13, 14 and 15 gives

$$\Delta\dot{Q}_I = \frac{\beta_1\beta_3}{1 + \beta_2\beta_3 + \beta_4\beta_5}\,\Delta\dot{Q}_s \qquad 21.$$

In the open-loop condition (where $k_{eq} = \infty$ so that $\dot{Q}_I(\infty)$ is no longer influenced by $C_I(\infty)$ and β_4 is thus zero),

$$\Delta\dot{Q}_I = \frac{\beta_1\beta_3}{1 + \beta_2\beta_3}\,\Delta\dot{Q}_s \qquad 22.$$

From equations 21 and 22,

$$G_{fb,II} = \frac{\left.\dfrac{d\dot{Q}_I}{d\dot{Q}_s}\right|_{open}}{\left.\dfrac{d\dot{Q}_I}{d\dot{Q}_s}\right|_{closed}} - 1 = \frac{\beta_4\beta_5}{1 + \beta_2\beta_3} \qquad 23.$$

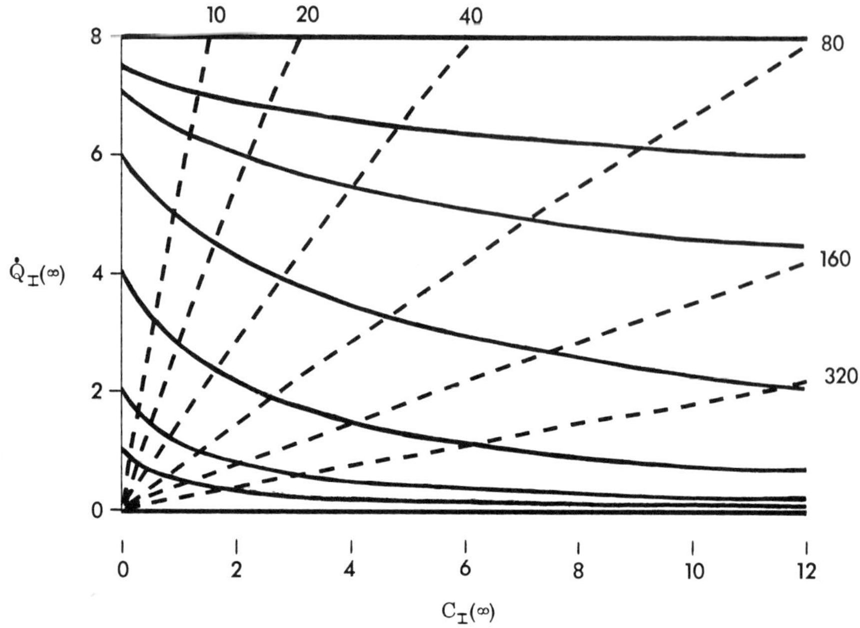

Figure 4. Broken lines, $C_I(\infty)$ as a function of $\dot{Q}_I(\infty)$ (equation 9) for the values of k_R indicated in the top and right margins. Solid lines, $\dot{Q}_I(\infty)$ as a function of $C_I(\infty)$ (equations 2 and 7) for various values of $\dot{Q}_{P,max}$. Reading in order from the top line where $\dot{Q}_I(\infty) = 8\ \mu$mole/min to the bottom line where $\dot{Q}_I(\infty) = 0\ \mu$mole/min, $\dot{Q}_{P,max} = 0, 3, 6, 15, 44, 133, 310$, and $\infty\ \mu$mole/min.

Substituting in equation 23 the values of the β's from equations 17 through 20 evaluated at the operating point specified in the example,

$$G_{fb,II} = \frac{[0.4725]\,[1.4025]}{1 + [0.8992]\,[0.2712]} = 0.533 \qquad 24.$$

In Figure 4, equation 9 and an equation obtained by combining 2 and 7 with the elimination of $C_S(\infty)$, have been used to construct curves showing the relationship between $\dot{Q}_I(\infty)$ and $C_I(\infty)$ for various values of $\dot{Q}_{P,max}$ and k_R, the other parameters retaining their listed values. From the slopes of the curves at the various operating points corresponding to the intersections, it is evident that for any given k_R, there is some optimum value of $\dot{Q}_{P,max}$ at which $G_{fb,II}$ is maximal. From that maximum, $G_{fb,II}$ decreases towards zero as $\dot{Q}_{P,max}$ approaches either zero (no conversion of S to P via the alternate pathway; see comment 3) or infinity (all of the S is converted to P via the alternate pathway).

Comments

1. Since I do not know of any experimental observations of the kind of system here described, the exemplary values are completely arbitrary. More precise analysis of such an artificial system is hardly warranted. However, in a series of somewhat casual attempts, I was unable to find reasonable values of the parameters which would give high gain. It seems likely that this kind of feedback inhibition would not be very effective as a regulatory mechanism.

2. In the present system, only loop II can be ascribed to the inhibitory effect of the product, I, upon the enzyme, F. If the same series of steps is applied to a system without any feedback inhibition (i.e., with $k_{eq} = \infty$), Loop I will still be present

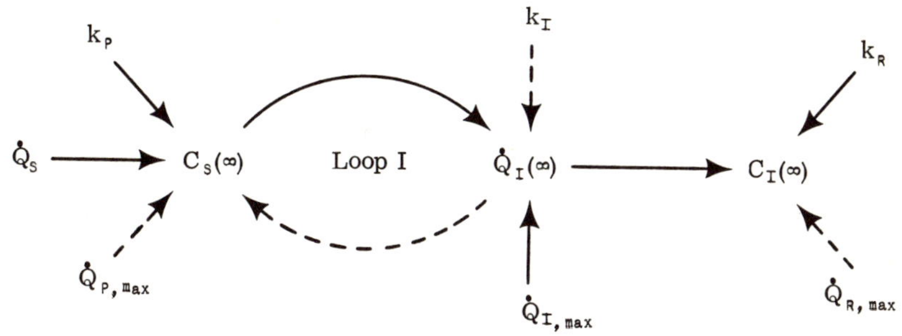

Figure 5.

3. The present model system is unrealistically simple. In most actual synthetic pathways where feedback inhibition has been identified, the inhibitory product is the last small molecule in the synthetic chain, and is itself used as a building block for some kind of macromolecule. The enzyme which it inhibits is often several irreversible steps back in the synthetic chain, and the inhibitory effect is not usually of the reversible competitive type illustrated here but is allosteric (See Comment 4). Nevertheless, the present naive model illustrates one of the essential characteristics of feedback control of a synthetic pathway: the enzyme which is inhibited by the end-product is always located at a branch point (usually the last one before a series of irreversible steps) so that the substrate has an alternate pathway available to it. If an alternate pathway were not available, substrate would accumulate until its concentration became high enough so that $\dot{Q}_I(\infty)$ would equal $\dot{Q}_s$ in spite of the inhibitory effect of I. Feedback control would than be of the pure derivative type (Section 4.5) and would therefore not influence the steady-state values, $\dot{Q}_I(\infty)$ and $\dot{Q}_R(\infty)$.

4. In allosteric inhibition, the last small molecule synthesized by an irreversible series of steps inhibits a multiunit enzyme at a branch point not by competing with substrate (which it usually does not resemble structurally) for the same site on the enzyme but by combining with the enzyme, independently of substrate, at a different site, so as to reduce the rate at which the enzyme forms product from substrate. The kinetic behavior of such a system is quite complex (2, 3, 4).

References

1. For a derivation of the Michaelis-Menten equations used here, see Sections 11-2 and 11-5 in Riggs, D.S.: The Mathematical Approach to Physiological Problems. Williams and Wilkins, Baltimore, 1963.

2. Monod, J., Wyman, J., and Changeux, J.P.: On the nature of allosteric transitions: a plausible model. J. molec. Biol., 12, 88-118, 1965.

3. Changeux, J.P., and Rubin, M.M.: Allosteric interactions in aspartate transcarbamylase. III interpretation of experimental data in terms of the model of Monod, Wyman and Changeux. Biochemistry, 7, 553-561, 1968. (See also the two papers preceding this one.)

4. Sweeny, J.R., and Fisher, J.R.: An alternative to allosterism and cooperativity in the interpretation of enzyme kinetic data. Biochemistry 7, 561-565, 1968.

For a descriptive account of feedback control of metabolic pathways, see

Changeux, J.P.: The control of biochemical reactions. Scientific American, 212, 36-45, April 1965.

REGULATION OF BODY TEMPERATURE

NEGATIVE AND POSITIVE

Type	Symbol	Definition	Example (Adult human)
Ind.	T_{amb}	Ambient temperature	27.69° C
	v_{air}	Velocity of the air	0.192 km/hr
Par.	$\dot{q}_{ex}$	Heat production by exercise (= all non-shivering, non-resting heat production)	22 kcal/hr
	$\dot{q}_{basal, 37}$	Basal heat production when $T_{core}(\infty) = 37°$ C	70 kcal/hr
	$\dot{q}_{lung}$	Rate of heat loss in expired air	6 kcal/hr
	$\dot{q}_{insen}$	Rate of heat loss due to "insensible perspiration"	6 kcal/hr
	A_{surf}	Surface area of body	1.86 m^2
	A_{rad}	Area contributing to heat loss by radiation	1.50 m^2
	A_{conv}	Area contributing to heat loss by convection	1.70 m^2
	k_{rad}	Radiative heat loss constant	7.00 kcal/m^2•hr•° C
	k_{cond}	Thermal conductivity, core to skin, when $\dot{V}_{skin}(\infty) = 0$	8.71 kcal/m^2•hr•° C
	k_{blood}	[Thermal capacity] [Density] for blood	1.00 kcal/liter • °C
	F_{cc}	Fraction of $\dot{V}_{skin}(\infty)$ rendered ineffective by countercurrent heat exchange between arteries and veins	zero
	$T_{c, crit}$	Critical core temperature. (= "SET POINT")	37.00° C
	$T_{sk, crit}$	Critical skin temp. for maximum $\dot{q}_{shiv}(\infty)$	21.00° C
	C_{core}	Thermal capacitance of core	58 kcal/° C
	C_{skin}	Thermal capacitance of skin	

Type	Symbol	Definition	Example (Adult human)
	k_{conv}	Depend. par. (Eq. 15). Convective heat loss constant	5.59 kcal/m^2•hr•° C
	$R_{sk,a}$	Depend. par. (Eq. 6) Resistance to heat flow between skin and ambient media	0.05 hr•° C/kcal
Dep.	$T_{core}(\infty)$	Core temperature (measured at tympanic membrane)	37.00° C
	$T_{skin}(\infty)$	Skin temperature	31.69° C
	$\dot{q}_{prod}(\infty)$	Total heat production	92.0 kcal/hr
	$\dot{q}_{basal}(\infty)$	Basal heat production	70.0 kcal/hr
	$\dot{q}_{shiv}(\infty)$	Production of heat by shivering	0 kcal/hr
	$\dot{q}_{evap}(\infty)$	Evaporative heat loss from skin	6.0 kcal/hr
	$\dot{q}_{sweat}(\infty)$	Heat loss due to evaporation of sweat	0 kcal/hr
	$\dot{V}_{skin}(\infty)$	Blood flow to skin <u>in excess</u> of minimal flow	0 liters/hr
	$R_{c,sk}(\infty)$	Total resistance to heat flow between core and skin	0.0617 hr•° C/kcal
	$a(\infty)$	Empirical variable defined by equation 11	kcal/hr
	$k_{shiv}(\infty)$	Empirical variable defined by equation 12	° C

Model

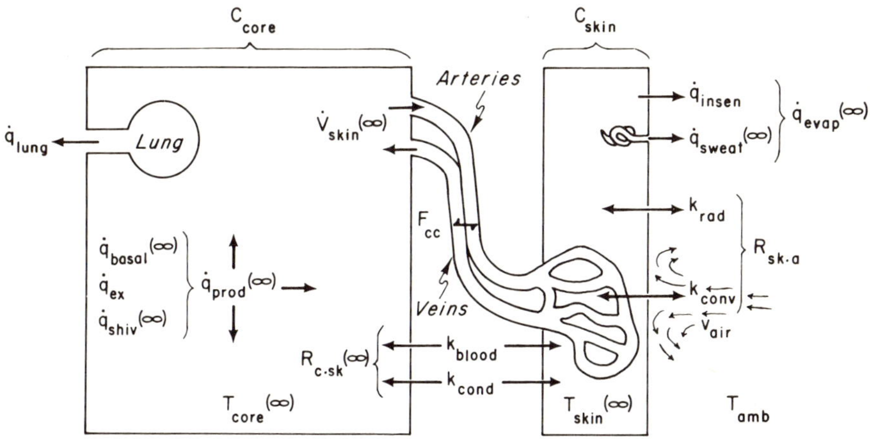

Figure 1.

The body is assumed to consist of two heat capacitances: a relatively large central core, surrounded by a relatively small amount of skin. Heat flow between core and skin is due either to conduction (which is here defined to include heat transfer by the minimal blood flow which remains during maximal cutaneous vasoconstriction) or to convection due to heat transfer by the supraminimal flow, $\dot{V}_{skin}(\infty)$. The fractional effectiveness of blood flow in heat transfer between core and skin may be reduced by counter-current heat exchange between arteries and veins (F_{cc}). Heat flow between skin and the environment is due to convection (which depends on air velocity) and to radiation. The direction of heat transfer, and the quantities of heat transferred by these mechanisms depend upon the temperature differences between core and skin, and between skin and air. In contrast, heat produced in the core by metabolic processes ($\dot{q}_{prod}(\infty)$), heat lost from the core by vaporization of water in the respiratory passages ($\dot{q}_{lung}$) and heat lost from the skin by vaporization of "insensible water" ($\dot{q}_{insen}$) and by sweat ($\dot{q}_{sweat}(\infty)$) are regarded as independent of temperature gradients. In the steady state, $T_{core}(\infty)$ and $T_{skin}(\infty)$ remain constant so that there is no net change in the quantity of heat stored in C_{core} or C_{skin}. These heat capacitances may therefore be neglected and the system may be treated as a dissipative network. (See Figure 2.).

Not illustrated in Figure 1 are the feedback control mechanisms which are activated principally by: (1.) a region in the anterior hypothalamus which is sensitive to $T_{core}(\infty)$, and (2.) cutaneous thermoreceptors which are sensitive to $T_{skin}(\infty)$. When $T_{core}(\infty)$ is higher than $T_{c,\,crit}$, the critical (= "set-point") temperature, the hypothalamic sensor initiates excitatory sudomotor impulses to the sweat glands, and decreases vasoconstrictor impulses to the cutaneous blood vessels. In contrast, the intensity of shivering depends in a complicated way (see equations 10, 11 and 12) upon a reduction of both core temperature and skin temperature. This simplified account neglects a variety of other effects whose contributions to thermal regulation in man have not been studied quantitatively.

Simplifying Assumptions :

1. All sweat evaporates, and $\dot{q}_{lung}$ and $\dot{q}_{insen}$ are constant; (i.e., the effects of changes in humidity are assumed to be negligible.)

2. $T_{c, crit}$ is assumed to be 37° C, and, for economy of notation, "37" will be used in place of $T_{c, crit}$ in the derivation.

3. Heat loss from skin to environment by conduction is negligible.

4. The effects of local heating or cooling of the skin upon local blood flow are negligible.

5. $\dot{q}_{prod}(\infty) - \dot{q}_{lung}$ is always positive. (See comment 6).

6. $T_{skin}(\infty) - T_{amb}$ is the same for radiation as for convection.

7. The increment of basal heat production per unit increment of T_{core} is assumed to be constant. Actually, the Q_{10} for man (i.e., the factor by which $q_{basal}(\infty)$ would be increased by a 10° C rise in body temperature) has been estimated as 2.3 (1). Over a wide range of body temperatures, therefore, equation 9 should be replaced by the following version of the Arrhenius equation,

$$\dot{q}_{basal}(\infty) = \dot{q}_{basal, 37}\, \epsilon^{8000\left[\frac{1}{310} - \frac{1}{T_{core}(\infty) + 273}\right]} \quad \text{kcal/hr}$$

8. Radiative heat loss is directly proportional to the temperature difference, $T_{skin}(\infty) - T_{amb}$. (Actually, according to the Stefan-Boltzmann law of black-body radiation, it should be directly proportional to $[T_{skin}(\infty) + 273]^4 - [T_{amb} + 273]^4$, i.e. to the difference between the fourth powers of the absolute temperatures of radiating body and environment.)

Derivation

The first part of the derivation is most easily understood in terms of an electrical analog of the model described above. (Figure 2). Although electrical elements are depicted in Figure 2, the variables shown are the analogous thermal quantities, so the equations derived will be directly applicable to the present system. For simplicity, disregard the thermal capacitances of core and skin which are indicated by broken lines in Figure 2, and consider at once the steady-state situation. Since the three source elements are all T-variable (here, heat) sources it is clear that for given thermal resistances and a given ambient temperature the dependent variables are $T_{core}(\infty)$ and $T_{skin}(\infty)$. By inspection, the node equation at the $T_{core}(\infty)$ node is,

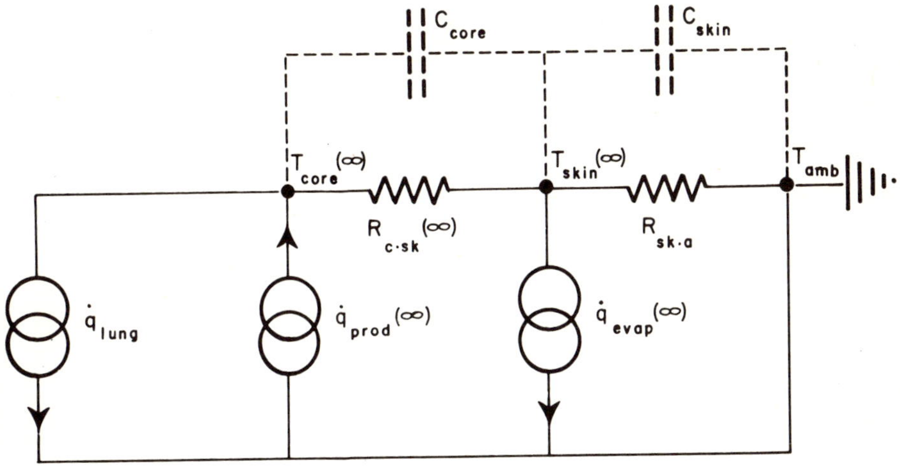

Figure 2.

$$\dot{q}_{prod}(\infty) = \dot{q}_{lung} + [T_{core}(\infty) - T_{skin}(\infty)]\, R^{-1}_{c,\,sk}(\infty) \qquad 1.T$$

The node equation at the $T_{skin}(\infty)$ node is

$$[T_{core}(\infty) - T_{skin}(\infty)]\, R^{-1}_{c,\,sk}(\infty) = \dot{q}_{evap}(\infty) + [T_{skin}(\infty) - T_{amb}]\, R^{-1}_{sk,\,a} \qquad 2.T$$

Solving equations 1 and 2 simultaneously for $T_{core}(\infty)$ and $T_{skin}(\infty)$,

$$T_{core}(\infty) = T_{amb} + R_{sk,\,a}[\dot{q}_{prod}(\infty) - \dot{q}_{lung} - \dot{q}_{evap}(\infty)] + R_{c,\,sk}(\infty)[\dot{q}_{prod}(\infty) - \dot{q}_{lung}] \qquad 3.$$

$$T_{skin}(\infty) = T_{amb} + R_{sk,\,a}[\dot{q}_{prod}(\infty) - \dot{q}_{lung} - \dot{q}_{evap}(\infty)] \qquad 4.$$

By definition,

$$R^{-1}_{c,\,sk}(\infty) = A_{surf}\{k_{cond} + k_{blood}[1 - F_{cc}]\dot{V}_{skin}(\infty)\} \qquad 5.T$$

where $k_{cond}A_{surf}$ is the thermal conductance between core and skin when cutaneous blood flow is minimal ($\dot{V}_{skin}(\infty) = 0$), and $k_{blood}A_{surf}[1 - F_{cc}]\dot{V}_{skin}(\infty)$ is the thermal conductance between core and skin due to blood flow greater than minimal. Also by definition,

$$R^{-1}_{sk,\,a} = [k_{rad}A_{rad} + k_{conv}A_{conv}] \qquad 6.T$$

where $k_{rad}A_{rad}$ is the thermal conductance between skin and environment due to transfer of heat by radiation, and $k_{conv}A_{conv}$ is the thermal conductance between skin and ambient fluid due to transfer of heat by convection. Further by definition,

$$\dot{q}_{prod}(\infty) = \dot{q}_{basal}(\infty) + \dot{q}_{shiv}(\infty) + \dot{q}_{ex} \qquad 7.T$$

$$\dot{q}_{evap}(\infty) = \dot{q}_{insen} + \dot{q}_{sweat}(\infty) \qquad 8.T$$

The basal rate of heat production is observed to increase by roughly 10% per °C increase in $T_{core}(\infty)$ above 37.0° C (1). By assumption 7,

$$\dot{q}_{basal}(\infty) = \dot{q}_{basal,37}\{1 + 0.1[T_{core}(\infty) - 37]\} \text{ kcal/hr} \qquad 9.E$$

Heat produced by shivering (in man) is observed to depend upon both skin temperature and core temperature. The following empirical equations are suggested by the data in Reference 2 , Figure 13, page 649.

$$\dot{q}_{shiv}(\infty) = a\left\{1 - \left[\frac{T_{skin}(\infty) - T_{skin,crit}}{k_{shiv}}\right]\right\}^2 \text{ kcal/hr; } \dot{q}_{shiv}(\infty) \geq 0 \qquad 10.E$$

where

$$a = 252\left\{1 - \epsilon^{-2.36[37.0 - T_{core}(\infty)]}\right\} \text{kcal/hr} \qquad 11.E$$

and

$$k_{shiv} = 11.25\,[37.0 - T_{core}(\infty)] + 2.5\left\{1 - \epsilon^{-20[37.0 - T_{core}(\infty)]}\right\} {}^\circ C \qquad 12.E$$

It is also observed empirically (Reference 2 , Figure 15, p. 651) that

$$\dot{q}_{sweat}(\infty) = 485[T_{core}(\infty) - 37.0] \text{ kcal/hr; } T_{core}(\infty) \geq 37.0^\circ C \qquad 13.E$$

and (Reference 2 , Figure 20, page 654) that

$$k_{blood}[1 - F_{cc}]\dot{V}_{skin}(\infty) = 113[T_{core}(\infty) - 37.0] \text{ kcal/m}^2\cdot\text{hr}\cdot{}^\circ\text{C}; \quad T_{core}(\infty) \geq 37.0^\circ C \qquad 14.E$$

or,

$$\dot{V}_{skin}(\infty) = 113\,\frac{T_{core}(\infty) - 37.0}{k_{blood}[1 - F_{cc}]}\;\frac{L}{m^2\cdot hr} \qquad 14.a$$

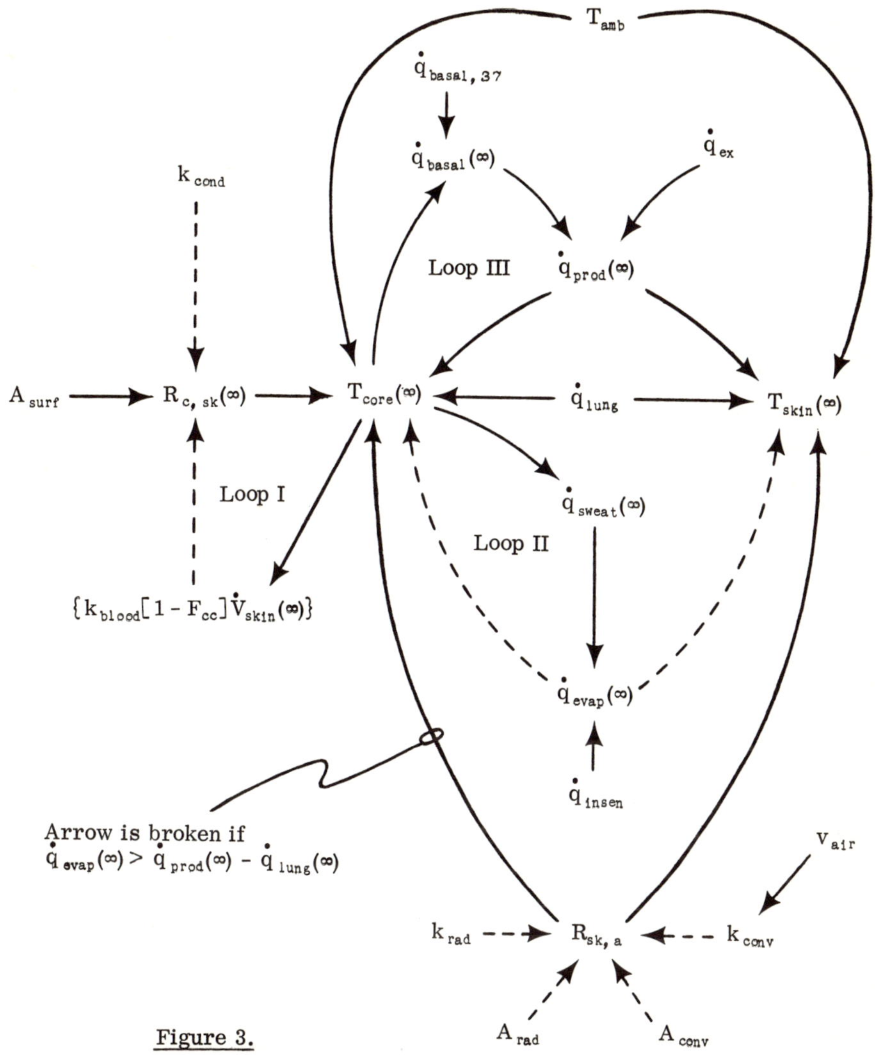

Figure 3.

Finally, it has been observed (reference 3, Figure 4, page 1053) that

$$k_{conv} = 4 + 4.28\, v_{air}^{0.6} \;\; \text{kcal/m}^2 \cdot \text{hr} \cdot {}^\circ\text{C} \qquad 15.\text{E}$$

Consider first the steady-state response to a warm environment where $T_{core}(\infty) \geq 37^\circ$. Under these circumstances, $\dot{q}_{shiv}(\infty)$ = zero, and the symbol-and-arrow diagram corresponding to equations 3 through 9, and 13 through 15 is shown in Figure 3.

Three feedback loops are evident: Loop I (negative) which depends upon regulation of cutaneous blood flow, Loop II (negative) which depends upon regulation of sweating, and Loop III (positive) which depends upon the increase in metabolism caused by an increase in core temperature.

For simplicity, neglect $T_{skin}(\infty)$ (a wholly dependent variable in Figure 3), and let all parameters except T_{amb} and v_{air} assume the values listed in the example. Combine equations 3, 5, 6, 8 and 15, deleting $R_{c,sk}(\infty)$, $\dot{q}_{evap}(\infty)$, $R_{sk,a}$, and k_{conv}.

$$T_{core}(\infty) = T_{amb} + \frac{\dot{q}_{prod}(\infty) - \dot{q}_{sweat}(\infty) - 12}{17.3 + 7.28\, v_{air}^{0.6}} + \frac{\dot{q}_{prod}(\infty) - 6}{16.2 + 1.86\, \dot{V}_{skin}(\infty)} \ ^\circ C \qquad 16.$$

Combine equations 7 and 9, deleting $\dot{q}_{basal}(\infty)$.

$$\dot{q}_{prod}(\infty) = 7\, T_{core}(\infty) - 167 \text{ kcal/hr} \qquad 17.$$

The symbol-and-arrow diagram redrawn according to equations 13, 14, 16 and 17 is shown in Figure 4.

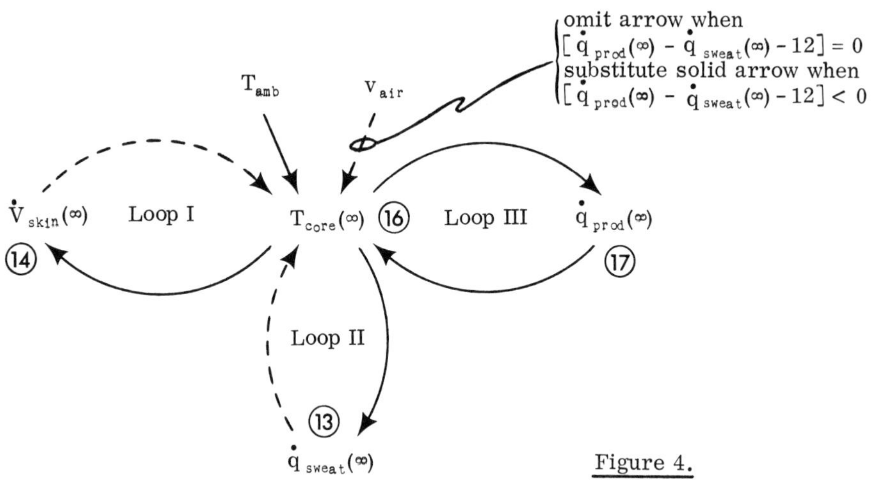

Figure 4.

The corresponding linear equations for the tangent hyperplanes about an operating point are,

$$\Delta T_{core} = \beta_1\, \Delta T_{amb} - \beta_2\, \Delta v_{air} - \beta_3\, \Delta \dot{V}_{skin} - \beta_4\, \Delta \dot{q}_{sweat} + \beta_5\, \Delta \dot{q}_{prod} \ ^\circ C \qquad 18.$$

$$\Delta \dot{V}_{skin} = \beta_6\, \Delta T_{core} \quad L/m^2 \cdot hr \qquad 19.$$

$$\Delta \dot{q}_{sweat} = \beta_7\, \Delta T_{core} \quad \text{kcal/hr} \qquad 20.$$

$$\Delta \dot{q}_{prod} = \beta_8 \, \Delta T_{core} \quad \text{kcal/hr} \qquad 21.$$

where:

$$\beta_1 = \left. \frac{\partial T_{core}(\infty)}{\partial T_{amb}} \right|_{op} = 1$$

$$\beta_2 = - \left. \frac{\partial T_{core}(\infty)}{\partial v_{air}} \right|_{op} = \frac{4.37[\dot{q}_{prod} - \dot{q}_{sweat}(\infty) - 12]}{v_{air}^{0.4}[17.3 + 7.28\, v_{air}^{0.6}]^2} \quad \frac{°C \cdot hr}{km}$$

$$\beta_3 = - \left. \frac{\partial T_{core}(\infty)}{\partial \dot{V}_{skin}(\infty)} \right|_{op} = \frac{1.86[\dot{q}_{prod}(\infty) - 6]}{[16.2 + 1.86\, \dot{V}_{skin}(\infty)]^2} \quad \frac{m^2 \cdot hr \cdot °C}{L}$$

$$\beta_4 = - \left. \frac{\partial T_{core}(\infty)}{\partial \dot{q}_{sweat}(\infty)} \right|_{op} = \frac{1}{17.3 + 7.28\, v_{air}^{0.6}} \quad \frac{hr \cdot °C}{kcal}$$

$$\beta_5 = \left. \frac{\partial T_{core}(\infty)}{\partial \dot{q}_{prod}(\infty)} \right|_{op} = \frac{1}{17.3 + 7.28\, v_{air}^{0.6}} + \frac{1}{16.2 + 1.86\, \dot{V}_{skin}(\infty)} \quad \frac{hr \cdot °C}{kcal}$$

$$\beta_6 = \left. \frac{d\dot{V}_{skin}(\infty)}{dT_{core}(\infty)} \right|_{op} = 113 \ L/m^2 \cdot hr \cdot °C$$

$$\beta_7 = \left. \frac{d\dot{q}_{sweat}(\infty)}{dT_{core}(\infty)} \right|_{op} = 485 \ \frac{kcal}{hr \cdot °C}$$

$$\beta_8 = \left. \frac{d\dot{q}_{prod}(\infty)}{dT_{core}(\infty)} \right|_{op} = 7 \ \frac{kcal}{hr \cdot °C} \qquad 22.$$

The block diagram corresponding to equations 18 through 21 is depicted in Figure 5. Solving equations 18 through 21 simultaneously for ΔT_{core},

$$\Delta T_{core} = \frac{\beta_1 \Delta T_{amb} - \beta_2 \Delta v_{air}}{1 + \beta_3 \beta_6 + \beta_4 \beta_7 - \beta_5 \beta_8} \quad °C \qquad 23.$$

where $\beta_3 \beta_6$ is the [OLG](j0) for Loop I (vasodilation), $\beta_4 \beta_7$ is the [OLG](j0) for Loop II (sweating), and $\beta_5 \beta_8$ is the [OLG](j0) for the positive Loop III. The gain due to Loops I and II together is

$$G_{fb,I+II} = \left| \frac{dT_{core}/dT_{amb} \,|_{I,II,\ open}}{dT_{core}/dT_{amb} \,|_{closed}} - 1 \right| ; \quad \Delta v_{air} = 0$$

or, differentiating the appropriate open-loop and closed-loop versions of equation 23,

$$G_{fb,I+II} = \frac{\beta_3 \beta_6 + \beta_4 \beta_7}{1 - \beta_5 \beta_8} ; \quad 0 \le \beta_5 \beta_8 < 1 \qquad 24.$$

In Figure 8, the values of the various gains have been plotted against T_{amb}, the other independent variables and parameters retaining the values listed in the example.

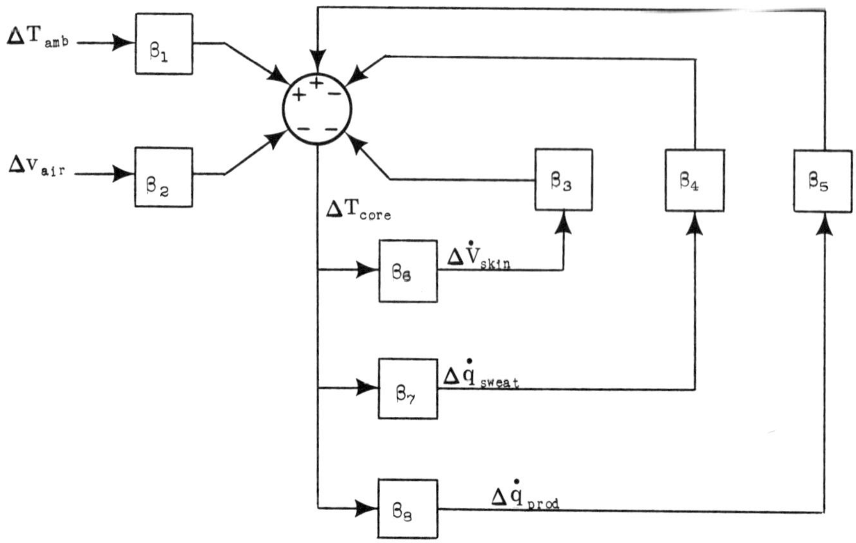

Figure 5.

Consider second the steady-state response to a cool environment where $T_{core}(\infty) \leq 37^\circ$ C. Under these circumstances, $\dot{q}_{sweat}(\infty)$ and $\dot{V}_{skin}(\infty)$ are both zero, and the symbol-and-arrow diagram of Figure 3 would have to be redrawn so that it would correspond to equations 3 through 10, and equation 15. Since $T_{skin}($ is now a member of one of the two feedback loops concerned with shivering, it cannot be disregarded. To simplify further analysis, let all parameters except T_{amb} and v_{air} assume the values listed in the example. Combine equations 3, 5, 6, 8 and 15, eliminating $R_{c,sk}(\infty)$, $R_{sk,a}$, k_{conv}, $\dot{q}_{evap}(\infty)$.

$$T_{core}(\infty) = T_{amb} + \frac{\dot{q}_{prod}(\infty) - 12}{17.3 + 7.28\, v_{air}^{0.6}} + \frac{\dot{q}_{prod}(\infty) - 6}{16.2} \ ^\circ C \qquad 25.$$

Combine equations 4, 6, 8 and 15, eliminating $R_{sk,a}$, $\dot{q}_{evap}(\infty)$, and k_{conv}.

$$T_{skin}(\infty) = T_{amb} + \frac{\dot{q}_{prod}(\infty) - 12}{17.3 + 7.28\, v_{air}^{0.6}} \ ^\circ C \qquad 26.$$

Combine equations 7 and 9, deleting $\dot{q}_{basal}(\infty)$.

$$\dot{q}_{prod}(\infty) = 7T_{core}(\infty) + \dot{q}_{shiv}(\infty) - 167 \text{ kcal/hr} \qquad 27.$$

For convenience, retain equations 10, 11 and 12 as separate entities. The symbol-and-arrow diagram corresponding to equations 10, 11, 12, 25, 26 and 27 is shown in Figure 6.

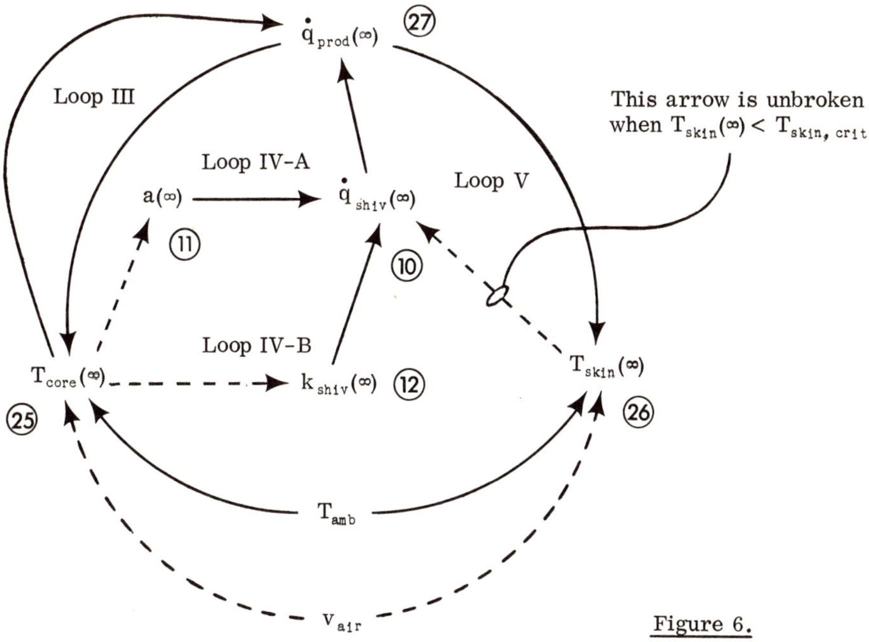

Figure 6.

The equations for the tangent hyperplanes about an operating point are:

$$\Delta T_{core} = \beta_1 \Delta T_{amb} - \beta_9 \Delta v_{air} + \beta_{10} \Delta \dot{q}_{prod} \ ^\circ C \qquad 28.$$

$$\Delta T_{skin} = \beta_1 \Delta T_{amb} - \beta_9 \Delta v_{air} + \beta_{11} \Delta \dot{q}_{prod} \ ^\circ C \qquad 29.$$

$$\Delta \dot{q}_{shiv} = \beta_{12} \Delta a - \beta_{13} \Delta T_{skin} + \beta_{14} \Delta k_{shiv} \quad \text{kcal/hr} \qquad 30.$$

$$\Delta a = -\beta_{15} \Delta T_{core} \quad \text{kcal/hr} \qquad 31.$$

$$\Delta k_{shiv} = -\beta_{16} \Delta T_{core} \ ^\circ C \qquad 32.$$

$$\Delta \dot{q}_{prod} = \beta_8 \Delta T_{core} + \beta_{17} \Delta \dot{q}_{shiv}, \quad \text{kcal/hr} \qquad 33.$$

where

$$\beta_9 = -\left.\frac{\partial T_{core}(\infty)}{\partial v_{air}}\right|_{op} = \frac{4.37[\dot{q}_{prod}(\infty) - 12]}{v_{air}^{0.4}[17.3 + 7.28\, v_{air}^{0.6}]^2} \ \frac{^\circ C}{\text{km/hr}}\ ; \ v_{air} > 0$$

$$\beta_{10} = \left.\frac{\partial T_{core}(\infty)}{\partial \dot{q}_{prod}(\infty)}\right|_{op} = \frac{1}{17.3 + 7.28\, v_{air}^{0.6}} + 0.0617 \ \frac{^\circ C}{\text{kcal/hr}}$$

$$\left.\begin{aligned}
\beta_{11} &- \frac{\partial T_{skin}(\infty)}{\partial \dot{q}_{prod}(\infty)}\bigg|_{op} = \frac{1}{17.3 + 7.28\, v_{air}^{0.6}} \quad \frac{{}^{\circ}C}{kcal/hr} \\
\beta_{12} &= \frac{\partial \dot{q}_{shiv}(\infty)}{\partial a(\infty)}\bigg|_{op} = 1 - \left\{\frac{T_{skin}(\infty) - T_{skin,\,crit}}{k_{shiv}(\infty)}\right\}^{2} \\
\beta_{13} &= -\frac{\partial \dot{q}_{shiv}(\infty)}{\partial T_{skin}(\infty)}\bigg|_{op} = \frac{2a\left[T_{skin}(\infty) - T_{skin,\,crit}\right]}{k^{2}_{shiv}(\infty)} \quad kcal/hr\cdot{}^{\circ}C \\
\beta_{14} &= \frac{\partial \dot{q}_{shiv}(\infty)}{\partial k_{shiv}(\infty)}\bigg|_{op} = \frac{2a\left[T_{skin}(\infty) - T_{skin,\,crit}\right]^{2}}{k^{3}_{shiv}(\infty)} \quad kcal/hr\cdot{}^{\circ}C \\
\beta_{15} &= -\frac{\partial a(\infty)}{\partial T_{core}(\infty)}\bigg|_{op} = 595\, \epsilon^{-2.36[37.0 - T_{core}(\infty)]} \quad kcal/hr\cdot{}^{\circ}C \\
\beta_{16} &= -\frac{\partial k_{shiv}(\infty)}{\partial T_{core}(\infty)}\bigg|_{op} = 11.25 + 50\, \epsilon^{-20[37.0 - T_{core}(\infty)]} \\
\beta_{17} &= \frac{\partial \dot{q}_{prod}(\infty)}{\partial \dot{q}_{shiv}(\infty)}\bigg|_{op} = 1
\end{aligned}\right\} \quad 34.$$

For convenience, combine equations 30, 31 and 32 by deleting Δa and Δk_{shiv}.

$$\dot{q}_{shiv} = -\left[\beta_{12}\beta_{15} + \beta_{14}\beta_{16}\right]\Delta T_{core} - \beta_{13}\,\Delta T_{skin} \quad kcal/hr \qquad 35.$$

The block diagram corresponding to equations 28, 29, 33 and 35 is shown in Figure 7. In solving equations 28, 29, 33 and 35 simultaneously for ΔT_{core}, it will be convenient to define

$$\left.\begin{aligned}
\beta_{18} &= \beta_{12}\beta_{15} + \beta_{14}\beta_{16} \\
I &= \beta_{1}\Delta T_{amb} - \beta_{9}\Delta v_{air} \\
\beta_{0} &= \beta_{10}\beta_{13}\beta_{17} \\
\beta_{III} &= \beta_{8}\beta_{10} \\
\beta_{IV} &= \beta_{10}\beta_{17}\beta_{18} \\
\beta_{V} &= \beta_{11}\beta_{13}\beta_{17}
\end{aligned}\right\} \quad 36.$$

Then the equations to be solved for ΔT_{core} may be written

$$\left.\begin{aligned}
\Delta T_{core} &= \beta_{10}\,\Delta\dot{q}_{prod} + I \\
\Delta T_{skin} &= \beta_{11}\,\Delta\dot{q}_{prod} + I \\
\Delta\dot{q}_{prod} &= \beta_{8}\,\Delta T_{core} + \beta_{17}\,\Delta\dot{q}_{shiv} \\
\Delta\dot{q}_{shiv} &= -\beta_{18}\,\Delta T_{core} - \beta_{13}\,\Delta T_{skin}
\end{aligned}\right\} \quad 37.$$

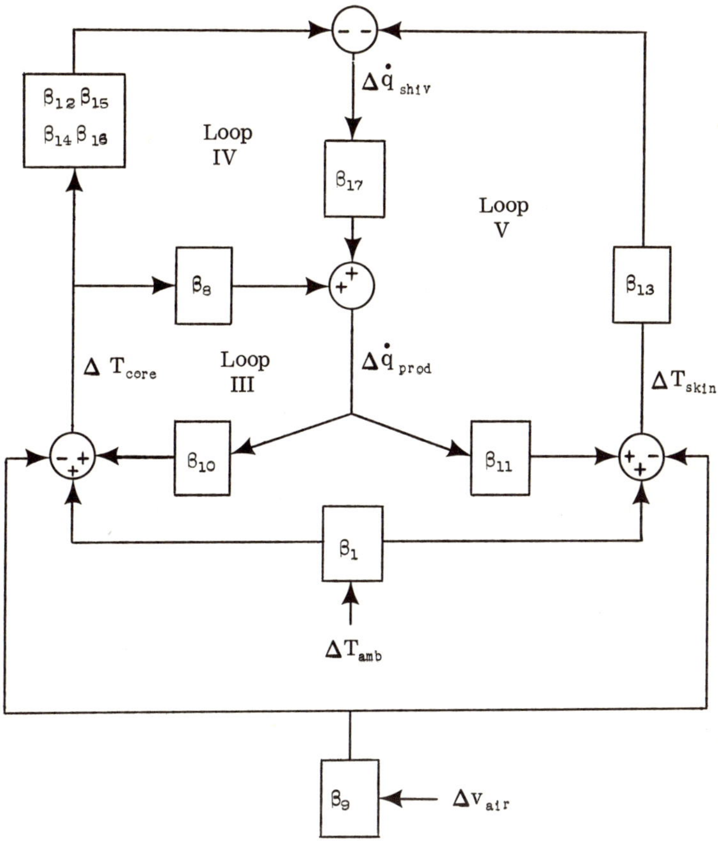

Figure 7.

Whence, eliminating first $\Delta\dot{q}_{shiv}$,

$$\Delta\dot{q}_{prod} = [\beta_8 - \beta_{17}\beta_{18}]\ \Delta T_{core} - \beta_{13}\beta_{17}\ \Delta T_{skin} \qquad 38.$$

then ΔT_{skin},

$$\Delta\dot{q}_{prod} = [\beta_8 - \beta_{17}\beta_{18}]\ \Delta T_{core} - \beta_{V}\ \Delta\dot{q}_{prod} - \beta_{13}\beta_{17} I \qquad 39.$$

or,

$$\dot{q}_{prod} = \frac{\beta_8 - \beta_{17}\beta_{18}}{1 + \beta_{V}}\ \Delta T_{core} - \frac{\beta_{13}\beta_{17}}{1 + \beta_{V}}\ I \qquad 40.$$

and, finally, eliminating $\Delta\dot{q}_{prod}$ and solving for ΔT_{core},

$$\Delta T_{core} = \frac{\beta_{III} - \beta_{IV}}{1 + \beta_{V}}\ \Delta T_{core} + \left\{1 - \frac{\beta_0}{1 + \beta_{V}}\right\}\ I \qquad 41.$$

or,

$$\Delta T_{core} = \left\{\frac{1 - \beta_0 + \beta_V}{1 + \beta_V + \beta_{IV} - \beta_{III}}\right\} [\beta_1 \, \Delta T_{amb} - \beta_9 \, \Delta v_{air}] \qquad 42.$$

which specifies the effect of the inputs, ΔT_{amb} and Δv_{air} on ΔT_{core} with all loops closed.

Let $\Delta v_{air} = 0$. Then by differentiating equation 42 with respect to ΔT_{amb}, we obtain

$$\left.\frac{dT_{core}}{dT_{amb}}\right|_{closed} = \frac{\beta_1[1 - \beta_0 + \beta_V]}{1 + \beta_V + \beta_{IV} - \beta_{III}} \qquad 43.$$

To calculate the influence of shivering on T_{core}, open loops IV and V by letting $\beta_{13} = \beta_{18} = 0$. Then $\beta_0 = \beta_{IV} = \beta_V = 0$, and

$$\frac{dT_{core}}{dT_{amb}}\bigg|_{IV, V\, open} = \frac{\beta_1}{1 - \beta_{III}} \qquad 44.$$

$RD_{IV,V}$, the ratio of these two derivatives, is

$$RD_{IV,V} = \frac{1}{1 - \beta_0 + \beta_V}\left\{1 + \frac{\beta_{IV} + \beta_V}{1 - \beta_{III}}\right\} \qquad 45.$$

In Figure 9 this ratio has been plotted as a function of T_{amb}, the other parameters retaining the values listed in the example. Notice that throughout a considerable range of T_{amb}, the effect of feedback is reversive. (See Comment 3).

Comments

1. The multiple feedback loops involved in temperature regulation constitute by all odds the most complicated system discussed in the present book. The ability of birds and mammals to maintain a practically constant body temperature in spite of wide fluctuations in ambient temperature depends upon an astonishing variety of behavioral and physiological mechanisms, only a few of which are fully exploited by any given species. The present analysis deals exclusively with human physiology; it is based primarily upon the elegant studies of Benzinger and his collaborators, (2, 4). A quite different set of feedback equations would be required to describe thermal homeostasis in other species, or, for that matter, in man under conditions less simple than those of Benzinger's experiments.

I have made no attempt to justify the "normal" values of the parameters exemplified here; I claim no more than that, by present criteria, they are reasonable. Most of them have been taken from, or calculated from standard texts.

2. In Figure 8, the negative of the open-loop transfer function of each of the five feedback loops, considered as isolated entities, has been plotted against ambient temperature. While the high gains due to sweating, cutaneous vaso-

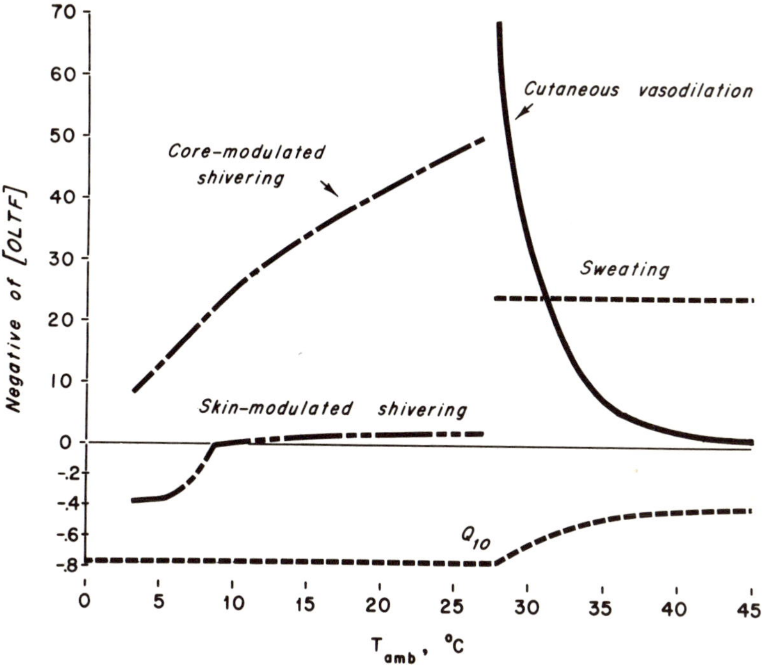

Figure 8. The negative of the open-loop transfer function of each individual loop plotted against ambient temperature.

Notice that below zero the values of -[OLTF] are plotted on a much larger scale than above zero. These values represent positive feedback.

---

dilation, and shivering are impressive enough by themselves, they indicate but dimly the extraordinary capacity of the system as a whole to defend body temperature against changes in ambient temperature. Nevertheless, they clearly show that von Euler's previous estimate of 8-10 for the gain of the hypothalamic temperature regulating mechanism (5) is far too low.

3. The overall regulatory capacity of the present system is more realistically displayed in Figure 9, where RD, defined as

$$RD = \left.\frac{dT_{core}}{dT_{amb}}\right|_{\text{negative loops open}} \Bigg/ \left.\frac{dT_{core}}{dT_{amb}}\right|_{\text{negative loops closed}}$$

has been plotted as a function of T_{amb}. The key feature of the graph is the very large magnitude of the RD over most of the range. The large absolute values of the RD show how small the closed-loop change is compared to the open-loop change. (See also comment 4.) In region A, the combined effects of cutaneous hyperemia and sweating reduce the change in core temperature caused by a

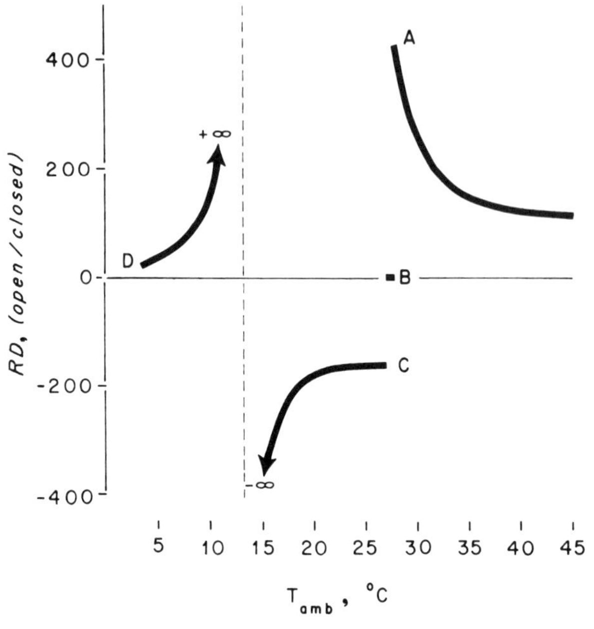

Figure 9. The ratio of derivatives for the negative loops only as a function of ambient temperature.

---

change in ambient temperature to minute fractions of what they would be without homeostatic control. In the small (and certainly artefactual) region B there is, by present assumptions, no homeostasis. In region C, body temperature actually rises slightly as ambient temperature falls, ("reversive feedback"), while in region D, body temperature falls faster and faster as T_{amb} decreases. Yet in both region C and region D, shivering serves to maintain a nearly normal body temperature. The truth is that in a normal unanesthetized human subject feedback control of body temperature is so extremely effective that precise quantitative characterization of the mechanisms involved is well-nigh impossible!

4. The large magnitudes of the ratios of derivatives in Figure 9 are partly due to the presence of Loop III, the positive "Q_{10}" feedback loop. The open-loop gain for that loop, being less than unity, at first seems to be trivial compared with the high gains for the negative loops (Figure 8). But in equation 24 for the ratio of derivatives in a warm environment, the open-loop transfer functions $\beta_3\beta_6$ and $\beta_4\beta_7$ for the negative loops I and II are multiplied by the factor $1/[1 - \beta_5\beta_8]$ which is, in fact, the magnification due to the positive loop III. For the exemplary values listed, this magnification is far from trivial, being about 4.6. This means that if the negative loops only are opened the change in

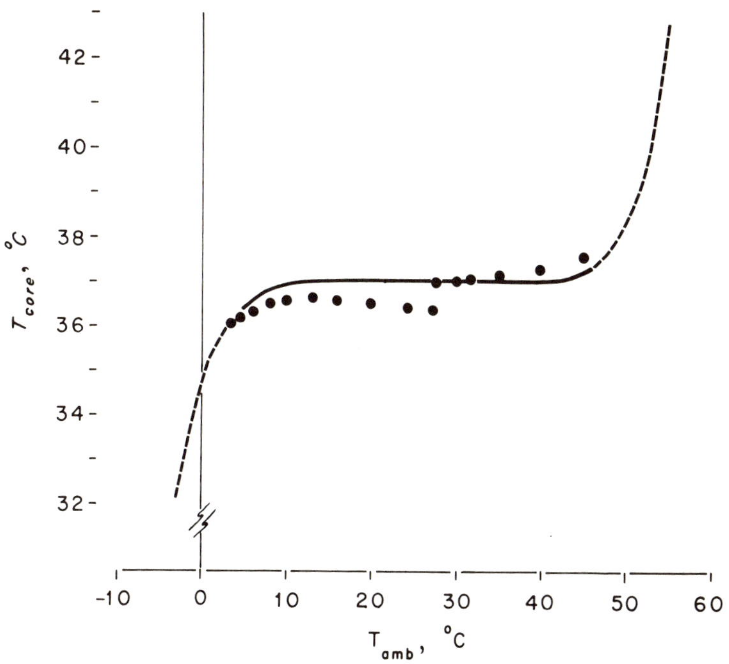

Figure 10. The solid dots for T_{core} have been calculated from T_{amb} using the feedback equations derived in this section. The practically horizontal solid curve for T_{core} as a function of T_{amb} fits the points within a few tenths of a degree over an extraordinarily wide range of ambient temperatures. (The broken-line extensions show that the regulatory mechanisms must fail at very high and very low ambient temperatures.) Notice that essentially perfect regulation can be achieved in spite of having no physically-real set point or reference standard.

---

body temperature due to a change in ambient temperature is 4.6 times as large as it would be if there were no feedback at all! Similar remarks apply to equation 45 for the ratio of derivatives during exposure to cold. The fact is that the gain of the "Q_{10}" loop by itself is uncomfortably close to the critical value of unity. Normally, this is of no consequence, since the homeostatic effect of the negative loops dominates the behavior of the system as a whole. But if the function of the homeostatic loops is impaired,the positive feedback may establish a vicious circle, leading to the rapid development of hyperthermia (for example in heat stroke) or hypothermia (for example during surgical anesthesia.)

5. "Normal" body temperature serves as a quasi-"set point." There is, of course, no physically-real set-point temperature maintained anywhere in the body. (see Section 4.6) The normal preservation of a virtually constant "errorless" core temperature over a wide range of ambient temperatures is due to the interplay of the various feedback loops here described. (See Figure 10).

6. In some animals (e.g. dogs) who pant in response to high temperatures, $\dot{q}_{lung}$ may actually exceed $\dot{q}_{prod}(\infty)$, so that, by equation 1, $T_{core}(\infty) < T_{skin}(\infty)$. Under these circumstances, a large resistance to heat transfer between core and skin would be advantageous; i.e. cutaneous vasodilation would be counter-productive as far as maintaining normal core temperature is concerned!

References

1. Cooper, K. E.: Quantitative observations in disordered temperature regulation. Scient. Basis Med. Ann. Rev.,239-258, 1965.

2. Benzinger, T. H., Kitzinger, C., and Pratt, A. W.: The human thermostat. Chap. 56, pp 637-665, in: Herzfeld, C. M., editor,: Temperature. Its Measurement and Control in Science and Industry, Part 3. Reinhold Publishing Corp., New York, 1963.

3. Brengelmann, G. and Brown, A. C.: Temperature regulation. Chap. 54, pp 1050-1069, in Ruch, T. C. and Patton, H. D.: Physiology and Biophysics, 19th edition. W. B. Saunders, Philadelphia, 1965.

4. Benzinger, T. H.: Heat regulation: homeostasis of central temperature in man. Physiol. Rev., 49, 671-759, 1969.

5. von Euler, C.: Physiology and pharmacology of temperature regulation. Pharmacol. Rev., 13, 361-398, 1961.

See also

Cornew, R. W., Houk, J. C., and Stark, L.: Fine control in the human temperature regulation system. J. theor. Biol., 16, 406-426, 1967.

NEGATIVE AND POSITIVE

Type	Symbol	Definition	Example (Human)
Ind.	$\dot{Q}_{O_2,\,other}$	Oxygen consumption other than that due to work of breathing.	14.47 mM/min
	$p_{O_2,I}$	Partial pressure of oxygen in inspired air.	149.6 mm Hg
	$p_{CO_2,I}$	Partial pressure of carbon dioxide in inspired air.	0.0 mm Hg
	$[HCO_3^-]$	Concentration of bicarbonate in plasma.	26.0 mEq/L
Par.	F_{RQ}	Respiratory quotient.	0.8
	$p_{CO_2,\,thr}$	Threshold value of $p_{CO_2,A}(\infty)$ below which $p_{CO_2,A}(\infty)$ does not influence $\dot{V}_A(\infty)$.	39.7 mm Hg
	C	Value of $p_{O_2,A}(\infty)$ for "infinite" $S(\infty)$.	32.0 mm Hg
	D	Minimum value of $S(\infty)$ approached as $p_{O_2,A}(\infty) \rightarrow \infty$.	4.0[L/min] per mm Hg
	K	An empirical parameter in equation 9.	68.0 L/min
	G	An empirical parameter in equation 7.	2.07 L/min
	H	Increase in $\dot{V}_A(\infty)$ per unit increase in $\dot{V}_{tot}(\infty)$.	0.91
	α	An empirical parameter in equation 10.	11.8 mm Hg
	β	Increase in B per unit increase in $[HCO_3^-]$	1.008 mm Hg/ [mEq/L]
	a	An empirical parameter in equation 14.	1.10 mM/min
	b	Value approached by $\dot{V}_{tot}(\infty)$ as $\dot{Q}_{O_2,\,vent}(\infty) \rightarrow \infty$.	120 L/min

Type	Symbol	Definition	Example (Human)
Dep. Par.			
	B	A dependent parameter defined by equation 10.	38.0 mm Hg
Const.	RT	Product of ideal gas law constant and body temperature.	19.34 mm Hg·L/mM
Dep.	$\dot{Q}_{O_2,in}(\infty)$	Quantity of 0_2 inspired into alveoli per min.	
	$\dot{Q}_{O_2,ex}(\infty)$	Quantity of 0_2 expired from alveoli per min.	
	$\dot{Q}_{CO_2,in}(\infty)$	Quantity of $C0_2$ inspired into alveoli per min.	
	$\dot{Q}_{CO_2,ex}(\infty)$	Quantity of $C0_2$ expired from alveoli per min.	
	$\dot{Q}_{O_2,vent}(\infty)$	Oxygen consumption due to the work of lung ventilation	0.08 mM/min
	$\dot{Q}_{O_2,tot}(\infty)$	Total oxygen consumption	14.55 mM/min
	$\dot{Q}_{CO_2,prod}(\infty)$	Carbon dioxide production	11.64 mM/min
	$p_{O_2,A}(\infty)$	Partial pressure of 0_2 in alveolar air.	100 mm Hg
	$p_{CO_2,A}(\infty)$	Partial pressure of $C0_2$ in alveolar air.	39.7 mm Hg
	$\dot{V}_A(\infty)$	Alveolar ventilation	5.67 L/min
	$\dot{V}_{tot}(\infty)$	Total ventilation	8.50 L/min
	$S(\infty)$	Increase in $\dot{V}_{tot}(\infty)$ per unit increase in $p_{CO_2,A}$.	5.00 [L/min] per mm Hg
	$[H^+](\infty)$	Hydrogen ion concentration in arterial blood.	$3.63 \cdot 10^{-5}$ mM/L
	$u(p_{CO_2,A}(\infty) - p_{CO_2,thr})$	The unit step function: u = 0 for negative values of its argument; u = 1 for non-negative values of its argument.	

Model

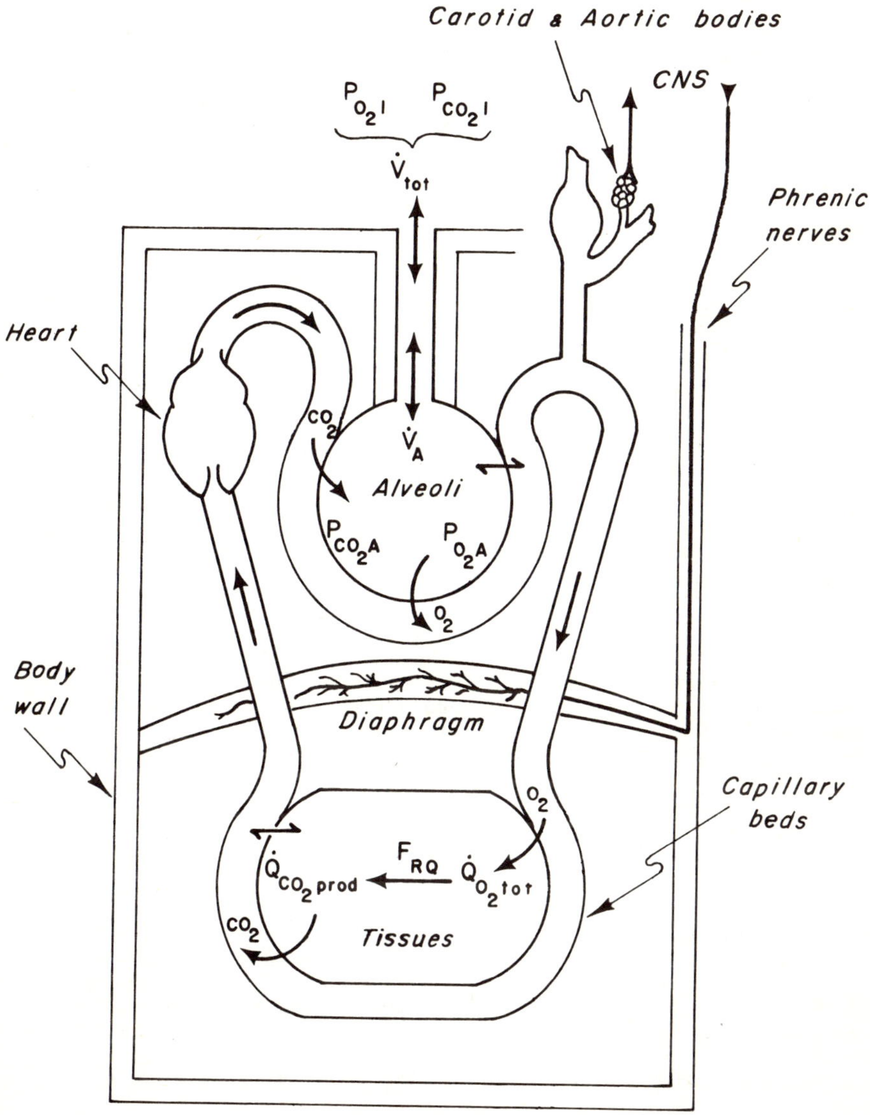

Figure 1.

The blood stream, greatly aided by its hemoglobin, carries 0_2 from the alveoli of the lungs to the tissues where the 0_2 is consumed by metabolic processes. The $C0_2$ produced thereby is proportional to the 0_2 consumed, the proportionality constant being the respiratory quotient, F_{RQ}. The blood stream, again aided by its hemoglobin, carries the $C0_2$ back to the lungs. In the steady state, the rate of oxygen consumption must equal the difference between the rate at which 0_2 enters the alveoli in the inspired air and the rate at which it leaves the alveoli in the expired air. There must be a similar balance for $C0_2$. It is assumed that both 0_2 and $C0_2$ have time to reach diffusion equilibrium with the alveolar air in the lungs while blood is flowing past the alveoli so that for each of these gases the partial pressure is the same in alveolar air and arterial blood.

An increase in arterial p_{CO_2} will increase $\dot{V}_A(\infty)$ by stimulating central receptors whose precise nature and location are still under investigation. A decrease in arterial p_{O_2} will stimulate chemoreceptors in the aortic and carotid bodies, impulses from which will also increase $\dot{V}_A(\infty)$ via central mechanisms. The increase in $\dot{V}_A(\infty)$ will tend to reduce $p_{CO_2, A}(\infty)$ and increase $p_{O_2, A}(\infty)$, thereby opposing the initial changes.

The muscular work expended in ventilating the lungs will itself require the consumption of oxygen and the production of carbon dioxide. Though normally small, $\dot{V}_{O_2, vent}(\infty)$, the oxygen consumption due to lung ventilation, does tend to increase the stimuli for breathing. This kind of <u>positive</u> feedback becomes increasingly important as total ventilation increases, particularly in patients with pulmonary disease (see Figure 8).

Although the central links of the feedback loops are still uncertain and cannot be studied quantitatively, the overall behavior of the system can be described by empirical equations. (See derivation.) Since arterial pH depends in part on arterial p_{CO_2}, this system may be regarded as regulating hydrogen ion concentration, changes of which in key receptor regions may indeed be the fundamental stimulus for increased activity of the respiratory "center."

<u>Simplifying assumptions</u>

1. The inspired air is already at body temperature and saturated with water vapor.

2. The respiratory quotient is near enough to unity so that the difference between volume of air inspired and volume of air expired can be neglected.

3. The changes described are acute; not those which would be seen after adaptation to an altered environment.

4. Gas flow through the lungs is assumed to be continuous, not intermittent.

5. Diffusion equilibrium is attained between alveolar air and pulmonary venous blood.

6. There are no arteriovenous shunts across the lungs.

7. See statement above equation 11.

8. $p_{CO_2,thr}$ is assumed to be an independent parameter. Actually it depends upon B so that $p_{CO_2,thr}$ = B + some constant, here assumed to be 1.7 mm Hg.

Derivation

In the steady state, the rate at which oxygen enters the alveoli must equal the rate at which it leaves.

$$\dot{Q}_{O_2,in}(\infty) = \dot{Q}_{O_2,ex}(\infty) + \dot{Q}_{O_2,tot}(\infty) \qquad 1.T.$$

Similarly, for carbon dioxide,

$$\dot{Q}_{CO_2,in}(\infty) + \dot{Q}_{CO_2,prod}(\infty) = \dot{Q}_{CO_2,ex}(\infty) \qquad 2.T.$$

For any gas, the ideal gas law states that

$$p_{gas}V_{gas} = RTQ_{gas} \qquad 3.T$$

Dividing both sides of equation 3 by time, and rearranging,

$$\dot{Q}_{gas} = \frac{1}{RT}\dot{V}_{gas}p_{gas} \qquad 3.a$$

Equation 3.a gives the quantity of gas transported per unit time by a flow of $\dot{V}_{gas}$. By assumption 2, $\dot{V}_A(\infty)$ is the same for air inspired into the aveoli as for air expired from the alveoli. We can therefore write a particular version of equation 3.a for each of the four quantities, $\dot{Q}_{O_2,in}(\infty)$, $\dot{Q}_{O_2,ex}(\infty)$, $\dot{Q}_{CO_2,in}(\infty)$ and $\dot{Q}_{CO_2,ex}(\infty)$, in $\dot{V}_A(\infty)$. We can then use the particular versions to eliminate these quantities from equations 1 and 2; thus obtaining

$$p_{O_2,A}(\infty) = p_{O_2,I} - RT\frac{\dot{Q}_{O_2,tot}(\infty)}{\dot{V}_A(\infty)} \qquad 4.$$

$$p_{CO_2,A}(\infty) = p_{CO_2,I} + RT\frac{\dot{Q}_{CO_2,prod}(\infty)}{\dot{V}_A(\infty)} \qquad 5.$$

By definition,

$$\dot{Q}_{CO_2,prod}(\infty) = F_{RQ}\,\dot{Q}_{O_2,tot}(\infty) \qquad 6.T$$

Gray and his collaborators (1) have observed that

$$\dot{V}_A(\infty) = -G + H\dot{V}_{tot}(\infty) \qquad 7.E$$

When $p_{CO_2,A}(\infty) \geq p_{CO_2,thr}$, Lloyd and Cunningham (2) have observed that

$$\dot{V}_{tot}(\infty) = S(\infty)\,[p_{CO_2,A}(\infty) - B]\ ;\ p_{CO_2,A}(\infty) \geq p_{CO_2,thr} \qquad 8.E$$

where

$$S(\infty) = D + \frac{K}{p_{O_2,A}(\infty) - C}\ ;\ p_{O_2,A}(\infty) > C \qquad 9.E$$

(In equation 9, "K" is the parameter defined by Lloyd and Cunningham (2) as "AD.") Furthermore, Cunningham et al. (3) have observed that the parameter B decreases when metabolic acidosis is produced by administering ammonium chloride. Their data can be described by the empirical equation,

$$B = \alpha + \beta[HCO_3^-] \qquad 10.E$$

However, we shall assume that when $p_{CO_2,A}(\infty)$ falls below $p_{CO_2,thr}$, $\dot{V}_{tot}(\infty)$ is no longer influenced by $p_{CO_2,A}(\infty)$ but retains the value appropriate for $p_{CO_2,A}(\infty) = p_{CO_2,thr}$ at the prevailing partial pressure of oxygen (4).

$$\dot{V}_{tot}(\infty) = S(\infty)\,[p_{CO_2,thr} - B]\ ;\ p_{CO_2}(\infty) < p_{CO_2,thr} \qquad 11.A$$

By introducing the unit step function (see Section 3.8), we can write equations 8 and 11 as the single equation,

$$\dot{V}_{tot}(\infty) = S(\infty)\left\{[p_{CO_2,A}(\infty) - p_{CO_2,thr}]\,u(p_{CO_2,A}(\infty) - p_{CO_2,thr}) + p_{CO_2,thr} - B\right\} \qquad 12.$$

We shall now define $\dot{Q}_{O_2,tot}(\infty)$, the total consumption of oxygen, as the sum of $\dot{Q}_{O_2,vent}(\infty)$, the oxygen consumption required by the work of breathing, and $\dot{Q}_{O_2,other}(\infty)$.

$$\dot{Q}_{O_2,tot}(\infty) = \dot{Q}_{O_2,vent}(\infty) + \dot{Q}_{O_2,other} \qquad 13.T$$

It has been observed (5) that

$$\dot{Q}_{O_2,vent}(\infty) = \frac{a\,\dot{V}_{tot}(\infty)}{b - \dot{V}_{tot}(\infty)}\ ;\ \dot{V}_{tot}(\infty) < b \qquad 14.E$$

Finally, the nonlogarithmic version of the Henderson-Hasselbalch equation,

$$[H^+](\infty) = 2.38 \cdot 10^{-5} p_{CO_2,A}(\infty)/[HCO_3^-] \qquad 15.T$$

shows how $[H^+]$ depends upon $[HCO_3^-]$, kept constant as a parameter, and $p_{CO_2,A}(\infty)$, determined by equation 5.

The symbol-and-arrow diagram corresponding to the working equations 4, 5, 6, 7, 9, 10, 12, 13, 14 and 15 is shown in Figure 2.

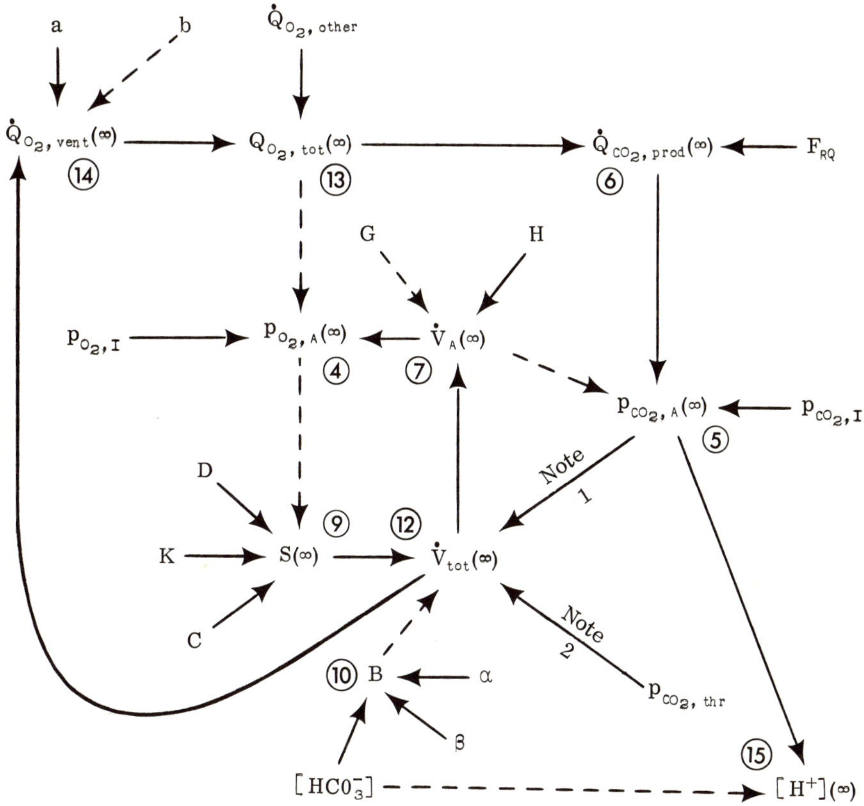

Note 1. Omit this arrow when $p_{CO_2, A}(\infty) < p_{CO_2, thr}$

Note 2. Omit this arrow when $p_{CO_2, A}(\infty) \geq p_{CO_2, thr}$

Figure 2.

In order to simplify further analysis, combine equations 5 and 6, eliminating $\dot{Q}_{CO_2, prod}(\infty)$.

$$p_{CO_2, A}(\infty) = p_{CO_2, I} + RTF_{RQ} \frac{\dot{Q}_{O_2, tot}(\infty)}{\dot{V}_A(\infty)} \qquad 16.$$

Combine equations 9, 10, and 12, eliminating $S(\infty)$ and B,

$$\dot{V}_{tot}(\infty) = \left\{D + \frac{K}{p_{O_2,A}(\infty) - C}\right\} \left\{[p_{CO_2,A}(\infty) - p_{CO_2,thr}]\, u(p_{CO_2,A}(\infty) - p_{CO_2,thr}) \right.$$

$$\left. p_{CO_2,thr} - \alpha - \beta\,[HCO_3^-]\right\} \quad ; p_{O_2,A}(\infty) > C \qquad 17.$$

Combine equations 13 and 14, eliminating $\dot{Q}_{O_2,vent}(\infty)$.

$$\dot{Q}_{O_2,tot}(\infty) = \frac{a\,\dot{V}_{tot}(\infty)}{b - \dot{V}_{tot}(\infty)} + \dot{Q}_{O_2,other} \qquad 18.$$

Equations 4, 7, 15, 16, 17 and 18 constitute the smallest set of working equations which will preserve all of the feedback loops and show their relationships to each other and to $[H^+](\infty)$.

For simplicity, rewrite these equations using the exemplary values except for the six remaining variables and the four inputs.

$$p_{O_2,A}(\infty) = p_{O_2,I} - 19.34\,\frac{\dot{Q}_{O_2,tot}(\infty)}{\dot{V}_A(\infty)} \text{ mm Hg,} \qquad 19.$$

$$p_{CO_2,A}(\infty) = p_{CO_2,I} + 15.47\,\frac{\dot{Q}_{O_2,tot}(\infty)}{\dot{V}_A(\infty)} \text{ mm Hg,} \qquad 20.$$

$$\dot{V}_A(\infty) = -\,2.07 + 0.91\,\dot{V}_{tot}(\infty) \quad \text{L/min} \qquad 21.$$

$$\dot{Q}_{O_2,tot}(\infty) = \dot{Q}_{O_2,other} + \frac{1.1\,\dot{V}_{tot}(\infty)}{120 - \dot{V}_{tot}(\infty)} \quad \text{mM/min} \qquad 22.$$

$$[H^+](\infty) = 2.38\cdot 10^{-5}\,p_{CO_2,A}(\infty)/[HCO_3^-] \text{ mM/L} \qquad 23.$$

$$\dot{V}_{tot}(\infty) = \left\{4 + \frac{68}{p_{O_2,A}(\infty) - 32}\right\}\left\{p_{CO_2,A}(\infty) - 11.8 - 1.008[HCO_3^-]\right\} \text{ L/min;}$$

$$p_{CO_2,A}(\infty) \geq p_{CO_2,thr}$$

$$= \left\{4 + \frac{68}{p_{O_2,A}(\infty) - 32}\right\}\left\{27.9 - 1.008[HCO_3^-]\right\} \text{ L/min;}$$

$$p_{CO_2,A}(\infty) < p_{CO_2,thr} \qquad 24.$$

The symbol-and-arrow diagram for equations 19 through 24 is shown in Figure 3. The corresponding <u>linear</u> equations for the tangent hyperplanes at any operating point are

$$\Delta p_{O_2,A} = \Delta p_{O_2,I} - \beta_1 \Delta \dot{Q}_{O_2,tot} + \beta_2 \Delta \dot{V}_A \text{ mm Hg} \qquad 25.$$

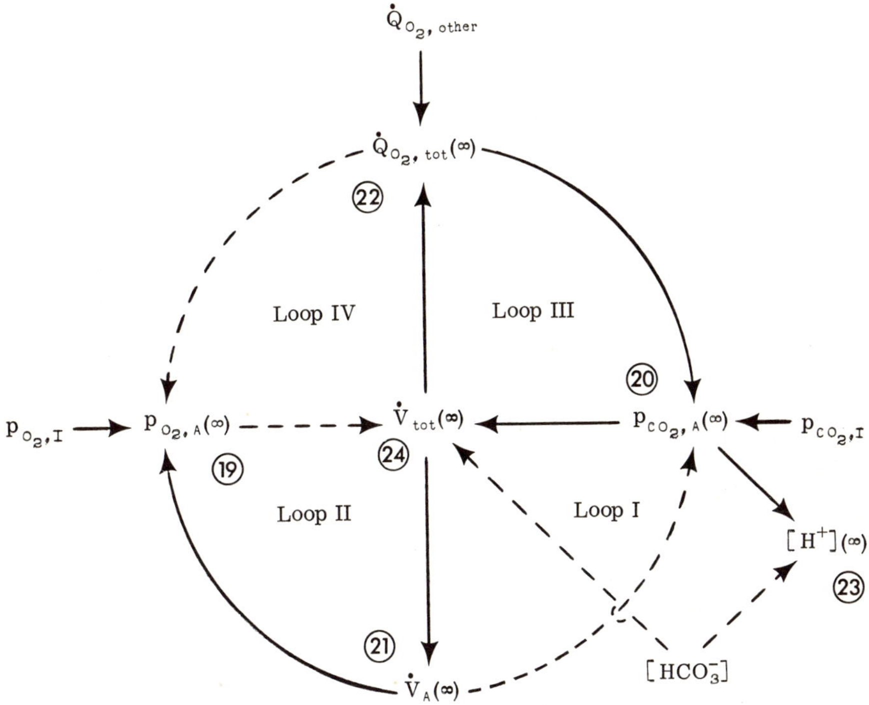

Figure 3.

$$\Delta p_{CO_2,A} = \Delta p_{CO_2,I} + \beta_3 \Delta \dot{Q}_{O_2,\,tot} - \beta_4 \Delta \dot{V}_A \quad \text{mm Hg} \qquad 26.$$

$$\Delta \dot{V}_A = \beta_5 \Delta \dot{V}_{tot} \quad \text{L/min} \qquad 27.$$

$$\Delta \dot{Q}_{O_2,\,tot}(\infty) = \Delta \dot{Q}_{O_2,\,other} + \beta_6 \Delta \dot{V}_{tot} \quad \text{mM/min} \qquad 28.$$

$$\Delta [H^+] = \beta_7 \Delta p_{CO_2,A} - \beta_8 \Delta [HCO_3^-] \ \text{mM/L} \qquad 29.$$

$$\Delta \dot{V}_{tot} = -\beta_9 \Delta p_{O_2,A} + \beta_{10} \Delta p_{CO_2,A} - \beta_{11} \Delta [HCO_3^-] \quad \text{L/min};$$

$$p_{CO_2,A}(\infty) \geq p_{CO_2,\,thr}$$

$$= -\beta_{12} \Delta p_{O_2,A} - \beta_{11} \Delta [HCO_3^-] \quad \text{L/min};$$

$$p_{CO_2,A}(\infty) < p_{CO_2,\,thr} \qquad 30.$$

where,

$$\beta_1 = -\left.\frac{\partial p_{O_2,A}(\infty)}{\partial \dot{Q}_{O_2,\,tot}}\right|_{op} = \frac{19.34}{\dot{V}_A(\infty)} \qquad \text{mm Hg/[mM/min]}$$

$$\beta_2 = \left.\frac{\partial p_{O_2,A}(\infty)}{\partial \dot{V}_A(\infty)}\right|_{op} = \frac{19.34\,\dot{Q}_{O_2,\,tot}}{\dot{V}_A^2(\infty)} \qquad \text{mm Hg/[L/min]}$$

$$\beta_3 = \left.\frac{\partial p_{CO_2,A}(\infty)}{\partial \dot{Q}_{O_2,\,tot}}\right|_{op} = \frac{15.47}{\dot{V}_A(\infty)} \qquad \text{mm Hg/[mM/min]}$$

$$\beta_4 = -\left.\frac{\partial p_{CO_2,A}(\infty)}{\partial \dot{V}_A(\infty)}\right|_{op} = \frac{15.47\,\dot{Q}_{O_2,\,tot}}{\dot{V}_A^2(\infty)} \qquad \text{mm Hg/[L/min]}$$

$$\beta_5 = \left.\frac{\partial \dot{V}_A(\infty)}{\partial \dot{V}_{tot}(\infty)}\right|_{op} = 0.91$$

$$\beta_6 = \left.\frac{\partial \dot{Q}_{O_2,\,tot}(\infty)}{\partial \dot{V}_{tot}(\infty)}\right|_{op} = \frac{132}{[120 - \dot{V}_{tot}]^2} \qquad \text{mM/L}$$

$$\beta_7 = \left.\frac{\partial [H^+](\infty)}{\partial p_{CO_2,A}(\infty)}\right|_{op} = \frac{2.38 \cdot 10^{-5}}{[HCO_3^-]} \qquad \text{[mM/L]/mm Hg}$$

$$\beta_8 = -\left.\frac{\partial [H^+](\infty)}{\partial [HCO_3^-]}\right|_{op} = \frac{2.38 \cdot 10^{-5}\, p_{CO_2,A}}{[HCO_3^-]^2}$$

$$\beta_9 = -\left.\frac{\partial \dot{V}_{tot}(\infty)}{\partial p_{O_2,A}(\infty)}\right|_{op} = \frac{68\{p_{CO_2,A} - 11.8 - 1.008[HCO_3^-]\}}{[p_{O_2,A} - 32]^2} \qquad \text{[L/min]/mm Hg}$$

$$\beta_{10} = \left.\frac{\partial \dot{V}_{tot}(\infty)}{\partial p_{CO_2,A}(\infty)}\right|_{op} = 4 + \frac{68}{p_{O_2,A} - 32} \qquad \text{[L/min]/mm Hg}$$

$$\beta_{11} = -\left.\frac{\partial \dot{V}_{tot}(\infty)}{\partial [HCO_3^-]}\right|_{op} = 1.008\left\{4 + \frac{68}{p_{O_2,A} - 32}\right\} \qquad \text{[L/min]/[mM/L]}$$

$$\beta_{12} = -\left.\frac{\partial \dot{V}_{tot}(\infty)}{\partial p_{O_2,A}(\infty)}\right|_{op} = \frac{68\{27.9 - 1.008[HCO_3^-]\}}{[p_{O_2,A} - 32]^2} \qquad \text{[L/min]/mm Hg}$$

For the situation where $p_{CO_2, A}(\infty) \geq p_{CO_2, thr}$, the corresponding block diagram is shown in Figure 4. When $p_{CO_2, A}(\infty) < p_{CO_2, thr}$, the diagram must be suitably modified.

We can now answer several questions of physiological interest. For example, to find out how effectively these feedback loops minimize a change in $p_{CO_2, A}(\infty)$ caused by a change in $\dot{Q}_{O_2, other}$, we let $\Delta p_{O_2, I} = \Delta p_{CO_2, I} = \Delta[HCO_3^-] = 0$, and solve the corresponding versions of equations 25 through 28 and equation 30 simultaneously for $\Delta p_{CO_2, A}/\Delta\dot{Q}_{O_2, other} = dp_{CO_2, A}/d\dot{Q}_{O_2, other}\Big|_{closed}$.

$$\left.\frac{dp_{CO_2, A}}{d\dot{Q}_{O_2, other}}\right|_{closed} = \frac{\beta_3 + \beta_2\beta_3\beta_5\beta_9 - \beta_1\beta_4\beta_5\beta_9}{1 + \beta_2\beta_5\beta_9 + \beta_4\beta_5\beta_{10} - \beta_1\beta_8\beta_9 - \beta_3\beta_8\beta_{10}} \qquad 31.$$

From Figure 4, it is obvious that we can open all of the feedback loops without disturbing the primary influence of $\Delta\dot{Q}_{O_2, other}$ upon $\Delta p_{CO_2, A}$ by letting $\beta_1 = \beta_4 = \beta_8 = 0$. Equation 31 then reduces to

$$dp_{CO_2, A}/d\dot{Q}_{O_2, other}\Big|_{\substack{all\\open}} = \beta_3 \qquad 32,$$

which is, in fact, obvious from mere inspection of Figure 4. The magnification by all of the feedback loops of the effect of a change in $\dot{Q}_{O_2, other}$ upon $p_{CO_2, A}(\infty)$ is the quotient of the closed-loop change, given by equation 31, divided by the open-loop change, given by equation 32.

$$Mag = \frac{\beta_3 + \beta_2\beta_3\beta_5\beta_9 - \beta_1\beta_4\beta_5\beta_9}{\beta_3[1 + \beta_2\beta_5\beta_9 + \beta_4\beta_5\beta_{10} - \beta_1\beta_8\beta_9 - \beta_3\beta_8\beta_{10}]} \qquad 33.$$

Therefore the gain due to the joint effect of all of the feedback loops is

$$G_{fb} = \frac{1}{Mag} - 1 = \frac{\beta_4\beta_5\beta_{10} - \beta_1\beta_8\beta_9 - \beta_3\beta_8\beta_{10} + [\beta_1\beta_4\beta_5\beta_9/\beta_3]}{1 + \beta_2\beta_5\beta_9 - [\beta_1\beta_4\beta_5\beta_9/\beta_3]} \qquad 34.$$

Equation 34 has been used to calculate the upper curve in Figure 5.

Comments

1. The present analysis leans heavily upon the careful quantitative studies of Lloyd and his collaborators. (2, 3, 6) I am indebted to Mr. Lloyd for reviewing the present section, and I apologize for my stubbornness in rejecting most of his thoroughly reasonable suggestions for revision.

2. In Figure 5, the gain due to the operation of all of the feedback loops has been plotted against the output, $p_{CO_2, A}(\infty)$. When the input is $\dot{Q}_{O_2, other}$ (upper curve), the gain at the normal resting operating point, specified by the exemplary values, is about 32. As $\dot{Q}_{O_2, other}$ (and hence $p_{CO_2, A}(\infty)$) increases, the gain decreases, rapidly at first, then more slowly. However, even when a ten-fold increase in $\dot{Q}_{O_2, other}$ has elevated the $p_{CO_2, A}(\infty)$ to 48 mm Hg, the gain is still about 5, indicating that the system is still able to offer substantial opposition to further increases in $p_{CO_2, A}(\infty)$. In contrast, when the input is

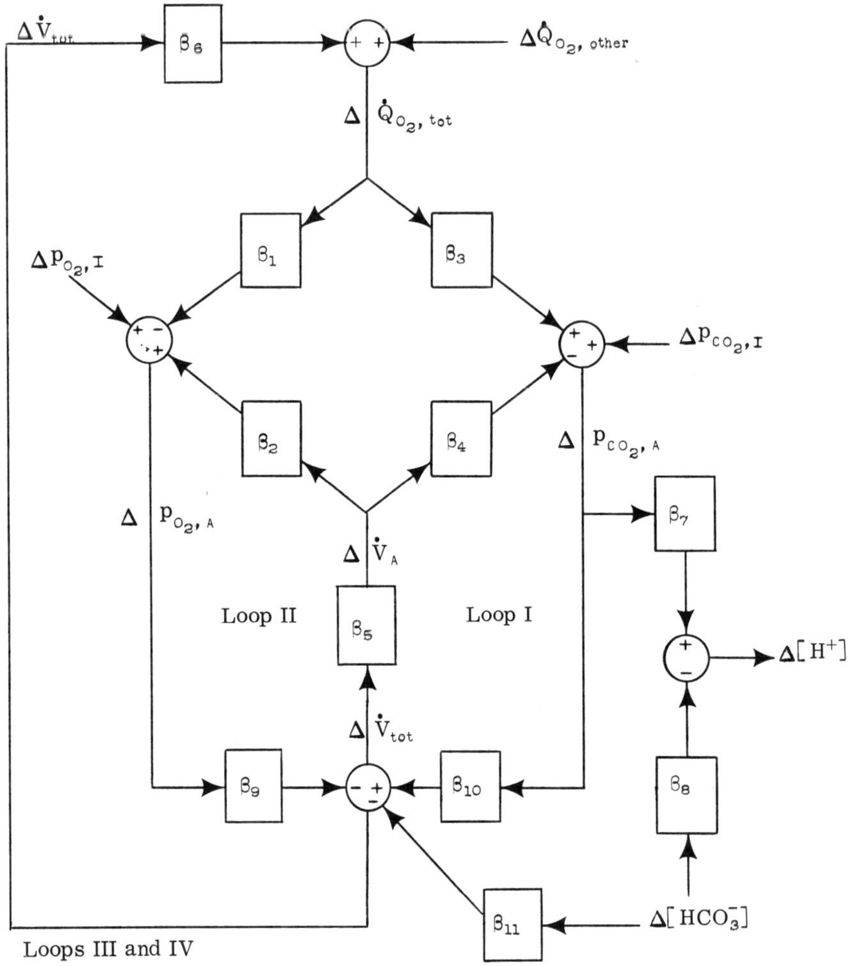
$\Delta\dot{V}_{tot}$
β_6
$\Delta\dot{Q}_{O_2,\ other}$
$\Delta\ \dot{Q}_{O_2,\ tot}$
β_1
β_3
$\Delta p_{O_2,\ I}$
$\Delta p_{CO_2,\ I}$
β_2
β_4
$\Delta\ p_{CO_2,\ A}$
$\Delta\ p_{O_2,\ A}$
β_7
$\Delta\ \dot{V}_A$
Loop II
β_5
Loop I
$\Delta[H^+]$
$\Delta\ \dot{V}_{tot}$
β_9
β_{10}
β_8
β_{11}
$\Delta[HCO_3^-]$
Loops III and IV

Figure 4.

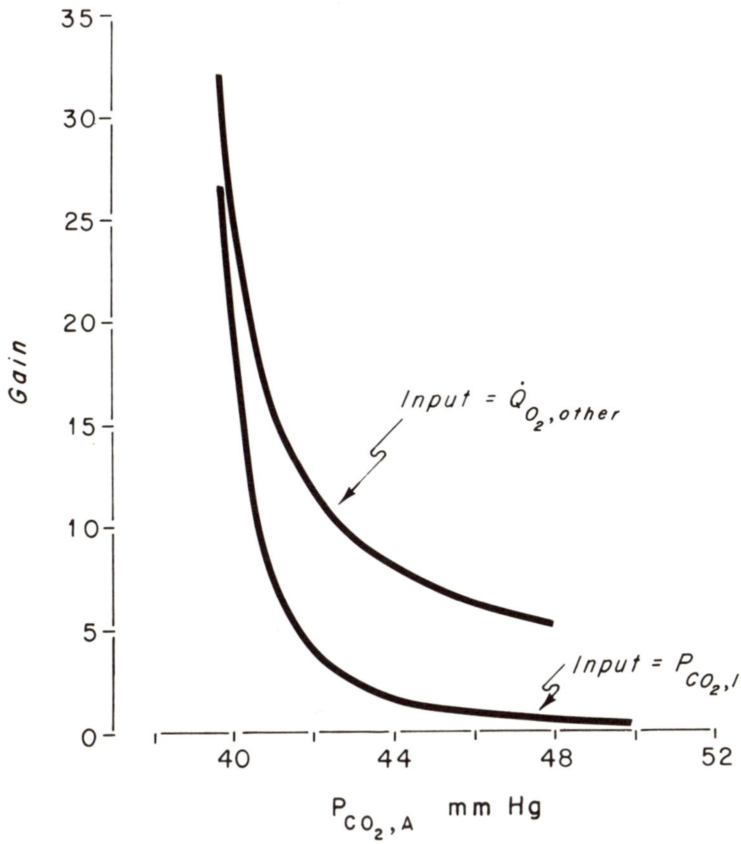

Figure 5.

$p_{CO_2,I}$ (lower curve) the gain at any particular value of $p_{CO_2,A}(\infty)$ is always smaller than when the input is $\dot{Q}_{O_2,other}$, and by the time $p_{CO_2,A}(\infty)$ has risen to about 45.5 mm Hg (at a $p_{CO_2,I}$ of about 38 mm Hg) the gain has already fallen to 1, so that the ability of the system to combat further increases in $p_{CO_2,A}$ is very limited. This relatively poor regulation when the system is loaded with CO_2 via the inspired air is easily accounted for. It is obvious that hyperventilation always tends to decrease the difference between $p_{CO_2,I}$ and $p_{CO_2,A}(\infty)$ so that as $\dot{V}_A \to \infty$, $p_{CO_2,A}(\infty) \to p_{CO_2,I}$. If a $p_{CO_2,A}(\infty)$ of 45.5 is due to metabolic CO_2 production while the subject is breathing room air, a given increase in $\dot{V}_A(\infty)$ would decrease $p_{CO_2,A}(\infty)$ by some fraction, F, of the difference between $p_{CO_2,I} = 0$ and $p_{CO_2,A}$; i.e., by $[45.5 - 0]F = 45.5F$ mm Hg.

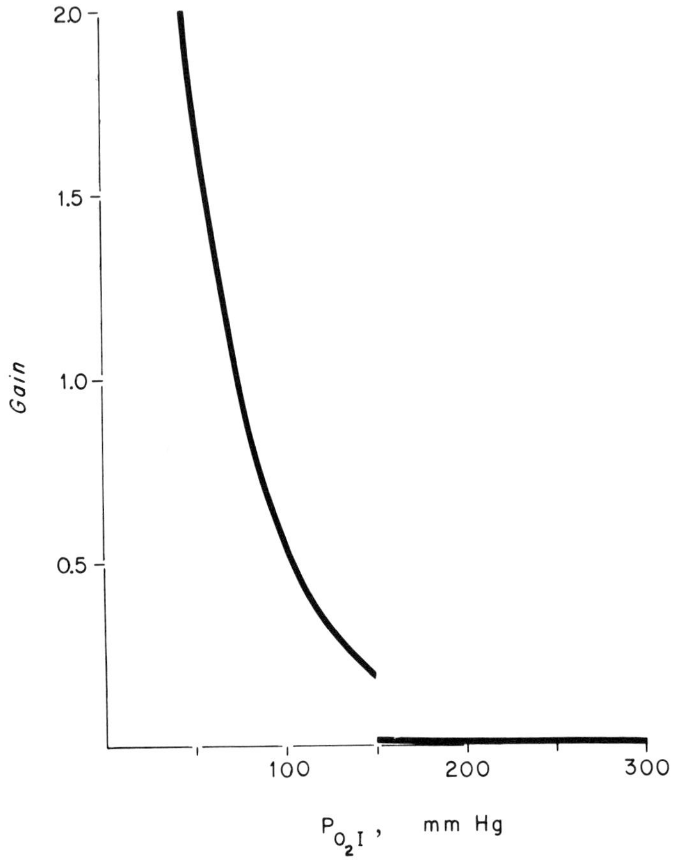

Figure 6.

But if the same $p_{CO_2,A}(\infty)$ is due to inspiring a gas mixture in which $p_{CO_2,I} = 38$ mm Hg, the same increase in $\dot{V}_A(\infty)$ would decrease $p_{CO_2,A}(\infty)$ by only $[45.5 - 38.0]F = 7.5\ F$ mm Hg. So the effectiveness of the system in regulating against increased CO_2 production would be about six times as great as against increased CO_2 inhalation. (Actually, the difference will be slightly larger than this because, with increased O_2 consumption, the decrease in $p_{O_2,A}(\infty)$ will also activate Loop II). Hence the response to inspired CO_2 is a very poor index of the ability of this system to compensate for increases in CO_2 production.

3. Figure 6 illustrates the ability of this system to protect $p_{O_2,A}(\infty)$ against changes caused by increases or decreases in $p_{O_2,I}$. Significant regulation occurs only in the region where $p_{O_2,I}$ is below normal; in the region above normal, the gain is practically zero. There is a discontinuity between these

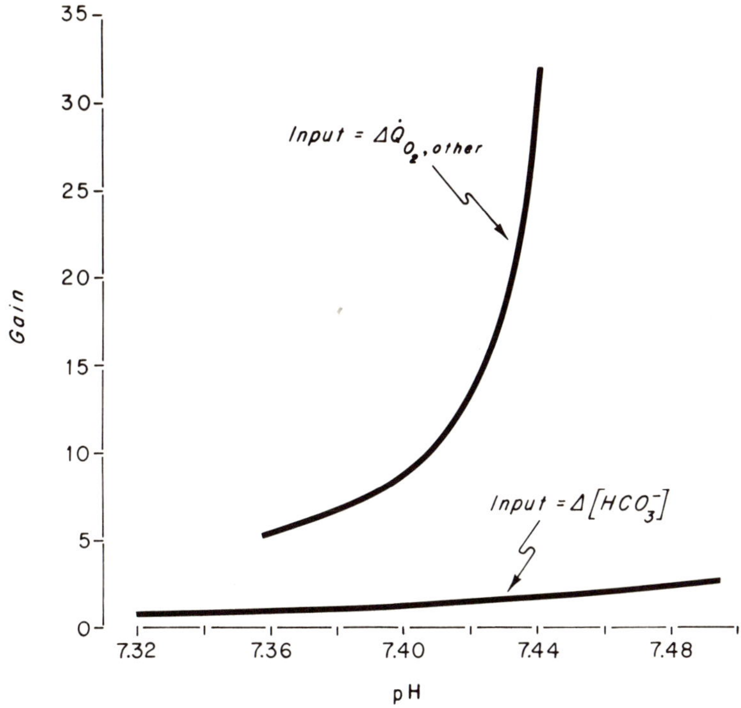

Figure 7.

two regions because we are assuming an abrupt change in the behavior of the system at the point where $p_{CO_2, A}(\infty) = p_{CO_2, thr}$ (equation 30). A discontinuity of this kind is usually an artifact. In all likelihood, the curve on the left should actually descend somewhat further and somewhat more rapidly as $p_{O_2, I}$ increases so as to join smoothly with the curve on the right.

4. Figure 7 shows how effectively the present system opposes changes in $[H^+]$. The gain due to all of the feedback loops has been plotted against pH, both for changes in $\dot{Q}_{O_2, other}$ and for changes in $[HCO_3^-]$. Since $[H^+](\infty)$ is directly proportional to $p_{CO_2, A}(\infty)$ (equation 15), changes in $[H^+](\infty)$ due to changes in $\dot{Q}_{O_2, other}$ are minimized just as effectively as are changes in $p_{CO_2, A}(\infty)$ itself (Figure 5). But if $\dot{Q}_{O_2, other}$ is kept constant and $[HCO_3^-]$ is altered, this system is much less able to regulate $[H^+]$, though the gain of about 2 near the "normal" operating point is by no means negligible. More precise regulation of bicarbonate is attained through renal mechanisms which are not described here.

5. The positive feedback loops which depend upon the oxygen consumption and carbon dioxide production obligated by the work of breathing are normally of negligible importance. But in a pulmonary disease such as emphysema they

may set an absolute limit to the amount of exercise which can be tolerated in the steady state. As a measure of the importance of these positive feedback loops, we compare $dp_{CO_2,A}/d\dot{Q}_{O_2,other}\Big|_{closed}$, calculated from equation 31, with

$$\frac{dp_{CO_2,A}}{d\dot{Q}_{O_2,other}}\Bigg|_{\substack{\text{III and IV} \\ \text{open}}} = \frac{\beta_3 + \beta_2\beta_3\beta_5\beta_9 - \beta_1\beta_4\beta_5\beta_9}{1 + \beta_2\beta_5\beta_9 + \beta_4\beta_5\beta_{10}} \qquad 35,$$

which represents the situation after loops III and IV have been opened by letting $\beta_8 = 0$. The ratio of the closed-loop change (equation 31) to the open-loop change (equation 35) gives the magnification, and the corresponding gain will be $\left|\frac{1}{\text{Mag}} - 1\right| = 1 - \frac{1}{\text{Mag}}$.

$$G_{fb,III,IV} = \frac{\beta_1\beta_8\beta_9 + \beta_3\beta_8\beta_{10}}{1 + \beta_2\beta_5\beta_9 + \beta_4\beta_5\beta_{10}} \qquad 36.$$

In Figure 8, gains for a normal subject and for a patient with emphysema have been calculated using equation 36 and values of a and b calculated from empirical curves fitted to observed data (5). For the normal subject, a was 1.10 mM/min, b was 120 L/min. For the patient with emphysema, a was 2.94 mM/min, b was 33.2 L/min. All other parameters for both subjects were assumed to be the exemplary values listed above. At rest, ($\dot{Q}_{O_2,other}$ = 14.47 mM/min) the gain for the patient with emphysema, though much higher than for the normal subject, was less than 0.1. But with even small increases in $\dot{Q}_{O_2,other}$, the gain rose precipitously, reaching the critical value of unity at a $\dot{Q}_{O_2,other}$ of only about 50 mM/min. It is obvious that the operation of the positive feedback loops would make it impossible for such a patient to maintain a steady state during even moderately heavy exercise.

6. It is well known that the ventilatory response to exercise cannot be completely accounted for by the operation of the "chemical" feedback system described above. Neural signals from the exercising muscles also increase $\dot{V}_{tot}(\infty)$. In particular, they are presumably responsible for the immediate increase in ventilation at the onset of exercise. But in the steady state, a large portion of the total ventilatory response to exercise can be explained by the present system of feedback loops, at least in subjects breathing room air (6). For example, in heavy exercise ($\dot{Q}_{O_2,tot}$ = 134 mM/min), Holmgren (7) observed that [HCO_3^-] was decreased by roughly 3.5 mM/L. If we accordingly let [HCO_3^-] = 22.5 mM/L, keeping $p_{CO_2,I}$ at 0 mm Hg and $p_{O_2,I}$ at 150 mm Hg, the present equations predict an increase in $p_{CO_2,A}(\infty)$ of only about 5 mm Hg. This is of the same order of magnitude as the mean increase actually observed in Holmgren's male subjects (7). (See also Table 1 in Reference 6). But the feedback loops described here cannot explain the response to exercise during hypoxia (6).

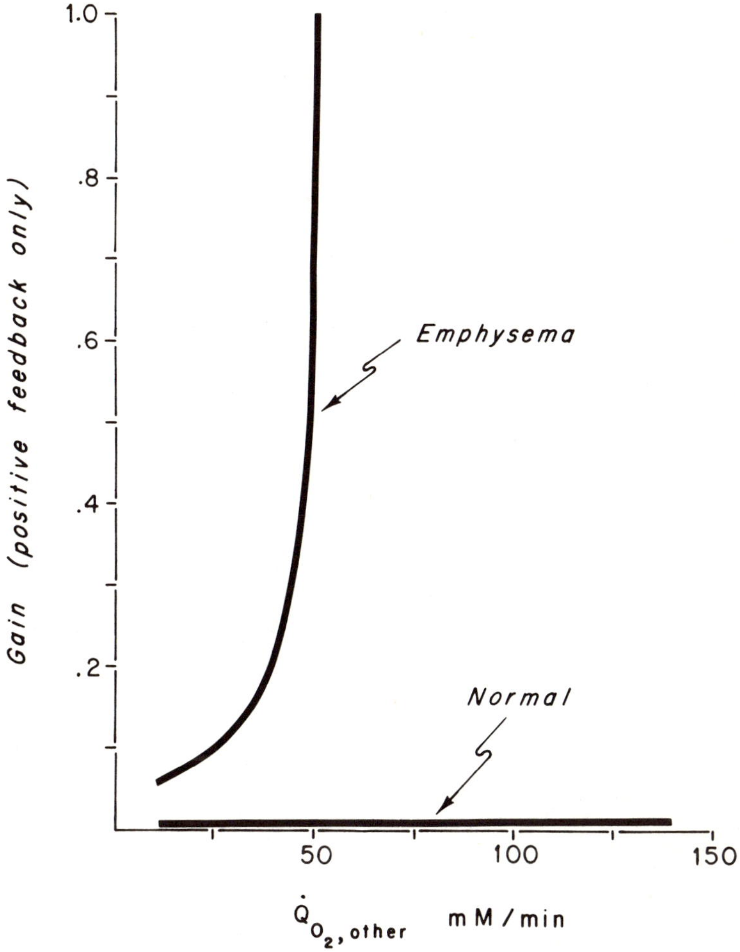

Figure 8.

Transient behavior

Many investigators have attempted to account for the transient changes in lung ventilation which occur in response to various inputs (8, 9, 10). In addition, Guyton, Crowell, and Moore (11) have demonstrated that cyclic fluctuations in breathing, which simulate Cheyne-Stokes respiration, can be produced experimentally by lengthening the circulatory path between lungs and central nervous system so as to introduce an artificial delay in signal transmission.

References

1. Gray, J.S., Grodins, F.S., and Carter, E.T.: Alveolar and total ventilation and the dead space problem. J. Appl. Physiol. 9, 307-320, 1956.

2. Lloyd, B.B., and Cunningham, D.J.C.: Aquantitative approach to the regulation of human respiration. In Cunningham, D.J.C. and Lloyd, B.B., editors, The Regulation of Human Respiration, pp 331-339, Blackwell Scientific Publications, Oxford, 1963.

3. Cunningham, D.J.C., Shaw, D.G, Lahiri, S., and Lloyd, B.B.: The effect of maintained ammonium chloride acidosis on the relation between pulmonary ventilation and alveolar oxygen and carbon dioxide in man. Q. Jl. exp. Physiol., 46, 323-334, 1961.

4. Nielsen, M. and Smith, H.: Studies on the regulation of respiration in acute hypoxia. Acta physiol. scand., 24, 293-313, 1952.

5. See Figure 21, page 756, in Ruch, T.C., and Patton, H.D., editors. Physiology and Biophysics. 19th edition. W.B. Saunders Co. Philadelphia, 1965.

6. Cunningham, D.J.C., Spurr, D., and Lloyd, B.B: The drive to ventilation from arterial chemoreceptors in hypoxic exercise. In Torrance, R.W., editor. Arterial Chemoreceptors. Blackwell Scientific Publications, Oxford, 1968.

7. Holmgren, A. Figures 50 and 51, pp 186-187 in Sjöstrand, T. editor, Clinical Physiology, J.B. Lippincott Co. Philadelphia, 1967.

8. Defares, J.G.: Principles of feedback control and their application to the respiratory control system. Chap. 26, pp 649-680 in Handbook of Physiology, Sec. 3 (Respiration), Vol. 1. Am. Physiol. Society, Washington, D.C. 1964.

9. Defares, J.G., Hara, H.H., and Billinghurst, E.M.: The stability of the respiratory servomechanism: an analog computer study. Progress in Biocybernetics 1, 109-123, 1964.

10. Grodins, F.S. and James, G.: Mathematical models of respiratory regulation. Ann N.Y. Acad. Sci., 109, 852-868, 1963.

11. Guyton, A.C., Crowell, J.W., and Moore, J.W. Basic oscillating mechanism of Cheyne-Stokes breathing. Am. J. Physiol. 187, 395-398, 1956.

CARDIAC FILLING PRESSURE; CARDIAC OUTPUT. I. STARLING HEART-LUNG PREPARATION.

NEGATIVE

Type	Symbol	Definition	Example (Dog)
Ind.	P_{res}	Pressure head between venous reservoir and heart.	11.0 mm Hg
Par.	R_{tube}	Resistance to flow through tube from reservoir to heart.	18.0 mm Hg per [L/min]
	a	Empirical parameter, equation 2	0.8 L/min
	b	Increase in $\dot{V}(\infty)$ per unit increase in $P_{RA}(\infty)$	0.7 [L/min] per mm Hg
Dep	$P_{RA}(\infty)$	Right atrial pressure. (= "Filling pressure.")	1.87 mm Hg
	$\dot{V}_{out}(\infty)$	Cardiac output	0.508 L/min
	$\dot{V}_{tube}(\infty)$	Blood flow from reservoir to heart.	0.508 L/min

Model

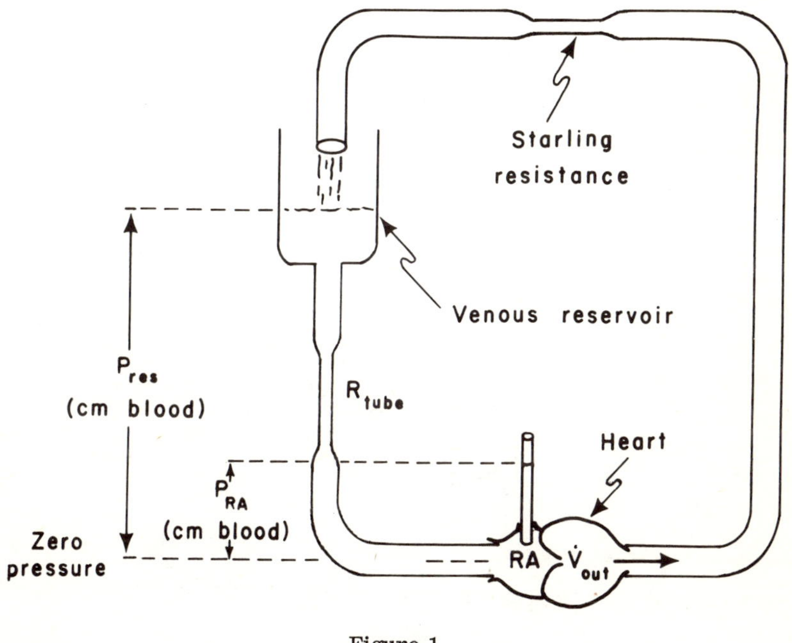

Figure 1.

Blood, pumped by the heart through an artificial "systemic resistance" (Starling resistance), flows into an open venous reservoir. The pressure head, P_{res}, from the surface of the blood in the reservoir to the heart is maintained constant by the investigator, who raises or lowers the reservoir to compensate for any changes in blood flow. (He provides essentially error-free integral control of P_{res} in the steady state.) Blood flows back to the right atrium, RA, through a tube whose resistance is R_{tube}. (For simplicity, the lesser circulation is disregarded.) The output of the heart is determined by how much the filling pressure stretches the ventricular muscle fibers, i.e., by the diastolic fiber length. (Starling's law of the heart.)

Assumptions

1. Flow from reservoir to right atrium obeys Ohm's law for fluid flow.

2. A change in filling pressure causes a proportional change in cardiac output within the range of interest.

Derivation

By assumption 1,

$$\dot{V}_{tube}(\infty) = \frac{P_{res} - P_{RA}(\infty)}{R_{tube}} \qquad 1.T$$

By assumption 2,

$$\dot{V}_{out}(\infty) = b\, P_{RA}(\infty) - a \qquad 2.E$$

In the steady state, conservation requires,

$$\dot{V}_{out}(\infty) = \dot{V}_{tube}(\infty) \qquad 3.T$$

Combine equations 1 and 3 and solve for the dependent variable, $P_{RA}(\infty)$.

$$P_{RA}(\infty) = P_{res} - R_{tube}\dot{V}_{out}(\infty) \qquad 4.$$

Equations 2 and 4 are the two feedback equations. The corresponding symbol-and-arrow diagram is,

From equation 2,

$$d\dot{V}_{out}(\infty)/dP_{RA}(\infty) = b \qquad 5.$$

From equation 4,

$$dP_{RA}(\infty)/d\dot{V}_{out}(\infty) = -R_{tube} \qquad 6.$$

Whence,

$$[OLG](j0) = -bR_{tube} \qquad 7.$$

which is 12.6 for the exemplary values.

Comments

1. The exemplary values used here are based upon actual measurements of the performance of a competent dog heart in an experiment previously used to illustrate the feedback relation between filling pressure and cardiac output (1). While equation 2 is a reasonably valid description of the observed dependence of $\dot{V}_{out}(\infty)$ upon $P_{RA}(\infty)$ over the range $1.4 < P_{RA}(\infty) < 2.4$ mm Hg, over a wider range the relationship was distinctly nonlinear, particularly when the heart was failing. And at high rates of blood flow even equation 4 (Ohm's law for fluid flow) had to be replaced by a non-linear modification. (See Section 5-11, in Reference 1).

2. The chief virtue of the Starling heart-lung preparation is its simplicity. Notice that according to equation 7, the effectiveness of the negative feedback loop as measured by the [OLG](j0) depends on both b and R_{tube}. R_{tube} has no biological significance; it is b, the change in cardiac output per unit change in right atrial pressure, which truly indicates the competence of the heart. Notice also that in the Starling heart-lung preparation, the circulation is artifically opened between the Starling resistance and the venous reservoir. Changes in blood volume are thus made irrelevant. This is not true of the normal closed circulation as the succeeding section will show.

Reference

1. Riggs, D.S.: The Mathematical Approach to Physiological Problems. Williams and Wilkins, Baltimore, 1963.

CARDIAC FILLING PRESSURE; CARDIAC OUTPUT. II INTACT CIRCULATION. (HETEROMETRIC REGULATION (1)).

NEGATIVE

Type	Symbol	Definition	Example (dog)
Ind.	V_{bl}	Total blood volume.	1400 ml
Par.	R	Total vascular resistance to flow.	0.0635 cm H_2O per [ml/min]
	α	Maximum work per stroke.	55 gm•m
	β	Parameter in equation 1.	230 gm•m •cm H_2O
	γ	Parameter in equation 1.	0.2 cm H_2O
	a	Parameter in equation 3.	43.3 cm H_2O
	b	Assumed value of dP_{art}/dV_{art}	0.722 cm H_2O/ml
	A	Parameter in equation 4.	6 cm H_2O
	B	Assumed maximal capacity of "veins."	2320 ml
	$\dot{N}$	Heart rate	150 beats/min
Dep.	$W(\infty)$	Work per stroke	16.13 gm•m
	$V_{art}(\infty)$	Volume of blood in "arteries."	240 ml
	$V_{ven}(\infty)$	Volume of blood in "veins."	1160 ml
	$P_{art}(\infty)$	Mean pressure in "arteries."	130 cm H_2O
	$P_{ven}(\infty)$	Mean pressure in "veins."	6 cm H_2O
	$\dot{V}_{out}(\infty)$	Cardiac output.	1953 ml/min
Const.	δ	Constant of proportionality in equation 2	0.01 gm• m per ml•cm H_2O

Model

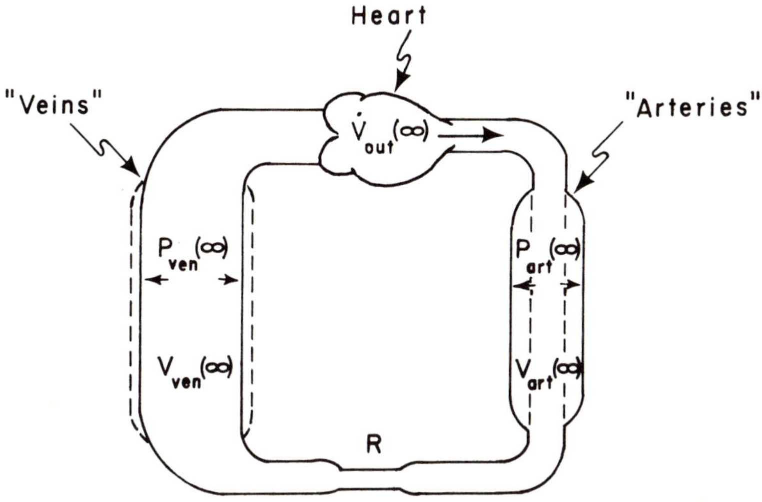

Figure 1. The broken lines are intended to suggest the situation when $\dot{V}_{out}(\infty)$ = 0 so that $P_{art}(\infty) = P_{ven}(\infty)$.

The circulation is assumed to consist of a relatively small high-pressure elastic reservoir ("Arteries") into which the heart pumps blood, a resistance, R, representing the total resistance to flow through all blood vessels, and a relatively large low-pressure elastic reservoir ("Veins") from which blood enters the heart. The lesser circulation is neglected. The entire blood volume is contained in the two reservoirs. The pressure in the venous reservoir is assumed equal to the atrial pressure and thus represents the filling pressure for the heart. As in the preceding section, the output of the heart depends upon the filling pressure (Starling's law of the heart.)

Assumptions

1. The work done by the heart in overcoming the inertance of blood is negligible.

2. See equations 3 and 4 below.

Derivation

In the steady state, the observed (1) dependence of stroke work upon filling pressure may be fitted by the empirical equation,

$$W(\infty) = \frac{\alpha P_{ven}(\infty) - \beta}{P_{ven}(\infty) + \gamma} \qquad 1.E$$

where, by assumption 1, $W(\infty)$ is proportional to the product of stroke volume and change in pressure.

$$W(\infty) = \delta \left\{ \frac{\dot{V}_{out}(\infty)}{\dot{N}} \right\} [P_{art}(\infty) - P_{ven}(\infty)] \qquad 2.T$$

The relation between pressure and volume is assumed to be of the form,

$$P_{art}(\infty) = b\, V_{art}(\infty) - a \qquad 3.S$$

for the "arteries," (Figure 2) and of the form,

$$P_{ven}(\infty) = \frac{A\, V_{ven}(\infty)}{B - V_{ven}(\infty)} = \frac{A}{\dfrac{B}{V(\infty)} - 1} \qquad 4.S$$

for the "veins," (Figure 2). Since the total resistance to flow is assumed to lie between the "arteries" and the "veins,"

$$\dot{V}_{out}(\infty) = \frac{[P_{art}(\infty) - P_{ven}(\infty)]}{R} \qquad 5.T$$

Conservation of blood volume requires that

$$V_{bl} = V_{art}(\infty) + V_{ven}(\infty) \qquad 6.T$$

Combine equations 1, 2 and 5 by eliminating $W(\infty)$ and $[P_{art}(\infty) - P_{ven}(\infty)]$, and solve for $\dot{V}_{out}(\infty)$.

$$\dot{V}_{out}(\infty) = \left\{ \frac{\dot{N}\,[\alpha\, P_{ven}(\infty) - \beta]}{\delta\, R[P_{ven}(\infty) + \gamma]} \right\}^{1/2} \qquad 7.$$

Combine equations 3, 4, 5 and 6 by eliminating $V_{art}(\infty)$, $P_{art}(\infty)$, and $V_{ven}(\infty)$, and <u>for convenience</u> solve for $\dot{V}_{out}(\infty)$.

$$\dot{V}_{out}(\infty) = \frac{1}{R} \left\{ b\, V_{bl} - a - P_{ven}(\infty) - \frac{b\, B\, P_{ven}(\infty)}{A + P_{ven}(\infty)} \right\} \qquad 8.$$

or

$$P_{ven}(\infty) = \frac{-J + \sqrt{J^2 - 4K}}{2} \qquad 8a.$$

where

$$J \equiv A + b\, B + R\, \dot{V}_{out}(\infty) + a - b\, V_{bl}$$

$$K \equiv A\, [R\, \dot{V}_{out}(\infty) + a - b\, V_{bl}]$$

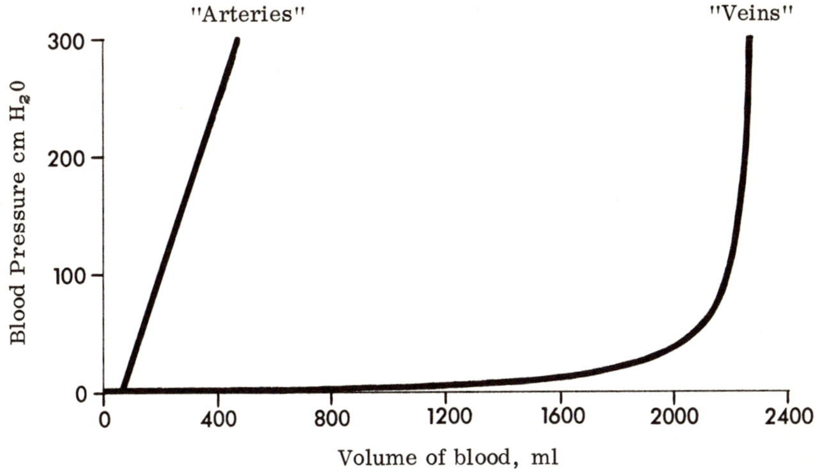

Figure 2. Assumed pressure-volume relationships

Equations 7 and 8a are the two feedback equations. The corresponding symbol-and-arrow diagram is

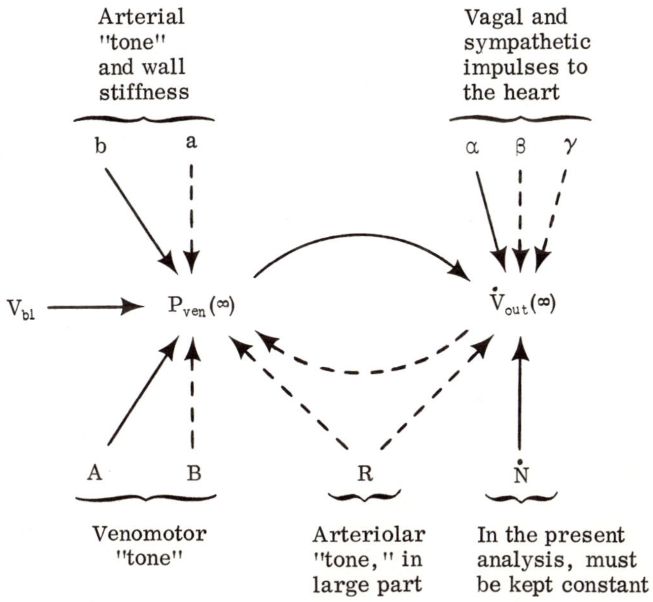

Figure 3.

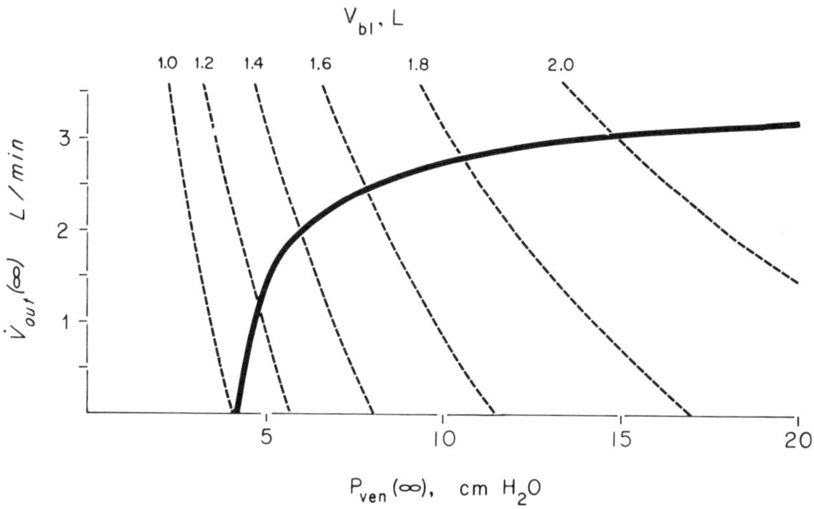

Figure 4.

From equation 7,

$$\frac{d\,\dot{V}_{out}(\infty)}{d\,P_{ven}(\infty)} = \frac{\frac{\dot{N}\alpha}{\delta} - R\,\dot{V}^2_{out}(\infty)}{2\,R\,\dot{V}_{out}(\infty)\,[P_{ven}(\infty) + \gamma]} \qquad 9.$$

From equation 8,

$$\frac{d\,\dot{V}_{out}(\infty)}{d\,P_{ven}(\infty)} = -\frac{1}{R}\left\{1 + \frac{b\,B\,A}{[A + P_{ven}(\infty)]^2}\right\} \qquad 10,$$

or, taking reciprocals on both sides of equation 10,

$$\frac{d\,P_{ven}(\infty)}{d\,\dot{V}_{out}(\infty)} = -\frac{R}{1 + \left\{\frac{b\,B\,A}{[A + P_{ven}(\infty)]^2}\right\}} \qquad 10a,$$

whence,

$$[OLG](j0) = \frac{\frac{\dot{N}\alpha}{\delta} - R\,\dot{V}_{out}(\infty)}{\left\{1 + \frac{b\,B\,A}{[A + P_{ven}(\infty)]^2}\right\}\left\{2\dot{V}_{out}(\infty)[P_{ven}(\infty) + \gamma]\right\}} \qquad 11.$$

Using the exemplary values, curves for various blood volumes have been plotted in Figure 4 according to equations 7 and 8a, and [OLG](j0) as a function of V_{bl} has been plotted in Figure 5.

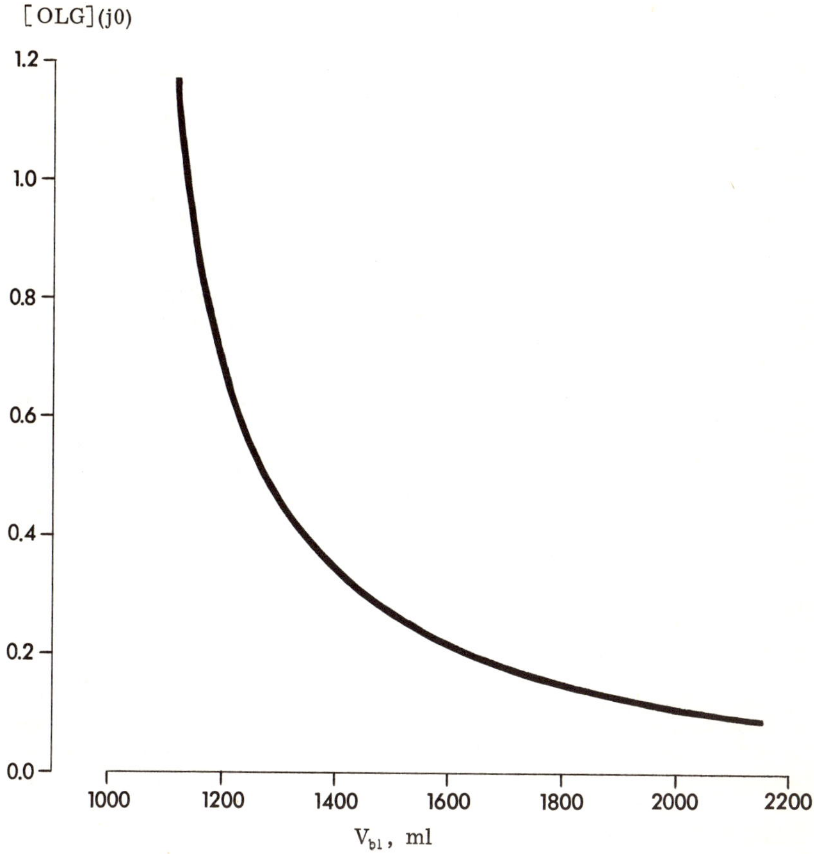

Figure 5.

Comments

1. The exemplary values have been chosen as reasonable for an 18 Kg dog. In this analysis I have relied heavily upon the careful quantitative studies of Sarnoff and his collaborators (1). Equation 1 describes one of their ventricular function curves (reproduced in Ruch and Patton's text (2) (Figure 4, p. 648)) obtained by varying V_{bl} in a dog whose cardiac accelerator nerves were being stimulated once every two seconds. Heart rate was kept constant at 150 per min by atrial stimulation.

2. In the present analysis no account is taken of the change in stroke work which would accompany a change in heart rate; hence, $\dot{N}$ must be regarded as fixed, and no conclusions about the effect of a change in $\dot{N}$ upon the values of $\dot{V}_{out}(\infty)$ or $P_{ven}(\infty)$ can be drawn.

3. The pressure-volume curves for "arteries" and "veins" (Figure 2) are patterned after the actual curves for aorta and vena cava in Figure 15, p. 539 of Ruch and Patton's text (3). But the numerical values have been chosen to fit the crude assumption made here that the entire blood volume is distributed between an arterial, aorta-like reservoir and a venous, vena-cava-like reservoir. Nevertheless, the exemplary values agree with the "mean circulatory filling pressure" of 8.6 cm H_20 in normal dogs, as estimated by Guyton et al. (4). By setting $P_{art}(\infty)$ (equation 3) equal to $P_{ven}(\infty)$(equation 4) with $\dot{V}_{bl} = 1400$ ml, we obtain 8.0 cm H_20 as the "mean circulatory filling pressure" in the present model system.

4. Equally crude is the assumption of a single-chambered heart, particularly since Sarnoff and Berglund (5) have documented the need to consider right ventricular function and left ventricular function separately.

5. In spite of these oversimplifications, the present model is sufficiently realistic to cast serious doubt upon the importance of the filling pressure-cardiac output feedback loop in the homeostatic regulation of cardiac output in the intact circulation.(Compare the low values of [OLG](j0) in Figure 5 with the value of 12.6 for the Starling heart-lung preparation in the previous section). This in no way detracts from the importance of filling pressure in the control of cardiac output via the Starling mechanism.

Related feedback loops

1. See the previous section.

2. The several physiological factors which control the parameters of the present system are indicated in Figure 3. All of these factors are themselves subject to feedback regulation, and some will be considered in the subsequent sections.

References

1. Sarnoff, S.J., Brockman, S.K., Gilmore, J.P., Linden, R.J., and Mitchell, J.H.: Regulation of ventricular contraction. Influence of cardiac sympathetic and vagal nerve stimulation on atrial and ventricular dynamics. Circulation Res., 8, 1108-1136, 1960.

2. Rushmer, R.F.: Control of cardiac output. Chap. 33, pp 644-659, in Ruch, T.C., and Patton, H.D.: Physiology and Biophysics. 19th edition. W.B. Saunders, Philadelphia, 1965.

3. Burton, A.C.: Hemodynamics and the physics of the circulation. Chap. 27, pp 523-542, in Ruch, T.C., and Patton, H.D.: Physiology and Biophysics, 19th edition. W.B. Saunders, Philadelphia, 1965.

4. Guyton, A.C., Polizo, D., and Armstrong, G.G.: Mean circulatory filling pressure measured immediately after cessation of heart pumping. Amer. J. Physiol., 179, 261-267, 1954.

References (continued)

5. Sarnoff, S.J., and Berglund, E.: Ventricular function. I. Starling's law of the heart studied by means of simultaneous right and left ventricular function curves in the dog. Circulation, 9, 706-718, 1954.

CAROTID SINUS PRESSURE; MEAN ARTERIAL BLOOD PRESSURE

NEGATIVE

Type	Symbol	Definition	Example
Ind.	$\dot{V}_{out}$	Cardiac output	
	P_{ven}	Venous pressure at right atrium	
Par.	A	Parameter of equation 3	15 mm Hg
	$\dot{N}_{max}$	Maximum value of $\dot{N}_{sinus}(\infty)$ for very large $P_{sinus}(\infty)$	50 per sec
	P_{thr}	Threshold value of $P_{sinus}(\infty)$	100 mm Hg
	P_{max}	Mean arterial blood pressure when $P_{sinus}(\infty) = 0$	210 mm Hg
	P_{min}	Mean arterial blood pressure when $P_{sinus}(\infty)$ is very large	90 mm Hg
	a	Parameter of equation 4	22
	b	Parameter of equation 4	0.133 per mm Hg
Dep.	$P_{art}(\infty)$	Mean arterial blood pressure	
	$P_{sinus}(\infty)$	Transmural pressure in carotid sinus	
	$R(\infty)$	Total peripheral vascular resistance	
	$\dot{N}_{sinus}(\infty)$	Rate of firing of sinus nerve fibers	
	$\dot{N}_{vaso}(\infty)$	Rate of firing of vaso-constrictor nerves	

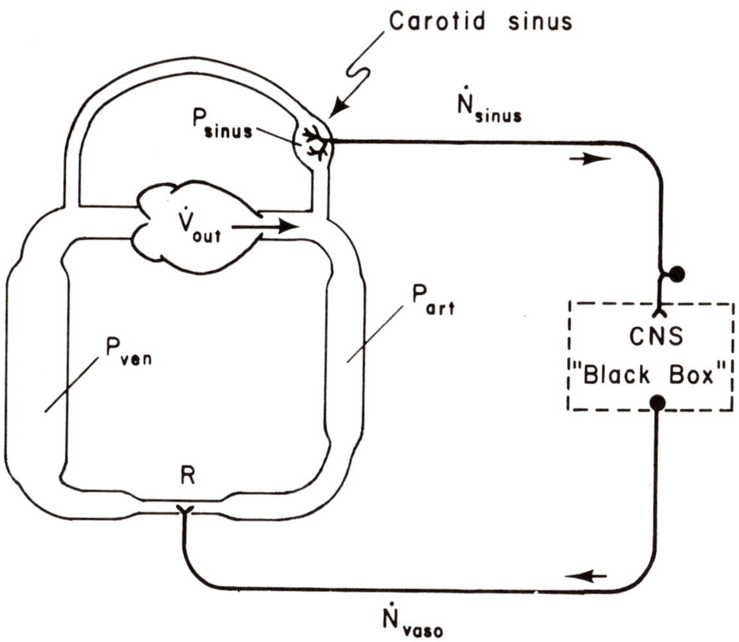

Figure 1.

The crude model of the circulation used here is much the same as the one used in the previous section.

An increase in $P_{art}(\infty)$ will increase $P_{sinus}(\infty)$, stretching the terminals of the sinus nerve fibers, and increasing $\dot{N}_{sinus}(\infty)$. Via central connections, an increase in the afferent $\dot{N}_{sinus}(\infty)$ will cause a decrease in the efferent $\dot{N}_{vaso}(\infty)$, which, in turn, will decrease vasomotor tone. The decrease in vasomotor tone will decrease R. The decrease in R will cause a decrease in $P_{art}(\infty)$, thereby opposing the initial change.

Derivation

If P_{ven} and $\dot{V}_{out}$ are kept constant as parameters,

$$P_{art}(\infty) = P_{ven} + \dot{V}_{out}\, R(\infty) \qquad 1.T$$

In the intact circulation

$$P_{sinus}(\infty) = P_{art}(\infty) \qquad 2.T$$

The observed (1) response of single fibers in the sinus nerve may be described by

$$\dot{N}_{sinus}(\infty) = \frac{\dot{N}_{max}[P_{sinus}(\infty) - P_{thr}]}{[P_{sinus}(\infty) - P_{thr}] + A}\,;\ P_{sinus}(\infty) \geq P_{thr} \qquad 3.E$$

$$\dot{N}_{sinus}(\infty) = 0 \qquad P_{sinus}(\infty) < P_{thr}$$

Exemplary values of the parameters for one such fiber are given above, but these parameters will differ considerably from fiber to fiber so that even the form of the total response is impossible to predict from observations of only a few members of the whole population of fibers.

An incomplete symbol-and-arrow diagram corresponding to the preceding description is given in Figure 2. Unfortunately, equations for the remaining parts of the loop are not available. We must therefore turn to a simpler analysis, based on the experimental observations of Scher and Young (2). In the vagotomized dog or cat under light chloralose anesthesia, the pressure in the carotid sinuses was controlled by isolating the common carotids from the systemic arterial system and perfusing them by a pump. With the feedback loop thus opened between P_{art} and P_{sinus}, the observed dependence of $P_{art}(\infty)$ upon $P_{sinus}(\infty)$ can be described by the empirical equation,

$$P_{art}(\infty) = P_{min} + \frac{P_{max} - P_{min}}{1 + \epsilon^{-a + bP_{sinus}(\infty)}} \qquad 4.E$$

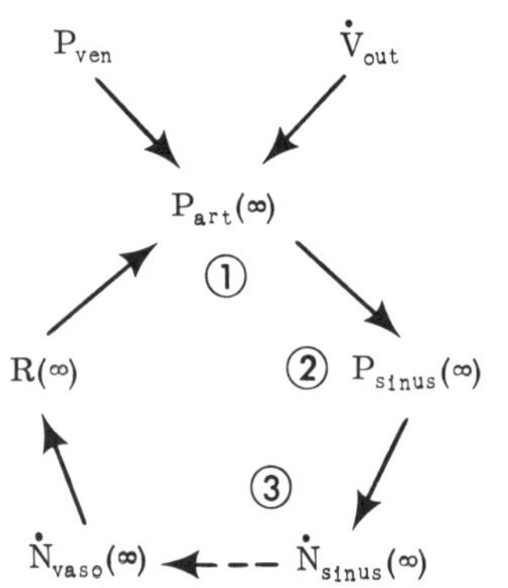

Figure 2.

Equations 2 and 4 are the two feedback equations corresponding to the symbol-and-arrow diagram of Figure 3.

From equation 2,

$$dP_{sinus}(\infty)/\ dP_{art}(\infty) = 1 \qquad 5.$$

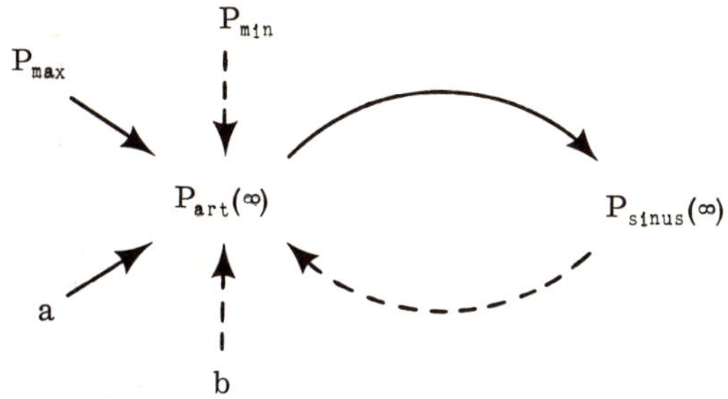

Figure 3.

From equation 4,

$$\frac{dP_{art}(\infty)}{dP_{sinus}(\infty)} = -\frac{b\,\epsilon^{-a + bP_{sinus}(\infty)}\left[P_{max} - P_{min}\right]}{\left[1 + \epsilon^{-a + bP_{sinus}(\infty)}\right]^2} \qquad 6.$$

The open-loop gain is the absolute value of the product of these two derivatives.

$$[OLG](j0) = \frac{b\,\epsilon^{-a + bP_{sinus}(\infty)}\left[P_{max} - P_{min}\right]}{\left[1 + \epsilon^{-a + bP_{sinus}(\infty)}\right]^2} \qquad 7.$$

Comment

In cats, the maximum value of [OLG](j0) ranged from 1 to 15 with a mean of about 5; in dogs from 0.8 to 8 with a mean of about 2. (2) The exemplary values listed here, and used for calculating the curves in Figures 4 and 5 give a maximum gain of 4.

Transient behavior

The stretch receptors in the carotid sinus are well-known (1, 2) to give a considerably greater response (larger $\dot{N}_{sinus}(t)$) when $P_{sinus}(t)$ is increasing rapidly than after they have adapted to a new steady-state value of $P_{sinus}(\infty)$. The lack of any pulsatile change in $P_{sinus}(\infty)$ may well account for the abnormally high blood pressures observed when $P_{sinus}(\infty)$ was kept constant in order to estimate [OLG](j0). Indeed, the response of $P_{art}(\infty)$ to a given mean value of $P_{sinus}(t)$ was greater when $P_{sinus}(t)$ was a sinusoidal function of time than when it was steady, even when ω was so large that [OLG]($j\omega$) was practically zero. For details of the responses to sinusoidal and square-wave changes in $P_{sinus}(t)$, Scher and Young's paper should be consulted (2).

Related feedback loops

Baroreceptors also exist in the aortic arch. Their effects were usually

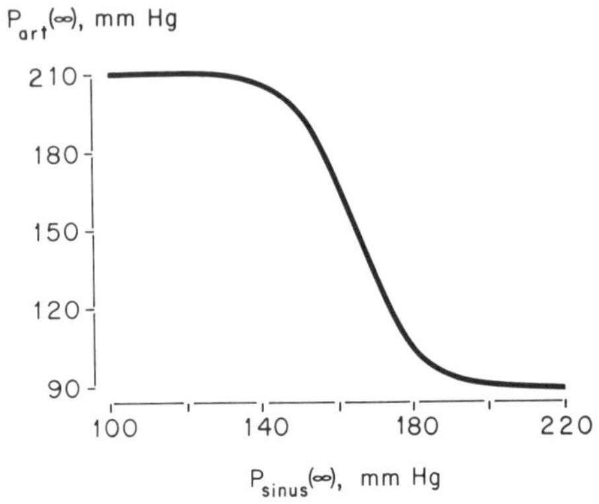

Figure 4.

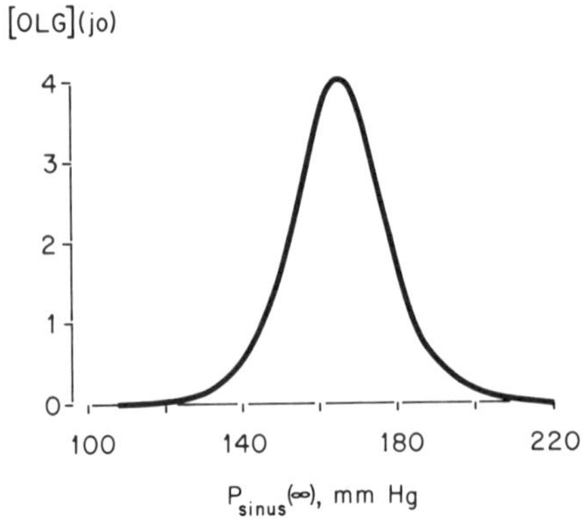

Figure 5.

eliminated in the experiments described above because the vagi were cut. Cutting the vagi also seriously impairs the baroreceptor-cardiac feedback loops which help to control heart rate and myocardial contractility (3).

The systemic renin-angiotensin loops described elsewhere in Part II, are important for the long-term regulation of blood pressure.

For a simple "artificial carotid sinus," see the following section.

References

1. Landgren, S.: On the excitation mechanism of the carotid baroreceptors. Acta physiol. scand., 26, 1-34, 1952.

2. Scher, A.M., and Young, A.C.: Servoanalysis of carotid sinus reflex effects on peripheral resistance. Circulation Res., 12, 152-162, 1963.

3. Sarnoff, S.J., Gilmore, J.P., Brockman, S.K., Mitchell, J.H., and Linden, R.J.: Regulation of ventricular contraction by the carotid sinus. Circulation Res., 8, 1123-1136, 1960.

See also,

Roston, S.: The cardiovascular effects of the carotid sinus mechanism. Bull. math. Biophys., 27, 167-173, 1965.

Kezdi, P., editor. Baroreceptors and Hypertension. Pergamon Press Inc., New York, 1967.

AN "ARTIFICIAL CAROTID SINUS": BLOOD PRESSURE; RATE OF INFUSION OF NOREPINEPHRINE

NEGATIVE

Type	Symbol	Definition	Example
Ind.	P_{res}	Pressure head of norepinephrine solution in reservoir.	105 mm Hg
Par.	P_{valve}	Pressure needed to open check valve.	17 mm Hg
	R	Resistance of infusion apparatus to flow.	12 mm Hg per [ml/min]
	C	Concentration of norepinephrine in infusion solution.	5 µg/ml
	P_{min}	Diastolic pressure when $\dot{Q}(\infty) = 0$	51 mm Hg
	P_{max}	Maximum diastolic pressure approached as $\dot{Q}(\infty) \rightarrow \infty$	191 mm Hg
	$\dot{Q}_{1/2}$	Rate of norepinephrine infusion for half-maximal response.	6.7 µg/min
Dep.	$P_{diast}(\infty)$	Mean diastolic blood pressure.	83.2 mm Hg
	$\dot{V}(\infty)$	Rate of infusion of solution from reservoir.	0.40 ml/min
	$\dot{Q}(\infty)$	Rate of infusion of norepinephrine.	2.00 µg/min

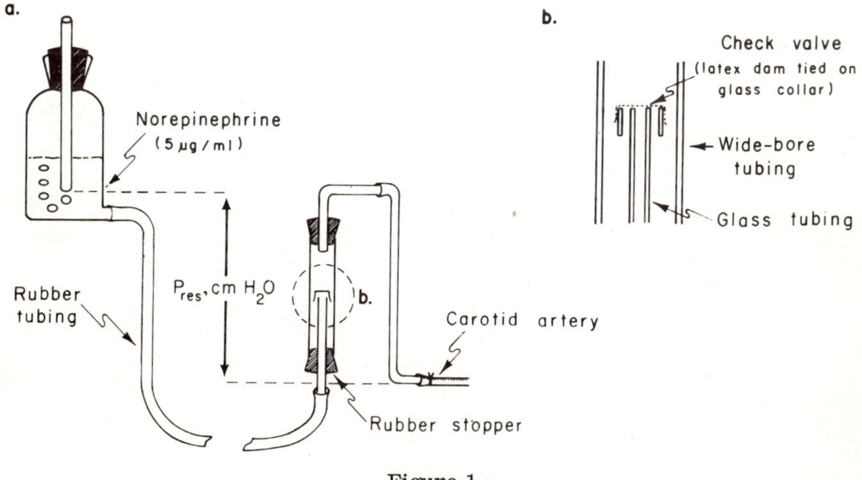

Figure 1.

A solution of norepinephrine is infused intra-arterially by gravity from a reservoir whose height above the point of infusion can be controlled. A check-valve in the infusion system permits the norepinephrine solution to flow into the artery whenever the net infusion pressure, $P_{res} - P_{valve}$, exceeds arterial pressure, but prevents any backflow of blood when arterial pressure exceeds net infusion pressure.

Simplifying Assumptions

1. Throughout the whole cardiac cycle the pressure which opposes $P_{res} - P_{valve}$, the infusion pressure, is the mean diastolic pressure.

2. The amplitude of any periodic fluctuation of diastolic pressure (limit cycle) is small compared with changes in mean diastolic pressure.

Derivation

By assumption 1,

$$\dot{V}(\infty) = \frac{P_{res} - P_{valve} - P_{diast}(\infty)}{R} \qquad 1.T$$

By definition,

$$\dot{Q}(\infty) = C\,\dot{V}(\infty) \qquad 2.T$$

When norepinephrine is infused intra-arterially at a constant rate, the observed dependence of diastolic pressure on rate of infusion can be described by

$$P_{diast}(\infty) = P_{min} + \frac{[P_{max} - P_{min}]\,\dot{Q}(\infty)}{\dot{Q}_{1/2} + \dot{Q}(\infty)} \qquad 3.E$$

Combine equations 1 and 2 by eliminating $\dot{V}(\infty)$ and solve for $\dot{Q}(\infty)$

$$\dot{Q}(\infty) = \frac{C}{R}\,[P_{res} - P_{valve} - P_{diast}(\infty)] \qquad 4.$$

Equations 3 and 4 are the two feedback equations. The corresponding symbol-and-arrow diagram is

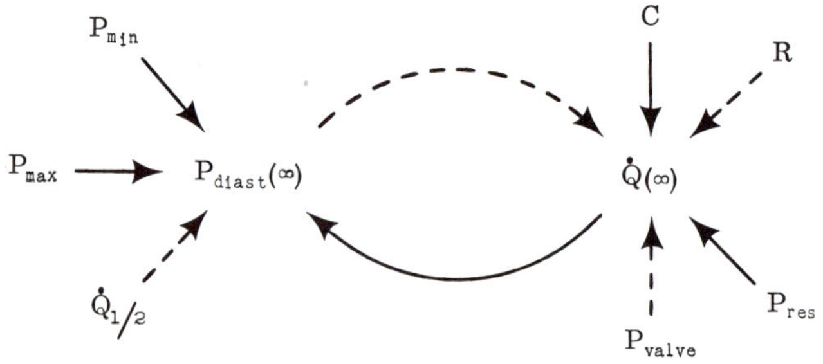

From equation 3,

$$\frac{dP_{diast}(\infty)}{d\dot{Q}(\infty)} = \frac{[P_{max} - P_{min}]\,\dot{Q}_{1/2}}{[\dot{Q}_{1/2} + \dot{Q}(\infty)]^2} \qquad 5.$$

From equation 4,

$$\frac{d\dot{Q}(\infty)}{dP_{diast}(\infty)} = -\frac{C}{R} \qquad 6.$$

The open-loop gain is the absolute value of the product of these derivatives.

$$[OLG](j0) = \frac{CQ_{1/2}[P_{max} - P_{min}]}{R[\dot{Q}_{1/2} + \dot{Q}(\infty)]^2} \qquad 7.$$

Comments

1. The exemplary values are the ones actually obtained in an experiment where the "artificial carotid sinus" was used to regulate the blood pressure of a pithed cat. (See Figure 2.) For any given reservoir pressure, the actual rate of infusion must have been somewhat less than the rate predicted on the assumption that the diastolic pressure was the pressure opposing infusion throughout the whole cardiac cycle. This is presumably why the observed mean diastolic pressures fall somewhat below the predicted pressures in Figure 2.

2. Figure 2 paints a misleadingly rosy picture of the regulatory talents of this simple apparatus, for it shows only the mean diastolic pressure. In fact, although the mean pressure was reasonably constant at a given reservoir pressure, assumption 2 was not really justified, for sinusoidal fluctuations about the mean were usually prominent, particularly at higher rates of infusion. Immediately after a step change in P_{res}, the fluctuations were often complicated, but later they tended to settle down to a monotonously regular sinusoidal pattern with a frequency of the order of 17 radians per min. (See Figure 3.) This limit cycle behavior was probably due in large part to the lag between the onset of norepinephrine infusion and the onset of the blood pressure response.

Transient behavior

See comment 2 above.

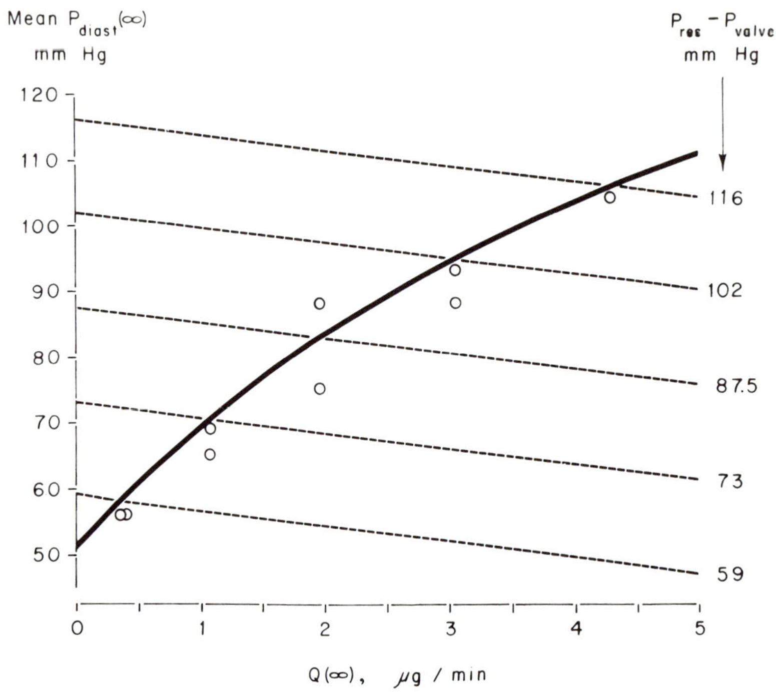

Figure 2. Regulation of mean diastolic blood pressure by the "artificial carotid sinus."

The diastolic blood pressure of a pithed cat during the intra-arterial administration of norepinephrine at various rates from a constant infusion pump could be described by the following specific version of equation 3. (Solid curve.)

$$P_{diast}(\infty) = 51 + \{140\dot{Q}(\infty)/[6.7 + \dot{Q}(\infty)]\} \text{ mm Hg}$$

The pressure-flow relationship for the infusion apparatus gave a value of 12 mm Hg per [ml/min] for R. Since C was held constant at 5 μg of norepinephrine per ml, the following specific version of equation 4,

$$\dot{Q}(\infty) = [5/12][P_{res} - P_{valve}] - [5/12]P_{diast}(\infty)\ \mu g/\text{min},$$

was used to calculate the broken lines at the values of $[P_{res} - P_{valve}]$ shown in the figure.

The points show the observed values of the mean diastolic blood pressure when the "artificial carotid sinus" was actually being used to regulate the cat's blood pressure. Each point was plotted at the $\dot{Q}(\infty)$ determined by the intersection corresponding to the $[P_{res} - P_{valve}]$ being used. Although the points tend to fall slightly below the intersections (see comment 1), they do follow the expected trend reasonably well. The [OLG](j0) varies from about 8.40 at the lowest rate of infusion used to about 3.25 at the highest rate of infusion used.

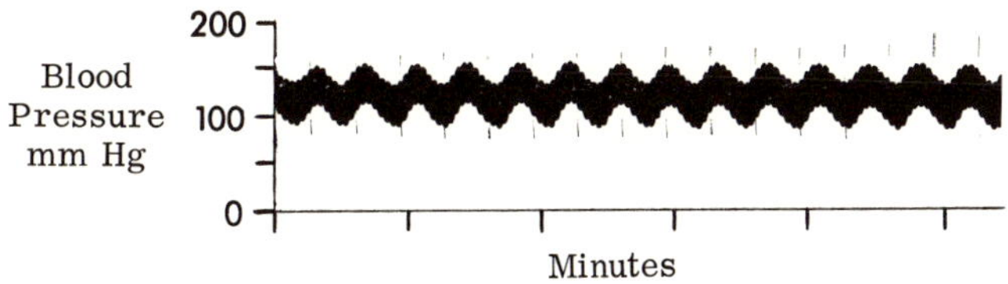

Figure 3. Pithed cat. Limit-cycle behavior of the blood pressure while being "regulated" by the "artificial carotid sinus."

The systolic blood pressure is the upper border, and the diastolic blood pressure is the lower border of the recording. The smaller fluctuations superimposed upon the larger ones are caused by the respiratory excursions of the thorax.

NEGATIVE

Type	Symbol	Definition	Example (Femoral artery flow. Dog)
Ind.	C_{art}	Concentration of oxygen in arterial blood.	0.2 ml (STPD)*/ml
Par.	P_{art}	Arterial perfusion pressure.	100 mm Hg
	P_{ven}	Venous blood pressure.	5 mm Hg
	$\dot{Q}_{max}$	Maximum O_2 consumption of perfused region.	1.5 ml (STPD)/ml
	$C_{1/2}$	$C_{ven}(\infty)$ for half-maximal oxygen consumption.	0.01 ml (STPD)/ml
Dep.	$C_{ven}(\infty)$	Concentration of O_2 in venous blood.	0.139 ml (STPD)/ml
	$p_{O_2,\ tiss}(\infty)$	Partial pressure of oxygen in perfused region.	mm Hg
	$R(\infty)$	Resistance to blood flow through perfused region.	4.13 mmHg/[ml/min]
	$\dot{V}(\infty)$	Blood flow through perfused region.	23 ml/min
	$\dot{Q}_{cons}(\infty)$	Oxygen consumption of perfused region.	1.40 ml (STPD)/min

[*STPD means "Standard Temperature and Pressure, Dry." i.e., the volume is volume of dry gas at 0° C and 760 mm Hg. One liter (STPD) of an ideal gas ≈ 44.6 mM.]

Model

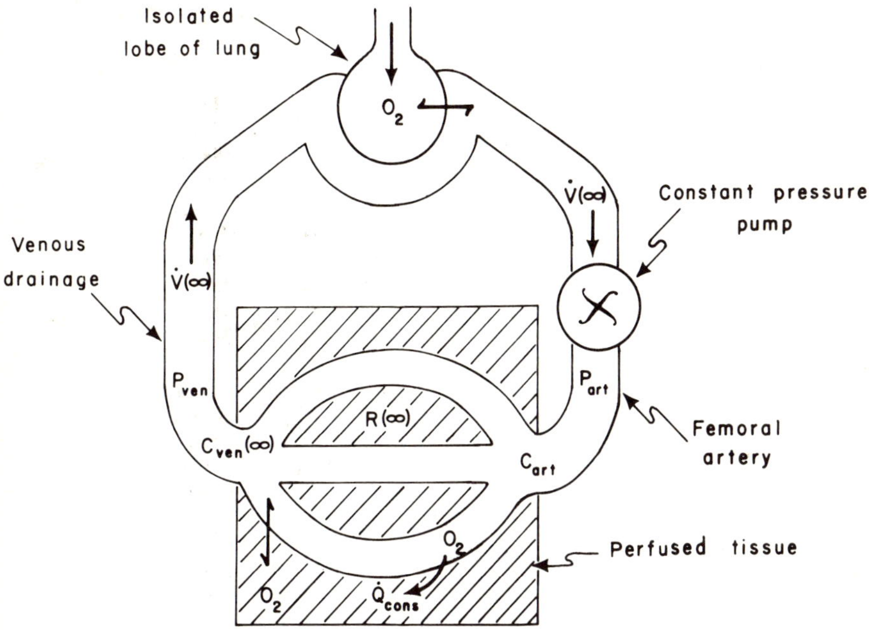

Figure 1.

This model, and some of the parameters, are based upon the experiments of Ross et al. (1). Blood used to perfuse the femoral artery of an anesthetized dog at constant pressure is oxygenated by passing it through an isolated lobe of lung where the composition of the inspired air can be controlled by the experimenter. By some unknown mechanism, hypoxia causes vasodilation, thus lowering $R(\infty)$ and increasing blood flow through the region.

Assumptions

1. The partial pressure of oxygen in the perfused tissues is equal to, or at least proportional to the partial pressure of oxygen in the venous blood.

Derivation

By Ohm's law for fluid flow,

$$\dot{V}(\infty) = \frac{P_{art} - P_{ven}}{R(\infty)} \qquad 1.T$$

In the steady state, conservation of oxygen requires that

$$C_{ven}(\infty) = C_{art} - \frac{\dot{Q}_{cons}(\infty)}{\dot{V}(\infty)} \qquad 2.T$$

By assumption 1,

$$p_{O_2,tiss}(\infty) = f(C_{ven}(\infty)) \qquad 3.E$$

where f is determined by the oxygen dissociation curve for venous blood.

Experimental observations (1) suggest that

$$R(\infty) = g(p_{O_2,tiss}(\infty)) \qquad 4.E.$$

Combining equations 3 and 4 by eliminating $p_{O_2,tiss}(\infty)$:

$$R(\infty) = g(f(C_{ven}(\infty))) \equiv h(C_{ven}(\infty)) \qquad 5$$

In Figure 2, equation 5 is represented by an empirical curve which was fitted by eye to points based upon experimental observations by Ross et al. (1).

On physiological grounds, (2), it seems likely that $\dot{Q}_{cons}(\infty)$ will itself be some function of $p_{O_2,tiss}(\infty)$ and hence, by equation 3, of $C_{ven}(\infty)$. As $C_{ven}(\infty)$ decreases from normal values, $\dot{Q}_{cons}(\infty)$ would probably change quite slowly at first, but as very low values of $C_{ven}(\infty)$ are approached $\dot{Q}_{cons}(\infty)$ would decrease more and more rapidly so as to become zero (as it must) when $C_{art} = C_{ven} = 0$. In accordance with this description, assume that

$$\dot{Q}_{cons}(\infty) = \frac{\dot{Q}_{max}\, C_{ven}(\infty)}{C_{1/2} + C_{ven}(\infty)} \qquad 6.S$$

Equations 1, 2, 5 and 6 describe the working relationships among the variables. The corresponding symbol-and-arrow diagram is shown in Figure 3.

The equations for the tangent hyperplanes at an operating point are

$$\Delta\dot{V} = \frac{1}{R(\infty)}\bigg|_{op} \Delta P_{art} - \frac{1}{R(\infty)}\bigg|_{op} \Delta P_{ven} - \frac{P_{art} - P_{ven}}{R^2(\infty)}\bigg|_{op} \Delta R \qquad 7.$$

$$\Delta C_{ven} = \Delta C_{art} - \frac{1}{\dot{V}(\infty)}\bigg|_{op} \Delta\dot{Q}_{cons} + \frac{\dot{Q}_{cons}(\infty)}{\dot{V}^2(\infty)}\bigg|_{op} \Delta\dot{V} \qquad 8.$$

$$\Delta R = \frac{\partial h(C_{ven}(\infty))}{\partial C_{ven}(\infty)}\bigg|_{op} \Delta C_{ven} + \text{terms in undefined parameters} \qquad 9.$$

$$\Delta\dot{Q}_{cons} = \frac{C_{ven}(\infty)}{C_{1/2} + C_{ven}(\infty)}\bigg|_{op} \Delta\dot{Q}_{max} + \frac{C_{1/2}\dot{Q}_{max}}{[C_{1/2} + C_{ven}(\infty)]^2}\bigg|_{op} \Delta C_{ven}$$

$$- \frac{\dot{Q}_{max}\, C_{ven}(\infty)}{[C_{1/2} + C_{ven}(\infty)]^2}\bigg|_{op} \Delta C_{1/2} \qquad 10.$$

The block diagram corresponding to these equations is shown in Figure 4.

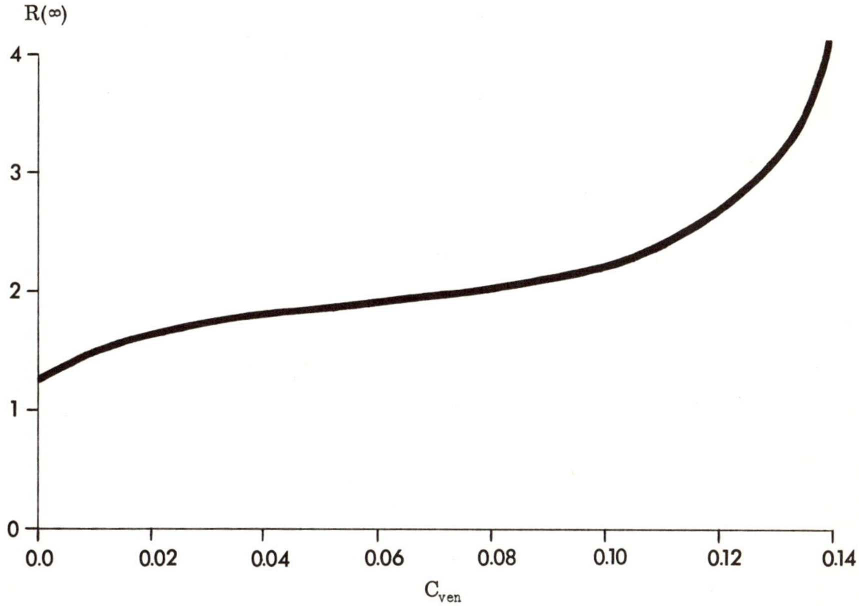

Figure 2.

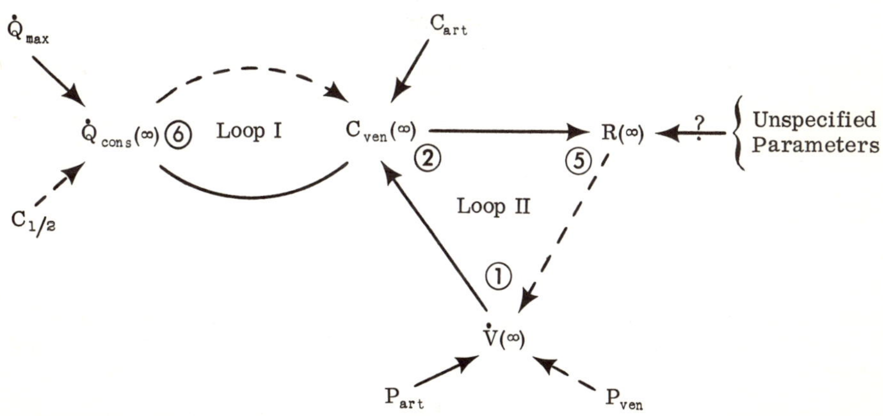

Figure 3.

For the response of $C_{ven}(\infty)$ to a change of C_{art}, the gain due to feedback from Loop I only is

$$G_{fb,I}(\infty) = [\mathrm{OLG}]_I(j0) = \frac{C_{1/2}\dot{Q}_{max}}{\dot{V}(\infty)\,[C_{1/2} + C_{ven}(\infty)]^2} \qquad 11.$$

Similarly, for Loop II only,

$$G_{fb,II}(\infty) = [OLG]_{II}(j0) = \frac{\partial h(C_{ven}(\infty))}{\partial C_{ven}(\infty)} \left\{ \frac{\dot{Q}_{cons}(\infty)[P_{art} - P_{ven}]}{[\dot{V}(\infty) R(\infty)]^2} \right\} \qquad 12.$$

Or, combining equation 12 with equation 1 so as to eliminate $\dot{V}(\infty)$ and $R(\infty)$,

$$[OLG]_{II}(j0) = \frac{\partial h(C_{ven}(\infty))}{\partial C_{ven}(\infty)} \left\{ \frac{\dot{Q}_{cons}(\infty)}{P_{art} - P_{ven}} \right\} \qquad 13.$$

For both loops together,

$$G_{fb,both}(\infty) = [OLG]_{both}(j0) = OLG_{I}(j0) + OLG_{II}(j0) \qquad 14.$$

In Figure 5 values of these gains calculated from the listed parameters and the empirical curve of Figure 2 have been plotted against C_{art}. In calculating the gain for Loop II, the partial derivative in equation 12 was estimated from the slope of tangent lines fitted by eye to the curve of Figure 2 at various points.

Comments

1. In this analysis I have leaned heavily upon the observed data of Ross et al (1) who measured C_{art} and $\dot{V}(\infty)$ in the experiments upon which the model here described is based. The curve in Figure 2 was calculated from the data in their Figure 3 by using the relationships stated by equations 1, 2 and 6.

2. Although I have tried to choose sensible values for the parameters not specified by Ross et al (i.e. P_{ven}, $\dot{Q}_{max}$ and $C_{1/2}$), equation 6 for Loop I is entirely suppositious. However, some such relationship must exist. If equation 6 is at all reasonable, Loop I has little effect on $C_{ven}(\infty)$ except in severe anoxia (Figure 5), and even then its ability to minimize further decreases in $C_{ven}(\infty)$ depends upon a decrease in $\dot{Q}_{cons}(\infty)$. Since $\dot{Q}_{cons}(\infty)$ rather than $C_{ven}(\infty)$ is presumably the important physiological variable, Loop I can hardly be regarded as a valid regulatory mechanism.

Related loops

1. Similar autoregulation of blood flow occurs in many regions, notably the myocardium.

2. Autoregulation of renal blood flow may depend upon a mechanism more specific than mere anoxia. (See section in Part II describing the feedback loop involving the macula densa.)

3. Regulation of cerebral blood flow depends largely upon the concentration of carbon dioxide. See, for example, Figure 8, p.955 in (3).

References

1. Ross, J.M., Fairchild, H.M., Weldy, J. and Guyton, A.C.: Autoregulation of blood flow by oxygen lack Amer. J. Physiol. 202, 21-24, 1962.

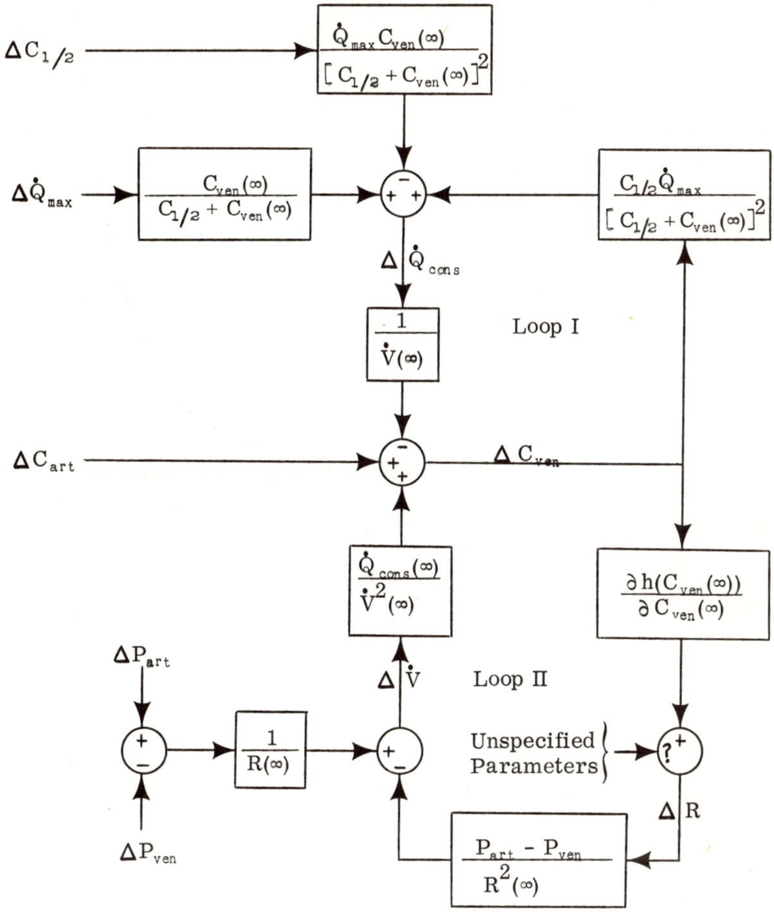

Figure 4.

References (continued)

2. Jöbsis, F. F.: Basic processes in cellular respiration. Chapter 2 in Handbook of Physiology, Section 3 (Respiration), Vol. I. American Physiological Society, Washington, D. C., 1964.

3. Patterson, J. L. Jr.: Circulation through the brain. In Ruch, T. C. and Patton, H. D., Physiology and Biophysics. 19th edition. W. B. Saunders Co. Philadelphia, 1965.

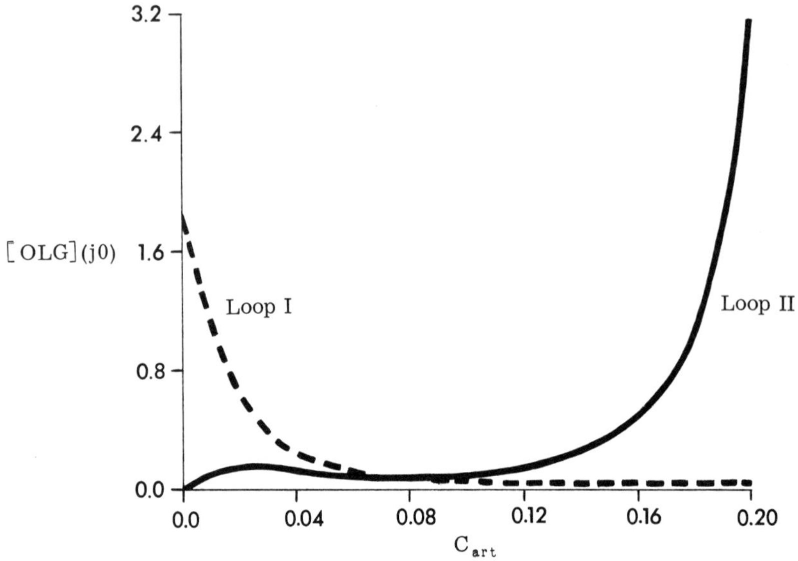

Figure 5.

RADIUS AND WALL THICKNESS DURING DISTENSION OF THIN-WALLED ELASTIC TUBES

POSITIVE

Type	Symbol	Definition	Example (latex tubing)
Ind.	P	Pressure in lumen of tube. (Net transmural pressure.)	dynes/cm^2
Par.	L	Length of tube.	4.00 cm
	r_o	Radius of unstressed tube.	0.33 cm
	w_o	Wall thickness of unstressed tube.	0.04 cm
	V	Volume of material of wall of tube.	0.332 cm^3
	Y	Young's modulus of elasticity.	$1.44 \cdot 10^7$ dynes/cm^2
	θ	A small angle between two radial planes.	radians
	P_{crit}	The critical blowout pressure. A dependent parameter defined by equation 19.	$4.36 \cdot 10^5$ dynes/cm^2 (327 mm Hg)
Dep.	$r(\infty)$	Radius of tube.	cm
	$w(\infty)$	Wall thickness.	cm
	$P_{stress}(\infty)$	Tangential stress in wall of tube.	dynes/cm^2
	$[Strain](\infty)$	Strain in wall of tube.	
	$F_p(\infty)$	Force exerted by P on segment of tube subtended by θ.	dynes
	$F_{tan}(\infty)$	Force exerted by $P_{stress}(\infty)$ on one lateral surface of the segment of tube subtended by θ.	dynes

Model

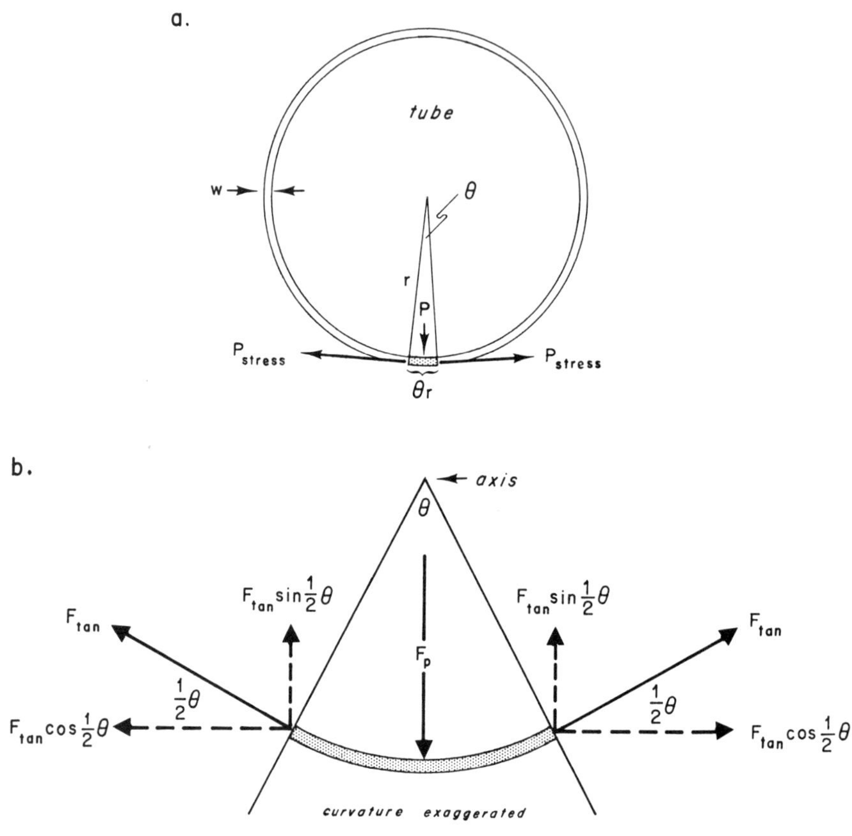

Figure 1.

A uniform, thin-walled tube of constant length, L, is distended by an intraluminal pressure, P. In the steady state, any increase in P must be counterbalanced by an equal increase in the inwardly-directed component of the stress in the wall of the tube caused by stretching. But stretching the tube also thins and weakens the wall, thereby demanding even greater stretching to produce the stress needed to balance the pressure.

Assumptions

1. $w_0 <<< r_0$, so that when the tube is distended all layers of the wall are subjected to essentially the same strain.

2. The wall of the tube is of uniform composition throughout, and is subject to no stresses or strains when P = 0.

Derivation

Consider a small segment of the wall of the tube (Figure 1). The length L, of the segment is measured parallel to the axis of the tube; its thickness $w(\infty)$ is the wall thickness; its width, measured along the circumference, is $\theta r(\infty)$, i.e. the length of arc subtended by the very small angle, θ. The outward force, $F_P(\infty)$, exerted upon this segment is the product of the intraluminal pressure, P, and the area of the inner surface of the segment on which it presses.

$$F_P(\infty) = P\, L\theta r(\infty) \qquad 1.T$$

At each lateral surface of the segment, the stress set up in the stretched wall exerts a tangential force, $F_{tan}(\infty)$, which is the product of $P_{stress}(\infty)$ and the area of the lateral surface of the segment on which it pulls (Figure 1b).

$$F_{tan}(\infty) = P_{stress}(\infty) L w(\infty) \qquad 2.T$$

Each of the two tangential forces may be resolved into two components. (See Figure 1b.) The cosine components, being opposite in direction but equal in magnitude, cancel each other. The sine components are equal in both magnitude and direction, and together constitute the inward force opposing $F_P(\infty)$. Hence in the steady state,

$$F_P(\infty) = 2\, F_{tan}(\infty) \sin(\theta/2) \qquad 3.T.$$

But θ may be taken as arbitrarily small. For very small θ,

$$\sin(\theta/2) \approx \theta/2 \qquad 4.T$$

Combining equations 3 and 4,

$$F_P(\infty) = F_{tan}(\infty)\theta \qquad 5.$$

Combining equations 1, 2 and 5, by eliminating $F_P(\infty)$ and $F_{tan}(\infty)$, and solving for $P_{stress}(\infty)$,

$$P_{stress}(\infty) = P\, r(\infty)/w(\infty) \qquad 6.$$

By definition, the circumferential strain is

$$[\text{Strain}](\infty) = \frac{2\pi r(\infty) - 2\pi r_0}{2\pi r_0} = \frac{r(\infty)}{r_0} - 1 \qquad 7.T$$

Hooke's law states that

$$P_{stress}(\infty) = Y[\text{Strain}](\infty) \tag{8.T}$$

Combining equations 6, 7 and 8 by eliminating $[\text{Strain}](\infty)$ and $P_{stress}(\infty)$, and solving for $r(\infty)$,

$$r(\infty) = \frac{1}{\dfrac{1}{r_o} - \dfrac{P}{Yw(\infty)}} \tag{9.}$$

Since the parameter V remains constant,

$$V = L\,w_o\,2\pi\,r_o = L\,w(\infty)\,2\pi\,r(\infty) \tag{10.T}$$

or,

$$w(\infty) = r_o\,w_o/r(\infty) \tag{10.a}$$

Equations 9 and 10a are the working equations for the two feedback variables, $r(\infty)$ and $w(\infty)$. The corresponding symbol-and-arrow diagram is

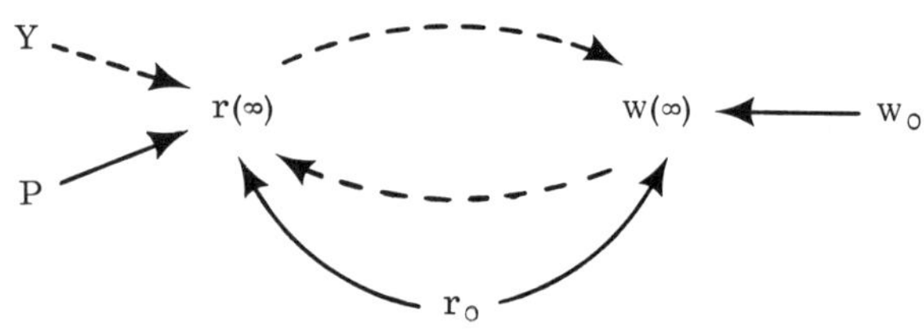

Figure 2.

From equation 9,

$$\frac{d\,r(\infty)}{d\,w(\infty)} = -\frac{P\,r^2(\infty)}{Y\,w^2(\infty)} \tag{11.}$$

From equation 10a,

$$\frac{d\,w(\infty)}{d\,r(\infty)} = -\frac{r_o\,w_o}{r^2(\infty)} = -\frac{w(\infty)}{r(\infty)} \tag{12.}$$

Hence,

$$[\text{OLG}](j0) = \frac{P\,r(\infty)}{Y\,w(\infty)} \tag{13.}$$

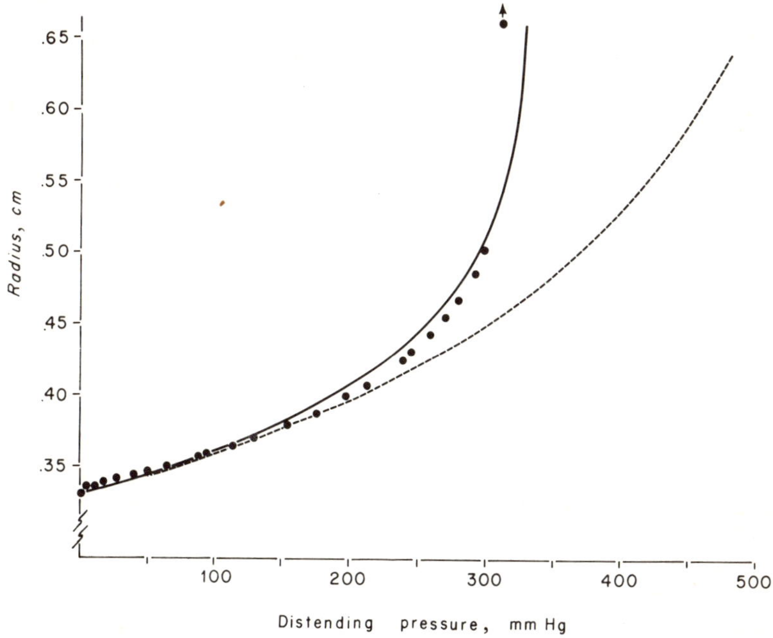

Figure 3. The steady-state radius of the thin-walled latex tube used as an example is plotted against the intraluminal pressure. Each point is an experimental observation. The uppermost point, (with arrow), does not represent a steady state but corresponds to the observed blowout pressure. Its position on the ordinate is therefore arbitrary. The solid line shows the theoretical relationship predicted by equations 9 and 10a from the dimensions of the unstressed tube and the measured value of Young's modulus. The broken line shows, for the same parameters, the relationship derived by Streeter et al. (2). It is obvious that the solid line fits the points better than the broken line.

Combining equations 6 and 8 by eliminating $P_{stress}(\infty)$,

$$[\text{Strain}](\infty) = \frac{P\ r(\infty)}{Y\ w(\infty)} \qquad 14.$$

Combining equations 13 and 14,

$$[\text{OLG}](j0) = [\text{Strain}](\infty) \qquad 15.$$

Since this is a positive feedback system, a stable steady state cannot be maintained if $[\text{OLG}](j0)$ equals or exceeds unity. Hence, from equation 15,

$$[\text{Strain}](\infty) < 1 \qquad 16.$$

Equation 16 predicts that the tube will become unstable, and will therefore "blow out" by rupture or ballooning when it has been stretched to twice its original radius. This prediction was borne out reasonably well by the actual behavior of a latex tube with the parameters listed in the example. (See Figure 3.)

The critical blowout pressure needed to produce unit strain may be expressed in terms of parameters by setting [Strain](∞) = 1 both in equation 14

$$P_{crit} = Y\frac{w(\infty)}{r(\infty)} \qquad 17.$$

and in equation 7 combined with 10a

$$\frac{r(\infty)}{r_o} = \frac{w_o}{w(\infty)} = 2 \qquad 18.$$

Combining equations 17 and 18 by eliminating $w(\infty)$ and $r(\infty)$,

$$P_{crit} = \frac{Yw_o}{4r_o} \qquad 19.$$

In Figure 4, curves have been plotted according to equations 9 and 10a for various values of P. At the critical pressure the curves for the two feedback equations do not actually intersect, but (as plotted on the graph) are tangent to each other at their one point of contact. This means that their derivatives at that point are, in fact, reciprocals of each other, so that their product is unity. At any higher pressure, the curves do not even touch, making it perfectly clear that no steady-state solution is possible.

Comments

1. With V and L fixed, equation 10a really expresses a kind of equilibrium between $w(\infty)$ and $r(\infty)$, and it could therefore be regarded as a valid working equation for $r(\infty)$ just as well as for $w(\infty)$. This is not true of equation 9, which confers upon the system the directionality shown in Figure 2.

2. In two publications (1,2), the critical blowout pressure has erroneously been given as the equivalent of $Yw_o/2r_o$, just double the value of equation 19. In deriving their equation, Streeter et al. (2) state that "the diameter change, on a percentage basis, is much less than the pressure change" and they therefore neglect the change in "tensile force" (and hence stress) with change in diameter. But as the critical blowout pressure is approached, the change in diameter becomes much larger than the change in pressure on a percentage basis; indeed, at the critical blowout pressure, an infinitesimal change in pressure causes an indefinitely large change in diameter. This explains the failure of their equation for $r(\infty)$ as a function of P (broken curve, Figure 3) to fit the observed points as P_{crit} is approached.

3. Since "blowing out" of the tube is determined by strain alone (equation 16), it should occur at twice the unstressed radius even if different layers of the tube are composed of different materials (as would be true of "biological" tubes) provided that for each layer the underlying assumptions hold true. But these assumptions include constancy of the parameters, and it is not likely that V, and particularly Y, would actually remain unchanged over a wide range of strains. The model analyzed here is therefore of highly dubious validity as a model of such things as blood vessels.

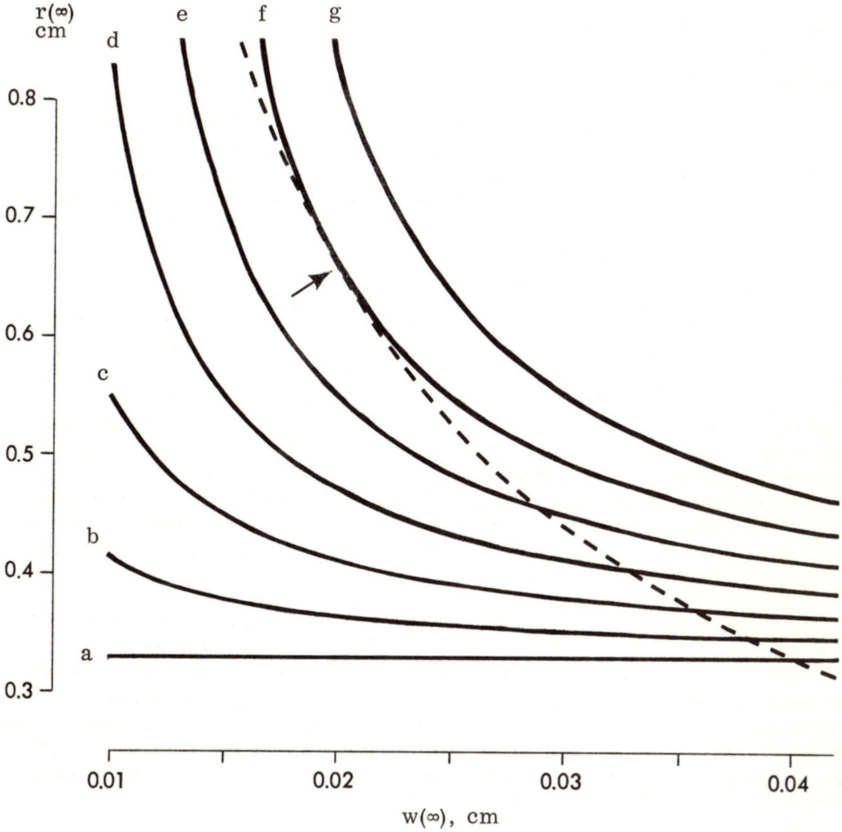

Figure 4. r(∞) as a function of w(∞) (solid curves) has been plotted according to equation 9 for the following pressures in dynes/cm^2:

a, 0.000; b, $0.872 \cdot 10^5$; c, $1.744 \cdot 10^5$; d, $2.616 \cdot 10^5$; e, $3.488 \cdot 10^5$; f, $4.360 \cdot 10^5 = P_{crit}$; g, $5.232 \cdot 10^5$.

w(∞) as a function of r(∞) (broken curve) has been plotted according to equation 10a. No curve for $P > P_{crit}$ will intersect the broken line, so no steady state could exist. The arrow indicates the critical point of tangency.

References

1. Burton, A.C. : Physical principles of circulatory phenomena: the physical equilibria of the heart and blood vessels. In Handbook of Physiology, Section 2, Vol. 1, pp. 85-106. American Physiological Society, Washington, D.C., 1962.

2. Streeter, V.L., Keitzer, W.F., and Bohr, D.R.: Pulsatile pressure and flow through distensible vessels. Circulation Res. 13, 3-20, 1963.

RENIN, ANGIOTENSIN, SYSTEMIC BLOOD PRESSURE

NEGATIVE

Qualitative description.

Figure 1 presents a purely descriptive variable-and-arrow diagram showing the causal relationships among the factors thought to be involved in the regulation of blood pressure via the release of renin into the systemic circulation. The possible role of angiotensin in the local regulation of glomerular blood flow and glomerular filtration is described elsewhere in Part II. (See "Macula densa chloride," etc.).

A decrease in mean arterial blood pressure of as little as five or ten mm Hg will promptly increase renin release from the granular cells in the walls of the afferent glomerular arterioles of the kidney (1). The immediate cause of the renin secretion may be a decreased stretching (= decreased strain) of the arteriolar wall (2). In the blood stream, renin acts upon its substrate to form angiotensin I, which, though itself inactive, is converted to the active angiotensin II by the converting enzyme in the lungs (3). Since "angiotensin is, on a wright basis, by far the most active pressor substance known" (4), it may well tend to counteract the initial fall in blood pressure by directly constricting small blood vessels and increasing peripheral vascular resistance (loop I in Figure 1). However, the importance of this loop in regulating blood pressure has been questioned (5).

The second loop (loop II in Figure 1), depends upon the ability of angiotensin II to stimulate secretion of aldosterone from the zona glomerulosa of the adrenal cortex. The aldosterone promotes the reabsorption of sodium (and, secondarily, water) from the glomerular filtrate, thereby expanding the volume of extracellular fluid (= ECF) including plasma. The resulting increase in blood volume increases cardiac output and hence mean arterial blood pressure. Loop II can be activated by a decrease in sodium intake. In normal dogs, four days of sodium depletion produced both a marked elevation of plasma renin and a three-fold increase in aldosterone secretion (6).

Loop III in Figure 1 does not involve renin, but depends upon the ability of a high concentration of potassium in plasma to cause the secretion of aldosterone, apparently by a direct effect upon the zona glomerulosa. The aldosterone increases potassium excretion thus tending to counteract the original increase in plasma potassium.

References

1. Skinner, S.L., McCubbin, J.W., and Page, I.H.: Control of renin secretion. Circulation Res., 15, 64-76, 1964.

2. Tobian L.: Relationship of juxtaglomerular apparatus to renin and angiotensin. Circulation, 25, 189-192, 1962.

3. Ng, K.K.F., and Vane, J.R.: Conversion of angiotensin I to angiotensin II. Nature, 216, 762-766, 1967.

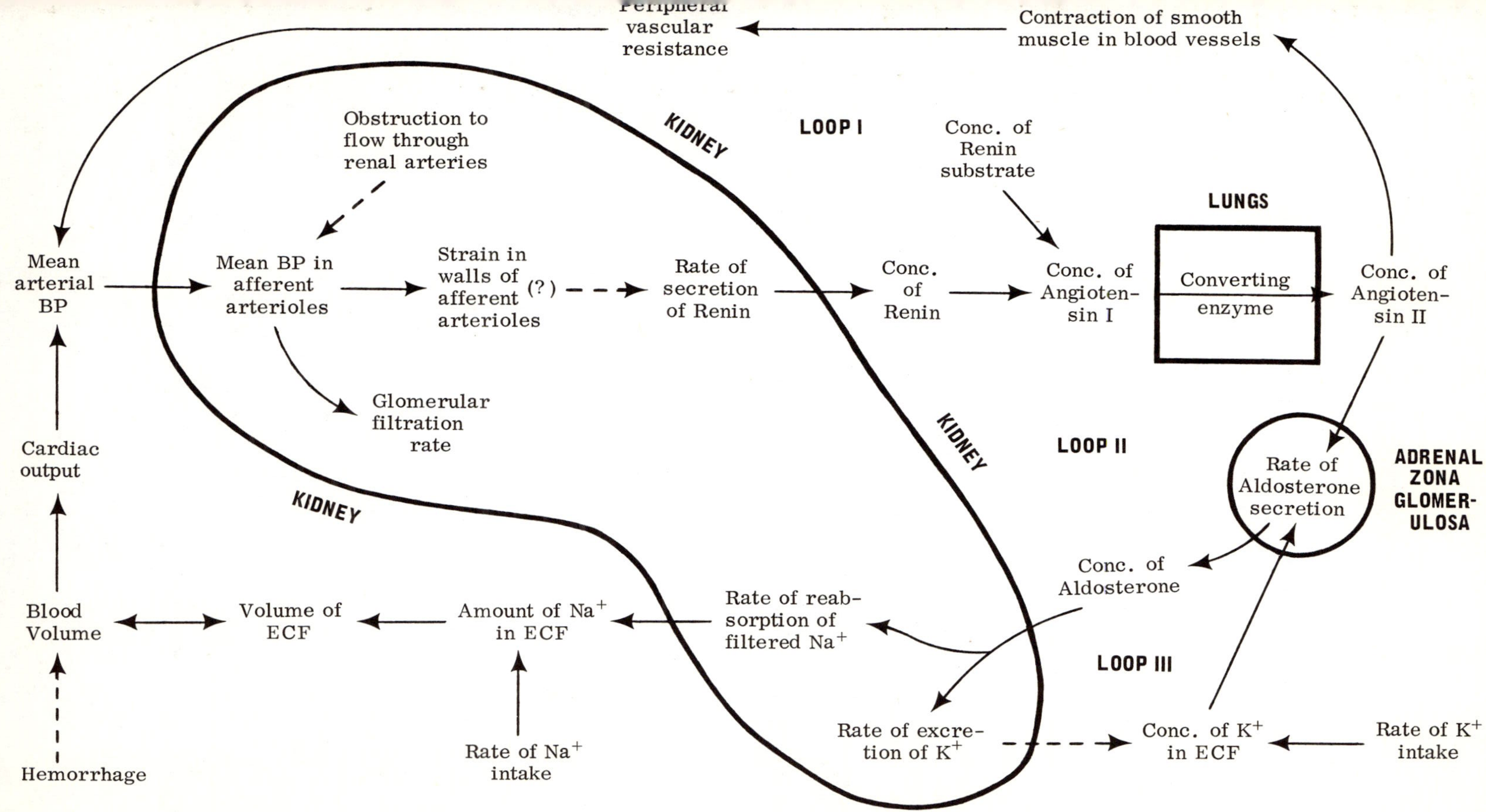
Peripheral vascular resistance
Contraction of smooth muscle in blood vessels
KIDNEY
LOOP I
Obstruction to flow through renal arteries
Conc. of Renin substrate
LUNGS
Mean arterial BP
Mean BP in afferent arterioles
Strain in walls of afferent arterioles
(?)
Rate of secretion of Renin
Conc. of Renin
Conc. of Angiotensin I
Converting enzyme
Conc. of Angiotensin II
Glomerular filtration rate
KIDNEY
LOOP II
Cardiac output
Rate of Aldosterone secretion
ADRENAL ZONA GLOMERULOSA
KIDNEY
Conc. of Aldosterone
Blood Volume
Volume of ECF
Amount of Na+ in ECF
Rate of reabsorption of filtered Na+
LOOP III
Rate of excretion of K+
Conc. of K+ in ECF
Rate of K+ intake
Hemorrhage
Rate of Na+ intake

Figure 1.

References (continued)

4. Page, I.H.: A story of hypertension. Federation Proc., 23, 693-704, 1964.

5. Peart, W.S.: The renin-angiotensin system. Pharm. Rev., 17, 143-182, 1965.

6. Binnion, P.F., Davis, J.O., Brown, T.C., and Olichney, M.J.: Mechanisms regulating aldosterone secretion during sodium depletion. Am. J. Physiol., 208, 655-661, 1965.

7. Cannon, P.J., Ames, R.P., and Laragh, J.H.: Relation between potassium balance and aldosterone secretion in normal subjects and in patients with hypertensive or renal tubular disease. J. clin. Invest., 45, 865-879, 1966.

See also

Tobian, L.: Renin release and its role in renal function and the control of salt balance and arterial pressure. Federation Proc., 26, 48-54, 1967.

Renin mechanisms and hypertension. Circulation Res., 21, Supplement II, pp. II-1 to II-226, 1967.

Vander, A.J.: Control of renin release. Physiol. Rev., 47, 359-382, 1967.

MACULA DENSA CHLORIDE, ANGIOTENSIN II, GLOMERULAR FILTRATION RATE. (A PRELIMINARY HYPOTHESIS)

NEGATIVE

Experimental Observations

1. The macula densa of a given nephron is always located at the vascular pole of the glomerulus of its own nephron. This strongly suggests that the macula densa has to do with the function of that nephron only.

2. At the vascular pole of its glomerulus, the macula densa is in intimate contact both with the afferent and efferent arterioles and with the layers of small, usually agranular cells ("lacis cells" (1); many earlier synonyms (2)) which occupy the space between the macula densa and the two arterioles, and which extend into the glomerulus as the "mesangial cells" (1) between the glomerular capillary loops. These small cells have been described as forming a "micromuscular sphincter" around the afferent arteriole (3). Like the adjacent granular, renin-containing cells of the afferent arteriole, they "contain fibrillar bundles resembling those of smooth muscle, and therefore" they "may be considered modified smooth muscle cells." (4) The macula densa is thus strategically placed to transmit some sort of signal from the post-loop-of-Henle fluid in the lumen of the nephron to the granular and agranular cells of the vascular pole of the glomerulus. (See Figure 1.)

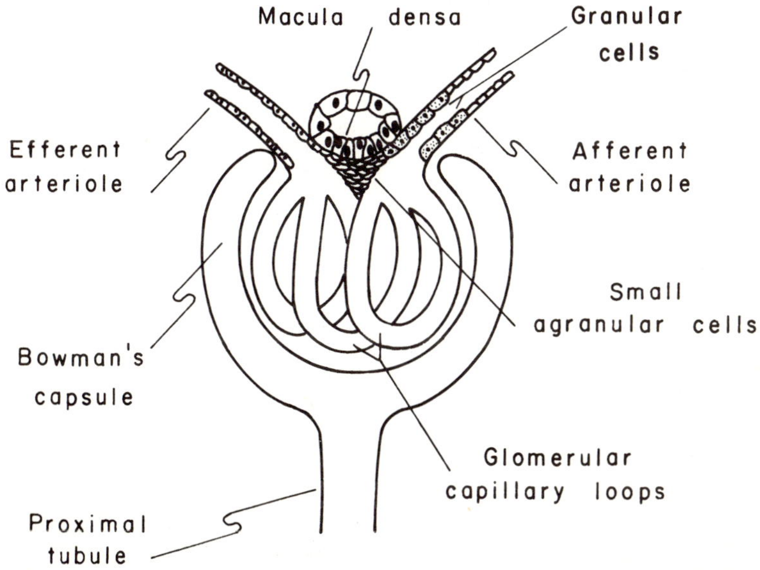

Figure 1. Diagram of the anatomical relationships among various juxtaglomerular structures.

The extension of the small agranular cells into the glomerulus as mesangium around the capillary loops is not shown.

3. Renin itself, being two stages away from the active Angiotensin II, is ideally suited to be a systemic, not a local hormone.

4. Exposure of the macula densa region of a particular nephron to a high concentration of sodium chloride or sodium bromide delivered by micropuncture causes almost immediate cessation of filtration at the glomerulus of that nephron (5).

5. The converting enzyme which converts the inactive Angiotensin I to the active Angiotensin II requires the chloride ion, or some similar anion such as Br^-, F^-, or NO_3^- (6).

Preliminary hypothesis

An increase in chloride concentration in the post-loop-of-Henle fluid causes an increase in chloride concentration in the cells of the macula densa region. Assume that converting enzyme is present in the juxtaglomerular region. Since renin is synthesized and stored in the granular cells which are found chiefly in the afferent arteriole, and since renin substrate is present in plasma (and may, in fact, be destroyed by the kidney), it seems logical to assume that Angiotensin I will also be present in the same region. Then the local increase in chloride concentration will increase the activity of the converting enzyme, so that more Angiotension II will be produced locally. The Angiotensin II will cause contraction of the adjacent smooth muscle cells, possibly including the renin-containing granular cells, or the "micromuscular sphincter", or both, thus constricting the afferent arteriole and decreasing the filtration pressure. Less chloride will be filtered, and the concentration of chloride in the post-loop-of-Henle fluid will decrease, thus tending to offset the original change.

The proposed feedback loop is depicted in the descriptive variable-and-arrow diagram of Figure 2.

Comments

1. As long ago as 1952, De Muylder deduced from the anatomical arrangement of the region "that in a normal functioning kidney, the role of the macula densa is very important in adapting the circulation of the nephron to the composition of the urine" (3). More recently Guyton et al. (7) have suggested that in response to an increase in osmotic pressure of the tubular fluid, the macula densa initiates a feedback signal which, in turn, produces arteriolar vasoconstriction. Like these and other (5, 8) previous theories, the present hypothesis, though plausible, lacks direct experimental confirmation.

2. Autoregulation of Renal Blood Flow and Glomerular Filtration Rate. A mechanism such as the one here postulated could account for the remarkable ability of the kidney to minimize changes in both blood flow and glomerular filtration rate caused by changes in blood pressure, particularly in the range above 70 mm Hg (9). As Figure 2 suggests, the direct effect of an increase in systemic blood pressure would be to increase both glomerular filtration and

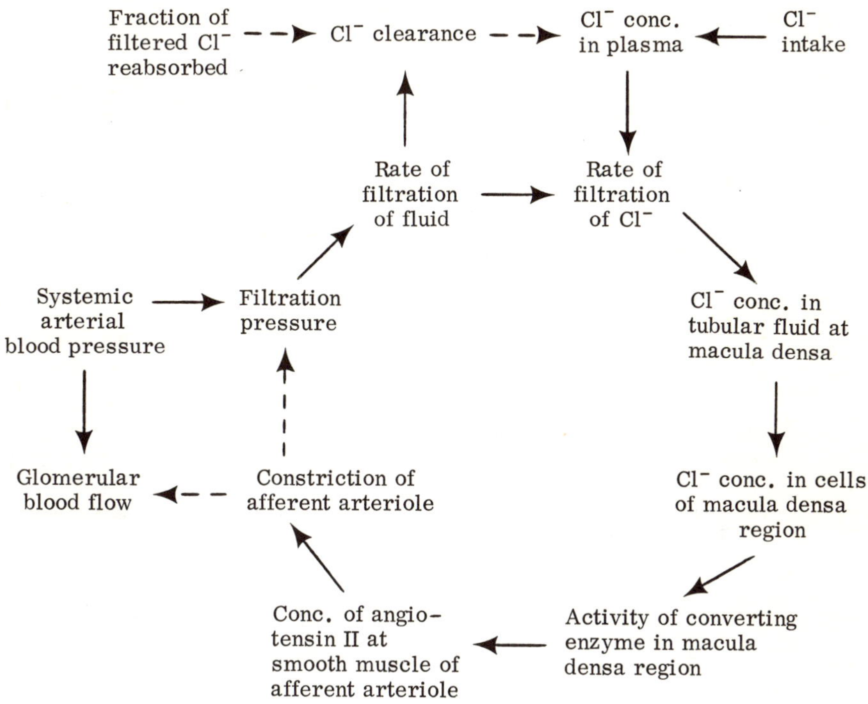

Figure 2.

renal blood flow, but these changes would be opposed by the operation of the feedback loop.

3. Because the activity of the converting enzyme is chloride-dependent, the present hypothesis focuses upon the renal handling of the chloride ion. But similar changes would also occur in the renal handling of sodium which is the chief cation accompanying chloride in the glomerular filtrate.

4. Figure 2 suggests that this system might act as a positive feedback system with respect to the effect of a change in chloride intake upon the chloride concentration in plasma. Any such action would presumably be offset by compensatory negative feedback loops, not shown here, which would decrease the fraction of filtered chloride reabsorbed when the concentration of chloride in plasma is increased.

References

1. Dalton, A. J., and Haguenau, F., editors. Ultrastructure of the Kidney. Academic Press, New York, 1967.

References (continued)

2. Cook. W. F.: Renin and the juxtaglomerular apparatus. In Hormones and the Kidney, pp. 247-254, Memoirs of the Society for Endocrinology, No. 13 Academic Press, New York, 1963.

3. De Muylder, C. G.: The Neurility of the Kidney. Blackwell, Oxford, 1952.

4. Barajas, L. and Latta, H.: Structure of the juxtaglomerular apparatus. Circulation Res., Supplement II, vol. 20 and 21, II-15 to II-28 1967.

5. Thurau, K., Schnermann, J., Nagel, W., Horster, M. and Wahl, M.: Composition of tubular fluid in the macula densa segment as a factor regulating the function of the juxtaglomerular apparatus. Circulation Res., Supplement II, vol. 20 and 21, II-79 to II-89, 1967.

6. Skeggs, L.T., Marsh, W.H., Kahn, J.R., and Shumway, N.P.: The existence of two forms of hypertension. J. exper. Med., 99, 275-282, 1954.

7. Guyton, A.C., Langston, J.B., and Navar, G.: Theory for renal autoregulation by feedback at the juxtaglomerular apparatus. Circulation Res., Supplement I, vol. 14 and 15, I-187 to I-197, 1964.

8. Britton, K.E.: Renin and renal autoregulation. Lancet, 2, 329-333, 1968.

9. Thurau, K.: Renal hemodynamics. Am. J. Med, 36, 698-719, 1964.

RENAL PLASMA CLEARANCE--PLASMA CONCENTRATION

POSITIVE

Type	Symbol	Definition	Example (Human)
Ind.	$\dot{Q}_{in}$	Rate of infusion of solute, S.	30 mg/min
Par.	V_{dist}	Volume of distribution of S.	12.5 L
	$\dot{V}_{filt}$	Glomerular filtration rate	0.125 L/min
	$\dot{V}_{ERP}$	Effective renal plasma flow. The value of $[\dot{V}cl](t)$ when $C(t) \leq C_{crit}$.	0.625 L/min
	$\dot{Q}_{sec, max}$	Maximum rate of tubular secretion of S.	15 mg/min
	C_{crit}	The critical value of C(t) at which the rate of delivery of S to the tubules in the post-glomerular plasma just equals $\dot{Q}_{sec, max}$. A dependent parameter defined by equation 14.	30 mg/L
Dep.	C(t)	Concentration of S in plasma	120 mg/L
	$\dot{Q}_{ex}(t)$	Rate of excretion of S.	30 mg/min
	$\dot{Q}_{filt}(t)$	Rate of filtration of S.	15 mg/min
	$\dot{Q}_{sec}(t)$	Rate of tubular secretion of S.	15 mg/min
	$[\dot{V}cl](t)$	Renal plasma clearance of S.	0.250 L/min

Steady-state values (Dep. examples)

Model

A solute, S, is infused at a constant rate, $\dot{Q}_{in}$, starting at $t = 0$. S is filtered through the glomeruli but not reabsorbed by the tubules. The S remaining in the post-glomerular plasma is completely removed by active tubular secretion until the tubular transfer maximum has been exceeded. Thereafter, $\dot{Q}_{sec}(t) = \dot{Q}_{sec, max}$, a constant parameter, so that $[\dot{V}cl](t)$ decreases as C(t) rises.

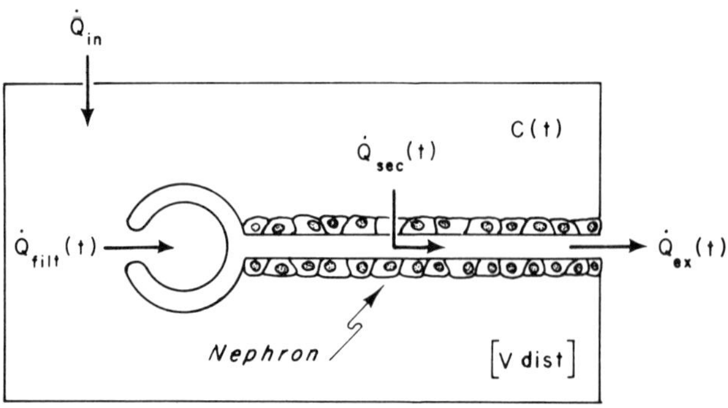

Figure 1.

Simplifying assumptions

1. $\dot{Q}_{in} \geq \dot{Q}_{sec,\,max}$
2. S is not bound to plasma protein.
3. S is not metabolized, stored, or excreted by non-renal routes.
4. The tubular transport mechanism becomes saturated abruptly when $C(t) = \dot{Q}_{sec,\,max}/[\dot{V}_{ERP} - \dot{V}_{filt}] \equiv C_{crit}$

Derivation

Under assumption 3, conservation of S requires that

$$V_{dist}\left[\frac{dC(t)}{dt}\right] = \dot{Q}_{in} - \dot{Q}_{ex}(t) \qquad 1.T$$

Since S is not reabsorbed by the tubules,

$$\dot{Q}_{ex}(t) = \dot{Q}_{filt}(t) + \dot{Q}_{sec}(t) \qquad 2.T$$

By assumption 2,

$$\dot{Q}_{filt}(t) = \dot{V}_{filt}C(t) \qquad 3.T$$

By assumption 4,

$$\dot{Q}_{sec}(t)\begin{cases} = C(t)[\dot{V}_{ERP} - \dot{V}_{filt}]; & 0 \leq C(t)[\dot{V}_{ERP} - \dot{V}_{filt}] \leq \dot{Q}_{sec,\,max} \qquad 4.T \\ = \dot{Q}_{sec,\,max} & ;\ \dot{Q}_{sec,\,max} \leq C(t)[\dot{V}_{ERP} - \dot{V}_{filt}] \qquad 5.T \end{cases}$$

where $C(t)[\dot{V}_{ERP} - \dot{V}_{filt}]$ is the rate at which S is delivered to the tubules in the post-glomerular plasma. By definition of the term "clearance",

$$\dot{Q}_{ex}(t) = C(t)[\dot{V}cl](t) \qquad 6.T$$

By assumption 1, equation 5 must be the equation describing $\dot{Q}_{sec}(t)$ as $t \to \infty$. Therefore, combine equations 2, 3, 5 and 6 eliminating $\dot{Q}_{ex}(t)$, $\dot{Q}_{filt}(t)$ and $\dot{Q}_{sec}(t)$,

$$[\dot{V}cl](t) = \dot{V}_{filt} + \frac{\dot{Q}_{sec,\,max}}{C(t)} \qquad 7.$$

or, in the steady state,

$$[\dot{V}cl](\infty) = \dot{V}_{filt} + \frac{\dot{Q}_{sec,\,max}}{C(\infty)} \qquad 7a.$$

Combine equations 1 and 6, eliminating $\dot{Q}_{ex}(t)$.

$$V_{dist}\left[\frac{dC(t)}{dt}\right] = \dot{Q}_{in} - [\dot{V}cl](t)C(t) \qquad 8.$$

In the steady state, $dC(t)/dt = 0$ and equation 8 becomes

$$C(\infty) = \dot{Q}_{in}/[\dot{V}cl](\infty) \qquad 9.$$

The positive feedback loop described by equations 7a and 9 is shown in the following symbol-and-arrow diagram:

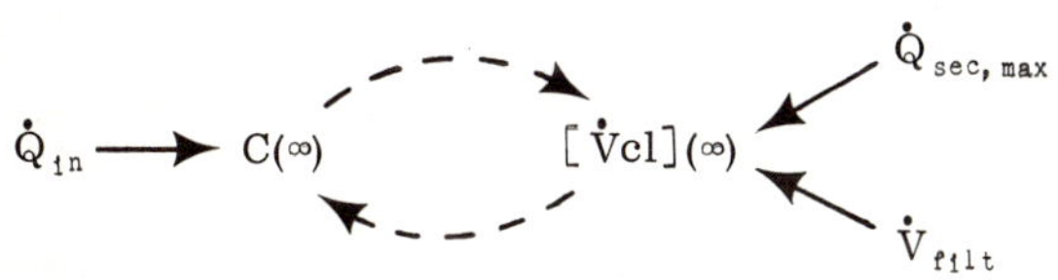

From equation 7a,

$$\frac{d[\dot{V}cl](\infty)}{dC(\infty)} = -\dot{Q}_{sec,\,max}/C^2(\infty) \qquad 10.$$

From equation 9,

$$\frac{dC(\infty)}{d[\dot{V}cl](\infty)} = -\dot{Q}_{in}/[\dot{V}cl]^2 \qquad 11.$$

Hence:

$$[OLG](j0) = \left|\frac{-\dot{Q}_{in}\dot{Q}_{sec,\,max}}{-[\dot{V}cl]^2(\infty)C^2(\infty)}\right| = \frac{\dot{Q}_{in}\dot{Q}_{sec,\,max}}{[\dot{V}cl]^2(\infty)C^2(\infty)} \qquad 12.$$

or, combining equations 9 and 12 so as to eliminate $[\dot{V}cl](\infty)$ and $C(\infty)$,

$$[OLG](j0) = \dot{Q}_{sec,max}/\dot{Q}_{in} \qquad 13.$$

Equation 13 shows that $[OLG](j0)$ decreases as $\dot{Q}_{in}$ increases. Hence, $[OLG](j0)$ will be maximal when $\dot{Q}_{in}$ is barely large enough to saturate the tubular transport mechanism for S as $t \to \infty$. But when the mechanism is <u>just</u> saturated, both equation 4 and equation 5 are valid. Combining them with the elimination of $\dot{Q}_{sec}(t)$, and solving for $C(t) = C_{crit}$,

$$C_{crit} = \frac{\dot{Q}_{sec,max}}{\dot{V}_{ERP} - \dot{V}_{filt}} \qquad 14.$$

where C_{crit} is the critical concentration which will be just large enough to saturat the transport mechanism. Combining equations 13 and 14,

$$[OLG]_{max}(j0) = \frac{\dot{V}_{ERP}C_{crit}}{\dot{Q}_{in}}\left[1 - \frac{\dot{V}_{filt}}{\dot{V}_{ERP}}\right] \qquad 15.$$

But by definition $\dot{V}_{ERP} = [\dot{V}cl](t)$ whenever equation 4 is valid. And for maximum $[OLG](j0)$, $C(t)$ will reach C_{crit} only as $t \to \infty$. Hence we may write equation 15 as

$$[OLG]_{max}(j0) = \frac{[\dot{V}cl](\infty)C_{crit}}{\dot{Q}_{in}}\left[1 - \frac{\dot{V}_{filt}}{\dot{V}_{ERP}}\right] \qquad 16.$$

Furthermore, under these circumstances equation 9 shows that $[\dot{V}cl](\infty)C_{crit} = \dot{Q}_{in}$. Combining this with equation 16,

$$[OLG]_{max}(j0) = 1 - \frac{\dot{V}_{filt}}{\dot{V}_{ERP}} \equiv 1 - [\text{Filtration fraction}] \qquad 17.$$

The filtration fraction in man is about 0.2; in the dog, about 0.3. Consequently, the maximum value of $[OLG](j0)$ would be about 0.8 for man; 0.7 for the dog. The corresponding maximum values for magnification would be 5 in man, 3.33 in the dog.

<u>Comments</u>

1. The same sequence of steps may be used to derive equations for the <u>negative</u> feedback loop which exists when a "threshold" substance (such as glucose) is delivered to the renal tubules at a rate which exceeds their <u>reabsorptive</u> capacity. Assuming that (unlike glucose!) the infused substance is not metabolize or excreted by other routes, the working equations would be equation 9 and

$$[\dot{V}cl](\infty) = \dot{V}_{filt} - \frac{\dot{Q}_{reab,max}}{C(\infty)} \qquad 18.$$

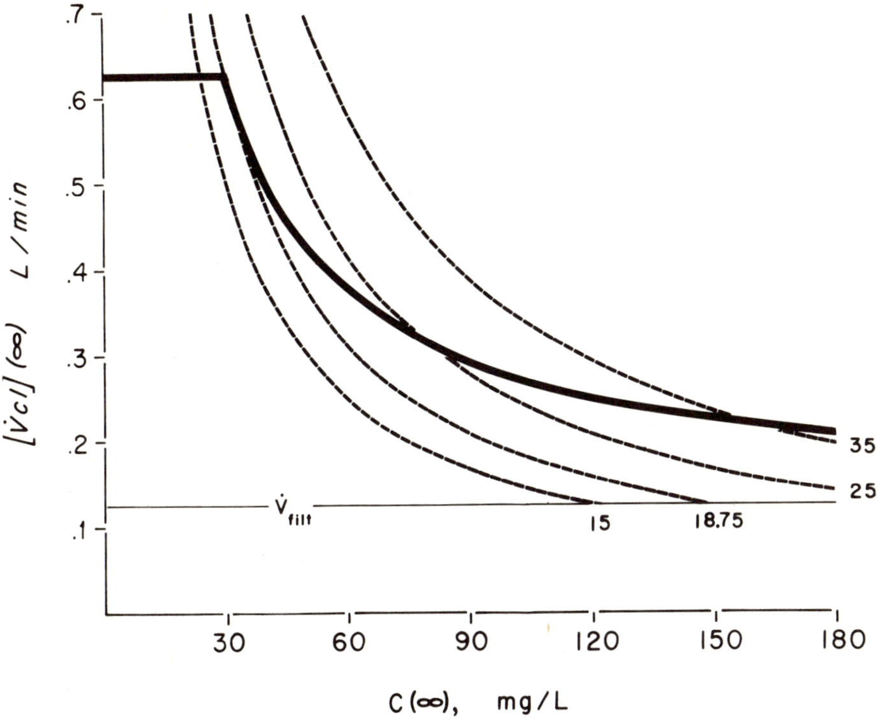

Figure 2. Simultaneous plot of the two feedback equations.

$[\dot{V}cl](\infty)$ as a function of $C(\infty)$ (equation 7a, for curved portion) is plotted as a solid line. Four curves for $C(\infty)$ as a function of $[\dot{V}cl](\infty)$ (equation 9) are plotted as broken lines, each representing a different rate of infusion as indicated in the graph. When $\dot{Q}_{in}$ is 15 mg/min, the transport mechanism is not saturated, the broken curve intersects the horizontal portion of the solid curve, and no feedback exists. When $\dot{Q}_{in} = 18.75$ mg/min, the broken curve intersects the solid curve just at the point where the transport mechanism becomes saturated so that the clearance starts to decrease as $C(\infty)$ increases further. At this point, or, more properly, for increases in $C(\infty)$ at this point, the $[OLG](j0)$ is maximum. The other broken lines illustrate still higher rates of infusion for which the $[OLG](j0)$ is smaller.

where $\dot{Q}_{reab,\,max}$ is the maximum rate of reabsorption of S(= tubular transfer maximum).

2. In the range governed by equation 4, (i.e. when $C(\infty) \leq C_{crit}$), the transfer mechanism can handle all of the S in the peritubular blood. Under these circumstances $[\dot{V}cl]$ is a constant (equal, in this example, to $\dot{V}_{ERP}$) and therefore does not depend on $C(\infty)$. Hence no feedback loop exists. (See leftmost broken curve in Figure 2.)

3. If a substance is cleared from plasma by metabolic alteration, a steady state cannot be achieved (and hence no steady-state feedback exists) if $\dot{Q}_{in}$ exceeds the capacity of the metabolic mechanism, unless S can be handled by some other non-saturated mechanism as well.

4. For an example of a feedback loop which does involve clearance by metabolism, see the section on "Thyroxine-thyrotropin" elsewhere in Part II.

Transient behavior

Combining equations 1, 2, 3 and 4, the linear differential equation for C(t) when the transport mechanism is not saturated is

$$\frac{dC(t)}{dt} + \frac{\dot{V}_{ERP}}{[Vdist]} C(t) = \frac{\dot{Q}_{in}}{[Vdist]} \qquad 19.$$

If we let C(0) = 0, the solution of equation 19 is

$$C(t) = \frac{\dot{Q}_{in}}{\dot{V}_{ERP}} \left[1 - \epsilon^{-\frac{\dot{V}_{ERP}}{[Vdist]} t} \right]; \; t < t_{crit} \qquad 20.$$

where t_{crit} is the time at which the plasma concentration just reaches C_{crit} (equation 14). For $t > t_{crit}$, we must combine equations 1, 2, 3 and 5, obtaining:

$$\frac{dC(t - t_{crit})}{dt} + \frac{\dot{V}_{filt}}{[Vdist]} C(t - t_{crit}) = \frac{\dot{Q}_{in} - \dot{Q}_{sec,max}}{[Vdist]} \qquad 21.$$

The solution of equation 21, subject to the condition $C(t_{crit}) = C_{crit} = \dfrac{\dot{Q}_{sec,max}}{\dot{V}_{ERP} - \dot{V}_{filt}}$ is

$$C(t - t_{crit}) = \frac{\dot{Q}_{in} - \dot{Q}_{sec,max}}{\dot{V}_{filt}} + \left[\frac{\dot{Q}_{sec,max}}{\dot{V}_{ERP} - \dot{V}_{filt}} - \frac{\dot{Q}_{in} - \dot{Q}_{sec,max}}{\dot{V}_{filt}} \right] \epsilon^{-\frac{\dot{V}_{filt}}{[Vdist]}[t - t_{crit}]}$$

$$t \geq t_{crit} \qquad 22.$$

In Figure 3, equations 20 and 22 have been used to calculate curves for C(t) as a function of time, using the exemplary values of the parameters. The solid curve shows the open-loop response which would occur if the transfer mechanism did not become saturated.

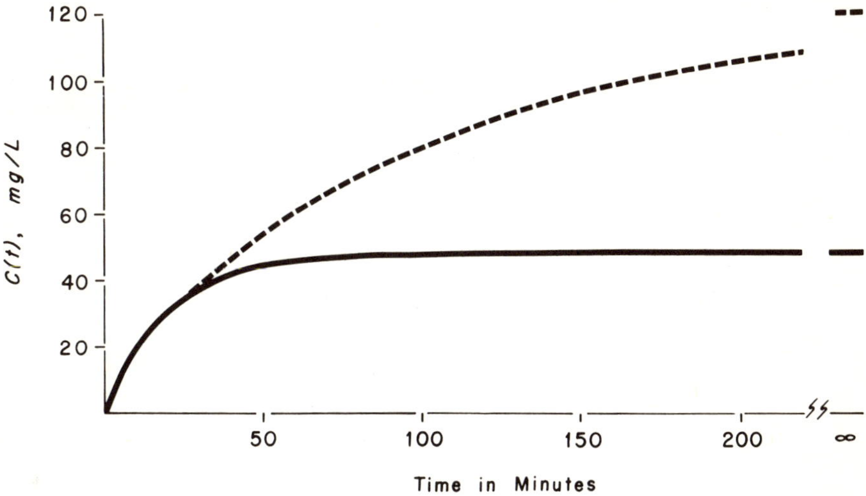

Figure 3. Plasma concentration as a function of time. Broken curve, closed-loop response; solid curve, open-loop response. The curves coincide when $C(t) \leq C_{crit} = 30$ mg/L.

NEGATIVE

Type	Symbol	Definition	Example (15 Kg. dog)
Ind.	$\dot{V}_{in}$	Rate of intake of water.	0.001 L/min
Par.	Q_{osm}	Total solute in extracellular fluid.	900 mOsm
	Q_{prot}	Total quantity of protein in plasma.	44.8 g
	P_{cap}	Mean blood pressure in glomerular capillaries.	70 mm Hg
	P_{Bow}	Mean pressure in Bowman's capsule.	10 mm Hg
	F_{ECF}	$V_{ECF}(t)/V_{H_2O}(t)$. Extracellular fluid volume as a fraction of total body water.	0.3
	F_{pl}	$V_{pl}(t)/V_{H_2O}(t)$. Plasma volume as a fraction of total body water.	0.07
	k	Mean oncotic pressure in glomerular capillaries per unit of protein concentration in systemic plasma.	0.4883 mm Hg/[g/L]
	α	Filtration rate per unit of net filtration pressure.	$0.001739 \frac{L/min}{mm\ Hg}$
	F_{max}	Max.value of $F_{reab}(\infty)$ approached as C_{osm} becomes very large.	0.997
	F_{min}	Min.value of $F_{reab}(\infty)$ approached as C_{osm} becomes very small.	0.910
	a	Empirical parameter in equation 10.	147
	b	Empirical parameter in equation 10.	0.5 L/mOsm
Dep.	$V_{H_2O}(t)$	Total volume of water in the body.	10.11 L
	$V_{ECF}(t)$	Volume of extracellular fluid.	3.033 L

Type	Symbol	Definition	Example (15 Kg. dog)
	$V_{pl}(t)$	Volume of plasma.	0.708 L
	$\dot{V}_{filt}(t)$	Glomerular filtration rate.	0.0506 L/min
	$\dot{V}_U(t)$	Urine flow.	0.001 L/min
	$C_{osm}(t)$	Osmotic concentration of solute in plasma.	296.9 mOsm/L
	$C_{prot}(t)$	Concentration of protein in systemic plasma.	63.3 g/L
	$P_{onc}(t)$	Mean oncotic pressure in glomerular capillaries.	30.9 mm Hg
	$P_{filt}(t)$	Net glomerular filtration pressure.	29.1 mm Hg
	$F_{reab}(t)$	Fraction of filtered water which is reabsorbed by the tubules.	0.9802

Model

A decrease in the rate of water intake, $\dot{V}_{in}$, will cause a decrease in the volume of, and an increase in the osmolal concentration of plasma and other extracellular fluids. Since most cells are freely permeable to water, similar changes will occur in intracellular fluids. The osmoreceptors thought to exist in the brain, perhaps in the supraoptic and paraventricular nuclei, will be stimulated and will increase the number of impulses per second passing down the supraopticohypophysial tract to the posterior pituitary. More antidiuretic hormone (ADH) will be released from the nerve terminals into the blood stream. The ADH will "open pores" in the distal portions of the nephron where facultative water reabsorption occurs. This allows water in the tubular fluid to diffuse back into the plasma in accordance with osmotic gradients already established by active solute reabsorption. Consequently, a larger proportion of the filtered water is reabsorbed, and a smaller proportion is excreted, so that urine flow soon equals $\dot{V}_{in}$ once more, thus minimizing the extensive changes in the volumes and osmolal concentrations of body fluids which, in the absence of ADH regulation, would have to occur before a decrease in glomerular filtration (secondary to an increased plasma protein concentration) could establish a new steady state.

Simplifying Assumptions

1. Absorption of water from the alimentary tract and urinary excretion of water are the only modes of water exchange.

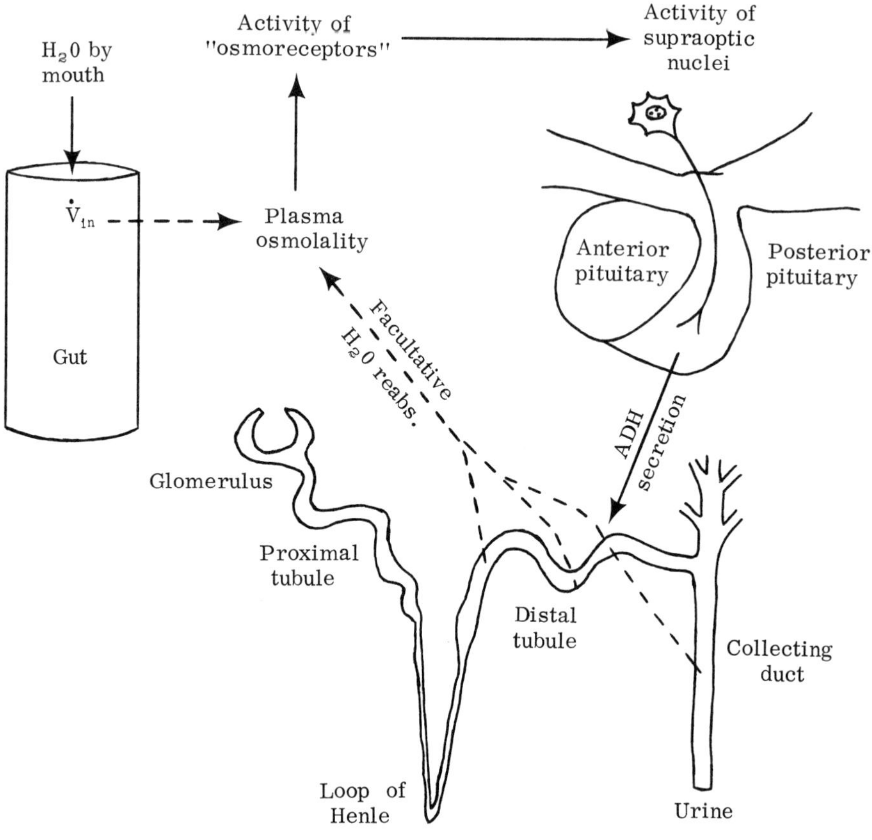

Figure 1.

2. Changes in the fraction of filtered water reabsorbed do not alter either the rate of solute reabsorption or the total quantity of solute in extracellular fluid.

3. The mean oncotic pressure in the glomerular capillaries is directly proportional to the concentration of protein in systemic plasma. (See comment 1.)

4. The changes which occur in the volume of plasma and other body fluids do not appreciably alter blood pressure in the glomerular capillaries. (See comment 2.)

Derivation

Under assumption 1, conservation of water requires that

$$\frac{dV_{H_2O}(t)}{dt} = \dot{V}_{in} - \dot{V}_U(t) \qquad 1.T$$

By definition,

$$\dot{V}_U(t) = \dot{V}_{filt}(t)[1 - F_{reab}(t)] \qquad 2.T$$

The glomerular filtration rate is proportional to filtration pressure.

$$\dot{V}_{filt}(t) = \alpha\, P_{filt}(t) \qquad 3.T$$

where $P_{filt}(t)$ is defined as

$$P_{filt}(t) = P_{cap} - P_{Bow} - P_{onc}(t) \qquad 4.T$$

with P_{cap} remaining constant (Assumption 4.) By assumption 3,

$$P_{onc}(t) = k\, C_{prot}(t) \qquad 5.T$$

By definition,

$$C_{prot}(t) = Q_{prot}/V_{pl}(t) \qquad 6.T$$

$$V_{pl}(t) = F_{pl}\, V_{H_2O}(t) \qquad 7.T$$

$$C_{osm}(t) = Q_{osm}/V_{ECF}(t) \qquad 8.T$$

$$V_{ECF}(t) = F_{ECF} V_{H_2O}(t) \qquad 9.T$$

In normal unanesthetized dogs, the observed dependence of $F_{reab}(t)$ on $C_{osm}(t)$ in the steady state may be described by the following empirical equation (1):

$$F_{reab}(\infty) = \frac{F_{max} - F_{min}}{1 + \epsilon^{a - bC_{osm}(t)}} + F_{min} \qquad 10.E$$

Combine equations 1 through 7, eliminating all variables except $V_{H_2O}(t)$, $F_{reab}(t)$, and t.

$$\frac{dV_{H_2O}(t)}{dt} = \dot{V}_{in} - \alpha[1 - F_{reab}(t)][P_{cap} - P_{Bow}]$$

$$+ \alpha[1 - F_{reab}(t)]\,\frac{kQ_{prot}}{F_{pl}}\left\{\frac{1}{V_{H_2O}(t)}\right\} \qquad 11.$$

In the steady state, $dV_{H_2O}(t)/dt = 0$, and equation 11 becomes

$$\frac{1}{V_{H_2O}(\infty)} = \frac{[P_{cap} - P_{Bow}]\,F_{pl}}{k\,Q_{prot}} - \frac{\dot{V}_{in}\,F_{pl}}{\alpha k\,Q_{prot}[1 - F_{reab}(\infty)]} \qquad 12.$$

Combining equations 8 and 9 by eliminating $V_{ECF}(t)$,

$$\frac{1}{V_{H_2O}(\infty)} = \frac{F_{ECF}}{Q_{osm}}\,C_{osm}(\infty) \qquad 13.$$

Combining equations 12 and 13 by eliminating $1/V_{H_2O}(\infty)$,

$$C_{osm}(\infty) = \frac{F_{pl}\,Q_{osm}}{k\,F_{ECF}\,Q_{prot}}\left\{P_{cap} - P_{Bow} - \frac{\dot{V}_{in}}{\alpha[1 - F_{reab}(\infty)]}\right\} \qquad 14.$$

Equations 10 and 14 describe the interrelationships between the loop variables $C_{osm}(\infty)$ and $F_{reab}(\infty)$. If we let the parameters Q_{osm}, Q_{prot}, F_{ECF}, and F_{pl} retain the values listed in the example, the symbol-and-arrow diagram showing the relationships among the remaining quantities is

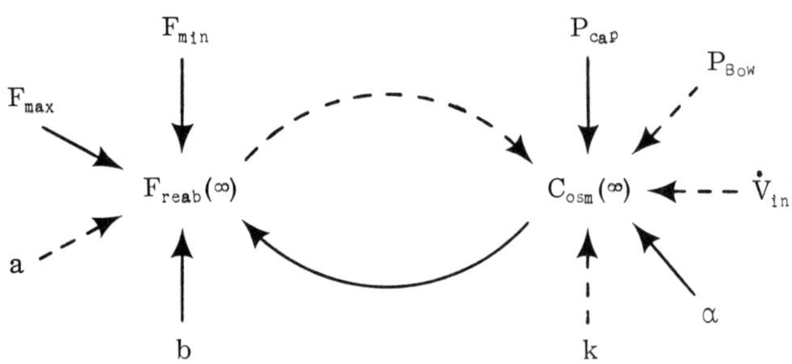

Figure 2.

From equation 10,

$$\frac{dF_{reab}(\infty)}{dC_{osm}(\infty)} = b[F_{max} - F_{min}]\,\frac{\epsilon^{a - bC_{osm}(\infty)}}{\left\{1 + \epsilon^{a - bC_{osm}(\infty)}\right\}^2} \qquad 15.$$

From equation 14,

$$\frac{dC_{osm}(\infty)}{dF_{reab}(\infty)} = -\frac{F_{pl}\,Q_{osm}\,\dot{V}_{in}}{\alpha\,k\,F_{ECF}\,Q_{prot}}\left\{\frac{1}{1 - F_{reab}(\infty)}\right\}^2 \qquad 16.$$

Whence

$$[OLG](j0) = \frac{b\,F_{pl}\,Q_{osm}\,\dot{V}_{in}[F_{max} - F_{min}]\,\epsilon^{a - b\,C_{osm}(\infty)}}{\alpha k\,F_{ECF}\,Q_{prot}[1 - F_{reab}(\infty)]^2\left\{1 + \epsilon^{a - b\,C_{osm}(\infty)}\right\}^2} \qquad 17.$$

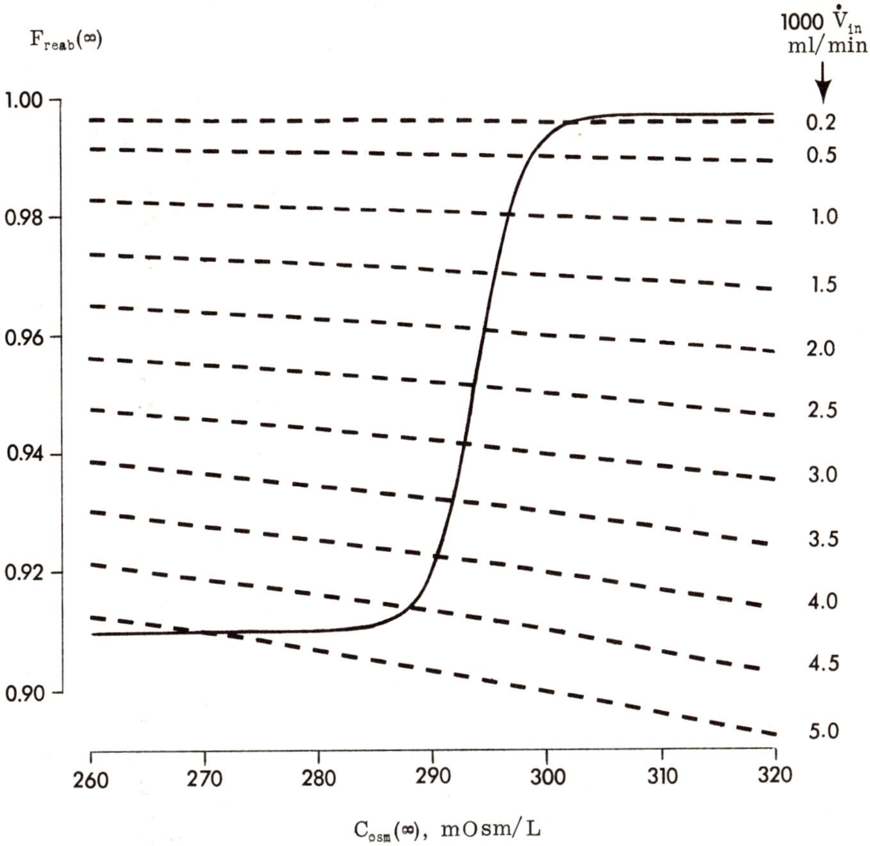

Figure 3. Broken curves: $C_{osm}(\infty)$ as a function of $F_{reab}(\infty)$ (equation 10) plotted for the designated values of 1000 $\dot{V}_{in}$. Solid curve: $F_{reab}(\infty)$ as a function of $C_{osm}(\infty)$. The figure shows that the feedback loop regulates $C_{osm}(\infty)$ quite precisely, but over a rather narrow range. (See also Figure 4.)

Using the exemplary values listed, the open-loop gain calculated from equation 17 is 94.

In Figure 3, curves corresponding to the two feedback equations 10 and 14 have been plotted for various values of $\dot{V}_{in}$. The influence of $\dot{V}_{in}$ on [OLG](j0) is shown in Figure 4.

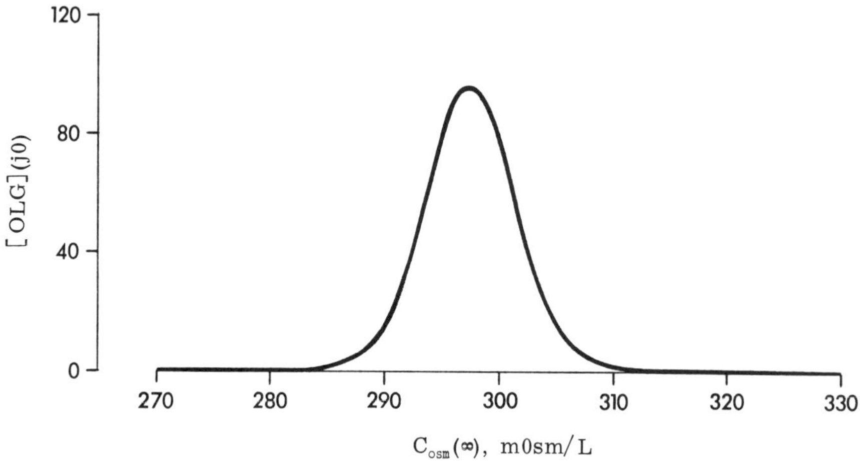

Figure 4. $[OLG](j0)$ as a function of $C_{osm}(\infty)$ (equation 17). The high gains within the normal range of $C_{osm}(\infty)$ show how very effective this system is in maintaining a reasonably constant plasma osmolality.

Comments

1. The effect of changes in the concentration of protein in plasma upon glomerular filtration rate is more complicated than is here assumed. (See Section 6.13).

2. The assumption that changes in plasma volume do not alter blood pressure may be reasonable for the small alterations which occur in the closed-loop system subject to moderate water loads. But this assumption can hardly be valid in the face of the large open-loop changes in plasma volume which are required by the simple dilution-of-plasma-protein mechanism here postulated. In fact, it is well known that severe dehydration can cause circulatory failure, and that in dogs with diabetes insipidus the glomerular filtration rate is often decreased more than could plausibly be ascribed to an increase in the oncotic pressure of the plasma in the glomerular capillaries. Effects of this kind would reduce the open-loop changes predicted by the simple theory presented here which therefore tends to over-estimate the open-loop gain.

Transient behavior

In addition to maintaining a reasonably constant plasma osmolality, the feedback loop greatly hastens the approach to a new steady state when $\dot{V}_{in}$ is changed. If there were no feedback, as would be true in a dog with diabetes insipidus produced by severing the supraopticohypophysial tract, $F_{reab}(\infty)$ would no longer be a function of $C_{osm}(\infty)$, and (according to present assumptions) the behavior of the system would be determined entirely by equation 11 with

$F_{reab}(t) = F_{min} = 0.91$. Separating the variables, t and $V_{H_2O}(t)$, in equation 11 and integrating,

$$\frac{1}{B}\left[V_{H_2O}(t) - V_{H_2O}(0)\right] + \frac{A}{B^2}\left\{\ln \frac{A + BV_{H_2O}(0)}{A + BV_{H_2O}(t)}\right\} = t \qquad 18.$$

where

$$A \equiv \alpha[1 - F_{reab}]kQ_{prot}/F_{pl}$$

$$B \equiv \dot{V}_{in} - \alpha[1 - F_{reab}][P_{cap} - P_{Bow}]$$

Suppose that at time zero $V_{H_2O}(0) = 10.000$ L and $\dot{V}_{in} = \dot{V}_U = 4.5 \cdot 10^{-3}$L/min so that, with $F_{reab} = 0.91$, $\dot{V}_{filt} = 50 \cdot 10^{-3}$L/min. (These steady-state values are in accord with the exemplified values of the other parameters.) If we now decrease $\dot{V}_{in}$ from $4.5 \cdot 10^{-3}$ to $4.0 \cdot 10^{-3}$L/min, the system will approach a new steady state where $V_{H_2O}(\infty) = 9.070$ L, and $\dot{V}_{filt}(\infty) = 44.44 \cdot 10^{-3}$L/min. By equation 18, the time needed for 90% of the required change (i.e. the time to go from $V_{H_2O} = 10.000$ liters to $V_{H_2O} = 9.163$ liters) would be about 4060 minutes or 68 hours. In contrast, when a normal dog is subjected to a step change of $\dot{V}_{in}$, a new steady state has practically been achieved within 3 hours (1).

Related loops

1. Thirst; water intake. The importance of this loop which regulates $\dot{V}_{in}$ is most obvious in diabetes insipidus, where ADH regulation of water excretion is absent or severely impaired, and in prolonged sweating, (roughly equivalent to negative values of $\dot{V}_{in}$) which cannot be compensated for by regulation of water excretion.

2. Blood volume; ADH secretion. A decrease in blood volume may cause ADH secretion (and thus water retention) even when plasma osmolality is normal or decreased. (2).

3. An increase in solute (especially sodium) intake may also increase C_{osm} and hence the rate of secretion of ADH. The excretion of sodium is regulated, in part at least, independently of facultative water excretion via aldosterone. (See elsewhere in Part II)

References

1. Millerschoen, N.R., and Riggs, D.S.: Homeostatic control of plasma osmolality in the dog, and the effect of ethanol. Am. J. Physiol., 217, 431-437, 1969.

2. Gauer, O.H.: Osmocontrol vs volume control. Fedn Proc., 27, 1132-1136, 1968.

BLOOD GLUCOSE; INSULIN SECRETION

NEGATIVE

Type	Symbol	Definition
Ind.	$\dot{Q}_{G,\,gut}$	Rate of entry of glucose from ingested carbohydrate.
	$\dot{Q}_{G,\,iv}$	Rate of intravenous infusion of glucose.
	$\dot{Q}_{I,\,iv}$	Rate of intravenous infusion of insulin.
Par.	$\dot{Q}_{G,\,CNS}$	Rate of glucose utilization by brain and other insulin-independent tissues.
	$C_{G,\,thr}$	Threshold concentration of glucose in blood above which glucosuria occurs.
	$\dot{V}_{filt}$	Glomerular filtration rate.
	$\dot{V}_{port}$	Blood flow in the portal vein.
	$[\dot{V}cl]_{I,\,liv}$	Hepatic clearance of insulin.
	$[\dot{V}cl]_{I,\,tiss}$	Clearance of insulin by other tissues.
Dep.	$C_{G,\,art}(\infty)$	Conc. of glucose in arterial blood.
	$[\dot{V}cl]_{G,\,kid}(\infty)$	Renal plasma clearance of glucose.
	$[\dot{V}cl]_{G,\,met}(\infty)$	Metabolic clearance of glucose.
	$\dot{Q}_{G,\,liv}(\infty)$	Rate of glucose production by the liver.
	$\dot{Q}_{I,\,sec}(\infty)$	Rate of secretion of insulin from the β-cells of the pancreatic islets.
	$C_{I,\,port}(\infty)$	Concentration of insulin in portal vein blood.
	$C_{I,\,hv}(\infty)$	Concentration of insulin in hepatic vein blood.
	$C_{I,\,art}(\infty)$	Concentration of insulin in arterial blood.

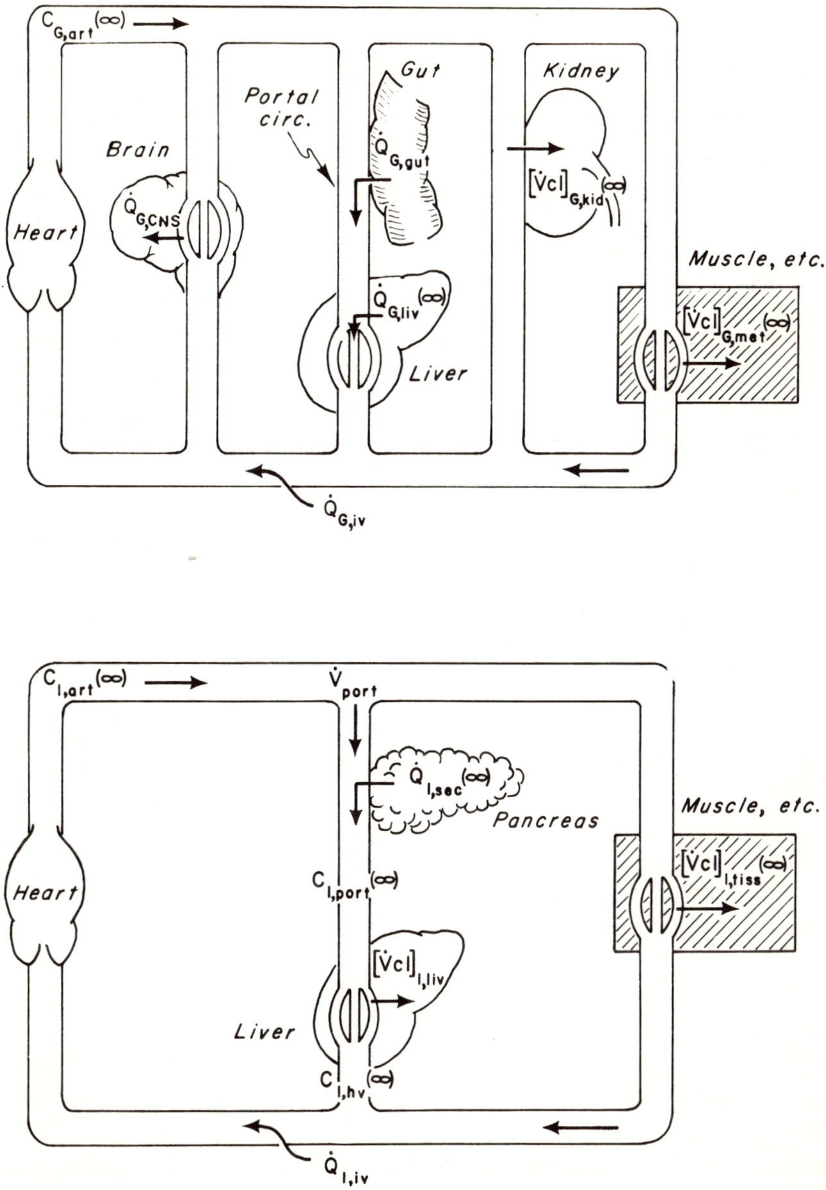

Figure 1. Routes for glucose (top) and for insulin (bottom).

Model

The avenues by which insulin and glucose enter and leave plasma are indicated in Figure 1. It is assumed that in the steady state there is no net change in the amount of glycogen stored. $\dot{Q}_{G,CNS}$ is treated as a parameter which is not influenced by glucose or insulin concentrations. Depending on circumstances, $\dot{Q}_{G,liv}(\infty)$ may be positive, zero, or negative. In experimental animals, it is both technically possible and physiologically reasonable to infuse exogenous insulin, or glucose, or both into the portal circulation rather than systemically as shown in the present model.

The rate of insulin secretion depends on the concentration of glucose in arterial blood. Insulin in turn reduces hepatic gluconeogenesis and increases the rate of utilization of glucose by insulin-sensitive tissues such as muscle and fat. Finally, glucose is excreted in the urine when the filtered load exceeds the tubular reabsorptive capacity for glucose. These feedback loops cannot be shown in the separate diagrams of Figure 1, but are depicted in the symbol-and-arrow diagram of Figure 2.

Simplifying assumptions

1. Saturation of the renal reabsorptive mechanism for glucose occurs abruptly when $C_{G,art}(\infty) = C_{G,thr}$.

2. Hepatic vein blood is, for convenience, assumed to be derived wholly from the portal circulation.

Derivation

In the steady state, the concentration of glucose in arterial blood is equal to the net rate of entry of glucose divided by the total plasma clearance of glucose.

$$C_{G,art}(\infty) = \frac{\dot{Q}_{G,gut} + \dot{Q}_{G,iv} + \dot{Q}_{G,liv}(\infty) - \dot{Q}_{G,CNS}}{[\dot{V}cl]_{G,met}(\infty) + [\dot{V}cl]_{G,kid}(\infty)} \qquad 1.T$$

Conservation of insulin in portal vein blood requires that

$$C_{I,art}(\infty)\,\dot{V}_{port} + \dot{Q}_{I,sec}(\infty) = C_{I,port}(\infty)\,\dot{V}_{port} \qquad 2.T$$

or, solving for $C_{I,port}(\infty)$

$$C_{I,port}(\infty) = C_{I,art}(\infty) + \frac{\dot{Q}_{I,sec}(\infty)}{\dot{V}_{port}} \qquad 2.a.$$

Conservation of insulin as $\dot{V}_{port}$ passes through the liver (see assumption 2) requires that

$$C_{I,port}(\infty)\,\dot{V}_{port} = C_{I,hv}(\infty)\,\dot{V}_{port} + C_{I,port}(\infty)[\dot{V}cl]_{I,liv} \qquad 3.T$$

In the systemic circulation, conservation of insulin requires that

$$\dot{Q}_{I,\,iv} + C_{I,\,hv}(\infty)\,\dot{V}_{port} = C_{I,\,art}(\infty)\,\dot{V}_{port} + C_{I,\,art}(\infty)[\dot{V}cl]_{I,\,tiss} \qquad 4.T$$

Combining equations 2, 3 and 4 by eliminating $C_{I,\,port}(\infty)$ and $C_{I,\,hv}(\infty)$, and solving for $C_{I,\,art}(\infty)$,

$$C_{I,\,art}(\infty) = \frac{\dot{Q}_{I,\,iv} + \dot{Q}_{I,\,sec}(\infty)\left\{1 - \dfrac{[\dot{V}cl]_{I,\,liv}}{\dot{V}_{port}}\right\}}{[\dot{V}cl]_{I,\,liv} + [\dot{V}cl]_{I,\,tiss}} \qquad 5.$$

By assumption 1,

$$[\dot{V}cl]_{G,\,kid}(\infty) = \dot{V}_{filt}\left\{1 - \frac{C_{G,\,thr}}{C_{G,\,art}(\infty)}\right\}\,u(C_{G,\,art}(\infty) - C_{G,\,thr}) \qquad 6.T$$

where, as always, u is defined as zero for negative values, and unity for non-negative values of its argument.

From qualitative observations, it is known that

$$[\dot{V}cl]_{G,\,met}(\infty) = f(C_{I,\,art}(\infty)) \qquad 7,$$

$$\dot{Q}_{G,\,liv}(\infty) = g(C_{I,\,port}(\infty)) \qquad 8,$$

$$\dot{Q}_{I,\,sec}(\infty) = h(C_{G,\,art}(\infty)) \qquad 9.$$

The symbol-and-arrow diagram corresponding to equations 1, 2a, 5, 6, 7, 8 and 9, is shown in Figure 2.

Unfortunately, the functions f, g, and h are not yet defined well enough to allow us to complete the derivation. But the prospects are favorable. For example, Vranic and Wrenshall (1) have devised a method for studying rates of glucose appearance and disappearance as a function of $\dot{Q}_{I,\,sec}(\infty)$ (actually, rate of intraportal infusion of insulin after acute pancreatectomy). Furthermore, the availability of accurate radioimmunoassay methods for measuring insulin concentrations in body fluids should facilitate evaluation of $\dot{Q}_{I,\,sec}(\infty)$ as a function of $C_{G,\,art}(\infty)$. With the techniques now available, it should be possible to characterize the behavior of this system in the steady state by analyzing arterial, hepatic vein, and portal vein blood for both glucose and insulin during continuous infusions of glucose and (in a separate series of experiments) insulin.

Comments

1. A good deal of effort has hitherto been devoted to studying fluctuations in plasma glucose and insulin after single large doses such as are used clinically in glucose tolerance and insulin tolerance tests. The complexities of transient behavior make it very difficult to interpret such studies quantitatively.

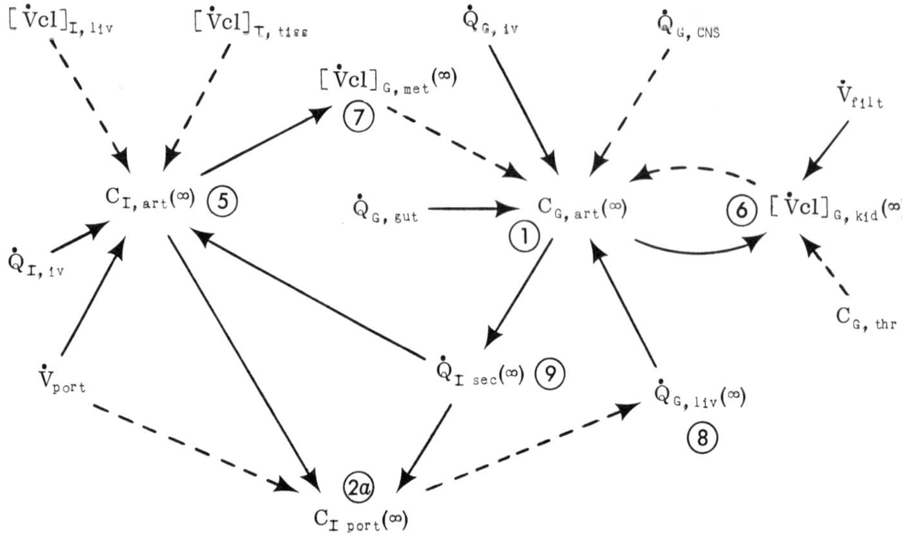

Figure 2.

2. Several investigators have tried to study the behavior of this system by simulating it with a computer (2, 3, 4, 5, 6). For example, Gatewood et al. (5) assumed a linear model, very much simpler than the one described above. Using a computer, they were able to adjust the five parameters of their linear differential equations so that actual observations of $C_{G,\,art}(t)$ as a function of time after giving a glucose load by mouth or intravenously could be fitted reasonably well. With five parameters available for adjustment, fair agreement between observed and calculated values is hardly surprising and does not prove the validity of their model. The steady-state behavior of Bolie's very similar model (2) was discussed in detail in Section 7.15. (See especially Figure 7.15.2. See also the discussion of the Galatea delusion in Section 1.14).

Related feedback loops

A decrease in blood glucose causes an increase in the rate of secretion of glucagon from the α-cells of the pancreatic islets. Glucagon causes glycogenolysis and increases gluconeogenesis in the liver, thereby opposing the initial change. The glucagon-blood glucose loop is important physiologically in protecting against hypoglycemia, for example during starvation (7, 8).

Hypoglycemia is also a very powerful stimulus for the secretion of growth hormone which (among other things) causes lipolysis and thereby increases the mobilization and utilization of fat. (9)

Glucocorticoids and catecholamines also participate in the regulation of carbohydrate, fat, and protein metabolism.

Kline et al. (10) have devised an artificial feedback system to maintain blood glucose at a relatively constant concentration. The rate of infusion of glucose and the injection of single doses of insulin were determined by the difference between a glucose concentration chosen as a standard and the actual blood glucose continuously monitored by a Technicon Auto-Analyzer.

References

1. Vranic, M., and Wrenshall, G.A.: Matched rates of insulin infusion and secretion and concurrent tracer-determined rates of glucose appearance and disappearance in fasting dogs. Canadian J. Physiol. and Pharmac., 46, 383-390, 1968.

2. Bolie, V.W.: Coefficients of normal blood glucose regulation. J. appl. Physiol., 16, 783-788, 1961.

3. Cerasi, E.: An analogue computer model for the insulin response to glucose infusion. Acta endocr., 55, 163-183, 1967.

4. Ceresa, F., Ghemi, F., Martini, P.F., Martino, P., Segre, G., and Vitelli, A.: Control of blood glucose in normal and in diabetic subjects. Diabetes, 17, 570-578, 1968.

5. Gatewood, L.C., Ackerman, E., Rosevear, J.W., and Molnar, G.D.: Tests of a mathematical model of the blood-glucose regulatory system. Comput. and biomed. Res., 2, 1-14, 1968.

6. Seed, J.C., Acton, F.S., and Stunkard, A.J.: A model for the appraisal of glucose metabolism. Clin. Pharmac. Ther., 3, 191-215, 1962.

7. Ohneda, A., Aguilar-Parada, E., Eisentraut, A.M., and Unger, R.H.: Control of pancreatic glucagon secretion by glucose. Diabetes, 18, 1-10, 1969.

8. Sokal, J.E.: Glucagon--an essential hormone. Am. J. Med., 41, 331-341, 1966.

9. Greenwood, F.C.: Growth Hormone. Chap. 10, pp 195-231, in Gray, C.H. and Bacharach, A.L.: Hormones in Blood, 2nd edition, vol. 1, Academic Press, London, 1967.

10. Kline, N.S., Shimano, E., Stearns, H., McWilliams, C., Kohn, M., and Blair, J.H.: Technique for automatic in vivo regulation of blood sugar. Techn. Q., 1, 10-16, 1969.

THYROXINE-THYROTROPIN (TSH)

NEGATIVE

Type	Symbol	Description	Exemplary values (Adult human)
Ind.	$\dot{Q}_{in,I}$	Iodide ingested daily	150 μg/day
	$\dot{Q}_{in,H}$	Thyroxine administered daily	0 μg/day
	$\dot{Q}_{in,TSH}$	TSH administered daily	0 μg/day
Par.	$[\dot{V}cl]_I$	Renal plasma clearance of iodide	48 L/day
	F_{tyr}	Fraction of iodine in thyroglobulin which is in iodotyrosines	2/3
	$[\dot{V}cl]_{TSH}$	Plasma clearance of TSH	
	K	Constant of proportionality, equation 9	
	J	Dissociation constant for protein-bound thyroxine	
	F_{org}	Fraction of hormone excreted in organic form (fecal excretion)	0.2
	$C_{pr,tot}$	Total conc. of protein binding sites for thyroxine, expressed as thyroxine iodine equivalents	
Dep.	$C_{TSH}(\infty)$	Concentration of TSH in plasma	
	$C_I(\infty)$	Concentration of iodide in plasma	2.92 μg/L
	$C_H(\infty)$	Concentration of free thyroxine in plasma	0.05 μg/L
	$C_{Hpr}(\infty)$	Concentration of protein-bound thyroxine in plasma	50 μg/L
	$C_{pr}(\infty)$	Concentration of unoccupied thyroxine binding sites on protein	
	$[\dot{V}cl]_H(\infty)$	Total plasma clearance of free thyroxine	1000 L/day
	$\dot{Q}_{TSH}(\infty)$	Daily secretion of TSH	
	$\dot{Q}_{lys}(\infty)$	Daily amount of I in amino acids of thyroglobulin released by proteolysis	100 μg/day

$\dot{Q}_{GH}(\infty)$	Daily secretion of thyroxine iodine	50 μg/day
$\dot{Q}_{IG}(\infty)$	Rate of transfer of iodide from blood to gland	
$\dot{Q}_{GI}(\infty)$	Rate of transfer of iodide from gland to blood	

Model

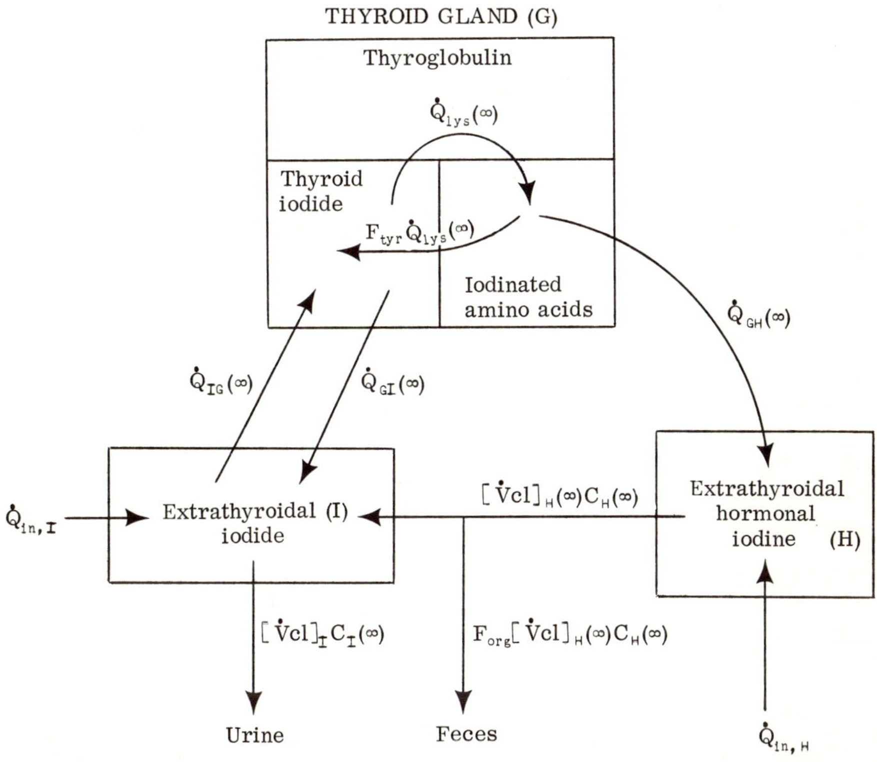

Figure 1. Pathways of iodine translocation assumed in the model described here.

Iodide collected by the thyroid gland from blood, or salvaged within the gland by deiodination of iodotyrosines released during the proteolysis of thyroglobulin (1), is used for the synthesis of new thyroglobulin. Thyroid stimulating hormone (= TSH) from the anterior pituitary gland controls $\dot{Q}_{lys}(\infty)$, the rate of proteolysis of thyroglobulin, and thereby regulates $\dot{Q}_{GH}(\infty)$, the rate of secretion of hormone, which is chiefly thyroxine. $\dot{Q}_{GH}(\infty)$ helps to determine $C_H(\infty)$, the concentration of free (= non-protein-bound) thyroxine. $C_H(\infty)$, in turn, controls the rate of secretion of TSH, an increase of $C_H(\infty)$ causing a decrease in $\dot{Q}_{TSH}(\infty)$. $C_H(\infty)$ and $\dot{Q}_{TSH}(\infty)$ are thus members of a negative feedback loop tending to stabilize the concentration of free thyroxine in plasma. An increase in $C_H(\infty)$ also causes some increase in $[\dot{V}cl]_H(\infty)$, the clearance of thyroxine by metabolic degradation and fecal excretion. The increase in $[\dot{V}cl]_H(\infty)$ tends to decrease $C_H(\infty)$. Hence $C_H(\infty)$ and $[\dot{V}cl]_H(\infty)$ are members of a second, but relatively unimportant negative feedback loop.

Simplifying assumptions

1. All of the thyroid hormone secreted is thyroxine; triiodothyronine is neglected.

2. All binding sites for thyroxine on plasma protein have the same dissociation constant, J, (see comment 2.).

3. $[\dot{V}cl]_H(\infty)$ is directly proportional to $C_H(\infty)$, the concentration of free thyroxine in plasma, (see comment 2.).

4. The participation of TRF, the thyrotropin releasing factor of the hypothalamus, and the participation of LATS, the "long-acting thyroid stimulator," in the regulation of thyroid function are neglected.

5. The sole effect of TSH is to increase the rate of secretion of thyroxine by increasing the rate of proteolysis of thyroglobulin, (see comment 1.).

Derivation

In the steady state, the iodide ion entering I, the extrathyroidal iodide pool, equals the iodide ion leaving I. (see Figure 1.)

$$\dot{Q}_{in,I} + \dot{Q}_{GI}(\infty) + [1 - F_{org}][\dot{V}cl]_H(\infty)C_H(\infty) = \dot{Q}_{IG}(\infty) + [\dot{V}cl]_I C_I(\infty) \qquad 1.T.$$

Similarly, steady-state iodine balance in the entire thyroid gland (see Figure 1.) requires that

$$\dot{Q}_{IG}(\infty) = \dot{Q}_{GI}(\infty) + \dot{Q}_{GH}(\infty) \qquad 2.T.$$

and steady-state balance in the pool of iodinated amino acids freed by proteolysis in the thyroid gland (see Figure 1.) requires that

$$\dot{Q}_{lys}(\infty) = F_{tyr}\dot{Q}_{lys}(\infty) + \dot{Q}_{GH}(\infty) \qquad 3.T$$

or, solving for $\dot{Q}_{GH}(\infty)$,

$$\dot{Q}_{GH}(\infty) = [1 - F_{tyr}]\,\dot{Q}_{lys}(\infty) \qquad 3.a.$$

Combine equations 1 and 2 by eliminating $\dot{Q}_{GI}(\infty) - \dot{Q}_{IG}(\infty)$, and solve for $C_I(\infty)$.

$$C_I(\infty) = \frac{\dot{Q}_{in,I} + [1 - F_{org}][\dot{V}cl]_H(\infty)C_H(\infty) - \dot{Q}_{GH}(\infty)}{[\dot{V}cl]_I} \qquad 4.$$

It has been observed (assumption 5) that

$$\dot{Q}_{lys}(\infty) = f(C_{TSH}(\infty)) \qquad 5.E.$$

By definition of "clearance,"

$$C_{TSH}(\infty) = \frac{\dot{Q}_{TSH}(\infty) + \dot{Q}_{in,TSH}}{[\dot{V}cl]_{TSH}} \qquad 6.T.$$

It has been demonstrated that

$$\dot{Q}_{TSH}(\infty) = g(C_H(\infty)) \qquad 7.E.$$

By definition of "clearance,"

$$C_H(\infty) = \frac{\dot{Q}_{GH}(\infty) + \dot{Q}_{in,H}}{[\dot{V}cl]_H(\infty)} \qquad 8.T.$$

By assumption 3,

$$[\dot{V}cl]_H(\infty) = KC_H(\infty) \qquad 9.E.$$

By assumption 2, and the law of mass action,

$$C_{Hpr}(\infty) = C_H(\infty)C_{pr}(\infty)/J \qquad 10.T.$$

By definition,

$$C_{pr,\,tot} = C_{pr}(\infty) + C_{Hpr}(\infty) \qquad 11.T$$

Combining equations 10 and 11 by eliminating $C_{pr}(\infty)$,

$$C_{Hpr}(\infty) = \frac{C_{pr,\,tot}\, C_H(\infty)}{J + C_H(\infty)} \qquad 12.$$

Equations 3a, 4, 5, 6, 7, 8, 9 and 12 constitute a set of working equations for this system. Accordingly, we may construct the symbol-and-arrow diagram of Figure 2. That diagram shows two feedback loops whose identity will not be compromised by combining equations 3a, 5, 6, and 7 to obtain an equation for $\dot{Q}_{GH}(\infty)$ as a function of $C_H(\infty)$.

$$\dot{Q}_{GH}(\infty) = [1 - F_{tyr}]f\left(\frac{g(C_H(\infty)) + \dot{Q}_{in,TSH}}{[\dot{V}cl]_{TSH}}\right) \qquad 13.$$

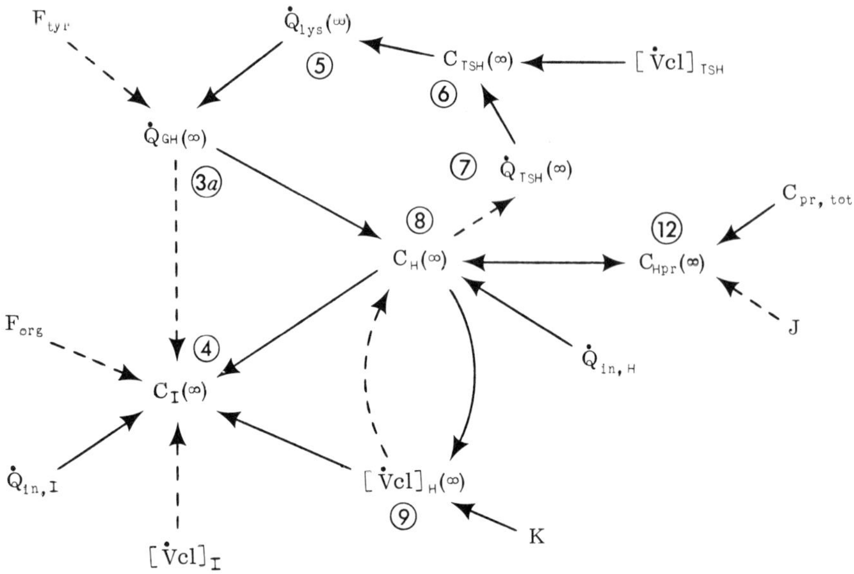

Figure 2.

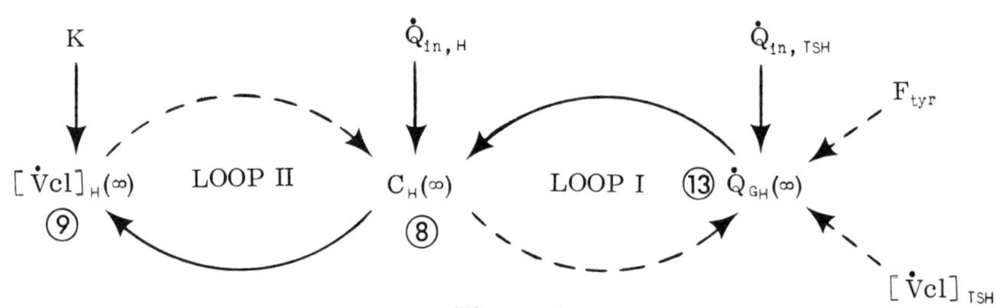

Figure 3.

The symbol-and-arrow diagram corresponding to equations 8, 9 and 13 is shown in Figure 3.

Unfortunately, the functions f and g are not known. However, at least a vague idea about the dependence of $\dot{Q}_{GH}(\infty)$ upon $C_H(\infty)$ can be obtained from studies of the ability of $\dot{Q}_{in,H}$, the rate of supply of exogenous hormone, to depress $\dot{Q}_{GH}(\infty)$. Riggs (Table 4, p. 350, Reference 2) calculated that, around the normal operating point, an increase of $C_{Hpr}(\infty)$ from 53 to 66 μg/L, caused by the administration of 195 mg of dried thyroid daily to euthyroid humans, would decrease $\dot{Q}_{GH}(\infty)$ from 82.5 to zero μg per day. Assuming that

$C_H(\infty) = 0.001\ C_{Hpr}(\infty)$ (3, 4) this would imply that $d\dot{Q}_{GH}(\infty)/dC_H(\infty)$ is of the order of $-82.5/0.013 = -6.35 \cdot 10^3$ L/day. However, Greer (5) was able to decrease the uptake of radioactive iodine (and hence, presumably, the rate of hormone synthesis and secretion) in normal human subjects to about 1/3 of its initial value by administering only 65 mg of dried thyroid daily, equivalent to perhaps 40 μg of endogenously-secreted thyroxine iodine. This suggests that $\dot{Q}_{GH}(\infty)$ might decrease from a normal value of 50 μg/day (4) to 50/3 μg/day in response to an increase in $C_H(\infty)$ of only about 0.004 μg/L. Using these values, $d\dot{Q}_{GH}(\infty)/dC_H(\infty) = -\dfrac{[2/3]50}{0.004} = -8.33 \cdot 10^3$ L/day. I believe that even this is a considerable underestimate, and I shall therefore make a wild and unsupported guess that, in the vicinity of the normal operating point,

$$\dot{Q}_{GH}(\infty) = 1050 - 20 \cdot 10^3\ C_H(\infty)\ \mu g/day \qquad 14.S.$$

From equation 8, with $[\dot{V}cl]_H(\infty) = 1000$ L/day, and $\dot{Q}_{in,H} = 0$,

$$dC_H(\infty)/d\dot{Q}_{GH}(\infty) = 10^{-3}\ \text{day/L} \qquad 15.$$

From equation 14,

$$d\dot{Q}_{GH}(\infty)/dC_H(\infty) = -20 \cdot 10^{-3}\ \text{L/day} \qquad 16.$$

Hence, for loop I,

$$[OLG]_{I}(j0) = \left| [10^{-3}][-20 \cdot 10^3] \right| = 20 \qquad 17.$$

a value which, it is to be emphasized, is highly suppositious but probably of the correct order of magnitude.

From equation 8, with $\dot{Q}_{in,H} = 0$,

$$\frac{d\,C_H(\infty)}{d[\dot{V}cl]_H(\infty)} = -\dot{Q}_{GH}(\infty)/[\dot{V}cl]_H^2(\infty) = -C_H(\infty)/[\dot{V}cl]_H(\infty) \qquad 18.$$

From equation 9,

$$d[\dot{V}cl]_H(\infty)/d\,C_H(\infty) = K = [\dot{V}cl]_H(\infty)/C_H(\infty) \qquad 19.$$

Hence, for loop II,

$$[OLG]_{II}(j0) = \left| -\frac{C_H(\infty)}{[\dot{V}cl]_H(\infty)} \cdot \frac{[\dot{V}cl]_H(\infty)}{C_H(\infty)} \right| = 1 \qquad 20,$$

but the true value of $[OLG]_{II}(j0)$ is probably less than this. (See comment 2.)

Comments

1. The effects of TSH on the thyroid gland are actually multiple and complex. In particular, TSH increases $\dot{Q}_{IG}(\infty)$ as well as $\dot{Q}_{GH}(\infty)$. Under the present naive

assumption that TSH influences only $\dot{Q}_{lys}(\infty)$, we must correspondingly assume that, in the steady state, the value of $\dot{Q}_{IG}(\infty) - \dot{Q}_{GI}(\infty)$ is dependent upon (indeed, equal to) $\dot{Q}_{GH}(\infty)$. This assumption makes it impossible to combine the present model with the one discussed in Section 5.6, where $\dot{Q}_{IG}(\infty)$ was taken to be directly proportional to $C_I(\infty)$. The model of section 5.6 was primarily intended to explain the distribution of radioactive iodine during any particular steady state. The present model emphasizes the homeostatic control of $C_H(\infty)$ by $C_{TSH}(\infty)$ as we move from one steady state to another. There is no real discrepancy between these models, each being correct within a particular, highly limited, set of assumptions. Mere ignorance prevents us from bringing them together.

2. In blood plasma, thyroxine is primarily bound to TBG, thyroxine-binding globulin, secondarily to TBPA, thyroxine-binding prealbumin, and only slightly to albumin itself. These proteins differ from each other both in concentration and in affinity for thyroxine (3). With three binding proteins in the picture, complications naturally arise.

A direct proportionality between $C_{Hpr}(\infty)$ and the clearance of protein-bound iodine has been observed empirically (equation 68, Reference 2), but at least some of the increase in the clearance of protein-bound iodine must be due simply to the larger proportion of free thyroxine present when $C_{Hpr}(\infty)$ is elevated (6) (Cf. equation 12.) The true [OLG](j0) for loop II must therefore be even less than the value of unity given by equation 20.

References

1. Rosenberg, I.N., Athans, J.C., and Isaacs, G.H.: Studies on thyroid iodine metabolism. Recent Prog. Horm. Res., 21, 33-72, 1965.

2. Riggs, D.S.: Quantitative aspects of iodine metabolism in man. Pharmac. Rev. 4, 284-370, 1952.

3. Robbins, J. and Rall, J.E.: The iodine-containing hormones. Chapter 15 in: Gray, C.H. and Bacharach, A.L., editors. Hormones, in Blood. 2nd edition, Vol. 1, Academic Press, New York, 1967.

4. Sterling, K.: Thyroxine in blood. Proc.Staff Meet. Mayo Clin. 39, 586-608, 1964.

5. Greer, M.A.: The effect on endogenous thyroid activity of feeding desiccated thyroid to normal human subjects. New Engl. J. Med. 244, 385-390, 1951.

6. Schussler, G.C., and Vance, V.K.: Effect of thyroid-suppressive doses of triiodothronine on thyroxine turnover and on the free thyroxine fraction. J. Clin. Invest., 47, 720-728, 1968.

See Also,

Distefano, J.J.: A model of the regulation of circulating thyroxin unbound and bound to plasma proteins and its response to pregnancy, drugs, long-acting thyroid stimulator, and temperature stress. Math. Biosci. 4, 137-152, 1969.

PARATHYROID HORMONE-PLASMA CALCIUM

NEGATIVE

Type	Symbol	Definition	Example (15 Kg. dog)
Ind.	$\dot{Q}_{exog}$	Rate of infusion of calcium	mg/hr
Par.	[Vdist]	Volume of distribution of calcium in soft tissucs	3.87 L
	$\dot{V}_{bone}$	Effective plasma flow through passively equilibrating portions of bone	1.37 L/hr
	$[\dot{V}cl]_{PH}$	Plasma clearance of parathyroid hormone	
	C_{norm}	Normal plasma calcium concentration	99 mg/L
	C_{eq}	Calcium concentration in plasma which is in passive equilibrium with bone	58 mg/L
	$\dot{Q}_{reab, norm}$	Normal rate of active re-absorption of calcium from bone caused by parathyroid hormone when $C(\infty) = C_{norm}$	56.1 mg/hr
	$\dot{Q}_{bone, ven}$	A dependent parameter which is defined by equation 5	
	a	An empirical parameter in equation 12	0.054 L/mg
Dep.	C(t)	Concentration of calcium in plasma	mg/L
	$\dot{Q}_{depos}(t)$	Rate of calcium deposition in bone due to passive equilibration	mg/hr
	$\dot{Q}_{reab}(t)$	Rate of active calcium reabsorption from bone due to parathyroid hormone	mg/hr
	$\dot{Q}_{bone, art}(t)$	Quantity of calcium per unit time entering bone in arterial blood	mg/hr
	$\dot{Q}_{in}(t)$	Total rate of entry of calcium into plasma except by passive exchange with bone	mg/hr

Type	Symbol	Definition	Example (15 Kg. dog)
Dep.	$\dot{Q}_{PH}(t)$	Rate of secretion of parathyroid hormone	
	$C_{PH}(t)$	Concentration of parathyroid hormone in plasma	

Model

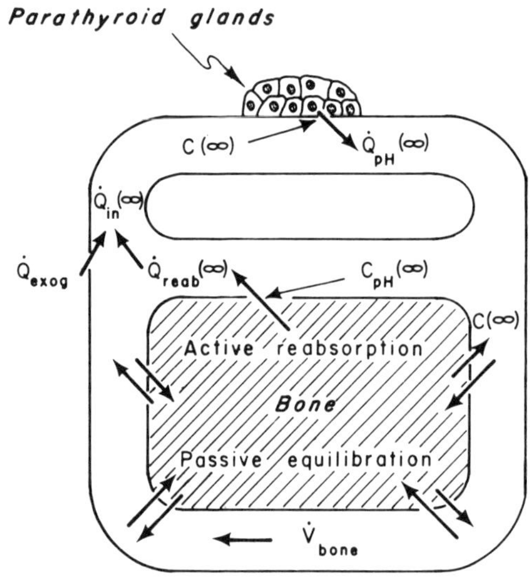

Figure 1.

A decrease in plasma calcium stimulates the secretion of parathyroid hormone, thereby increasing the concentration of hormone in plasma. This increased concentration of parathyroid hormone causes increased active reabsorption of calcium from bone, thus raising plasma calcium and opposing the original disturbance.

Simplifying assumptions

1. Bone is an infinitely extensive source of, or sink for calcium. (See comment 1.)

2. Non-bone avenues of calcium exchange (e.g., via the kidneys or the gastrointestinal tract) have a negligible influence upon the steady state reached during an intravenous infusion of calcium. (See comment 1.)

3. Compared with parathyroid hormone, thyrocalcitonin has a negligible influence upon the steady state reached during an intravenous infusion of calcium. (See comment 2.)

4. In the absence of parathyroid hormone, and with no exogenous source of calcium, plasma calcium remains at an abnormally low concentration, C_{eq}, determined by passive equilibrium between plasma and bone.

5. In the absence of parathyroid hormone, but with an exogenous source of calcium, bone passively removes from the plasma flowing past it enough calcium to establish C_{eq} in the venous effluent.

6. In the presence of parathyroid hormone, passive deposition of calcium in bone still occurs at the rate described in assumption 5; but parathyroid hormone causes active reabsorption of calcium at a rate of $\dot{Q}_{reab}(t)$ which tends to offset passive deposition, thus maintaining $C(\infty) > C_{eq}$.

Derivation

With respect to the calcium in soft tissue, conservation requires that

$$[Vdist]dC(t)/dt = \dot{Q}_{in}(t) - \dot{Q}_{depos}(t) \qquad 1.T$$

where

$$\dot{Q}_{in}(t) = \dot{Q}_{exog} + \dot{Q}_{reab}(t) \qquad 2.T$$

With respect to passive calcium exchange with bone,

$$\dot{Q}_{depos}(t) = \dot{Q}_{bone,\,art}(t) - \dot{Q}_{bone,\,ven} \qquad 3.T$$

where

$$\dot{Q}_{bone,\,art}(t) = \dot{V}_{bone}C(t) \qquad 4.T$$

By assumption 5,

$$\dot{Q}_{bone,\,ven} = \dot{V}_{bone}C_{eq} \qquad 5.T$$

Combining equations 1 through 5,

$$[Vdist]\frac{dC(t)}{dt} + \dot{V}_{bone}C(t) = \dot{V}_{bone}C_{eq} + \dot{Q}_{reab}(t) + \dot{Q}_{exog} \qquad 6.$$

Solving equation 6 for $C(\infty)$ in the steady state,

$$C(\infty) = C_{eq} + \frac{\dot{Q}_{reab}(\infty) + \dot{Q}_{exog}}{\dot{V}_{bone}} \qquad 7.$$

With respect to the role of parathyroid hormone,

$$\dot{Q}_{PH}(\infty) = f(C(\infty)) \qquad 8.$$

$$C_{PH}(\infty) = \dot{Q}_{PH}(\infty)/[\dot{V}cl]_{PH} \qquad 9.T$$

$$\dot{Q}_{reab}(\infty) = g(C_{PH}(\infty)) \qquad 10.$$

or, combining equations 8, 9 and 10,

$$\dot{Q}_{reab}(\infty) = h(C(\infty)) \qquad 11.$$

where the function h must be found empirically.

A partly-descriptive symbol-and-arrow diagram for the above relationships is:

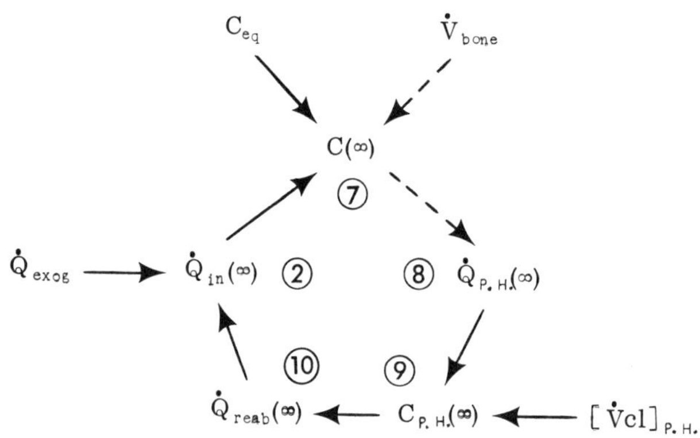

Figure 2.

In order to find the function, h, calcium was administered to normal dogs by intravenous infusion at a constant rate, $\dot{Q}_{exog}$, and $C(\infty)$ was measured when a new steady state had been reached (1). Similar experiments were conducted in thyroparathyroidectomized dogs. Assuming that $\dot{Q}_{reab}(\infty)$ = zero after parathyroidectomy, C_{eq} was estimated from equation 7 as the mean plasma calcium in the thyroparathyroidectomized dogs when $\dot{Q}_{exog} = 0$. With C_{eq} known, $\dot{V}_{bone}$ could also be estimated by applying equation 7 to the thyroparathyroidectomized dogs when $\dot{Q}_{exog} > 0$. Equation 7 was used yet a third time to construct the broken lines in Figure 3, each line representing one of the rates of calcium infusion used in the normal dogs. Finally, a point was plotted on each such theoretical line at the mean value of $C(\infty)$ which was actually observed experimentally when the corresponding rate of calcium infusion was used in normal dogs. The experimental points were thus constrained to obey equation 7 for $C(\infty)$ as a function of $\dot{Q}_{reab}(\infty)$. But by the argument detailed in Section 7.11, they should also conform to equation 11 for $\dot{Q}_{reab}(\infty)$ as a function of $C(\infty)$. Hence, the unknown function, h, of equation 11 was approximated (2) by the following

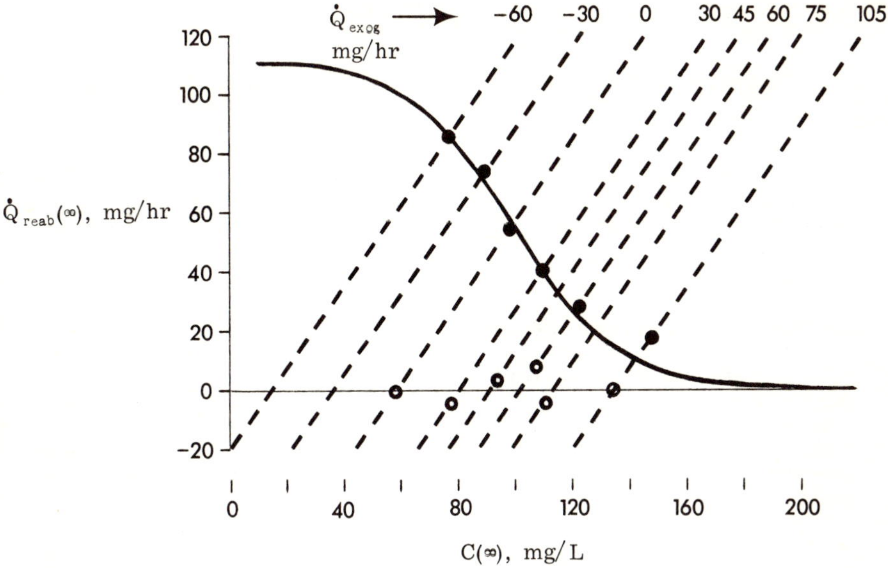

Figure 3. Broken lines, $C(\infty)$ as a function of $\dot{Q}_{reab}(\infty)$ (equation 7) for the designated values of $\dot{Q}_{exog}$. Solid dots, observed values of $C(\infty)$ in normal dogs plotted on the relevant broken lines for $\dot{Q}_{exog}$. Solid curve, equation 12 for $\dot{Q}_{reab}(\infty)$ as a function of $C(\infty)$ fitted to the solid points. Open dots, observed values of $C(\infty)$ in thyroparathyroidectomized dogs plotted on the relevant broken line for $\dot{Q}_{exog}$. Horizontal solid line, $\dot{Q}_{reab}(\infty) = 0$.

---

empirical equation (a symmetrical logistic curve) fitted to the points of Figure 3.

$$\dot{Q}_{reab}(\infty) = \frac{2\,\dot{Q}_{reab,\,norm}}{1 + e^{-a[C_{norm} - C(\infty)]}} \qquad 12.E$$

Equations 7 and 12 are a pair of working equations which describe the relationships between two members of the feedback loop. The derivative of equation 7,

$$\frac{dC(\infty)}{d\dot{Q}_{reab}(\infty)} = \frac{1}{\dot{V}_{bone}} \qquad 13,$$

and the derivative of equation 12,

$$\frac{d\dot{Q}_{reab}(\infty)}{dC(\infty)} = -\frac{2\dot{Q}_{reab,\,norm}\, a\, \epsilon^{-a[C_{norm} - C(\infty)]}}{\left\{1 + \epsilon^{-a[C_{norm} - C(\infty)]}\right\}^2}, \qquad 14,$$

may therefore be used to calculate [OLG](j0).

$$[OLG](j0) = \frac{2\,a\,\dot{Q}_{reab,\,norm}}{\dot{V}_{bone}} \left[\frac{\epsilon^{-a[C_{norm} - C(\infty)]}}{\left\{1 + \epsilon^{-a[C_{norm} - C(\infty)]}\right\}^2} \right] \qquad 15.$$

The open-loop gain for various values of $C(\infty)$ produced by infusion of calcium in normal dogs has been calculated using equation 15 and the exemplary values of the parameters listed above. The resulting curve is shown in Figure 4.

Comments

1. Several of the assumptions upon which the present model is based are crude and unrealistic. For example, bone is not an infinite source of, or sink for calcium (assumption 1), so that the steady states here discussed are, in fact, only quasi-steady states. A true steady state would be reached only when calcium intake (in the diet and by infusion) was equal to the rate of calcium loss from the body. In a true steady state, therefore, we could not neglect (assumption 2) calcium exchange via the kidney and via the gastrointestinal tract.

2. There is no longer any doubt that calcitonin (= thyrocalcitonin) is a genuine hormone which is secreted in response to elevation of the concentration of calcium in plasma. (See, for example, 3). The role of calcitonin in normal physiology is less clear, but recent evidence (4) suggests that it may indeed be important, at least in some species. In the experiments on which the present analysis is based, the feedback loop was opened by totally excising the thyroid and parathyroid glands, so that the sources of both parathyroid hormone and calcitonin were removed. The gains, therefore, represent the joint action of both hormones. The results of studying them separately should be interesting.

3. The rather small open-loop gain of this system suggests that "normally the most important function of the parathyroid glands is not to regulate plasma calcium in the face of fluctuations in calcium supply but rather to maintain plasma calcium at a concentration far higher than would prevail in the absence of any parathyroid hormone." (1).

4. According to the present model, (2), the rate of deposition of calcium in bone at a given plasma concentration is limited by plasma flow through bone (equations 3, 4 and 5) while the rate of reabsorption, which depends upon the concentration of parathyroid hormone (equation 10), is not flow-limited. A local decrease in $\dot{V}_{bone}$ would therefore decrease calcium deposition without affecting reabsorption. If $\dot{V}_{bone}$ decreases during immobilization, this hypothesis would help to explain the demineralization which occurs in an immobilized limb. Direct studies of how $\dot{V}_{bone}$ influences calcium metabolism are needed.

5. Of the several parameters in the empirical equation 12, only "a" was actually evaluated from the six observed points in Figure 3. The others were calculated independently or fixed by hypothesis. (Estimating three or four parameters from six observed points is almost never justified!)

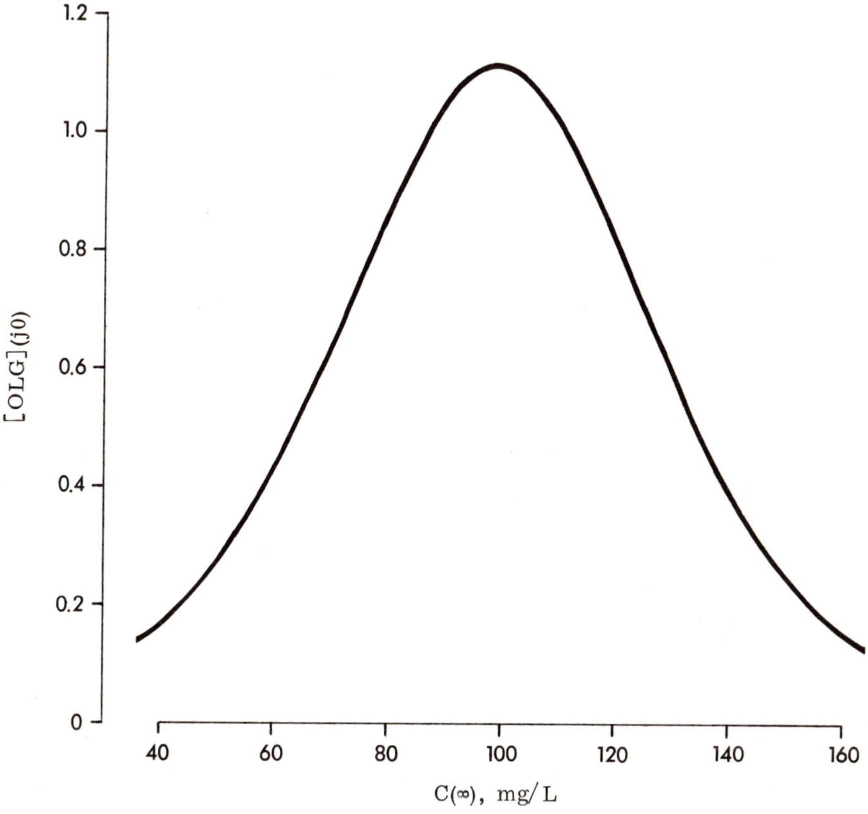

Figure 4.

Transient behavior

The midportion of the solid curve (equation 12) in Figure 3 can be represented reasonably well by a straight line corresponding to

$$\dot{Q}_{reab}(t) = 188.2 - 1.335\ C(t)\ \text{mg/hr} \qquad 16.$$

Combining this equation with equation 6 by eliminating $\dot{Q}_{reab}(t)$, inserting the values of the parameters listed in the example, and solving the result for C(t) subject to the initial condition, $C(0) = C_{norm} = 99$ mg/L, we obtain

$$C(t) = 99 + \frac{\dot{Q}_{exog}}{2.706}\left[1 - \epsilon^{-0.699t}\right]\ \text{mg/L} \qquad 17,$$

When the feedback loop has been opened surgically, we assume that $\dot{Q}_{reab}(t) = 0$. Then integration of equation 6, subject to the initial condition C(0) = 58 mg/L, yields

$$C(t) = 58 + \frac{\dot{Q}_{exog}}{1.37}\left[1 - e^{-0.354t}\right] \text{ mg/L} \qquad 18.$$

The exponential rate constants of -0.699 per hour for the closed loop situation and -0.354 per hour for the open-loop situation compare reasonably well with the values of -0.571 per hour and -0.355 per hour estimated experimentally (1). The observation that the approach to a new steady state was no more than about twice as fast in the normal dogs as in the thyroparathroidectomized dogs agrees with the calculated gains of about 1.0 in the neighborhood of the normal operating point, and supports the conclusion that this is not a very effective homeostatic mechanism.

References

1. Hausmann, E., and Riggs, D.S.: The effectiveness of negative feedback in regulating plasma calcium in the dog. J. Theoret.Biol., 12, 350-363, 1966.

2. Riggs, D.S.: A quantitative hypothesis concerning the action of the parathyroid hormone. J. Theoret. Biol., 12, 364-372, 1966.

3. Copp, D.H.: Parathyroid hormone, calcitonin and calcium homeostasis. Excerpta med. Int. Congr. Ser, 159, 25-39, 1968.

4. Gray, T.K., and Munson, P.L.: Thyrocalcitonin: Evidence for physiological function. Science, 166, 512-513, 1969.

PUPILLARY REACTION TO LIGHT

NEGATIVE

Type	Symbol	Definition
Ind.	B	Brightness of illuminated screen facing human subject in an otherwise dark room.
Par.	k	Proportionality constant in equation 1.
Dep.	$\dot{q}_{ret}(\infty)$	Quantity of visible radiant energy per unit time incident on retina.
	$A_{pupil}(\infty)$	Area of pupil of eye.

Model

When the quantity of light falling upon the retina increases, reflex contraction of the circular muscle of the iris occurs. This reduces the area of the pupil, and decreases the illumination of the retina, thus tending to reduce the magnitude of the original change.

Derivation

Crawford (1) studied the steady-state diameter of the pupil in 10 normal human subjects at 16 different values of B, the brightness of a wide illuminated screen facing the subject. Both $A_{pupil}(\infty)$ and $\dot{q}_{ret}(\infty)$ (measured in convenient arbitrary units) can be calculated from his data. $\dot{q}_{ret}(\infty)$ is directly proportional to the product of B, the screen brightness, and $A_{pupil}(\infty)$, the pupillary area.

$$\dot{q}_{ret}(\infty) = k\,B\,A_{pupil}(\infty) \qquad 1.\text{T}$$

where k is a constant of proportionality with appropriate units. The observed dependence of $A_{pupil}(\infty)$ upon $\dot{q}_{ret}(\infty)$ can be fitted by the following empirical equation (reference 2, p.109):

$$A_{pupil}(\infty) = \frac{40\,\dot{q}_{ret}^{-0.4}(\infty) + 23.7}{\dot{q}_{ret}^{-0.4}(\infty) + 3.95} \qquad 2.\text{E}$$

Equations 1 and 2 are the two working equations describing the steady-state behavior of this system. The derivative of equation 1 is

$$\frac{d\dot{q}_{ret}(\infty)}{dA_{pupil}(\infty)} = kB = \frac{\dot{q}_{ret}(\infty)}{A_{pupil}(\infty)} \qquad 3.$$

The derivative of equation 2 is

$$\frac{dA_{pupil}(\infty)}{d\dot{q}_{ret}(\infty)} = -\frac{53.72\,\dot{q}_{ret}^{-1.4}(\infty)}{[\dot{q}_{ret}^{-0.4}(\infty) + 3.95]^2} \qquad 4.$$

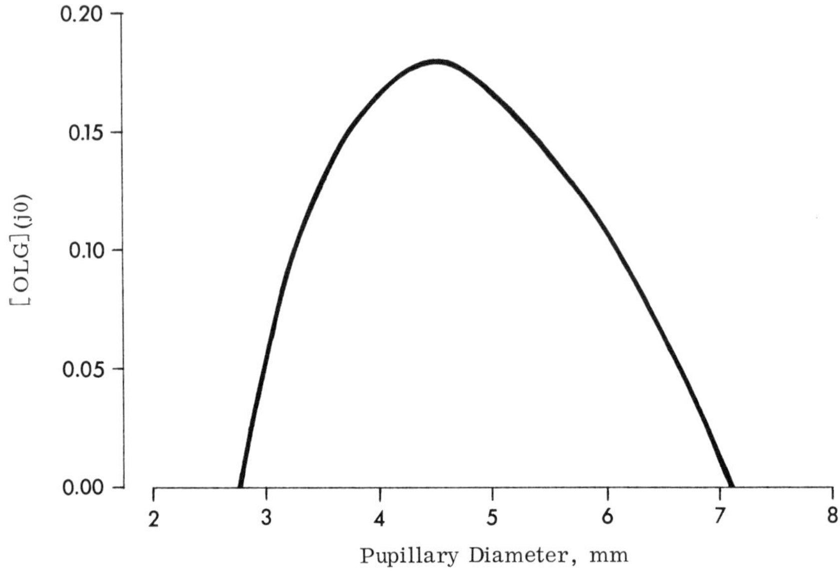

Figure 1.

The [OLG](j0) is the absolute value of the product of these two derivatives.

$$[OLG](j0) = \frac{53.72\, \dot{q}_{ret}(\infty)^{-0.4}}{A_{pupil}(\infty)\left[\dot{q}_{ret}(\infty)^{-0.4} + 3.95\right]^2} \qquad 5.$$

In Figure 1, [OLG](j0) calculated from equation 5 has been plotted against $2\sqrt{A_{pupil}(\infty)/\pi}$, the pupillary diameter in millimeters. Even at its maximum the gain is very small because a substantial change in $\dot{q}_{ret}(\infty)$ causes only a small change in the size of the pupil.

Comment

The pupillary reaction to light illustrates how a feedback loop which is not very effective at any particular operating point can nevertheless exert some homeostatic effect over an enormous range. In Crawford's experiments the largest value of B was almost 2,000,000,000 times the smallest value of B!

Transient behavior

Because the transient response of the pupil can be measured quite accurately under a variety of circumstances in intact man, this system has served as a leading example of how control theory can be used to analyze a biological feedback mechanism. For a detailed discussion of its transient behavior, the reader should consult the papers of Stark and his collaborators (see 3 and 4 for references) and of Clynes and his collaborators (see 5 for references). The depth and elegance of their studies contrast sharply with the rather uninteresting steady-state behavior of the system which is described here.

References

1. Crawford, B.H.: The dependence of pupil size upon external light stimulus under static and variable conditions. Proc. roy. Soc. Ser. B, 121, 376-395, 1937.

2. Riggs, D.S.: The Mathematical Approach to Physiological Problems. The Williams and Wilkins Co., Baltimore, 1963.

3. Stark, L.: Pupillary control system: its nonlinear adaptive and stochastic engineering design characteristics. Fedn. Proc., 28, 52-64, 1969.

4. Stark, L.: Neurological Control Systems, Plenum, N.Y., 1968.

5. Clynes, M.: The non-linear biological dynamics of unidirectional rate sensitivity illustrated by analog computer analysis, pupillary reflex to light and sound, and heart rate behavior. Ann. N.Y. Acad. Sci., 98, 806-845, 1962.

VICIOUS CIRCLES IN DISEASE

POSITIVE

In medicine, a vicious circle is a positive feedback loop whose operation tends to intensify the pathophysiological changes responsible for a particular syndrome. Vicious circles often account for acute exacerbations in a chronic disease. If the organ system affected is essential for life, and if the gain of the positive feedback loop equals or exceeds unity, the patient is likely to die unless the vicious circle can be interrupted promptly. When a vicious circle underlies a pathological process, serial measurements of the open-loop gain would warn of impending disaster, and would serve as a quantitative index of the effectiveness of treatment. Unfortunately, the variables in the positive feedback loop are often difficult to identify and even more difficult to measure accurately enough so that feedback equations describing the interrelations between two members of the loop could be written. In this section, therefore, I shall merely present a few descriptive symbol-and-arrow diagrams to illustrate the role played by vicious circles in disease.

A. PAROXYSMAL NOCTURNAL DYSPNEA DUE TO ACUTE PULMONARY EDEMA

In left heart failure, acute episodes of pulmonary edema characteristically occur at night, some time after the patient has gone to bed. Unless prompt treatment is given, a severe attack may be fatal presumably because the overall gain of the vicious circles equals or exceeds unity.

In the following symbol-and-arrow diagram, the steps directly involved in the pathogenesis of pulmonary edema and the consequent anoxia have been identified by heavy arrows. In left heart failure, $R_{L\ vent}$, the resistance of the left ventricle to the inflow of blood is increased, so that $P_{L\ atr}$, the pressure in the left atrium is abnormally elevated, necessitating an elevation of $P_{lung\ capill}$ and hence of P_{filt}, the net transcapillary filtration pressure in the lungs, which causes movement of fluid from blood plasma into the alveoli. During the day, while the patient is "up and around," the force of gravity tends to limit venous return, (= $\dot{V}_{R\ vent}$, the output of the right heart) by pooling fluid in dependent parts of the body, and by vasoconstriction (mediated via the baroreceptors) which increases peripheral resistance, R_{periph}. (Venomotion, however, tends to maintain venous return by decreasing the volume of blood pooled in the veins. When the patient goes to bed, the angle which the long axis of the body makes with the horizontal is decreased to or towards zero; gravitational forces no longer hinder the return of blood to the heart, and cardiac output increases, particularly if there is any dependent edema which can be reabsorbed into the blood during recumbency. The increased output of the right heart increases $P_{lung\ capill}$ and P_{filt}. As P_{filt} increases, $\dot{V}_{filt}$, the rate of fluid transudation into the alveoli becomes more rapid and the volume of edema fluid in the lungs increases. Fluid accumulation increases the resistance to air flow into the alveoli (thus reducing $\dot{V}_{alv}$, the alveolar ventilation), and decreases the diffusing capacity of the lungs. Both effects reduce $p_{O_2,art}$, the partial pressure of oxygen in arterial blood, and cause anoxia. Both effects also increase $p_{CO_2,art}$, the partial pressure of carbon dioxide in arterial blood.

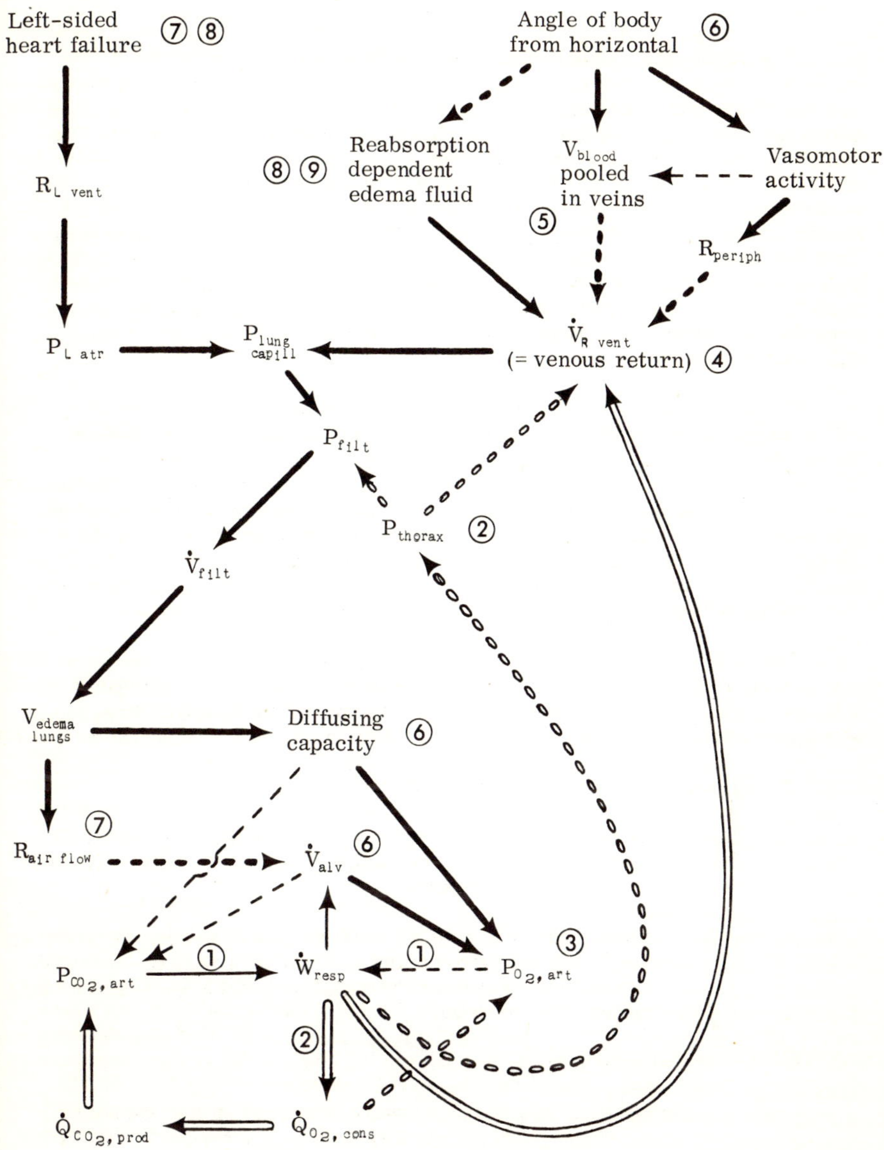
Left-sided heart failure ⑦ ⑧
Angle of body from horizontal ⑥
Reabsorption dependent edema fluid ⑧ ⑨
V_{blood} pooled in veins ⑤
Vasomotor activity
$R_{L\ vent}$
R_{periph}
$P_{L\ atr}$
$P_{lung\ capill}$
$\dot{V}_{R\ vent}$ (= venous return) ④
P_{filt}
P_{thorax} ②
$\dot{V}_{filt}$
$V_{edema\ lungs}$
Diffusing capacity ⑥
$R_{air\ flow}$ ⑦
$\dot{V}_{alv}$ ⑥
③
$P_{CO_2,art}$
①
$\dot{W}_{resp}$
①
$P_{O_2,art}$
②
$\dot{Q}_{CO_2,prod}$
$\dot{Q}_{O_2,cons}$

In the lower part of the diagram is shown the multiple-loop system, described elsewhere in Part II, which is concerned with the chemical control of breathing.

Key portions of positive feedback loops are emphasized by thick, but unfilled arrows. All of these positive loops depend upon the ability of hypercapnia and hypoxia to increase $\dot{W}_{resp}$, the rate at which the muscles of respiration do work. The positive feedback loops which involve increased O_2 consumption and CO_2 production by the respiratory muscles were described in the section on the chemical control of breathing. In pulmonary edema, two other positive feedback paths establish additional vicious circles. First, the labored, gasping respiration (= dyspnea) decreases the mean intrathoracic pressure. This decrease not only tends to promote return of blood to the heart but also decreases the pressure within the alveoli, thereby increasing P_{filt}. Second, as is true of any kind of exercise, the increased work of the respiratory muscles will cause an increase in cardiac output via mechanisms which are still not entirely clear.

The preponderance of positive feedback in pulmonary edema confronts us with a paradox: the rapidly worsening tissue anoxia would seem to demand measures to increase delivery of oxygen from the air to the lungs via increased lung ventilation, and from the lungs to the tissues via increased cardiac output. Yet increased cardiac output and increased ventilatory efforts are two of the main factors accounting for the initiation and perpetuation of the pulmonary edema! In fact, as clinicians have long recognized, in pulmonary edema it is essential to interrupt the greater-than-unity-gain positive feedback loops even if we do so by depressing respiration and decreasing cardiac output.

The following list shows what therapeutic measures are available to reverse the rapidly progressive deterioration characteristic of untreated pulmonary edema. Each circled number in the symbol-and-arrow diagram shows the point of attack of the correspondingly numbered therapeutic measure.

I. To interrupt the vicious circles.

1. Morphine sulfate intravenously. Lowers $\dot{W}_{resp}$ by decreasing the sensitivity of the respiratory center to stimuli caused by hypercapnia and hypoxia.

2. Assisted respiration. (Gas mixture is supplied at greater-than-atmospheric pressure during inspiration.) In part, substitutes exogenous for endogenous respiratory work. Also raises mean intrathoracic pressure.

II. To combat anoxia directly.

3. Increasing the partial pressure of oxygen in the inspired air.

III. To decrease venous return and cardiac output.

4. Bleeding (= venesection).

5. Pooling blood in limbs by applying tourniquets to obstruct venous outflow (= "bloodless venesection").

6. Elevation of head of bed, or sitting posture, to promote gravitational pooling. This may also favor redistribution of fluid in the lungs so that at least the upper portions can be aerated, thus increasing diffusing capacity and decreasing resistance to air flow.

IV. To combat the heart failure.

7. Aminophylline slowly intravenously, for its positive inotropic effect on the heart, and to relax bronchiolar smooth muscle thus reducing $R_{air\ flow}$.

8. Digitalization with a rapidly-acting cardiac glycoside, to relieve the heart failure.

V. Longer-term preventive measures.

9. Maintain digitalization with a suitable drug.

10. Administer diuretics if needed to prevent accumulation of dependent edema.

B. HEMORRHAGIC SHOCK

In experimental animals, a prolonged decrease in blood pressure due (for example) to hemorrhage, ultimately produces a state of irreversible circulatory "shock" which causes death in spite of restoration of blood volume. The following symbol-and-arrow diagram, patterned upon the work of Guyton and Crowell (1) shows three positive feedback loops which help to explain the pathogenesis of acute hemorrhagic shock. Decreased arterial blood pressure caused by decreased cardiac output is common to all three loops.

Loop 1. Decreased coronary blood flow restricts myocardial oxygenation, thus producing heart failure and causing still further decrease in cardiac output.

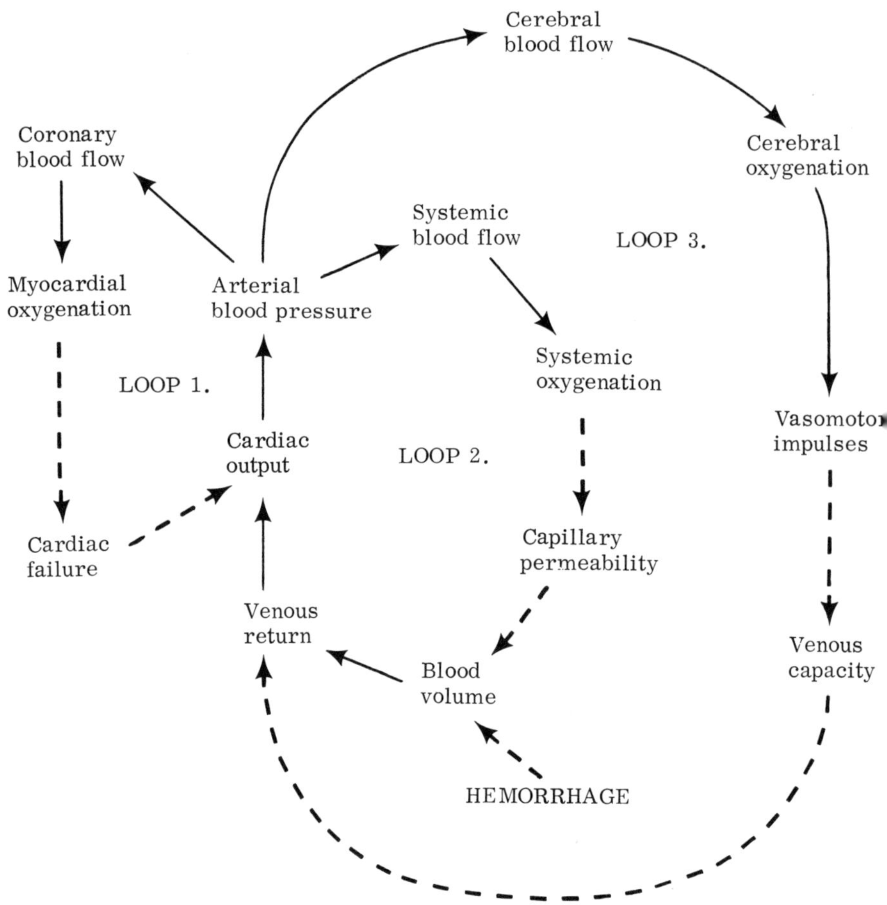

Loop 2. The decrease in oxygenation of peripheral tissues caused by decreased blood flow leads to an increase in capillary permeability. As a result, blood volume, venous return, and cardiac output all decrease.

Loop 3. Long-lasting cerebral anoxia may depress the vasomotor center. The consequent loss of venomotor tone may increase the capacity of the veins to hold blood at low pressure, thereby decreasing venous return and cardiac output. (N.B. It is well known that acute anoxia increases vasomotor outflow from the brain.)

Reference

1. Guyton, A.C. and Crowell, J.W.: Cardiac deterioration in shock: I. Its progressive nature. Pp 1-12 in Hershey, S.G., editor. Shock, Little, Brown, and Co., Boston, 1964.

C. MALIGNANT HYPERTENSION

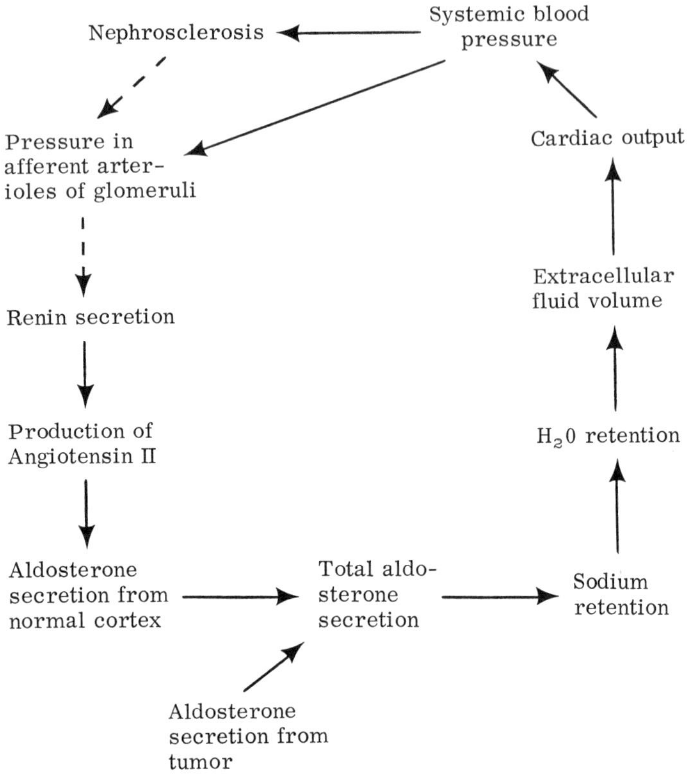

Persistent hypertension, from whatever cause, may eventually produce renal arteriolosclerosis (= nephrosclerosis) which impedes blood flow and decreases blood pressure in many of the afferent arterioles of the glomeruli. Via the mechanism described elsewhere in Part II, increased secretion of renin occurs which still further increases systemic blood pressure. This sequence of events helps to account for the sudden onset of a rapidly-progressing "malignant" phase in chronic hypertensive disease. A vicious circle of this kind may occur even in primary aldosteronism (= Conn's syndrome) caused by primary oversecretion of aldosterone from an adrenocortical adenoma --"It is likely that primary aldosteronism and renovascular hypertension occur not infrequently in the same patient. In this situation one should expect to observe a rapidly progressive form of hypertension with papilledema. Plasma renin activity levels would rise and the presence of primary aldosteronism could be completely obscured by the overriding effect of the plasma renin activity . . . One should expect to find at autopsy, not only evidence of a single discrete adenoma, but also evidence of secondary aldosteronism in the form of diffuse nodular hyperplasia of the nontumerous adrenal tissue. . . . Considering the degree of arteriolosclerosis which is frequently observed in the operative renal biopsies from our own patients with primary aldosteronism, it would be surprising if some of them did not develop a compromised renal circulation sufficient to increase renin output. It is very likely that this occurs more often than we presently believe but, by the time we observe the patient, the renovascular element as the cause of the hypertension appears to be obvious." (1.).

Reference

1. Conn, J.W., Cohen, E.L., and Rovner, D.R.: Plasma renin activity in primary aldosteronism. J. Amer. med. Assoc., 190, 213-221, 1964.

D. ACUTE GOUTY ARTHRITIS

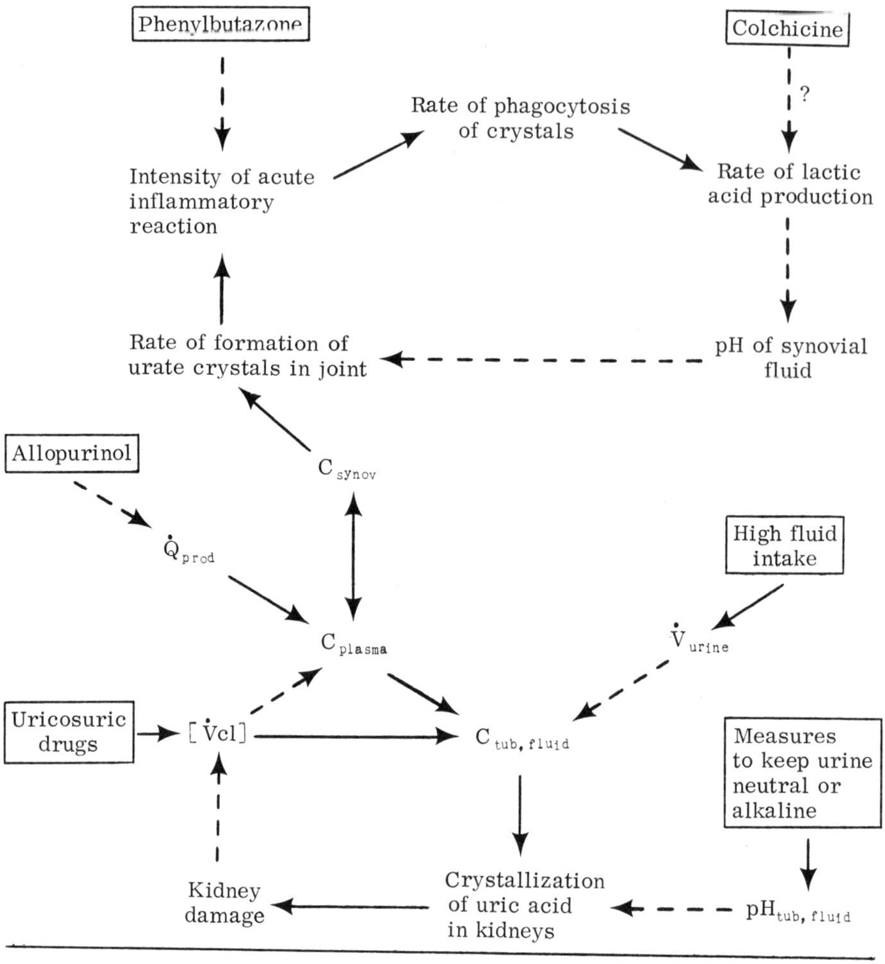

According to Seegmiller et al. (1), an acute attack of gouty arthritis is initiated by the formation of crystals of monosodium urate in the synovial fluid of the joint. Crystallization is favored by a high C_{synov}, the concentration of uric acid in synovial fluid, which is increased by an increase in C_{plasma}, the concentration of uric acid in plasma. The crystals cause a local inflammatory reaction, possibly by activating Hageman factor which, in turn, may lead to the formation of "kinin-like inflammatory mediators" (2). As part of the acute inflammatory response, increased numbers of leucocytes invade the joint fluid and phagocytose many of the crystals. Phagocytosis is accompanied by an increase in the metabolic activity of the leucocytes with the production of increased lactic acid. The lactic acid lowers the pH of the synovial fluid which decreases the solubility of monosodium urate and increases the rate of crystal formation. This completes the vicious circle (1). Colchicine may interrupt the cycle by decreasing the metabolic activity of the leucocytes and the rate at which they produce lactic acid. Phenylbutazone may decrease the intensity of the acute inflammatory reaction.

In the lower part of the diagram is shown a second vicious circle, which, though not involved in the acute attack of arthritis, does help to perpetuate the hyperuricemia. When C_{plasma} is elevated and $\dot{V}_{urine}$, the urine flow, is low, $C_{tub, fluid}$, the concentration of urate in the distal tubular fluid may be high enough to cause crystal formation, particularly when the pH of the fluid is low. The resulting renal damage reduces $[\dot{V}cl]$, the clearance of urate from plasma, and thus tends to elevate C_{plasma} still further. While uricosuric drugs such as probenecid decrease C_{plasma} by increasing $[\dot{V}cl]$, they may also increase $C_{tub, fluid}$ and hence cause further kidney damage, unless a high fluid intake and measures to decrease urinary acidity reduce the danger of crystallization of uric acid. Allopurinol reduces C_{plasma} by decreasing $\dot{Q}_{prod}$, the rate at which uric acid is formed from precursors.

References

1. Seegmiller, J.E., Laster, L, and Howell, R.R.: Biochemistry of uric acid and its relation to gout. New Engl. J. Med. 268, 712-716, 764-773, 821-827, 1963.

2. Kellermeyer, R.W., and Breckenridge, R.T.: The inflammatory process in acute gouty arthritis. I. Activation of Hageman factor by sodium urate crystals. J. lab. clin. Med., 65, 307-315, 1965.

APPENDIX

DIFFERENTIATING IMPLICIT FUNCTIONS

Suppose that we have an equation $f(x, y(x)) = 0$, and that we want to find dy/dx. If the equation can be solved for $y(x)$, so that $y(x) = g(x)$ where $g(x)$ is some reasonably simple function of x alone, we can find dy/dx in the usual way by differentiating $g(x)$ directly. Often, however, it is difficult or impossible to solve $f(x, y(x)) = 0$ explicitly for $y(x)$ and to differentiate the function of x thus obtained. Under these circumstances, we shall often find it easier to work with the original equation which describes $y(x)$ as an implicit function of x. To differentiate an implicit function, we need only remember that the total differential of the left side of an equation must equal the total differential of the right side. Hence, given $f(x, y) = 0$,

$$\frac{\partial f}{\partial x} dx + \frac{\partial f}{\partial y} dy = 0, \text{ or } \frac{\partial f}{\partial x} dx = -\frac{\partial f}{\partial y} dy$$

solving for the derivative,

$$\frac{dy}{dx} = -\frac{\partial f/\partial x}{\partial f/\partial y}$$

Example 1. Given $f(x, y) = x^2 + y^2 = 0$, find dy/dx.

A. By solving for y, then differentiating.

$$y = \sqrt{1 - x^2} = u^{1/2} \quad \text{where } u \equiv 1 - x^2$$

Using the chain rule, $[dy/dx] = [dy/du][du/dx]$,

$$dy/du = [1/2]/u^{1/2}; \quad du/dx = -2x$$

Hence,

$$dy/dx = \left[\frac{1}{2u^{1/2}}\right][-2x] = -\frac{x}{\sqrt{1 - x^2}}$$

B. By differentiating the implicit function.

$$dy/dx = -[\partial f/\partial x]/[\partial f/\partial y] = -x/y = -\frac{x}{\sqrt{1-x^2}}$$

Example 2. Given $f(x, y) = xy^2 - xy - 3x - 5 = 0$, find dy/dx.

A. If we try to solve explicitly for y, the best we can do is to get

$$y^2 - y = [5 + 3x]/x = 5x^{-1} + 3 \qquad (1.)$$

Define $u = y^2 - y$ so that

$$\frac{du}{dy} = 2y - 1; \quad \frac{dy}{du} = \frac{1}{2y - 1}$$

$$\frac{du}{dx} = -5/x^2$$

Using the chain rule, $dy/dx = [dy/du][du/dx]$,

$$\frac{dy}{dx} = -\frac{5}{2yx^2 - x^2} \qquad (2.)$$

B. By attacking the given implicit function directly,

$$\frac{dy}{dx} = -\frac{\partial f/\partial x}{\partial f/\partial y} = -\frac{y^2 - y - 3}{2xy - x}$$

which, if desired, can be transformed into (2.) by using (1.) to eliminate y^2-y from the numerator.

FITTING EMPIRICAL EQUATIONS TO EXPERIMENTAL DATA

Suppose that we have studied a dependent variable, y(x), as a function of an independent variable, x, which is either a space-time coordinate, or some other variable which we have manipulated experimentally over a certain domain. Assume that we have thus obtained n experimental points, $[x_i, y_i]$, i = 1, 2, . . . n, and that we have plotted the points on arthmetical coordinate graph paper with x as abscissa, y(x) as ordinate. We wish to find an empirical equation to describe the trend of the points.

General remarks on curve fitting

1. Choosing the general kind of equation which is to be fitted to the points. Selecting the most appropriate form of equation depends in large part upon common sense and experience. However, three rule-of-thumb guides are useful: a.) Make the most of whatever theoretical information is available about how y(x) would be expected to depend on x. In particular, consider what values y(x) would approach as x becomes very large or very small, and try to find an equation which will make sense at both extremes. b.) Choose the simplest equation which will describe the data adequately, i.e., the equation with the fewest parameters. c.) If two or more equations with the same number of parameters fit the observations equally well, choose the one which is most easily handled mathematically.

2. Estimating parameters by fitting an n-parameter curve to n points, or by using a linearizing transformation. It is not always easy to see by mere inspection what form of equation will satisfactorily describe the trend of the experimental observations, and it may then be desirable to try several forms before choosing one of them for more careful fitting by the method of least squares.

To test the ability of an equation with n parameters to fit the observations, choose n points which fairly represent the full range of the available data and which also represent any curvature or points of inflection. (The n points are chosen to show the trend of the data and need not coincide with any of the actual observations). For each point, write an equation in the unknown parameters by substituting the coordinates of the point for x and y(x) in the general equation. Solve the resulting n equations simultaneously for the n parameters. Finally, plot the curve determined by those parameters to see whether or not it represents the data satisfactorily over the whole range of observations.

Estimates of parameters may also be obtained by subjecting the data to an appropriate linearizing transformation. But see NOTE 2 below.

Example: The following points are representative of the relationship between oxygen consumption required by the work of breathing, and ventilation rate in normal human subjects, as shown by a curve in Figure 21, p. 756 of Ruch and Patton's textbook (1). (Oxygen consumption is here given in mM/min instead of in ml/min as in the original figure.)

Total ventilation (liters expired/min)	$\dot{V}$	= 0	20	40	60	80	100
oxygen consumption ascribed to work of breathing (mM/min)	$\dot{Q}(\dot{V})$	= 0.00	0.22	0.44	1.10	2.31	5.50

Notice that the oxygen consumption is necessarily zero for zero ventilation, so that the curve is constrained to pass through the origin.

Trial of power function:

$\dot{Q}(\dot{V}) = a\dot{V}^n$

$\log_{10}\dot{Q}(\dot{V})$ was plotted against $\log_{10}\dot{V}$, but the resulting points did not fall on the straight line required by the logarithmic transformation,

$\log_{10}\dot{Q}(\dot{V}) = \log_{10}a + n \log_{10}\dot{V}$

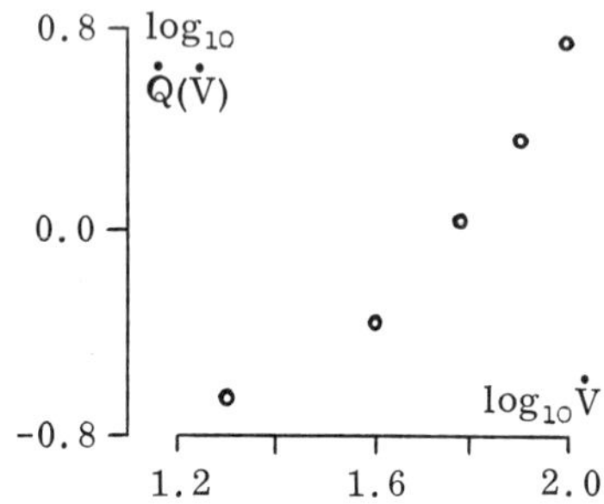

Trial of polynomial:

$\dot{Q}(\dot{V}) = a\dot{V} + b\dot{V}^2$

Try fitting to the points [60, 1.1] and [100, 5.5].

$$\left.\begin{aligned} 1.1 &= 60a + 3600b \\ 5.5 &= 100a + 10{,}000b \end{aligned}\right\} \begin{aligned} a &= -3.67 \cdot 10^{-2} \\ b &= 9.17 \cdot 10^{-4} \end{aligned}$$

But $\dot{Q}(\dot{V}) = -3.67 \cdot 10^{-2}\dot{V} + 9.17 \cdot 10^{-4}\dot{V}^2$ does not fit the remaining points (solid dots) satisfactorily.

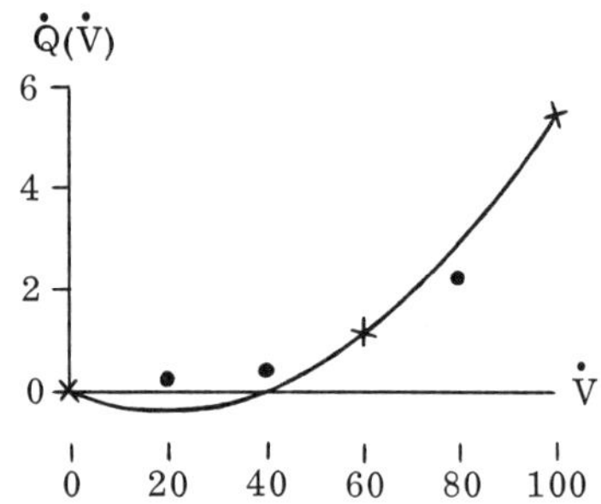

Trial of hyperbola:

$\dot{Q}(\dot{V}) = \dot{Q}_{asym}\dot{V}/[\dot{V} - \dot{V}_{asym}]$ or,

$\dot{V}\dot{Q}(\dot{V}) = \dot{Q}_{asym}\dot{V} + \dot{V}_{asym}\dot{Q}(\dot{V})$

which is more convenient for calculating parameters from the points [60, 1.1] and [100, 5.5]:

$$\left.\begin{aligned} 66 &= 60\dot{Q}_{asym} + 1.1\dot{V}_{asym} \\ 550 &= 100\dot{Q}_{asym} + 5.5\dot{V}_{asym} \end{aligned}\right\} \begin{aligned} \dot{Q}_{asym} &= -1.1 \\ \dot{V}_{asym} &= 120 \end{aligned}$$

The resulting equation,
$\dot{Q}(\dot{V}) = 1.1\dot{V}/[120 - \dot{V}]$
fits the remaining points very well.

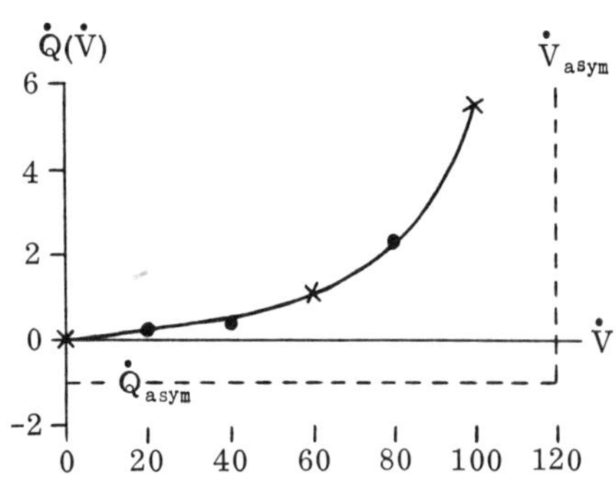

3. Fitting a specific equation to the points by the method of least squares. Having selected the general form of the equation to be used, we next evaluate its parameters so that the resulting specific equation will best fit the n experimental points. For sound statistical reasons, the "best" parameters are usually taken to be the ones which minimize the sum of the squares of the deviations of the observed y's from the curve determined by the fitted equation, the deviations being measured parallel to the y axis. The procedure employed for finding those parameters is called the method of least squares.

Let $y_{i,\,calc}(x_i)$ be the value of y calculated from the ith experimental value of x, using the general form of the chosen equation, i.e., the form in which the parameters, a, b, c, . . . , are still merely symbols. Let $y_{i,\,obs}$ be the experimentally observed value of y when $x = x_i$. Then SS(a, b, c, · · ·), the sum of squared deviations (= "sum of squares") which is to be minimized is

$$SS(a, b, c, \cdots) = \sum_{i=1}^{n} \left[y_{i,\,obs} - y_{i,\,calc}(x_i) \right]^2$$

The values of a, b, c, · · · which make SS a minimum are found by setting the partial derivatives of SS with respect to each parameter equal to zero, and solving the resulting normal equations,

$$\partial SS/\partial a = 0$$

$$\partial SS/\partial b = 0$$

$$\partial SS/\partial c = 0$$

$$\vdots$$

simultaneously for a, b, c, . . .

Example: Suppose we have studied the rate of disappearance of a drug from plasma, and that we have decided to fit the observed points

i =	1	2	3	4	5	6	
t_i =	0	1	2	3	4	5	days
C_i =	3.10	1.70	1.20	0.65	0.40	0.20	mg/L of plasma

by an exponential equation of the general form

$$C(t) = C_0 \epsilon^{-kt}$$

where C_0, the initial concentration, and k, the rate constant for disappearance are the parameters whose "best" values are to be found by the method of least squares. Each of the six deviations is of the form

$$C_{i,\,obs} - C_{i,\,calc}(t_i) = C_{i,\,obs} - C_0 \epsilon^{-kt_i}$$

Squaring,

$$[C_{i,\,obs} - C_{i,\,calc}(t_i)]^2 = C^2_{i,\,obs} - 2C_{i,\,obs}\,C_0\epsilon^{-kt_i} + C_0^2\epsilon^{-2kt_i}$$

Sum all such squared deviations from i = 1 to i = 6,

$$
\begin{aligned}
& C^2_{1,\,obs} - 2\,C_{1,\,obs}C_0\epsilon^{-kt_1} + C_0^2\epsilon^{-2kt_1} \\
+\ & C^2_{2,\,obs} - 2\,C_{2,\,obs}C_0\epsilon^{-kt_2} + C_0^2\epsilon^{-2kt_2} \\
& \quad\vdots \\
+\ & C^2_{6,\,obs} - 2\,C_{6,\,obs}C_0\epsilon^{-kt_6} + C_0^2\epsilon^{-2kt_6} \\
\hline
SS =\ & \Sigma\, C^2_{i,\,obs} - 2\,C_0\,\Sigma\, C_{i,\,obs}\epsilon^{-kt_i} + C_0^2\,\Sigma\,\epsilon^{-2kt_i}
\end{aligned}
$$

Taking partial derivatives with respect to each of the two parameters, C_0 and k, and setting them equal to zero, we get the normal equations,

$$\partial SS/\partial C_0 = -2\Sigma\, C_{i,\,obs}\epsilon^{-kt_i} + 2\,C_0\Sigma\,\epsilon^{-2kt_i} = 0$$

$$\partial SS/\partial k = 2C_0\Sigma\, t_i C_{i,\,obs}\,\epsilon^{-kt_i} - 2\,C_0^2\Sigma\, t_i\epsilon = 0$$

Combine the two normal equations by solving each for C_0 and equating the two results

$$\frac{\Sigma\, C_{i,\,obs}\,\epsilon^{-kt_i}}{\Sigma\,\epsilon^{-2kt_i}} = \frac{\Sigma t_i C_{i,\,obs}\,\epsilon^{-kt_i}}{\Sigma t_i\,\epsilon^{-2kt_i}}$$

To find the least-squares value for each parameter, solve the above equation graphically using the method detailed elsewhere in this Appendix. Using slide-rule calculations, I estimated C_0 = 3.06 mg/liter of plasma, k = 0.518 per day.

Additional specific examples of how to use the method of least squares are given in Exercise 4, p. 72 and the discussion thereof on pp. 373-374 of reference 2).

NOTE 1. In the preceding description it was assumed that the observed values of y would tend to deviate from the fitted curve as much at one point on the curve as at another, i.e., that the error variance of y was independent of x. But if we know from sound theoretical or mathematical arguments (NOT from our observed values of y) that the error variance of y depends on x, the values of y_i must be appropriately weighted when we apply the method of least squares (See Section 3.15 in reference 2).

NOTE 2. When y(x) is a curvilinear function of x, it may be possible to find a linearizing transformation such that the transformed data can be fitted by a straight line. This often facilitates graphical estimation of the parameters. However, certain linearizing transformations (e.g., the undeservedly popular Lineweaver-Burk (= "double reciprocal") method of plotting data from studies of enzyme kinetics) are disadvantageous because they may misleadingly distort the original data (3). Fortunately, a digital computer can be programmed to calculate the least-square values of parameters directly from the observed data, even when y(x) is a curvilinear function of x. I strongly recommend that such computer programs be used whenever possible.

4. Moving the origin of coordinates to a given point. Applying an equation--whether theoretical or empirical--to a given problem is often facilitated by shifting the origin of coordinates to some particular point of interest, e.g., the point whose coordinates are the mean of x and the mean of y, or the point where a time-dependent variable starts to change in response to an imposed input, or a point of equilibrium where all time derivatives vanish.

Given an equation for y as a function of x,

$$y = f(x) \tag{1}$$

we can shift the origin of coordinates to any desired point, $[x_p, y_p]$ on the x, y plane, simply by defining new variables, X and Y, as the deviations of x and y from x_p and y_p respectively.

$$X \equiv x - x_p \tag{2}$$

$$Y \equiv y - y_p \tag{3}$$

Combine equations 1, 2, and 3 so as to eliminate x and y, and solve for Y.

$$Y = f(X + x_p) - y_p \tag{4}$$

which is the desired relationship. The analysis can now be carried forward in terms of X and Y instead of the original x and y. It is easy to transform any intermediate or final result back to an expression in the original variables by replacing X by $x - x_p$ (equation 2) and Y by $y - y_p$ (equation 3).

Example: Given

$$y(x) = 2x^2 - 8x + 5$$

place the origin of coordinates at the point where $dy(x)/dx = 0$.

$$dy(x)/dx = 4x - 8 = 0; \; x_p = 2, \; y_p = -3$$

$$X \equiv x - 2; \; x = X + 2$$

$$Y \equiv y + 3; \; y = Y - 3$$

Rewriting the original equation by substituting Y - 3 for y and X + 2 for x,

$$Y - 3 = [2X^2 + 8X + 8] - [8X + 16] + 5 = 2X^2 - 3$$

Hence,

$$Y = 2X^2$$

which is the desired equation. (CHECK: By substituting y + 3 for Y and x - 2 for X in this equation and then solving for y we obtain the original equation.)

The effect of changing variables from x and y to X and Y in this example is illustrated graphically in the following figure:

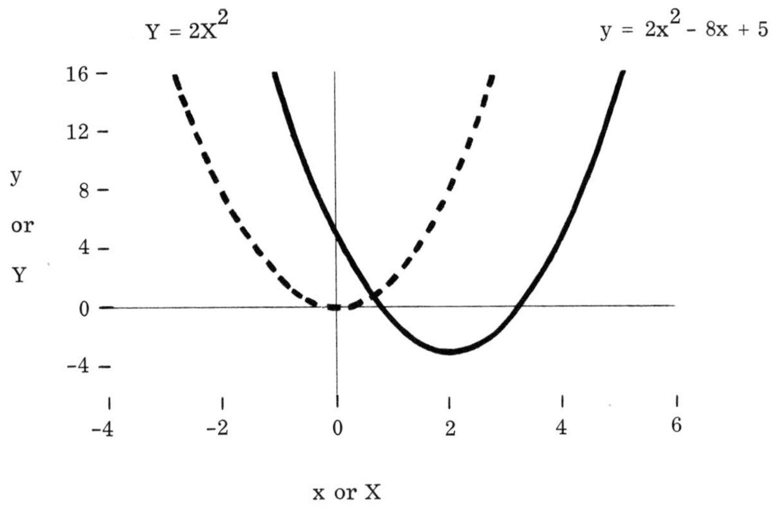

General forms of equations for fitting experimental observations

The following catalog of equations is by no means complete. Additional general equations can be found in references 4 and 5. In particular, Worthing and Geffner (reference 5, pp 58-64) give criteria for deciding whether a particular form of equation will fit a given set of experimental data satisfactorily

Biological observations are rarely numerous enough or accurate enough to require description by an empirical equation in which more than four parameters are to be calculated from the data. Hence equations with five or more parameters are not considered here.

Even an empirical equation may be subject to theoretical constraints which reduce the number of parameters to be calculated from the experimental observations. For example, the two-parameter equation for a general straight line is $y(x) = a + bx$. But if we know a priori that $y(0) = 0$, i.e., that the line must pass through the origin, the equation to be fitted, $y(x) = bx$, has only one parameter.

1. Polynomial equations.

One parameter:
$y(x) = a$

Example:
$y(x) = 2$

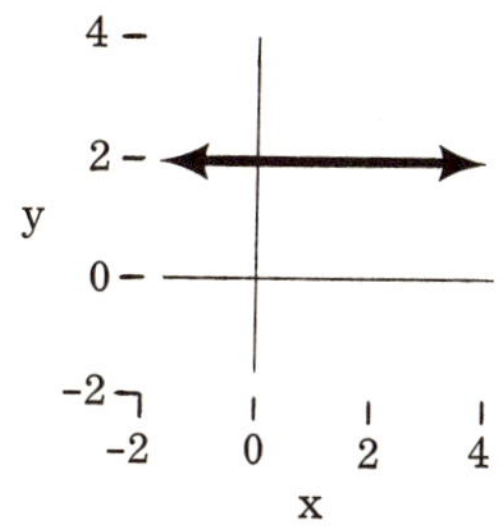

Two parameters:
$y(x) = a + bx$

Example:
$y(x) = 3 - 2x$

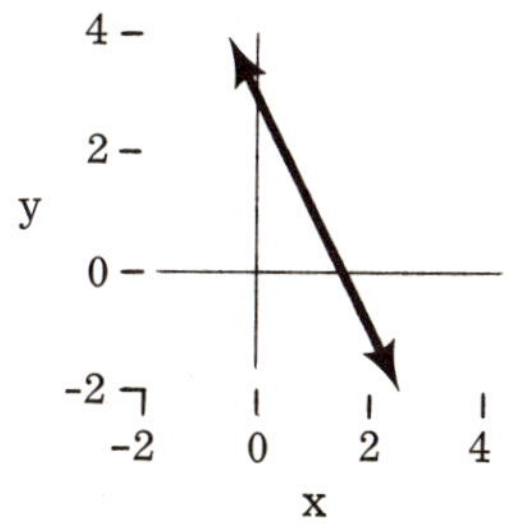

Three parameters:
$y(x) = a + bx + cx^2$

Example:
$y(x) = 5 + 2x - x^2$

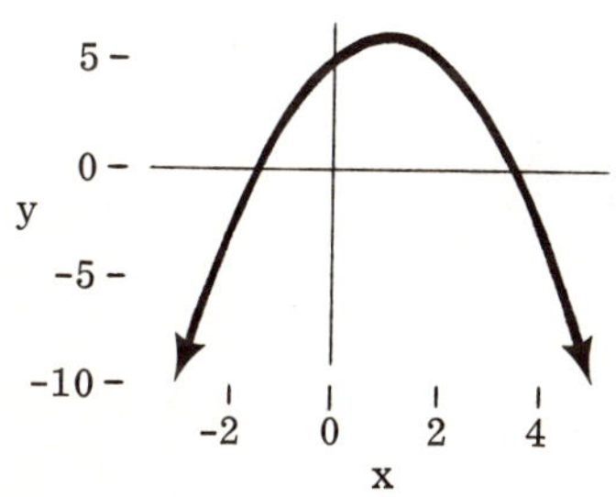

Four parameters:
$y(x) = a + bx + cx^2 + dx^3$

Example:
$y(x) = 2.2 - 3.6x - 0.3x^2 + 0.2x^3$

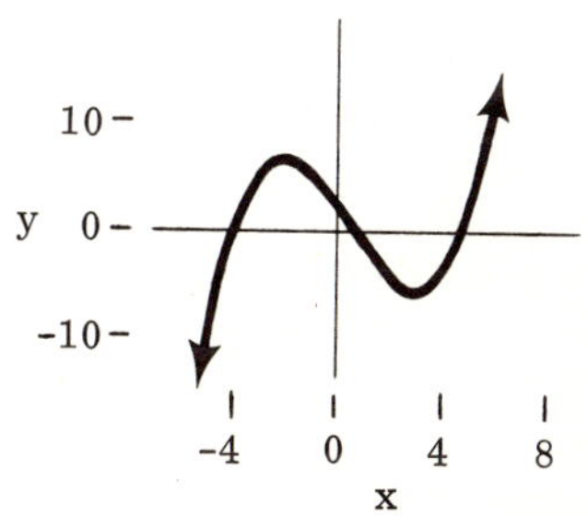

2. Rectangular hyperbolae with asymptotes parallel to the axes of coordinates.

General characteristics: Symmetrical curves approaching two asymptotes, x_{asym} and y_{asym}. Regardless of what numerical scales are chosen for x and y(x), when plotted on rectangular coordinate graph paper the hyperbola is symmetrical about a line with a slope of ± 45° drawn through the intersection of the asymptotes. The asymptotes together with lines drawn perpendicular to them from any point on the hyperbola bound a rectangle of constant area, A.

General equations:
Solid curves:

$$[x - x_{asym}]\,[y(x) - y_{asym}] = A$$

$$y(x) = y_{asym} + \frac{A}{x - x_{asym}}$$

Broken curves:

$$[x - x_{asym}]\,[y_{asym} - y(x)] = A$$

$$y(x) = y_{asym} - \frac{A}{x - x_{asym}}$$

Usually, only one limb of the curve is of any biological interest.

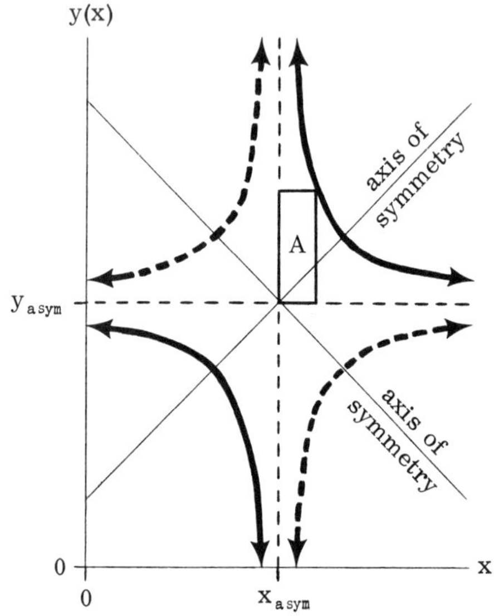

Example:
Parameters,
$A = 4$; $x_{asym} = 2$; $y_{asym} = 3$

$$y(x) = 3 + \frac{4}{x - 2}\ ;\ x > 2$$

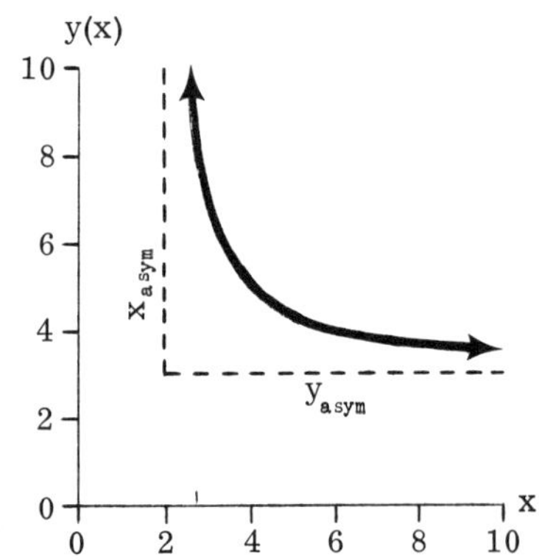

SPECIAL CASE: Michaelis-Menten eq. for initial rate of enzyme reaction

$v(C) = v(\infty)C/[k + C]$

C = concentration of substrate
v(C) = initial rate of reaction
$v(\infty)$ = maximum initial rate
k = Michaelis constant

Note that v(0) = 0. Hence curve is constrained to pass through origin, and there are only two parameters

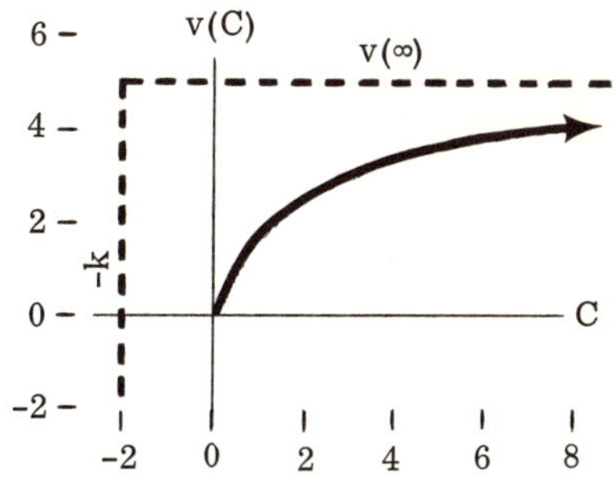

3. Power functions:

General equation:

$y(x) = ax^n + b$

Linearized form:

$\log[y(x) - b] = \log a + n \log x$

For simplicity, we shall let a = 1, b = 0. This does not alter the shape of the power-function curves, but (except for negative n) requires them to pass through the points [0, 0] and [1, 1] instead of the points [0, b] and [1, a + b]. The equation for each of the following curves is therefore of the form $y(x) = x^n$.

fractional powers

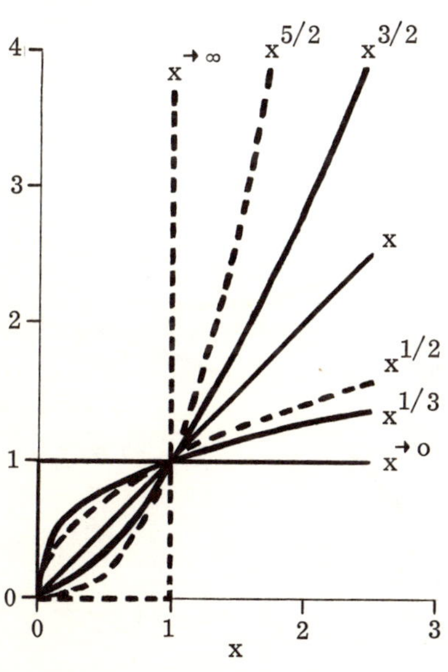

odd powers

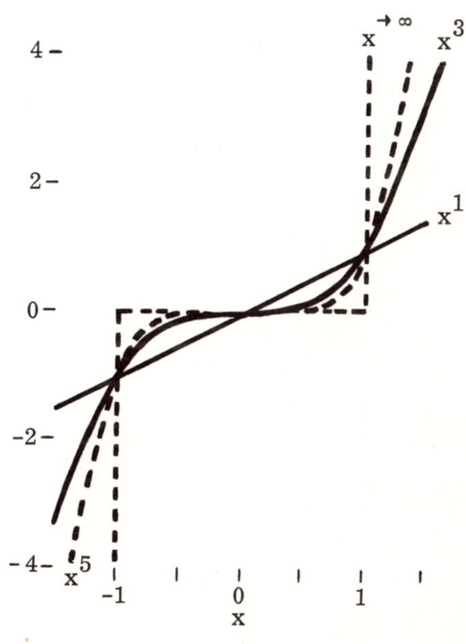

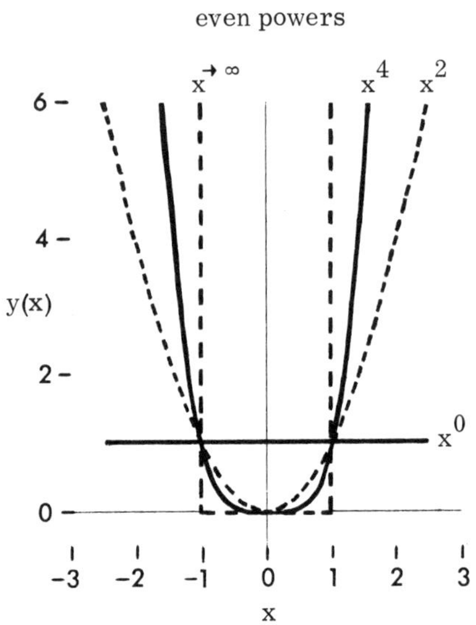

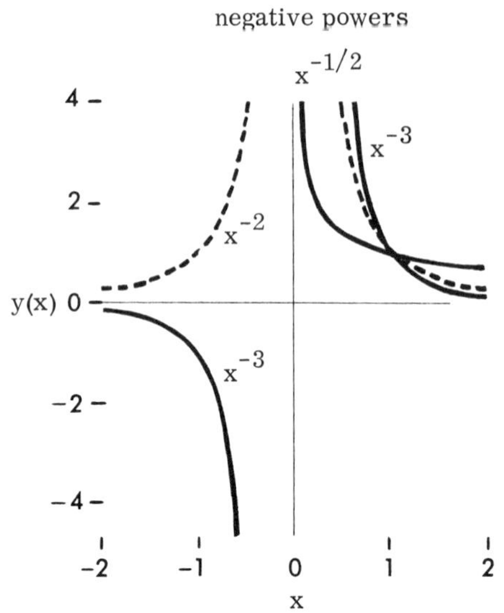

4. Exponential functions.

When linear systems respond either to initial conditions or to a constant input, their dependent variables very often change exponentially with time. It is not surprising therefore that exponential growth and decay are commonly encountered in nature. Moreover, the technique of fitting exponential functions to experimental data is deceptively easy. As a result, sums of exponentials are frequently used to describe experimental results even when there is little or no theoretical justification for choosing an exponential function. In particular, the disappearance of a radioactive isotope from plasma (due to distribution into other compartments, storage, metabolism, and excretion) is traditionally described by a sum of exponentials, all too often containing a preposterous number of parameters to be evaluated from the data. In a previous book (2), I have made extensive comments on exponential curve fitting. Only a few relatively simple examples will be given here.

$$y(x) = y(\infty) + [y_o - y_\infty]\epsilon^{-kx}$$

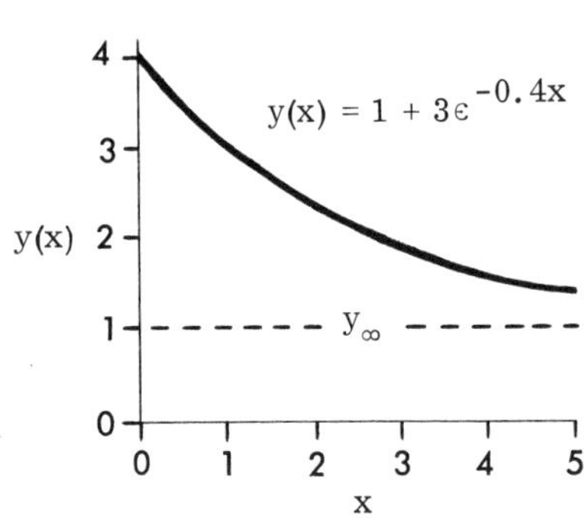

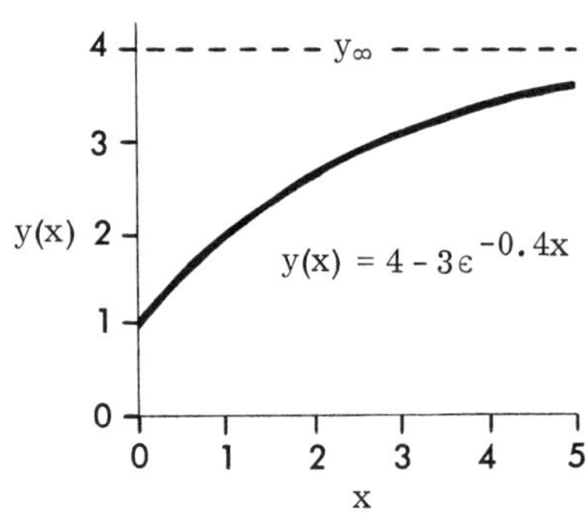

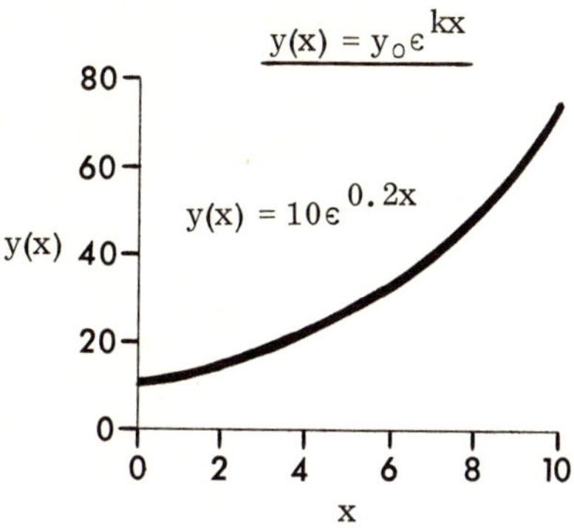

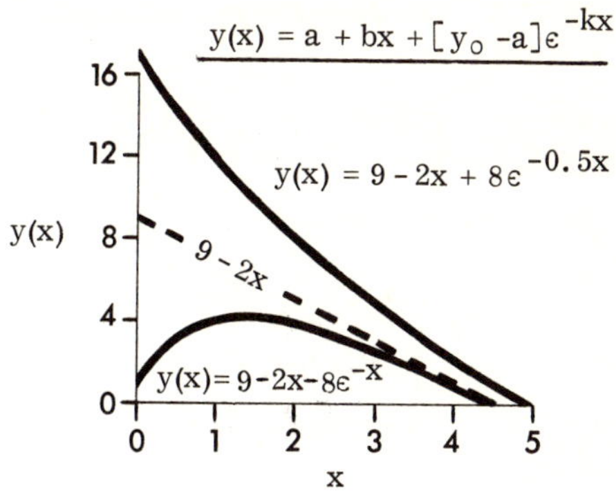

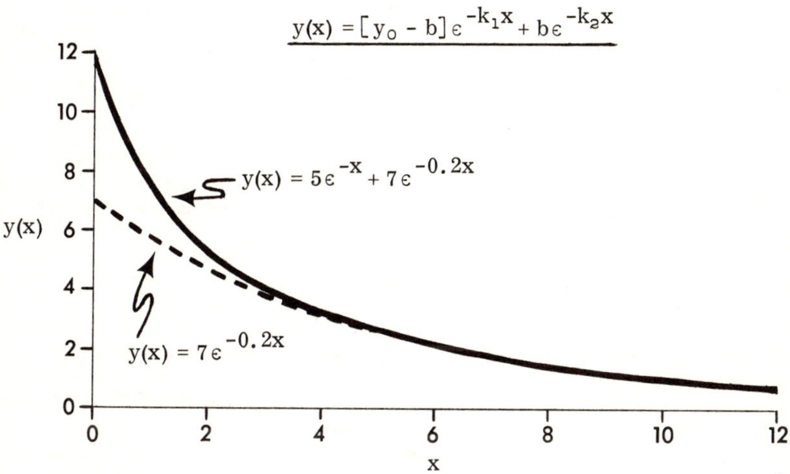

5. Sigmoid curves.

In studying biological systems it is quite common to encounter a dependent variable, y(x), which is a sigmoid function either of the independent variable, x, itself, or of log x. Sigmoid curves of this kind are familiar to the pharmacologist as log dose-response curves. (See Chapter 11 in Reference 2). Furthermore, when a regulatory component in a biological feedback system has a limited capacity, its output is likely to vary from some minimum value when the input is zero to some maximum value as the input approaches infinity. These minimum and maximum values are often approached asymptotically.

Two such sigmoid curves are described here. In addition, integration of the curve of normal (= Gaussian) distribution also generates a sigmoid curve. But the integration cannot be performed explicitly, so that the equation for the curve still contains an unevaluated integral. However, approximate values of the integral have been extensively tabulated. (6).

A. Consider the general equation,

$$y(x) = \frac{a + bx^k}{1 + cx^k} \quad ; k > 1, \; x \geq 0$$

or, more instructively,

$$y(x) = \frac{y_{min} + y_{max} x_{1/2}^{-k} x^k}{1 + x_{1/2}^{-k} x^k}, \text{ where}$$

y_{min} = minimum value approached by y(x)

y_{max} = maximum value approached by y(x)

$x_{1/2}$ = value of x at which y(x) is halfway between y_{max} and y_{min}. i.e., $y(x_{1/2}) = 1/2\,[y_{min} + y_{max}]$

k = $4x_{1/2}y'(x_{1/2}) / [y_{max} - y_{min}]$

It may be possible to obtain reasonably good estimates of y_{min}, y_{max}, and $x_{1/2}$ by inspection of the plotted points. In fact, y_{min} and y_{max} are often fixed by theory, for example as $y_{min} = 0\%$, $y_{max} = 100\%$. If $y'(x_{1/2})$, the slope where y(x) is halfway between y_{min} and y_{max}, can also be estimated from the trend of the points, k can be calculated by the equation given above.

<u>Example:</u> $y_{min} = 3$, $y_{max} = 9$, $x_{1/2} = 4$, $k = 2$

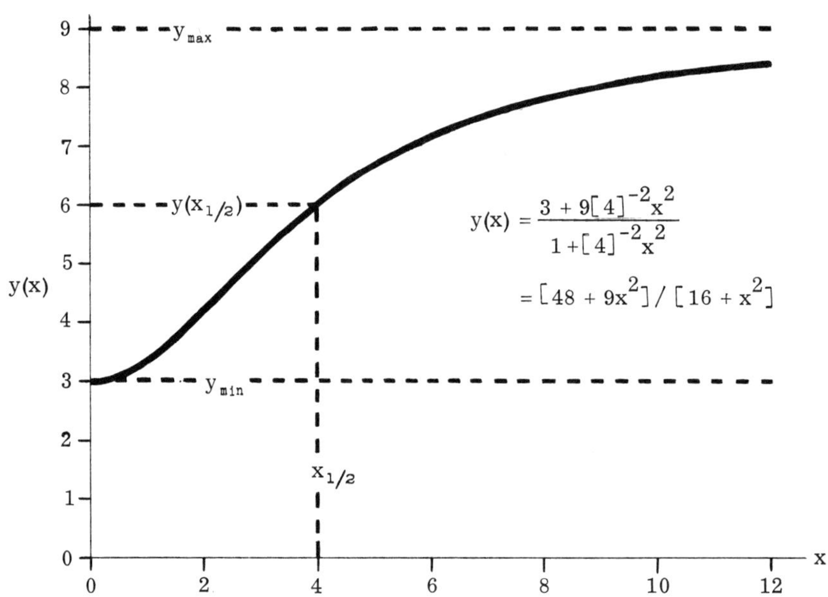

NOTE 1. (i.) If $y'(x)$ is positive, $y(x) \to y_{min}$ as $x \to 0$; $y(x) \to y_{max}$ as $x \to \infty$. (ii.) If $y'(x)$ is negative, $y(x) \to y_{min}$ as $x \to \infty$; $y(x) \to y_{max}$ as $x \to 0$. (iii.) The curve is not symmetrical when $y(x)$ is plotted against x on ordinary rectangular coordinate graph paper, but (iv.) the curve is symmetrical about its midpoint when $y(x)$ is plotted against $\log x$.

NOTE 2.

$$y'(x) = \frac{k x_{1/2}^{-k} [y_{max} - y_{min}] x^{k-1}}{\left\{1 + x_{1/2}^{-k} x^{k}\right\}^2}$$

$$= \frac{1}{x} \left[\frac{k}{y_{max} - y_{min}}\right] [y(x) - y_{min}] [y_{max} - y(x)]$$

B. "The logistic curve is a growth curve used to describe functions which continually increase, gradually at first, more rapidly in the middle growth period, and slowly again, reaching a maximum at the end of the growth." (7).

Although the actual growth of certain confined populations (e.g., bacteria growing in a finite volume of culture medium) is sometimes described reasonably well by the logistic curve, (see, for example, figures 5, 6, and 8 in Reference 8), all too frequently the curve fails to fit the observed growth of an individual organism or a group of organisms. It is best, therefore, to regard the logistic merely as a convenient empirical equation rather than as an equation specifically intended to describe rate of biological growth. It can then be used to describe monotonically decreasing sigmoid relationships as well as sigmoid "growth" curves.

The symmetrical logistic is a four-parameter equation of the general form

$$y(x) = y_{min} + \frac{y_{max} - y_{min}}{1 + \epsilon^{[a - bx]}}$$

or, more instructively,

$$y(x) = y_{min} + \frac{y_{max} - y_{min}}{1 + \epsilon^{k[x_{1/2} - x]}}$$

where

y_{min} = minimum value approached asymptotically by $y(x)$

y_{max} = maximum value approached asymptotically by $y(x)$

$x_{1/2}$ = value of x at which $y(x)$ is halfway between y_{max} and y_{min}. i.e., $y(x_{1/2}) = 1/2\,[y_{min} + y_{max}]$

k = $4y'(x_{1/2})/[y_{max} - y_{min}]$

As was true of the curve described under "A" above, approximate values for y_{min}, y_{max}, $x_{1/2}$ and $y'(x_{1/2})$ may be obtained by inspection of the plotted points. These values would then determine the appropriate value for k.

Example: $y_{min} = 2$, $y_{max} = 12$, $x_{1/2} = 40$, $k = 0.2$

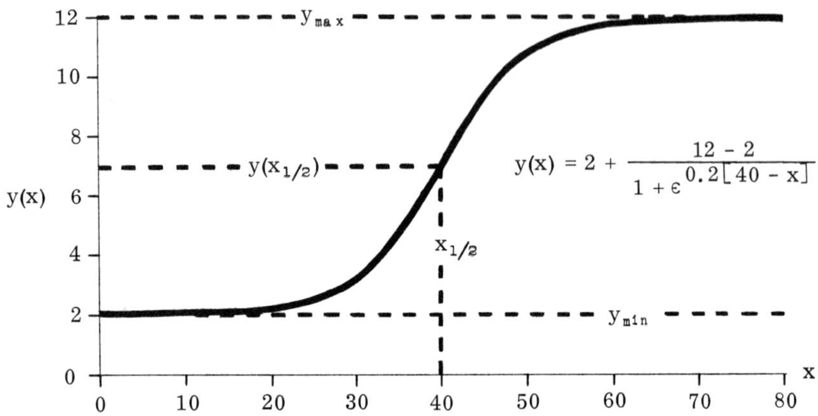

NOTE 1. (i.) If $y'(x)$ is positive, $y(x) \to y_{min}$ as $x \to -\infty$; $y(x) \to y_{max}$ as $x \to +\infty$. This means that if x is an intrinsically positive quantity, y(x) can never actually reach y_{min}. (ii.) If $y'(x)$ is negative, $y(x) \to y_{min}$ as $x \to +\infty$; $y(x) \to y_{max}$ as $x \to -\infty$. This means that if x is an intrinsically positive quantity, y(x) can never actually reach y_{max}. (iii.) This curve is symmetrical about its midpoint when y(x) is plotted against x on ordinary rectangular coordinate graph paper.

NOTE 2.

$$y'(x) = \frac{k[y_{max} - y_{min}]\epsilon^{k[x_{1/2} - x]}}{\left\{1 + \epsilon^{k[x_{1/2} - x]}\right\}^2}$$

$$= \left\{\frac{k}{y_{max} - y_{min}}\right\} [y(x) - y_{min}][y_{max} - y(x)]$$

6. Humped curves.

A. The normal (= Gaussian) distribution curve is a symmetrical, two-parameter curve extending from $x = -\infty$ to $x = +\infty$. The dependent variable y(x) approaches zero asymptotically at both extremes. The equation for the normal curve is written so that its integral between limits of $x = -\infty$ and $x = +\infty$ is unity. It can therefore be used as a probability density function.

$$y(x) = \frac{1}{\sigma\sqrt{2\pi}}\,\epsilon^{-\frac{1}{2}\left[\frac{x-\mu}{\sigma}\right]^2}$$

where

μ = mean of x

σ = standard deviation of x

$1/\sqrt{2\pi} \approx 0.3989$

If x is measured in units of σ about a mean of zero, the equation reduces to

$$y(x) = \frac{1}{\sqrt{2\pi}}\,\epsilon^{-x^2/2}$$

Values of this function, of its integral, and of its various derivatives are tabulated in The Handbook of Chemistry and Physics (6).

B. A curve with the same general appearance as the normal distribution curve is obtained by plotting the first derivative of the logistic curve as a function of x. The general equation is given in NOTE 2, Section 5, B above, and the following curve is plotted using the same parameters as were used for the example of that section.

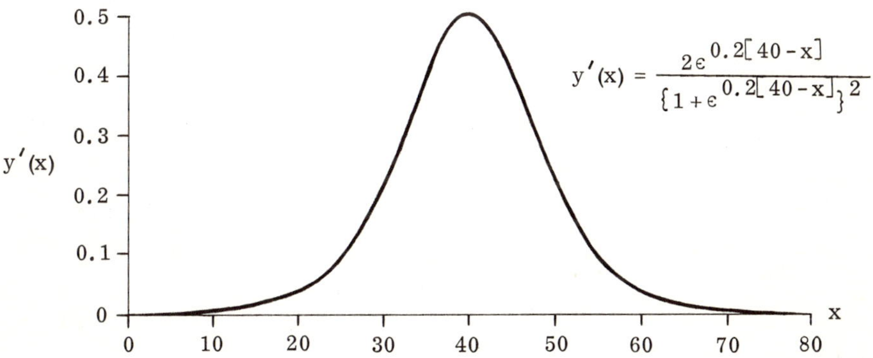

C. Asymmetrical humped curves. A variety of asymmetrical humped curves can be obtained by choosing different values for the four parameters a, b, k, and n in the general equation

$$y(x) = \frac{ax^{k-1}}{\{1 + bx^k\}^n} \quad k > 1,\ n \geq 1$$

In the example on the left, a = 15/14, b = 1/14, k = 3 n = 1. In the example on the right, Y(x) is the derivative of the curve used to illustrate Section 5.A above. a = 0.75, b = 1/16, k = 2, n = 2.

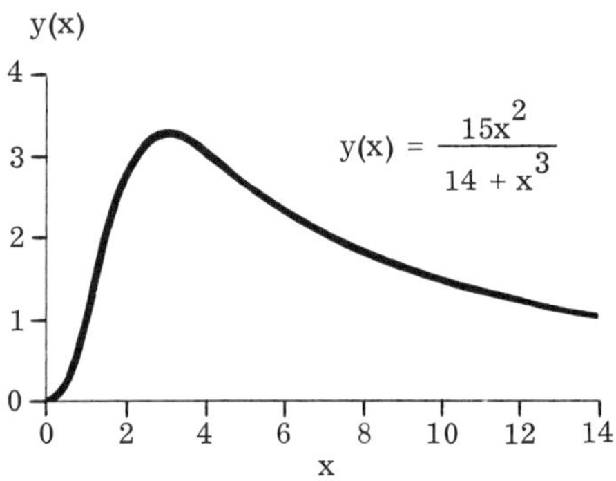

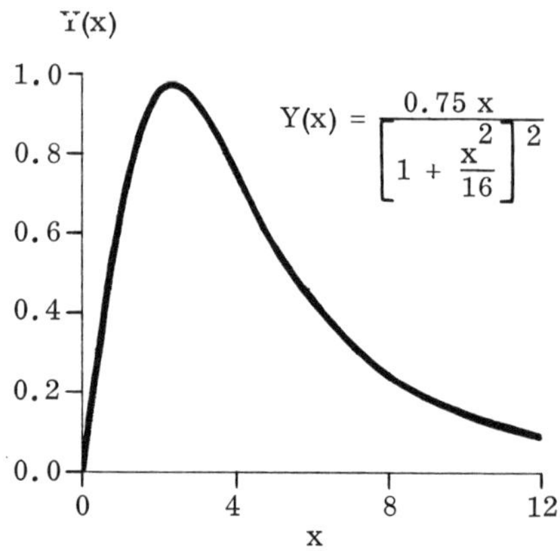

References

1. Ruch, T.C., and Patton, H.D.: Physiology and Biophysics, 19th edition. W.B. Saunders Co., Philadelphia, 1965.

2. Riggs, D.S.: The Mathematical Approach to Physiological Problems. The M.I.T. Press, Cambridge, 1970.

3. Dowd, J.E., and Riggs, D.S.: A comparison of estimates of Michaelis-Menten kinetic constants from various linear transformations. J. Biol. Chem., 240, 863-869, 1965.

4. Davis, D.S.: Nomography and Empirical Equations, 2nd edition. Reinhold Publishing Corp., New York, 1962.

5. Worthing, A.G., and Geffner, J.: Treatment of Experimental Data. John Wiley & Sons, New York, 1943.

References (continued)

6. Handbook of Chemistry and Physics, (any recent edition). Chemical Rubber Publishing Co., Cleveland.

7. The International Dictionary of Applied Mathematics. D. van Nostrand Co., Princeton, 1960.

8. Lotka, A.J.: Elements of Mathematical Biology. Dover Publications, New York, 1956.

SOLVING A SET OF SIMULTANEOUS LINEAR ALGEBRAIC EQUATIONS BY PUTTING THE CORRESPONDING AUGMENTED MATRIX IN ROW-ECHELON FORM

The following method is easier to remember and easier to use than the traditional method which depends upon Cramer's rule.

Consider the n simultaneous linear equations,

$$a_{11}x_1 + a_{12}x_2 + \cdots + a_{1n}x_n = b_1$$

$$a_{21}x_1 + a_{22}x_2 + \cdots + a_{2n}x_n = b_2$$

$$\vdots$$

$$a_{n1}x_1 + a_{n2}x_2 + \cdots + a_{nn}x_n = b_n$$

in the n unknowns, $x_1, x_2, \cdots, x_n$. We would like to evaluate the x_i's from the known values of the coefficients, a_{ij}, and the known values of the nonhomogeneous b_i terms. Begin by forming the augmented matrix,

Equation number	Variable x_1	$x_2 \cdots$	x_n	non-homog.
1	a_{11}	$a_{12} \cdots$	a_{1n}	b_1
2	a_{21}	$a_{22} \cdots$	a_{2n}	b_2
.	.	.	.	.
.	.	.	.	.
.	.	.	.	.
n	a_{n1}	$a_{n2} \cdots$	a_{nn}	b_n

In writing this matrix, the marginal labels have been included to remind us that each row in the matrix corresponds to one of the n equations, and that each of the first n columns consists of the coefficients multiplying a particular unknown. (The last column which "augments" the matrix of coefficients is simply the vector of nonhomogeneous terms.) Any such matrix represents a set of simultaneous equations and whatever operations we perform on the matrix must preserve its correspondence to a set of valid equations. The three row operations which we can use, singly or seriatim, are

(1.) Interchange rows.* This operation obviously does not impair the validity of the equations, and it allows us to arrange the matrix in the most convenient form.

(2.) Multiply all of the elements in a row by a constant. This is equivalent to multiplying both sides of an equation by a constant, an operation which does not destroy the validity of the equation.

(3.) Add the corresponding elements of two rows together. This is equivalent to adding two equations term by term, an operation which yields a new, but still valid equation.

[*Any of the n columns of coefficients may also be interchanged, provided that we remember to make the corresponding change in the order of the variables whose symbols appear above the columns.]

By repeated application of these three operations, we can put the original augmented matrix in the following row-echelon form,

$$\begin{array}{cccc|c} x_1 & x_2 & \cdots & x_n & \text{non-homog.} \\ 1 & 0 & \cdots & 0 & c_1 \\ 0 & 1 & \cdots & 0 & c_2 \\ \vdots & \vdots & & \vdots & \vdots \\ 0 & 0 & \cdots & 1 & c_n \end{array}$$

where the c's are numbers derived from the a's and b's by using the three operations. But this matrix is precisely equivalent to the solution which we seek, because the set of equations which it now represents is simply $x_1 = c_1$, $x_2 = c_2, \cdots, x_n = c_n$.

PROCEDURE.

1. If any rows have one or more initial zeros, arrange the rows so that the ones with the largest number of initial zeros appear nearest the bottom of the matrix, i.e., so that the first nonzero elements in successive rows appear as much as possible "in echelon" as in the final pattern shown above.

2. Multiply every element in the first row by the reciprocal of the first element in the row, thereby changing the first element to unity. Replace the original first row by the result so obtained. Multiply each element of the new first row by the negative of the first element in the second row, and sum the result with the second row so as to obtain a new second row whose first element is zero. Deal likewise with the third, fourth, · · ·, nth rows, thus "sweeping out" of the first column all nonzero elements except the "1" in the first row.

3. Repeat the second step for each successive row, first multiplying it by the factor which changes its initial nonzero element to unity, then using the result to "sweep out" the corresponding column above and below.

That's all there is to it! A couple of numerical examples will clarify the procedure.

Example 1. Solve

$$x_2 + 4x_3 = 2$$

$$[1/2]x_1 + x_2 + 2x_3 = 5/2$$

$$[1/3]x_1 - 4x_2 - 2x_3 = 0$$

By inspection, we see that the first equation should be placed last since it is the only one whose first coefficient is zero. The corresponding matrix is

$$\left[\begin{array}{ccc|c} 1/2 & 1 & 2 & 5/2 \\ 1/3 & -4 & -2 & 0 \\ 0 & 1 & 4 & 2 \end{array}\right]$$

Subsequent steps are indicated by the arrows in the following sequence of matrices. Multiplication of a row by a constant is indicated by an arrow pointing from the row to the result, with the constant labeling the arrow. Rows to be added are bracketed, with an unlabeled arrow pointing from the bracket to the sum. In each matrix, the row being used to sweep out a particular column is boxed for emphasis. In actual practice, no such arrows or boxes are needed; they are used here only as explanatory devices.

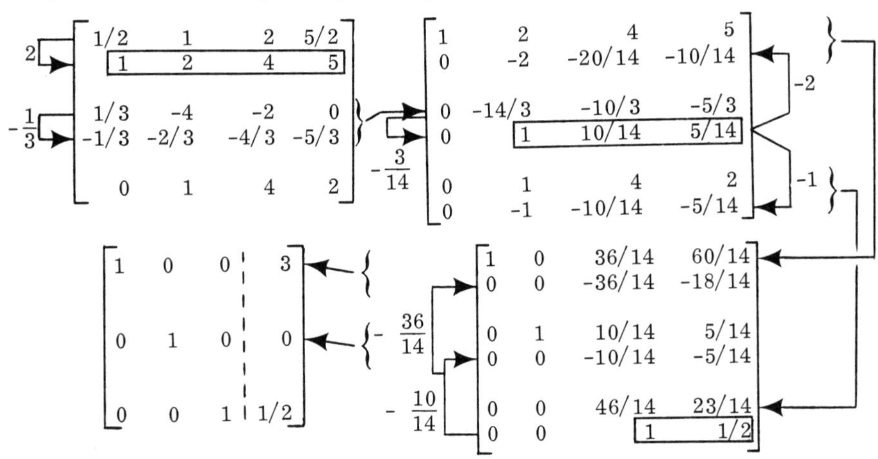

Therefore, $x_1 = 3$, $x_2 = 0$, $x_3 = 1/2$.

NOTE 1. ALWAYS check the solution by making sure that it satisfies the original equations.

NOTE 2. It is usually best to keep fractions in the lowest common denominator (here, 14) until a late stage of the calculations.

Example 2. This example shows what happens when the original set of simultaneous equations is not an independent set, i.e., when one or more of the equations can be derived from the remaining ones. This will be true whenever the number of equations (here 5) exceeds the number of unknowns (here 4). Under these circumstances, we use exactly the same procedure to put the matrix in row-echelon form, only remarking that somewhere along the way two of the rows will become identical as would be expected when one of the equations, being dependent upon the others, contributes nothing but redundant information.

As in the previous example, in each successive matrix the row being used to sweep out a particular column is boxed for emphasis.

1	**-1**	**0**	**1**	**6**
-3	2	1	4	1
3	-3	0	3	18
-2	-4	1	2	8
2	-2	0	2	12
0	-5	-2	-5	3
0	0	1	1	-1

⇒

1	-1	0	1	6
	1	-1	-7	-19
0	-1	1	7	19
	1	**-1**	**-7**	**-19**
0	-6	1	4	20
	6	-6	-42	-114
0	-5	-2	-5	3
	5	-5	-35	-95
0	0	1	1	-1

⇓

1	0	-1	-6	-13
		1	1	-1
0	1	-1	-7	-19
		1	1	-1
0	0	-5	-38	-94
		5	5	-5
0	0	-7	-40	-92
		7	7	-7
0	0	**1**	**1**	**-1**

Note 1

⇐

1	0	0	-5	-14
			5	15
0	1	0	-6	-20
			6	18
0	0	1	1	-1
			-1	-3
0	0	0	-33	-99
			1	**3**
0	0	0	-33	-99

Note 2

⇓

1	0	0	0	1
0	1	0	0	-2
0	0	1	0	-4
0	0	0	1	3
0	0	0	1	3

$$\Rightarrow \begin{cases} x_1 = 1 \\ x_2 = -2 \\ x_3 = -4 \\ x_4 = 3 \end{cases}$$

Note 1. It is easiest to "sweep out" the third column with the last row because its first non-zero element is already unity.

Note 2. Rows 4 and 5 are now identical. See the initial discussion of this example.

DETERMINANTS OF SQUARE MATRICES; CRAMER'S RULE

Determinants exist only for square matrices. Let $\underline{A}$ be a square matrix. We shall symbolize its determinant either by "det $\underline{A}$" or by $|\underline{A}|$. (There should be no confusion with the symbol for "the absolute value of" since the underline makes it clear that we are dealing with matrices.)

The determinant of a 1 by 1 matrix (i.e., a single number) is simply defined as the number itself with its proper sign. Thus det[2] = 2, det[-3] = -3, det[a_{11}] = a_{11}.

The determinant of an n by n matrix, n > 1, can be calculated by the method of expansion by minors. Let a_{ij} be any element of the matrix $\underline{A}$. Then the minor of a_{ij} is defined as the determinant of the matrix which remains after the ith row and the jth column have been deleted. For example

$$\underline{A} = \begin{bmatrix} a_{11} & a_{12} & a_{13} \\ a_{21} & a_{22} & a_{23} \\ a_{31} & a_{32} & a_{33} \end{bmatrix} \quad ; \quad \text{Minor of } a_{22} \equiv |\underline{A}_{22}| = \det \begin{bmatrix} a_{11} & a_{13} \\ a_{31} & a_{33} \end{bmatrix}$$

PROCEDURE. Choose any row or any column of $\underline{A}$. (For ease in calculation, it is best to choose one containing as many zeros as possible.) Let a_{ij} be any element in the chosen row or column. Then, by definition, the determinant of $\underline{A}$ is

$$\det \underline{A} = \sum_{1^{st}}^{n^{th}} [-1]^{i+j} a_{ij} \det \underline{A}_{ij}$$

where $\sum_{1^{st}}^{n^{th}}$ means that the summation is to extend over all of the n elements in the chosen row or column, and det $\underline{A}_{ij}$ is the minor of $\underline{A}_{ij}$.

Notice that each term in the sum is the product of three factors: first, the sign-designating $[-1]^{i+j}$ which will be +1 when i + j is even, -1 when i + j is odd; second, a_{ij}, the element (with its proper sign) for which the term is being calculated; third, the minor of a_{ij}. We calculate a term of this kind for each of the elements in the chosen row or column and add all n terms (some of which may be zero) to get det $\underline{A}$.

Example 1. Find the determinant of the general 2 by 2 matrix. To illustrate the calculation, let us choose to work with the second column.

$$\det \underline{A} = \det \begin{bmatrix} a_{11} & a_{12} \\ a_{21} & a_{22} \end{bmatrix} = [-1]^{[1+2]} a_{12} \det \underline{A}_{12} + [-1]^{[2+2]} a_{22} \det \underline{A}_{22}$$

But det $\underline{A}_{12}$, the minor of a_{12}, is simply a_{21}. Similarly, det $\underline{A}_{22} = a_{11}$. Hence det $\underline{A} = [-1] a_{12} a_{21} + [+1] a_{22} a_{11} = a_{11} a_{22} - a_{12} a_{21}$. This shows that the determinant of any 2 by 2 matrix is the product of the two elements on the principal diagonal (i.e., the diagonal from upper left to lower right) minus the product of the other two elements. This fact is worth memorizing, for ultimately the determinant of any higher-order matrix will be evaluated in terms of its various 2 by 2 minors.

Example 2. In the present book, determinants are of importance chiefly as a way of finding the characteristic equation for a square matrix representing a set of first-order linear differential equations. (See Section 5.4). For example, we might wish to find

$$\det \begin{bmatrix} s & -1 & 0 \\ 0 & s & -1 \\ 8 & 14 & s+7 \end{bmatrix}$$

where for simplicity we would do well to expand by minors with respect to the first or second row. Choosing the first row, we have

$$\det \underline{A} = [-1]^{1+1}[s] \det \begin{bmatrix} s & -1 \\ 14 & s+7 \end{bmatrix} + [-1]^{1+2}[-1] \det \begin{bmatrix} 0 & -1 \\ 8 & s+7 \end{bmatrix} + 0$$

making use of the result for a 2 by 2 matrix which was obtained in example 1 above, we see by inspection that

$$\det \begin{bmatrix} s & -1 \\ 14 & s+7 \end{bmatrix} = s^2 + 7s + 14; \ \det \begin{bmatrix} 0 & -1 \\ 8 & s+7 \end{bmatrix} = 8$$

Hence,

$$\det \underline{A} = s^3 + 7s^2 + 14s + 8$$

CRAMER'S RULE is a rule for solving a set of n simultaneous linear equations in n unknowns by using determinants. It is neither as easy to apply nor as general as the previously-described method of putting the augmented matrix in row-echelon form. Cramer's rule is given here for those readers who once had it drilled into them, and who would rather revive an old method than learn a new one.

Let $\underline{A}$ be the matrix of coefficients of the n equations in the n unknowns, $x_1, x_2, \cdots x_n$. Let $\underline{b}$ represent the column vector of n nonhomogeneous terms. Let det $\underline{A}_{\underline{b}i}$ be the determinant of the matrix $\underline{A}_{\underline{b}i}$ which is formed by substituting $\underline{b}$ for the ith column in $\underline{A}$. Then

$$x_i = \det \underline{A}_{\underline{b}i} / \det \underline{A}$$

Example. Find the value of x_3 in Example 1 of the section on the row-echelon method of solution.

$$\underline{A} = \begin{bmatrix} 1/2 & 1 & 2 \\ 1/3 & -4 & -2 \\ 0 & 1 & 4 \end{bmatrix} ; \underline{b} = \begin{bmatrix} 5/2 \\ 0 \\ 2 \end{bmatrix} ; \text{ hence } \underline{A}_{\underline{b}3} = \begin{bmatrix} 1/2 & 1 & 5/2 \\ 1/3 & -4 & 0 \\ 0 & 1 & 2 \end{bmatrix}$$

Expanding by minors with respect to the third row,

$$\det \underline{A} = [-1]^{3+2}[1][-5/3] + [-1]^{3+3}[4][-7/3] = -23/3$$

$$\det \underline{A}_{\underline{b}3} = [-1]^{3+2}[1][-5/6] + [-1]^{3+3}[2][-7/3] = -23/6$$

By Cramer's rule, $x_3 = [-23/6]/[-23/3] = 1/2$. This procedure would have to be repeated for each of the other unknowns.

INVERSION OF A NONSINGULAR SQUARE MATRIX

Given an n by n square matrix, $\underline{A}$, the easiest way to find its inverse, $\underline{A}^{-1}$, is to adjoin the n by n identity matrix, $\underline{I}$, to $\underline{A}$ on the right, and to put the entire n by 2n matrix thus obtained into row-echelon form. The result will be a matrix whose left half is $\underline{I}$ and whose right half is $\underline{A}^{-1}$, provided that $\underline{A}^{-1}$ exists. (Any square matrix whose determinant is zero is a singular matrix, and it has no inverse. Only non-singular matrices can be inverted.)

Example 1. Find the inverse of the matrix of coefficients for Example 1 of the previous section before it has been augmented by adjoining to it on the right the vector of nonhomogeneous terms.

$$\overbrace{\begin{matrix} 1/2 & 1 & 2 \\ 1/3 & -4 & -2 \\ 0 & 1 & 4 \end{matrix}}^{\underline{A}} \quad \overbrace{\begin{matrix} 1 & 0 & 0 \\ 0 & 1 & 0 \\ 0 & 0 & 1 \end{matrix}}^{\underline{I}}$$

$$\left[\begin{array}{cccccc} 1/2 & 1 & 2 & 1 & 0 & 0 \\ 1/3 & -4 & -2 & 0 & 1 & 0 \\ 0 & 1 & 4 & 0 & 0 & 1 \end{array}\right]$$

$$\downarrow$$

$$\left[\begin{array}{cccccc} 1 & 2 & 4 & 2 & 0 & 0 \\ 0 & -14/3 & -10/3 & -2/3 & 1 & 0 \\ 0 & 1 & 4 & 0 & 0 & 1 \end{array}\right]$$

$$\downarrow$$

$$\left[\begin{array}{cccccc} 1 & 0 & 36/14 & 24/14 & 6/14 & 0 \\ 0 & 1 & 10/14 & 2/14 & -3/14 & 0 \\ 0 & 0 & 46/14 & -2/14 & 3/14 & 1 \end{array}\right]$$

$$\downarrow$$

$$\left[\begin{array}{cccccc} 1 & 0 & 0 & 42/23 & 6/23 & -18/23 \\ 0 & 1 & 0 & 4/23 & -6/23 & -5/23 \\ 0 & 0 & 1 & -1/23 & 3/46 & 7/23 \end{array}\right]$$

$$\underbrace{\begin{matrix} 1 & 0 & 0 \\ 0 & 1 & 0 \\ 0 & 0 & 1 \end{matrix}}_{\underline{I}} \quad \underbrace{\begin{matrix} 42/23 & 6/23 & -18/23 \\ 4/23 & -6/23 & -5/23 \\ -1/23 & 3/46 & 7/23 \end{matrix}}_{\underline{A}^{-1}}$$

Check by showing that $\underline{AA}^{-1} = \underline{I}$, or that $\underline{A}^{-1}\underline{A} = \underline{I}$.

$$\underbrace{\begin{bmatrix} 1/2 & 1 & 2 \\ 1/3 & -4 & -2 \\ 0 & 1 & 4 \end{bmatrix}}_{\underline{A}} \underbrace{\begin{bmatrix} 42/23 & 6/23 & -18/23 \\ 4/23 & -6/23 & -5/23 \\ -1/23 & 3/46 & 7/23 \end{bmatrix}}_{\underline{A}^{-1}} = \underbrace{\begin{bmatrix} 1 & 0 & 0 \\ 0 & 1 & 0 \\ 0 & 0 & 1 \end{bmatrix}}_{\underline{I}}$$

NOTE. If we call the vector of unknowns $\underline{x}$, and the vector of nonhomogeneous terms $\underline{b}$, $\underline{x} = \underline{A}^{-1}\underline{b}$. For the example of the previous section,

$$\underbrace{\begin{bmatrix} x_1 \\ x_2 \\ x_3 \end{bmatrix}}_{\underline{x}} = \frac{1}{23} \underbrace{\begin{bmatrix} 42 & 6 & -18 \\ 4 & -6 & -5 \\ -1 & 3/2 & 7 \end{bmatrix}}_{\underline{A}^{-1}} \underbrace{\begin{bmatrix} 5/2 \\ 0 \\ 2 \end{bmatrix}}_{\underline{b}} = \frac{1}{23} \begin{bmatrix} 105-36 \\ 10-10 \\ -5/2+14 \end{bmatrix} = \begin{bmatrix} 3 \\ 0 \\ 1/2 \end{bmatrix}$$

which gives us the same solution as was obtained by putting the augmented matrix in row-echelon form.

Example 2.

$$\left[\underbrace{\begin{matrix} 1 & 2 & 4 \\ 5 & -4 & -2 \\ 3 & -1 & 1 \end{matrix}}_{\underline{A}} \quad \underbrace{\begin{matrix} 1 & 0 & 0 \\ 0 & 1 & 0 \\ 0 & 0 & 1 \end{matrix}}_{\underline{I}}\right] \Rightarrow \begin{bmatrix} 1 & 2 & 4 & 1 & 0 & 0 \\ 0 & -14 & -22 & -5 & 1 & 0 \\ 0 & -7 & -11 & -3 & 0 & 1 \end{bmatrix}$$

$$\Downarrow$$

$$\begin{bmatrix} 1 & 0 & 6/7 & 2/7 & 1/7 & 0 \\ 0 & 1 & 11/7 & 5/14 & -1/14 & 0 \\ 0 & 0 & 0 & -1/2 & -1/2 & 0 \end{bmatrix}$$

We cannot go any further because the third element of the third row is zero. Hence the third row cannot be used to "sweep out" the third column. The matrix, $\underline{A}$, in this example has no inverse. If we had calculated its determinant (see elsewhere in the Appendix) we would have found it to be equal to zero, showing that the matrix is singular. A singular matrix results when one or more of the equations which it represents can be derived from the remaining ones so that the equations do not form an independent set. (In this example, [row 2] = [-1] [row 1] + [2] [row 3].) Under these circumstances, we do not have enough information for a complete solution, and the inverse of such a square matrix does not exist.

NOTE (for lazy people). Any non-singular 2 by 2 or 3 by 3 matrix can be inverted by substituting specific values for the letters in the following general matrices.

If $\underline{A} = \begin{bmatrix} a & b \\ c & d \end{bmatrix}$, $\underline{A}^{-1} = \dfrac{1}{|\underline{A}|} \begin{bmatrix} d & -b \\ -c & a \end{bmatrix}$

where $|\underline{A}| = \det \underline{A} = ad - bc \neq 0$

If $\underline{A} = \begin{bmatrix} a & b & c \\ d & e & f \\ g & h & j \end{bmatrix}$

$$\underline{A}^{-1} = \frac{1}{|\underline{A}|}\begin{bmatrix} ej - fh & ch - bj & bf - ce \\ fg - dj & aj - cg & cd - af \\ dh - eg & bg - ah & ae - bd \end{bmatrix}$$

where $|\underline{A}| = \det \underline{A}$

$$= [aej + bfg + cdh] - [afh + bdj + ceg] \neq 0$$

L'HOSPITAL'S RULE

We sometimes encounter a ratio of functions, say f(x)/g(x) which approaches 0/0 or ∞/∞ as x approaches some particular value, a, where $-\infty \leq a \leq +\infty$. Since 0/0 and ∞/∞ are <u>indeterminate ratios,</u> $\lim\limits_{x \to a} \frac{f(x)}{g(x)}$ might actually be zero, or some other finite number, or $+\infty$, or $-\infty$. L'Hospital's rule is a method for evaluating such indeterminate ratios.

If f(x) and g(x) are differentiable in an open interval (x_1, x_2) except possibly at the point a in this interval, and if $g'(x) \neq 0$ for $x \neq a$, then

$$\lim_{x \to a} \frac{f(x)}{g(x)} = \lim_{x \to a} \frac{f'(x)}{g'(x)}$$

whenever the limit on the right is determinate. If the limit on the right is still indeterminate, and if f′(x) and g′(x) satisfy the conditions stated above for f(x) and g(x), the process can be repeated.

<u>Example 1.</u> Find $\lim\limits_{x \to \infty} \frac{x^3}{\epsilon^{bx}}$ where b is a finite positive constant. As it stands, the ratio approaches the indeterminate form ∞/∞ as $x \to \infty$. Applying L'Hospital's rule,

$$\lim_{x \to \infty} \frac{x^3}{\epsilon^{bx}} = \lim_{x \to \infty} \frac{3x^2}{b\epsilon^{bx}} = \lim_{x \to \infty} \frac{6x}{b^2 \epsilon^{bx}} = \frac{6}{b^3 \epsilon^{bx}} = 0$$

<u>Example 2.</u> Find $\lim\limits_{x \to 0} \frac{5\epsilon^{-3x} - 5}{4x}$ which, as it stands, approaches the indeterminate form, 0/0.

$$\lim_{x \to 0} \frac{5\epsilon^{-3x} - 5}{4x} = \lim_{x \to 0} \frac{-15\epsilon^{-3x}}{4} = -\frac{15}{4}$$

FACTORING POLYNOMIALS. LIN'S METHOD. GRAPHICAL METHOD.

Finding the factors of a high-order polynomial by hand is at best a tedious business. Whenever possible, it should be done by computer.

Consider the nth order polynomial in s,

$$P(s) = s^n + a_{n-1}s^{n-1} + a_{n-2}s^{n-2} + \cdots + a_1 n + a_0$$

[N.B. if the coefficient of the leading term (i.e., the s^n term) is not already unity, divide each term by that coefficient.]

We want to express P(s) in the factored form

$$P(s) = [s + b_1][s + b_2] \cdots [s + b_n]$$

where $-b_1, -b_2, \cdots, -b_n$ are the roots of the equation, $P(s) = 0$. However, we shall be quite content if we can find quadratic factors, particularly if they have complex roots, since we can easily factor a quadratic by applying the quadratic formula.

Lin's method. A.) n is even. If n is even, the polynomial might have only complex roots. To avoid working with complex numbers, therefore, we shall seek quadratic factors only.

1. Take $s^2 + \frac{a_1}{a_2}s + \frac{a_0}{a_2}$ as a first approximation to a quadratic factor. Divide P(s) by this approximation

$$\begin{array}{r|l} & s^{n-2} + q_{n-3}s^{n-3} + \cdots + q_1 s + q_0 \\ \hline s^2 + \frac{a_1}{a_2}s + \frac{a_0}{a_2} & s^n + a_{n-1}s^{n-1} + \cdots + a_2 s^2 + a_1 s + a_0 \\ & \cdots \qquad \vdots \\ & \qquad \cdots \qquad \vdots \\ & q_0 s^2 + b_1 s + a_0 \\ & q_0 s^2 + \frac{q_0 a_1}{a_2}s + \frac{q_0 a_0}{a_2} \\ \hline & R_1 s + R_0 \equiv \text{remainder} \end{array}$$

2. If the remainder is not negligibly small, take

$$s^2 + \frac{b_1}{q_0}s + \frac{a_0}{q_0} \text{ as a second approximation.}$$

3. Divide P(s) by the second approximation.

4. Continue repeating steps 2 and 3 until the remainder is negligibly small.

Example. Find the quadratic factors of

$$10s^4 + 31s^3 + 43s^2 + 24s + 2$$

Divide each term by the leading coefficient, 10. (Strictly speaking, [10] will then be one of our factors). This gives $s^4 + 3.1s^3 + 4.3s^2 + 2.4s + 0.2$.

Take $s^2 + [a_1/a_2]s + [a_0/a_2] = s^2 + [2.4/4.3]s + [0.2/4.3] = s^2 + 0.558s + 0.0465$ as a first approximation to a quadratic factor. The work proceeds as follows:

$$
\begin{aligned}
&\text{Divisor: } s^2 + 0.558s + 0.0465 \qquad \text{Quotient: } s^2 + 2.542s + 2.835\\
&s^4 + 3.100s^3 + 4.300s^2 + 2.400s + 0.200\\
&\underline{s^4 + 0.558s^3 + 0.046s^2}\\
&2.542s^3 + 4.254s^2 + 2.400s\\
&\underline{2.542s^3 + 1.419s^2 + 0.118s}\\
&2.835s^2 + 2.282s + 0.200\\
&\underline{2.835s^2 + 1.581s + 0.132}\\
&0.701s + 0.068 \quad \text{(Remainder)}
\end{aligned}
$$

Second approximation

$$s^2 + \frac{2.282}{2.835}s + \frac{0.200}{2.835}$$

$$
\begin{aligned}
&\text{Divisor: } s^2 + 0.805s + 0.0705 \qquad \text{Quotient: } s^2 + 2.295s + 2.381\\
&s^4 + 3.100s^3 + 4.300s^2 + 2.400s + 0.200\\
&\underline{s^4 + 0.805s^3 + 0.071s^2}\\
&2.295s^3 + 4.229s^2 + 2.400s\\
&\underline{2.295s^3 + 1.848s^2 + 0.162s}\\
&2.381s^2 + 2.238s + 0.200\\
&\underline{2.381s^2 + 1.918s + 0.168}\\
&0.320s + 0.032
\end{aligned}
$$

Third approximation

$$s^2 + \frac{2.238}{2.381} + \frac{0.200}{2.381}$$

$$
\begin{aligned}
&\text{Divisor: } s^2 + 0.940s + 0.084 \qquad \text{Quotient: } s^2 + 2.160s + 2.186\\
&s^4 + 3.100s^3 + 4.300s^2 + 2.400s + 0.200\\
&\underline{s^4 + 0.940s^3 + 0.084s^2}\\
&2.160s^3 + 4.216s^2 + 2.400s\\
&\underline{2.160s^3 + 2.030s^2 + 0.182s}\\
&2.186s^2 + 2.218s + 0.200\\
&\underline{2.186s^2 + 2.054s + 0.184}\\
&0.164s + 0.016
\end{aligned}
$$

Fourth approximation

$$s^2 + \frac{2.218}{2.186}s + \frac{0.200}{2.186}$$

$$\downarrow$$

$$\begin{array}{r|l}
 & s^2 + 2.085\,s \quad + 2.091 \\
s^2 + 1.015\,s + 0.0915 & s^4 + 3.100\,s^3 + 4.300\,s^2 + 2.400\,s + 0.200 \\
 & \underline{s^4 + 1.015\,s^3 + 0.0915\,s^2} \\
 & \qquad 2.085\,s^3 + 4.2085\,s^2 + 2.400\,s \\
 & \qquad \underline{2.085\,s^3 + 2.117\,s^2 + 0.191\,s} \\
 & \qquad\qquad\qquad 2.0915\,s^2 + 2.209\,s + 0.2000 \\
 & \qquad\qquad\qquad \underline{2.091\,s^2 + 2.122\,s + 0.1914}
\end{array}$$

which gives a reasonably small remainder $\longrightarrow 0.087\,s + 0.0086$

Check: Multiply together the two quadratic factors thus found

$$\begin{array}{l}
s^2 + 1.015\,s + 0.0915 \\
\underline{s^2 + 2.085\,s + 2.091} \\
s^4 + 1.015\,s^3 + 0.0915\,s^2 \\
\quad + 2.085\,s^3 + 2.117\,s^2 + 0.1908\,s \\
\quad\qquad\qquad \underline{+ 2.091\,s^2 + 2.123\,s + 0.1914} \\
s^4 + 3.100\,s^3 + 4.2995\,s^2 + 2.3138\,s + 0.1914
\end{array}$$

which is close to the true value.

(The true quadratic factors are $s^2 + 1.1\,s + 0.1$ and $s^2 + 2\,s + 2$).

B.) n is odd. If n is odd, there must be at least one linear factor whose root is real. It is best to dig this root out first, leaving us with a polynomial of even order which can be handled as described above. The procedure for finding the linear factor is quite analogous to the one already described for quadratic factors, except that our approximate divisors will be linear, not quadratic. A numerical example will suffice to show the steps.

Example. Factor $s^3 + 18\,s^2 + 87\,s + 70$

First approximation $= s + [a_0/a_1] = s + [70/87] = s + 0.805$

$$\begin{array}{r|l}
 & s^2 + 17.195\,s \quad + 73.16 \\
s + 0.805 & s^3 + 18.000\,s^2 + 87.000\,s + 70.000 \\
 & \underline{s^3 + \ \ 0.805\,s^2} \\
 & \quad 17.195\,s^2 + 87.000\,s \\
 & \quad \underline{17.195\,s^2 + 13.84\,s} \\
 & \qquad\qquad 73.16\,s + 70.000 \\
 & \qquad\qquad \underline{73.16\,s + 58.89} \\
 & \qquad\qquad\qquad\quad 11.11
\end{array}$$

Second approximation $= s + \dfrac{70.00}{73.16} = s + 0.957$

$$
\begin{array}{r|llll}
 & s^2 & + 17.043\,s & + 70.695 & \\
\hline
s + 0.957 & s^3 & + 18.000\,s^2 & + 87.000\,s & + 70.000 \\
 & s^3 & + \ \ 0.957\,s^2 & & \\
\cline{2-3}
 & & 17.043\,s^2 & + 87.000\,s & \\
 & & 17.043\,s^2 & + 16.305\,s & \\
\cline{3-4}
 & & & 70.695\,s & + 70.000 \\
 & & & 70.695\,s & + 67.66 \\
\cline{4-5}
 & & & & 2.34
\end{array}
$$

Third approximation $= s + \dfrac{70.000}{70.695} = s + 0.99$

$$
\begin{array}{r|llll}
 & s^2 & + 17.01\ s & + 70.16 & \\
\hline
s + 0.99 & s^3 & + 18.000\,s^2 & + 87.000\ s & + 70.000 \\
 & s^3 & + \ \ 0.990\,s^2 & & \\
\cline{2-3}
 & & 17.01\ s^2 & + 87.000\,s & \\
 & & 17.01\ s^2 & + 16.840\,s & \\
\cline{3-4}
 & & & 70.16\ s & + 70.000 \\
 & & & 70.16\ s & + 69.400 \\
\cline{4-5}
 & & & & 0.6
\end{array}
$$

which gives a reasonably small remainder. We therefore have,

$$s^3 + 18s^2 + 87s + 70 \approx [s + 0.99][s^2 + 17.01s + 70.16]$$

compared with the true values,

$$s^3 + 18s^2 + 87s + 70 = [s + 1][s^2 + 17s + 70]$$

Graphical approximation. If at least one of the roots of $P(s) = 0$ is known to be real, we can plot $P(s)$ against s on ordinary graph paper and estimate the real roots as the values of s where the curve crosses the $P(s) = 0$ axis.

Example. Estimate the factors of

$$P(s) = s^3 + 5.43\,s^2 + 8.31\,s + 3.33$$

graphically.

We begin by evaluating P(s) for various s, arranging the work in tabular form. (An appropriate range of s can usually be chosen by testing a few large values and a few small values of s.)

s	s^2	A s^3	B 8.31s	C $5.43s^2$	P(s) = A + B + C + 3.33
0	0	0	0	0	
-0.5	0.25	-0.125	-4.155	1.36	0.41
-1.0	1.00	-1.00	-8.31	5.43	-0.55
-1.5	2.25	-3.38	-12.47	12.22	-0.29
-2.0	4.00	-8.00	-16.62	21.72	0.43
-2.5	6.25	-15.63	-20.78	33.94	0.87
-3.0	9.00	-27.00	-24.93	48.87	0.27
-3.5	12.25	-42.87	-29.09	66.52	-2.11
-4.0	16.00	-64.00	-33.24	86.88	-7.03

Next, we plot P(s) against s as in the graph at the left below. Since the curve crosses the P(s) = 0 axis thrice, all three roots must be real, and they can be estimated as the values of s at the crossing points. But the scale of s is too

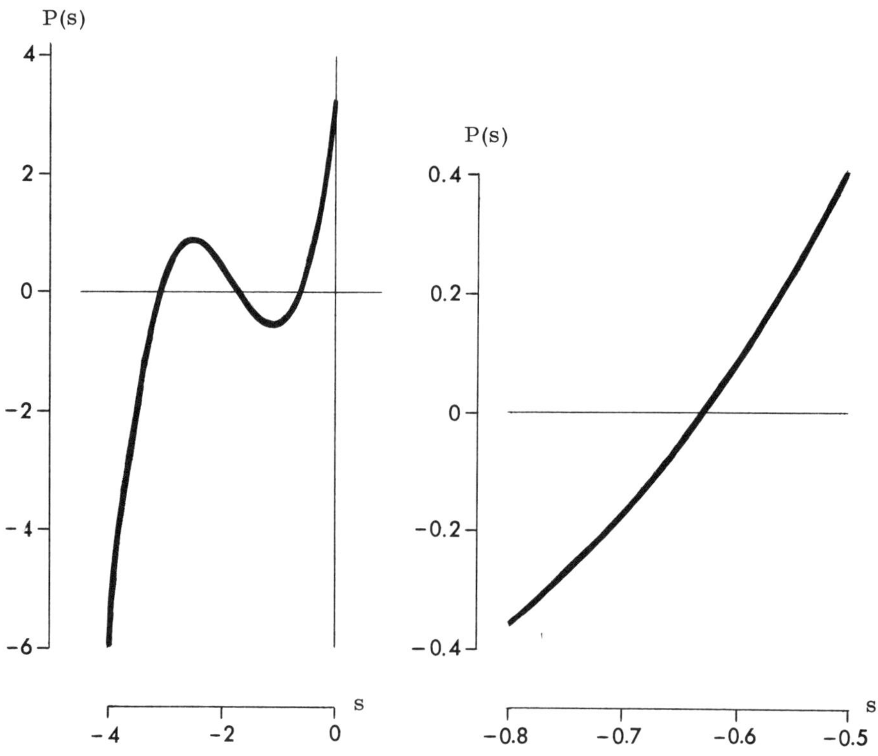

small for accuracy and this first graph must be regarded only as a range-finding device. To sharpen our estimate, we must repeat the procedure over a range of s which, though much narrower than before, is wide enough so that it will surely include the crossing point, due allowance being made for the crudity of the previous graph. For example, to improve our estimate of the root at roughly $s = -0.6$, we might calculate P(s) for $s = -0.50, -0.55, -0.60, \cdots, -0.80$ and replot this small segment of the curve on a much larger scale as shown in the graph on the right. This gives us a revised estimate of $s = -0.63$. Additional iterations could be used if still greater accuracy were needed.

In the present example, once we have estimated one of the factors with sufficient accuracy by this graphical technique, we would do best to divide the original polynomial by our known factor, $[s + 0.63]$, thus obtaining the second-order quotient, $s^2 + 4.80\,s + 5.28$ whose factors, $[s + 1.71]$ and $[s + 3.09]$ are easily obtained algebraically by using the quadratic formula.

GRAPHICAL SOLUTION OF SIMULTANEOUS EQUATIONS

The graphical method for finding roots which was described in the previous section is also useful for solving nonlinear simultaneous equations which cannot be solved analytically. As an example, consider equations 7.4.26, 7.4.27, and 7.4.28.

$$A(\infty) = \frac{WB(\infty)}{U + B(\infty)} + C(\infty)$$

$$B(\infty) = M/A(\infty)$$

$$C(\infty) = C_0 \epsilon^{-KB(\infty)}$$

It is easy enough to combine these equations algebraically so as to eliminate $A(\infty)$ and $C(\infty)$, leaving us with an expression in the single variable, $B(\infty)$,

$$\frac{M}{B(\infty)} - \frac{WB(\infty)}{U + B(\infty)} = C_0 \epsilon^{-KB(\infty)}$$

However, this equation cannot be solved explicitly for $B(\infty)$ because $B(\infty)$ appears in the exponent. By selecting suitable values of $B(\infty)$, evaluating each side of the equation separately for these values, and plotting the results against $B(\infty)$ on a single sheet of graph paper, we shall obtain two curves which will intersect at the desired value of $B(\infty)$. The accuracy of the estimate can be improved by calculating additional points for values of $B(\infty)$ close to the initial estimate, and plotting them on a graph with a much larger scale.

FOURIER SERIES REPRESENTATION OF PERIODIC FUNCTIONS

Let f(t) be a periodic function of t. Let 2T be the length of the period, so that the interval, T, is a half-period. Let t_o be the beginning of a period which ends at $t_o + 2T$. We may then assume that f(t) can be represented as the Fourier series,

$$f(t) = \frac{a_o}{2} + \sum_{n=1}^{\infty} \left\{ a_n \cos \frac{n\pi t}{T} + b_n \sin \frac{n\pi t}{T} \right\} \tag{1.}$$

where

$$a_o = \frac{1}{T} \int_{t_o}^{t_o + 2T} f(t)dt$$

$$a_n = \frac{1}{T} \int_{t_o}^{t_o + 2T} f(t) \cos \frac{n\pi t}{T} dt$$

$$b_n = \frac{1}{T} \int_{t_o}^{t_o + 2T} f(t) \sin \frac{n\pi t}{T} dt$$

and $a_o/2$ is the mean value of f(t) during the period.

Now although the above Fourier series representation of f(t) contains an infinite number of terms, f(t) can be approximately represented by the sum of a finite number of terms, the difference between the approximation and the actual function becoming smaller and smaller as the number of terms increases.

When dealing with symmetrical functions, we can simplify the calculations by choosing our coordinates so that 1.) the mean of f(t) is zero, and 2.) f(t) is symmetrical about t = 0 defined at the midpoint of the period. Then f(t) will be either an even function (i.e., f(-t) = f(t)) or an odd function (i.e., f(-t) = -f(t)). Even functions require only the cosine terms in the Fourier series, all b's equalling zero, while odd functions require only the sine terms, all a's equalling zero.

The periodic function, f(t), may be only piecewise continuous. This poses no analytical difficulty because the integrations needed to evaluate the a's and b's can be carried out correspondingly piecewise and then added over the whole period. The pieces are often segments of straight lines. Moreover, any arbitrary periodic curve, (for example the fluctuations of blood pressure during the cardiac cycle) can be approximated by a series of straight-line segments. We shall therefore begin by finding expressions for the coefficients, $a_{o,i}$, $a_{n,i}$ and $b_{n,i}$, for $f_i(t) = A_i + B_i t$, the ith straight-line segment, defined between a lower limit, $t = \alpha_i$, and an upper limit, $t = \beta_i$. These general expressions can be applied separately to each line segment within the period, and the resulting values can then be summed to obtain a_o, a_n, and b_n for the whole period. For the ith line segment, we have

$$a_{o,i} = \frac{1}{T} \int_{\alpha_i}^{\beta_i} [A_i + B_i t]dt = \frac{1}{T} \left[A_i t + \frac{1}{2} B_i t^2\right]_{\alpha_i}^{\beta_i}$$

$$= \frac{1}{T} \left\{A_i[\beta_i - \alpha_i] + \frac{1}{2} B \left[\beta_i^2 - \alpha_i^2\right]\right\}$$

(N.B. In this and the following equations, A_i, B_i, α_i, and β_i are defined separately for each line segment within the period, but T is one-half the total period, not one-half $[\beta_i - \alpha_i]$ except when the whole periodic function consists of a single line segment so that α_i and β_i are, in fact, the endpoints of the entire period.) By similar integrations we obtain

$$a_n = \frac{1}{n\pi}\left\{[A_i + \beta_i B_i]\sin\left(\frac{n\pi\beta_i}{T}\right) - [A_i + \alpha_i B_i]\sin\left(\frac{n\pi\alpha_i}{T}\right) + \frac{B_i T}{n\pi}\left[\cos\left(\frac{n\pi\beta_i}{T}\right) - \cos\left(\frac{n\pi\alpha_i}{T}\right)\right]\right\}$$

$$b_n = \frac{1}{n\pi}\left\{[A_i + \alpha_i B_i]\cos\left(\frac{n\pi\alpha_i}{T}\right) - [A_i + \beta_i B_i]\cos\left(\frac{n\pi\beta_i}{T}\right) + \frac{B_i T}{n\pi}\left[\sin\left(\frac{n\pi\beta_i}{T}\right) - \sin\left(\frac{n\pi\alpha_i}{T}\right)\right]\right\}$$

Example 1. Approximate the periodic function,

$$f(t) = \begin{cases} 4 + 1.5t & -2 \le t < 0 \\ 4 - 1.5t & 0 \le t < 2 \end{cases}$$

with period 2T = 4 by a finite number of terms of a Fourier series.

For the first segment of f(t), $A_1 = 4$, $B_1 = 1.5$, $\alpha_1 = -2$, $\beta_1 = 0$. Substituting these values into our general equations for a straight-line segment, we get

$$a_{o,1} = [1/2]\{4[0 + 2] + [1/2][1.5][0 - 4]\} = 5/2$$

$$a_{n,1} = \frac{1}{n\pi}\left\{4 \sin 0 - [4 - 3]\sin(-n\pi) + \frac{3}{n\pi}[\cos 0 - \cos(-n\pi)]\right\}$$

$$= 0 \text{ for all even n}$$

$= 6/[n\pi]^2$ for all odd n

$$b_{n,1} = \frac{1}{n\pi}\left\{[4-3]\cos(-n\pi) - [4+0]\cos 0 + \frac{3}{n\pi}[\sin 0 - \sin(-n\pi)]\right\}$$

$= -3/n\pi$ for all even n

$= -5/n\pi$ for all odd n

Similarly, for the second segment of f(t), $A_2 = 4, B_2 = -1.5, \alpha_2 = 0, \beta_2 = 2$. Substituting these values into our general equations, we get

$$a_{0,2} = [1/2]\,\{4[2-0] + [1/2][-1.5][4-0]\} = 5/2$$

$$a_{n,2} = \frac{1}{n\pi}\left\{[4-3]\sin(n\pi) - 4\sin 0 - \frac{3}{n\pi}[\cos(n\pi) - \cos 0]\right\}$$

$= 0$ for all even n

$= 6/[n\pi]^2$ for all odd n

$$b_{n,2} = \frac{1}{n\pi}\left\{4\cos 0 - [4-3]\cos(n\pi) - \frac{3}{n\pi}[\sin(n\pi) - \sin 0]\right\}$$

$= 3/n\pi$ for all even n

$= 5/n\pi$ for all odd n

Summing the corresponding results for both segments, we obtain

$$a_0 = a_{0,1} + a_{0,2} = 5$$

$$a_n = a_{n,1} + a_{n,2} = \begin{cases} 0 \text{ when n is even} \\ 12/[n\pi]^2 \text{ when n is odd} \end{cases}$$

$$b_n = b_{n,1} + b_{n,2} = 0 \text{ for all n}$$

Substituting these values into equation (1.) for f(t) expressed as a Fourier series,

$$f(t) = \frac{5}{2} + \sum_{n=1,3,5,\ldots}^{\infty} \left\{\frac{12}{\pi^2} \cdot \frac{1}{n^2} \cos \frac{n\pi t}{2}\right\}$$

or, since $12/\pi^2 \approx 1.2159$ is a constant coefficient of each of the summed terms,

$$f(t) \approx 2.5 + 1.2159 \sum_{n=1,3,5,\cdots}^{\infty} \left\{ \frac{1}{n^2} \cos \frac{n\pi t}{2} \right\} \qquad (2.)$$

NOTE 1. Check to make sure that $a_o/2 = 2.5$ is the mean of the function. (It is!)

NOTE 2. We could have shifted the zero of our scale of ordinates to the mean of f(t), so as to obtain the function

$$g(t) \equiv f(t) - 2.5 = \begin{cases} 1.5 + 1.5t & -2 \le t < 0 \\ 1.5 - 1.5t & 0 \le t < 2 \end{cases}$$

Then a_o would have been zero. Much more important, however, is our choice of the origin for the scale of abscissas so that our function is an even function. We thus avoid having to evaluate any of the sine terms in the Fourier series.

To illustrate the actual use of equation (2.), let us approximate f(t) at one or two selected values of t, using only the first four nonvanishing harmonics (i.e., those for n = 1, 3, 5 and 7.)

$$f(-2) \approx 2.5 + 1.2159 \left[\frac{-1}{1} + \frac{-1}{9} + \frac{-1}{25} + \frac{-1}{49} \right]$$

$$= 2.5 - [1.2159][1.17152] = 1.076$$

which we can compare with the true value, 1.000. (Parenthetically, it is interesting to note that by equating the infinite series with this true value, we obtain

$$2.5 - \frac{12}{\pi^2} \sum_{n=1,3,5,\cdots}^{\infty} \left\{ \frac{1}{n^2} = 1 \right\}$$

solving for the sum, $\sum_{n=1,3,5,\cdots}^{\infty} [1/n^2] = \pi^2/8$. Precise values for sums of certain other infinite series may be similarly calculated.)

$$f(0.4) \approx 2.5 + 1.2159 \left[\frac{0.809}{1} - \frac{0.309}{9} - \frac{1}{25} - \frac{0.309}{49} \right]$$

$$= 2.5 + [1.2159][0.7284] = 3.386$$

compared with the true value of 3.400.

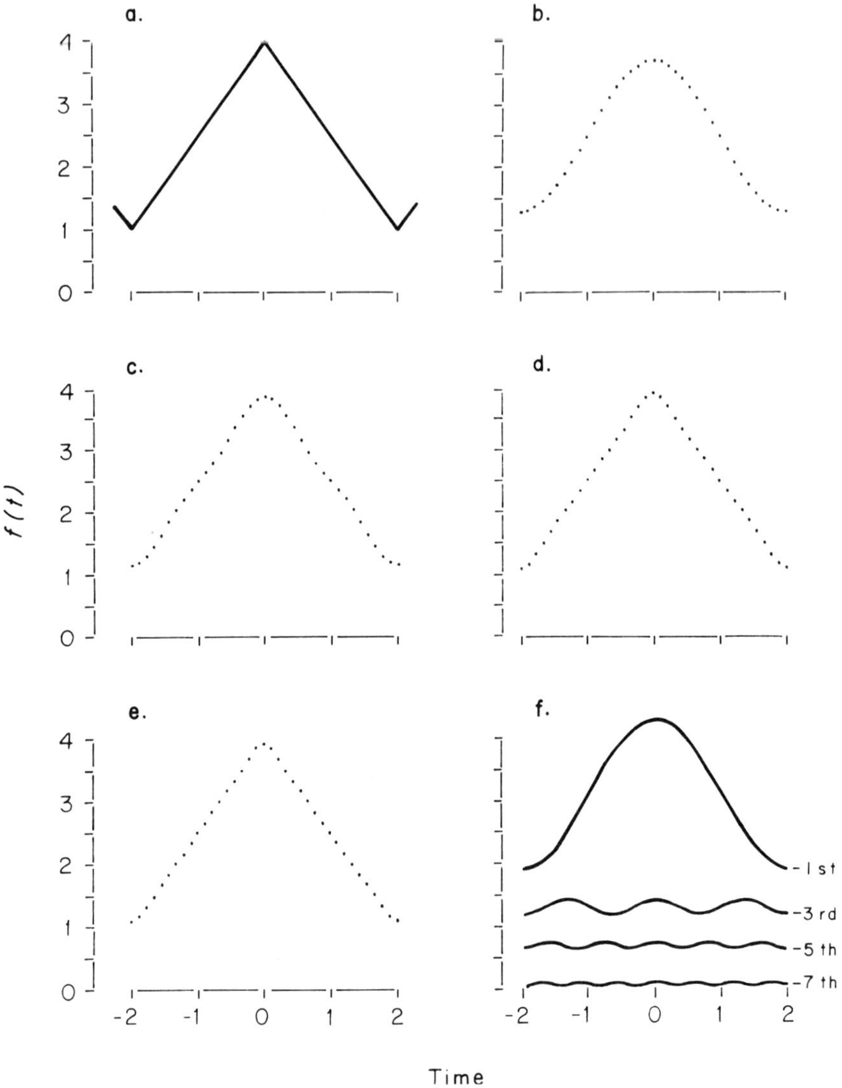
a.
b.
c.
d.
e.
f.
f (t)
0
1
2
3
4
-2
-1
0
1
2
-1 st
-3 rd
-5 th
-7 th
Time

In the figure, panel a. shows the function of the present example, while panel f. shows the first four non-vanishing harmonics of the Fourier series. Panels b., c., d., and e. show respectively the mean plus the first, the mean plus the first two, the mean plus the first three, and the mean plus all four of these harmonics, evaluated for the discrete values of t at which the points are plotted. It is obvious that a reasonably good approximation of this particular function can be achieved with only three or four harmonic terms. This is by no means always true, as the second example will show.

Example 2. Approximate the periodic function

$$f(t) = \begin{cases} 1 + [1/3]t & 0 \le t < 6 \\ 3 & 6 \le t < 8 \end{cases}$$

with period 2T = 8 by a finite number of terms of a Fourier series.

For the first segment of f(t), $A_1 = 1$, $B_1 = 1/3$, $\alpha_1 = 0$, $\beta_1 = 6$. Substituting these values into our general equation for a straight-line segment, we get

$$a_{0,1} = \frac{1}{4}\left\{[6] + \frac{1}{2} \cdot \frac{1}{3}[36-0]\right\} = 3$$

$$a_{n,1} = \frac{1}{n\pi}\left\{[1+2]\sin\frac{3}{2}n\pi - \sin 0 + \frac{4}{3n\pi}\left[\cos\frac{3}{2}n\pi - \cos 0\right]\right\}$$

$$= \frac{1}{n\pi}\left\{3\sin\frac{3}{2}n\pi + \frac{4}{3n\pi}\left[\cos\frac{3}{2}n\pi - 1\right]\right\}$$

$$b_{n,1} = \frac{1}{n\pi}\left\{\cos 0 - [1+2]\cos\frac{3}{2}n\pi + \frac{4}{3n\pi}\left[\sin\frac{3}{2}n\pi - \sin 0\right]\right\}$$

$$= \frac{1}{n\pi}\left\{1 - 3\cos\frac{3}{2}n\pi + \frac{4}{3\pi n}\left[\sin\frac{3}{2}n\pi\right]\right\}$$

Similarly, for the second segment of f(t), $A_2 = 3$, $B_2 = 0$, $\alpha_2 = 6$, $\beta_2 = 8$. Hence,

$$a_{0,2} = \frac{1}{4}\{6\} = 1.5$$

$$a_{n,2} = \frac{1}{n\pi}\left\{3\sin 2n\pi - 3\sin\frac{3}{2}n\pi\right\} = \frac{1}{n\pi}\left\{-3\sin\frac{3}{2}n\pi\right\}$$

$$b_{n,2} = \frac{1}{n\pi}\left\{3\cos\frac{3}{2}n\pi - 3\cos 2n\pi\right\} = \frac{1}{n\pi}\left\{3\cos\frac{3}{2}n\pi - 3\right\}$$

Summing the corresponding results for both segments, we obtain,

$$a_0 = a_{0,1} + a_{0,2} = 4.5$$

$$a_n = a_{n,1} + a_{n,2} = \frac{1}{n\pi}\left\{\frac{4}{3n\pi}\left[\cos\frac{3}{2}n\pi - 1\right]\right\} = \frac{4}{3[n\pi]^2}\left[\cos\frac{3}{2}n\pi - 1\right]$$

$$b_n = b_{n,1} + b_{n,2} = \frac{1}{n\pi}\left\{-2 + \frac{4}{3\pi n}\left[\sin\frac{3}{2}n\pi\right]\right\} = \frac{4}{3[n\pi]^2}\left[\sin\frac{3}{2}n\pi\right] - \frac{2}{n\pi}$$

Therefore,

$$f(t) = 2.25 + \sum_{n=1}^{\infty} \frac{4}{3[n\pi]^2}\left[\cos\frac{3}{2}n\pi - 1\right]\cos\frac{n\pi t}{4} + \left\{\frac{4}{3[n\pi]^2}\left[\sin\frac{3}{2}n\pi\right] - \frac{2}{n\pi}\right\}\sin\frac{n\pi t}{4}$$

As an example of the calculations needed to find just one pair of coefficients,

$$a_3 = \frac{4}{27\pi^2}[-1] = -\frac{4}{27}[0.10132] = -0.01501$$

$$b_3 = -0.01501 - \frac{2}{3\pi} = -0.01501 - 0.21221 = -0.1972$$

NOTE. Hand calculations of this kind are facilitated by using a table of numerical constants containing π, such as the one found in "The Handbook of Chemistry and Physics." But if Fourier analysis is to be used at all extensively, there is no substitute for a digital computer. Indeed, library programs for calculating Fourier coefficients are now available.

The coefficients for the first five sine and cosine terms for this example

n	1	2	3	4	5
a_n	-0.1351	-0.0675	-0.0150	0.0000	-0.0054
b_n	-0.7717	-0.3183	-0.1972	-0.1592	-0.1327

have been used to calculate the values of f(t) which are plotted as points in the following figure. The solid lines show the original function. It is at once obvious that this approximation is by no means as close as the one obtained, with far fewer terms, in the previous example. There are two reasons for this. First, the present function is not symmetrical. Consequently both sine and cosine terms are needed. Second, (and more important) at the beginning and end of each period there is a vertical "jump" between the line segments. Yet our approximating functions are sines and cosines which are everywhere continuous! When there is such a jump discontinuity in the function to be approximated, the Fourier series approximates the mean of the two function values (shown in the figure by a large dot) at the jump. But close to the discontinuity on either side of the jump, the sum of a finite number of smooth and continuous sines and cosines will have a hard time approximating the true function. This is quite evident from the figure. Nevertheless, the approximation will continue to become better and better as more terms are added so that the sum of the squares of the deviations of the approximating points from the true line will become less and less. So even when there are actual breaks such as this, f(t) will still be, for all practical purposes, precisely equal to the sum of an infinite number of terms in the Fourier series.

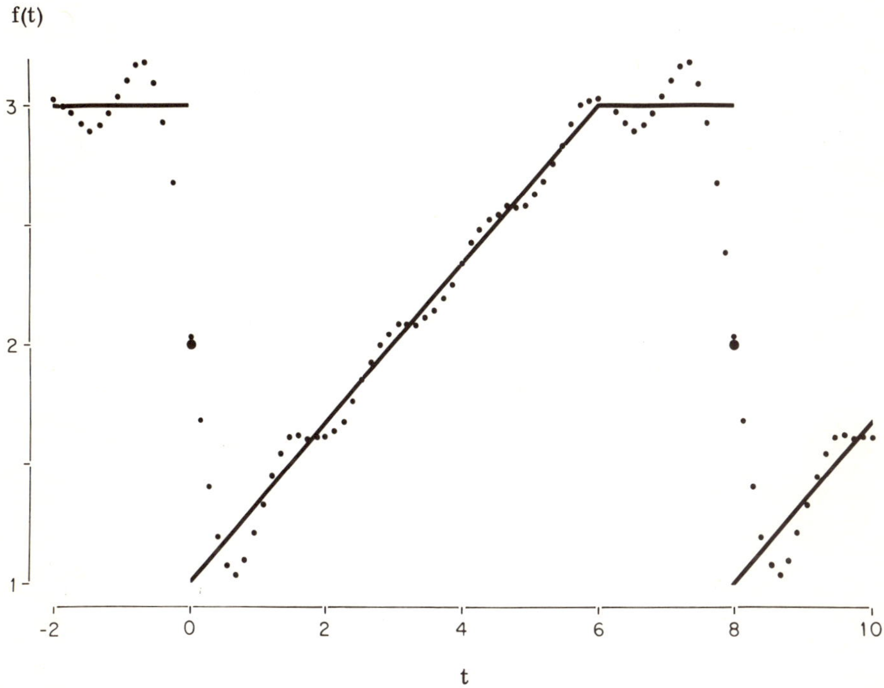

COMPLEX NUMBERS

The reader is undoubtedly familiar with the idea that any real number can be represented by a point on a straight line extending an infinite distance on either side of zero. We shall regard this line as the horizontal axis of coordinates on an infinite plane, and we shall call it the axis of reals, or, for short, the real axis.

In exactly the same way, any complex number, say $\alpha + j\beta$, $j \equiv \sqrt{-1}$, can be represented by a point on a plane called the complex plane. Keeping the horizontal axis as the real axis, we define the vertical axis as the axis of imaginaries, any point on which represents (or, in common parlance, is) a pure imaginary number. For example, j3 would be a point on the axis of imaginaries three units above the origin; $-j\sqrt{2}$ would be a point on the axis of imaginaries $\sqrt{2}$ units below the origin. For short, we call the axis of imaginaries the imaginary axis. But there is nothing 'imaginary' (in the non-mathematical sense) about it. We draw it, we see it, we work with it, precisely as we do with the real axis. With these coordinates defined, it is obvious that any complex number $\alpha + j\beta$ can be represented as the point $[\alpha, \beta]$ in the complex plane. This is the basis for all that follows. Initially, at least, you should always visualize a complex number as a point in the complex plane. E.g., you should visualize $-8-j\,0.5$ as the point $[-8, -0.5]$ in the third quadrant well to the left of the origin and just a bit below the real axis. But it is equally important to visualize the point $[\alpha, \beta]$ in the complex plane as the corresponding vector, extending from the origin to the point. By thinking of it as such a vector (after all, $[\alpha, \beta]$ is a row vector with two elements) we see at once that the position of the point (and therefore the corresponding complex number) can be specified just as well by giving the length of the vector (=its absolute magnitude) and the angle which it makes with the positive real axis. Magnitude and angle are called the polar coordinates of the complex number, just as α and β are called the rectangular coordinates. In phasor analysis in the frequency domain (See Section 3.16), we visualize the vector as rotating counterclockwise through all phases of the complete cycle of 2π radians which brings it full circle. The angle formed by the rotating vector with the positive real axis then represents the phase of the cycle which it has reached, and this angle is therefore called the phase angle. By extension, we often speak of the polar coordinates of any complex number as being the magnitude and the phase angle. It is customary to use the notation $M \angle \theta$ for the complex number whose magnitude is M and whose phase angle is θ radians.

For some purposes rectangular coordinates, and for other purposes polar coordinates are most convenient. We now derive the relationship between these two forms so that we can pass easily from one to the other.

Let C be any particular complex number. Let α= Re C (read "the real part of C"). Let β= Im C (read "the imaginary part of C"). Then,

$$C = \alpha + j\beta \qquad (1)$$

in rectangular coordinate form. Let the same number be,

$$C = M \angle \theta \qquad (2)$$

in polar coordinates. By reference to the figure at the end of this section it is easy to see that

$$\alpha = M\cos\theta \tag{3}$$

$$\beta = M\sin\theta \tag{4}$$

So if we know the magnitude and phase angle of any complex number we can find its rectangular coordinates by using equations 3 and 4.

Next, square both sides of equations 3 and 4 and add the results together,

$$\alpha^2 = M^2 [\cos\theta]^2$$

$$\beta^2 = M^2 [\sin\theta]^2$$

$$\alpha^2 + \beta^2 = M^2\{[\cos\theta]^2 + [\sin\theta]^2\} = M^2$$

Hence,

$$M = \sqrt{\alpha^2 + \beta^2} \tag{5}$$

Finally, from equations 3 and 4, we have,

$$\frac{\beta}{\alpha} = \frac{\sin\theta}{\cos\theta} = \tan\theta \tag{6}$$

Whence,

$$\theta = \text{Arctan}\,[\beta/\alpha] = \text{Tan}^{-1}\,[\beta/\alpha] \tag{7}$$

So if we know the rectangular coordinates of a complex number, we can find its polar coordinates by using equations 5 and 7.

CAUTION. Equation 7 is ambiguous in the following sense:

Consider $C_1 = 3 + j2$, a point in the <u>first</u> quadrant of the complex plane. $\beta/\alpha = 2/3$. To apply equation 7, we consult a table of trigonometric functions for angles in radians, thus finding $\theta = \text{Arctan}\ 0.6667 \approx 0.588$ radians which is correct. Now consider $C_2 = -3-j2$, a point in the <u>third</u> quadrant of the complex plane. By equation 7, we <u>again</u> find $\theta \approx 0.588$ radians which is wrong, the true angle for the point $[-3, -2]$ being $\approx 0.588 \pm \pi$ radians. The error is caused by the limited range of most trigonometric tables, and by our forgetting that tan $[\alpha/\beta]$ really <u>is</u> equal to tan $[-\alpha/-\beta]$ and tan $[-\alpha/\beta]$ to tan $[\alpha/-\beta]$. To avoid the error, we must consider the signs of α and β and use our common sense to make sure that our looked-up-in-a-table angle lies in the correct quadrant. If it doesn't we simply add $\pm\pi$ radians to the tabulated value.

By using the Euler identity,

$$\epsilon^{jx} = \cos x + j \sin x$$

$$\epsilon^{-jx} = \cos x - j \sin x \qquad (8)$$

we can derive still another very useful way of expressing a complex number. From equations 1, 3 and 4, we have,

$$C = M[\cos \theta + j \sin \theta] \qquad (9)$$

Whence, by equation 8,

$$C = M\epsilon^{j\theta} \qquad (10)$$

Since M and θ in equation 10 are the polar coordinates, the exponential form is rightly regarded as just another polar-coordinate way of expressing a complex number. But this exponential expression greatly facilitates certain mathematical computations.

At the end of this section the preceding discussion is summarized in a table, with specific examples, one for each quadrant of the complex plane.

In working with complex numbers, the following definitions, rules and notes are useful:

1. Definition The complex conjugate of $\alpha + j\beta$ is $\alpha - j\beta$.

   NOTE: Complex roots occur in conjugate pairs. E.g. the roots of $x^2 + 6x + 13$ are $-3 + j2$ and $-3 - j2$

   NOTE: $[\alpha + j\beta][\alpha - j\beta] = \alpha^2 + \beta^2 = M^2$

2. The magnitude of a complex number is always (as the name implies) intrinsically positive. If it is preceded by a minus sign, the minus must be the operator which means "subtract."

3. Two complex numbers are equal if, and only if the real part of the first is equal to the real part of the second, and the imaginary part of the first is equal to the imaginary part of the second,

   $C_1 = C_2 \Longrightarrow \text{Re } C_1 = \text{Re } C_2;\ \text{Im } C_1 = \text{Im } C_2$

4. In addition, complex numbers behave like two-element vectors: real parts are to be added only to other real parts, imaginary parts, only to other imaginary parts.

   $[-\alpha_1 + j\beta_1] - [\alpha_2 - j\beta_2] = -[\alpha_1 + \alpha_2] + j[\beta_1 + \beta_2]$

5. In multiplication, phase angles add; magnitudes multiply,

$[M_1\underline{/\theta_1}]\,[M_2\underline{/\theta_2}] = M_1 M_2 \underline{/\theta_1 + \theta_2}$

Proof: $M_1 \epsilon^{j\theta_1} M_2 \epsilon^{j\theta_2} = M_1 M_2 \epsilon^{j[\theta_1 + \theta_2]}$

6. The reciprocal of a complex number, $M\underline{/\theta}$, is $M^{-1}\underline{/-\theta}$.

Proof I.

Consider $C = M/\theta = \sqrt{\alpha^2 + \beta^2}\ \underline{/\mathrm{Tan}^{-1}[\beta/\alpha]} = \alpha + j\beta$

$$C^{-1} = \frac{1}{\alpha + j\beta} = \frac{1}{[\alpha + j\beta]} \cdot \frac{[\alpha - j\beta]}{[\alpha - j\beta]} = \frac{\alpha - j\beta}{\alpha^2 + \beta^2} =$$

$$\left[\frac{\alpha}{\alpha^2 + \beta^2}\right] - j\left[\frac{\beta}{\alpha^2 + \beta^2}\right]$$

Hence the magnitude of C^{-1} is,

$$\left\{\frac{\alpha^2}{[\alpha^2 + \beta^2]^2} + \frac{\beta^2}{[\alpha^2 + \beta^2]^2}\right\}^{1/2} = \frac{1}{\sqrt{\alpha^2 + \beta^2}} = M^{-1}$$

and the phase angle of C^{-1} is,

$$\tan^{-1}\left\{\frac{-\dfrac{\beta}{\alpha^2 + \beta^2}}{\dfrac{\alpha}{\alpha^2 + \beta^2}}\right\} = \text{Arctan } \frac{-\beta}{\alpha} = \mathrm{Tan}^{-1}\ -\theta$$

Hence,

$$C^{-1} = M^{-1}\underline{/-\theta}$$

as was to be shown.

Proof II.

Consider $C = M\underline{/\theta} = M\epsilon^{j\theta}$

$$C^{-1} = M^{-1}\epsilon^{-j\theta} = M^{-1}\epsilon^{j[-\theta]} = M^{-1}\underline{/-\theta}$$

as was to be shown.

(How much easier!)

7. Never leave a "j" in the denominator when reducing a fraction to a complex number.

   E.g. What is the complex number, $\frac{a + jb}{\alpha + j\beta}$?

   Multiply both the numerator and the denominator by the complex conjugate of the denominator,

$$\frac{a + jb}{\alpha + j\beta} = \frac{a + jb}{\alpha + j\beta} \cdot \frac{\alpha - j\beta}{\alpha - j\beta} = \frac{a\alpha + j[\alpha b - \beta a] + b\beta}{\alpha^2 + \beta^2}$$

$$= \left[\frac{a\alpha + b\beta}{\alpha^2 + \beta^2}\right] + j\left[\frac{\alpha b - \beta a}{\alpha^2 + \beta^2}\right]$$

The table below illustrates various ways of expressing the complex numbers which are indicated both by points and by the corresponding vectors shown as arrows in the complex plane of the opposite page.

quadrant	$ReC + jImC =$ $\alpha + j\beta$		$M\cos\theta$ α		$jM\sin\theta$ $j\beta$		$M\underline{/\theta}$ $\sqrt{\alpha^2+\beta^2}\underline{/\tan^{-1}\frac{\beta}{\alpha}}$		$M\epsilon^{j\theta}$ $\sqrt{\alpha^2+\beta^2}\,\epsilon^{j\tan^{-1}\frac{\beta}{\alpha}}$
		$=$		$+$		$=$		$=$	
I	$1 + j2$	$=$	$\sqrt{5}[0.447]$	$+$	$j\sqrt{5}[0.894]$	$=$	$\sqrt{1+4}\underline{/\tan^{-1}\frac{2}{1}}$	$=$	$\sqrt{5}\epsilon^{j1.106}$
II	$-1 + j2$	$=$	$\sqrt{5}[-0.447]$	$+$	$j\sqrt{5}[0.894]$	$=$	$\sqrt{1+4}\underline{/\tan^{-1}\frac{2}{-1}}$	$=$	$\sqrt{5}\epsilon^{j2.035}$
III	$-1 - j2$	$=$	$\sqrt{5}[-0.447]$	$+$	$j\sqrt{5}[-0.894]$	$=$	$\sqrt{1+4}\underline{/\tan^{-1}\frac{-2}{-1}}$	$=$	$\sqrt{5}\epsilon^{j4.248}$
IV	$1 - j2$	$=$	$\sqrt{5}[0.447]$	$+$	$j\sqrt{5}[-0.894]$	$=$	$\sqrt{1+4}\underline{/\tan^{-1}\frac{-2}{1}}$	$=$	$\sqrt{5}\epsilon^{j5.177}$

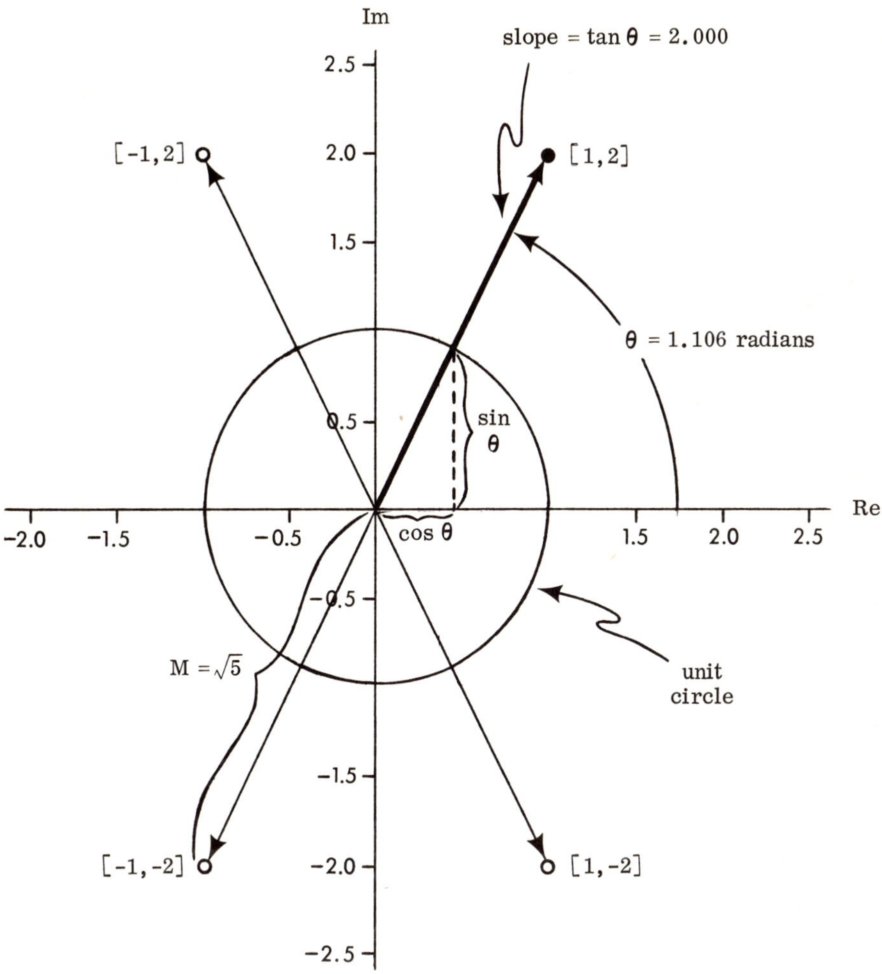

The Complex Plane

Any n^{th}-order linear differential equation may be arranged in the form,

$$\frac{d^n x(t)}{dt^n} + a_{n-1}\frac{d^{n-1}x(t)}{dt^{n-1}} + \cdot\cdot\cdot + a_1\frac{dx(t)}{dt} + a_0 x(t) = f(t)$$

where f(t) is some forcing function. (If the leading coefficient, i.e., the coefficient of the highest-order derivative, is not already unity, the equation can be put in the above form by dividing both sides by the leading coefficient.) When f(t) = 0, the equation is homogeneous and the system it describes is said to be unforced. If one or more of the coefficients (i.e., the a's) is not constant but varies with time, the system described by the differential equation is non-stationary, and the method of variation of parameters (or some equivalent method) would then have to be used. In this book we confine attention to stationary systems, so we shall assume that the a's are all constant. Then the easiest "classical" method of solution is usually the method of undetermined coefficients, the only one which will be considered here. Each step will be described, and then illustrated by a specific example.

METHOD of UNDETERMINED COEFFICIENTS

Specific example. Solve

$$\frac{d^2x(t)}{dt^2} + 5\frac{dx(t)}{dt} + 6x(t) = \epsilon^{-0.5t}$$

Subject to the initial conditions: $x(0) = 1$; $\dot{x}(0) = 2$

1. Define $D^n \equiv d^n/dt^n$ and rewrite the equation in the form,

$$P(D)\{x(t)\} \equiv [D^n + a_{n-1}D^{n-1} + \cdot\cdot\cdot + a_1 D + a_0]\{x(t)\} = f(t)$$

E.g., $P(D)\{x(t)\} = [D^2 + 5D + 6]\{x(t)\} = \epsilon^{-0.5t}$

2. Write the characteristic equation by substituting λ for D in P(D), the differential-operator polynomial, and setting the resulting characteristic polynomial equal to zero.

E.g., $P(\lambda) = \lambda^2 + 5\lambda + 6 = 0$

3. Find the eigenvalues which are simply the roots of the characteristic equation. Express P(λ) as a product of factors,

$$P(\lambda) = [\lambda + \alpha_n][\lambda + \alpha_{n-1}] \cdot\cdot\cdot [\lambda + \alpha_1]$$

Then the eigenvalues are the values of λ which make $P(\lambda) = 0$, namely, $\lambda_i = -\alpha_i$, $i = 1 \cdot\cdot\cdot n$. An nth-order equation has exactly n roots which may be distinct or multiple; zero, positive, or negative; real, imaginary, or complex.

E.g. $\lambda^2 + 5\lambda + 6 = \left[\lambda - \frac{-5 + \sqrt{25-24}}{2}\right]\left[\lambda - \frac{-5 - \sqrt{25-24}}{2}\right]$
$= [\lambda + 2][\lambda + 3]$

So the eigenvalues are $\lambda_1 = -2$, $\lambda_2 = -3$.

NOTE: To find the roots of higher-order polynomial equations, the polynomial can be factored by using Lin's method or a graphical method which are described elsewhere in the Appendix.

4. The complementary equation is the homogeneous equation obtained by substituting zero for f(t) in the original differential equation. If the eigenvalues are all distinct, its solution, $x_c(t)$ will be of the form,

$$x_c(t) = b_1 \epsilon^{\lambda_1 t} + b_2 \epsilon^{\lambda_2 t} + \ldots + b_n \epsilon^{\lambda_n t}$$

where the b's are constants to be evaluated from the initial conditions after the particular solution corresponding to f(t) has been added to the complementary solution.

E.g., $x_c(t) = b_1 \epsilon^{-2t} + b_2 \epsilon^{-3t}$

If there are any multiple roots, the solution will have terms of the form

$$b_{1,1}\epsilon^{\lambda_1 t} + b_{1,2} t\epsilon^{\lambda_1 t} + b_{1,3} t^2 \epsilon^{\lambda_1 t} + \ldots + b_{1,m} t^{m-1} \epsilon^{\lambda_1 t}$$

where m is the multiplicity of the λ_1 root. For example, if one of the factors of the characteristic equation is $[\lambda_1 + 5]^3$, the corresponding terms in the solution will be

$$b_{1,1}\epsilon^{-5t} + b_{1,2} t\epsilon^{-5t} + b_{1,3} t^2 \epsilon^{-5t}$$

NOTE: As is shown elsewhere in the Appendix, terms corresponding to a complex root and its conjugate, for example $b_1 \epsilon^{[-7 + j5]t}$ and $b_2 \epsilon^{[-7-j5]t}$ can equally well be written in the form $A\epsilon^{-7t}\sin 5t + B\epsilon^{-7t}\cos 5t$ where A and B are, like the b's, simply constants to be evaluated.

5. Consider the forcing function as representing additional roots, r_1, r_2 etc. which must be considered, together with the eigenvalues, when finding the terms of the full solution. For example, if $f(t) = 12 + t\epsilon^{-3t} - 4 \cos 5t$ we recognize that these terms correspond to factors of the general form $[r + 0]\,[r + 3]^2\,[r \pm j5]$ with roots (which are not eigenvalues) of 0, 3, 3, j5 and -j5.

The brief table below lists some kinds of terms commonly encountered in forcing functions and the roots to which they correspond. A, B, α, ω are constants, n is a non-negative integer.

Forcing function term	Corresponding root
A (i.e., any constant)	$r = 0$
$A\epsilon^{\alpha t}$	$r = \alpha$
$At^n\epsilon^{\alpha t}$	$r = \alpha$ repeated n + 1 times
A sin ωt or B cos ωt or A sin ωt + B cos ωt	$r = \pm j\omega$
$A\epsilon^{\alpha t}\sin \omega t$, or $B\epsilon^{\alpha t}\cos \omega t$ or $\epsilon^{\alpha t}[A \sin \omega t + B \cos \omega t]$	$r = \alpha \pm j\omega$
$At^n\epsilon^{\alpha t}\sin \omega t$ or $B t^n \epsilon^{\alpha t}\cos \omega t$ or $t^n\epsilon^{\alpha t}[A \sin \omega t + B \cos \omega t]$	$r = \alpha \pm j\omega$ repeated n + 1 times

NOTE: All of the forcing function terms tabulated could be represented by the single general form $At^n\epsilon^{[\alpha \pm j\omega]t}$ where n, α, and ω may each assume a value of zero.

6. Write the most general equation for x(t) which will include one term corresponding to each of the λ's and each of the r's, making no distinction between the λ's and r's when determining the multiplicity of the roots. For example, if the eigenvalues are 0 and -5 and the r's are 0, 3, 3, j5 and -j5, the general equation for x(t) would be

$$x_g(t) = b_1 + b_2\epsilon^{-5t} + b_3t + b_4\epsilon^{3t} + b_5t\epsilon^{3t} + b_6\epsilon^{j5t} + b_7\epsilon^{-j5t}$$

$$= b_1 + b_2\epsilon^{-5t} + b_3t + b_4\epsilon^{3t} + b_5t\epsilon^{3t} + A \sin 5t + B \cos 5t$$

where the subscript "g" is used to show that this is a perfectly general equation containing all of the kinds of terms--whether derived from the system's "own" roots (= eigenvalues) or from the roots forced upon it by f(t),--which could possibly occur in the specific solution for x(t). Notice that each root becomes a coefficient of time in the exponent of ϵ in the corresponding term of $x_g(t)$.

E.g., $x_g(t) = b_1\epsilon^{-2t} + b_2\epsilon^{-3t} + b_3\epsilon^{-0.5t}$

7. Allow the differential operator, P(D), to operate upon $x_g(t)$. The resulting expression must equal the forcing function for all values of t. We can therefore particularize $x_g(t)$ according to the given f(t) by setting coefficients of like terms on both sides of the equation

$$P(D)\{x_g(t)\} = f(t)$$

equal to each other. Simultaneous solution of the resulting equations will then allow us to determine the previously undetermined coefficients of those terms in $x_g(t)$ which belong to $x_p(t)$, the particular solution.

$$\text{E.g., } P(D)\{x_g(t)\} = [D^2 + 5D + 6]\{b_1\epsilon^{-2t} + b_2\epsilon^{-3t} + b_3\epsilon^{-0.5t}\} =$$

$$\begin{aligned} 4b_1\epsilon^{-2t} &+ 9b_2\epsilon^{-3t} + 0.25\, b_3\epsilon^{-0.5t} \\ -10b_1\epsilon^{-2t} &- 15b_2\epsilon^{-3t} - 2.50\, b_3\epsilon^{-0.5t} \\ +\, 6b_1\epsilon^{-2t} &+ 6b_2\epsilon^{-3t} + 6.00\, b_3\epsilon^{-0.5t} = \\ 0\epsilon^{-2t} &+ 0\epsilon^{-3t} + 3.75\, b_3\epsilon^{-0.5t} = \epsilon^{-0.5t} \\ & 3.75\, b_3 = 1 \\ & b_3 = 4/15 \end{aligned}$$

Therefore the particular integral, $x_p(t)$, is

$$x_p(t) = [4/15]\epsilon^{-0.5t}$$

NOTE: In finding $x_p(t)$, it is perfectly legitimate to drop the terms corresponding to the λ's alone before applying the differential operator, since operating on these terms (derived from the homogeneous equation) must yield zero. I prefer to carry them along as an additional check.

8. Now (not before) apply the initial conditions to the general solution, $x_g(t) = x_c(t) + x_p(t)$, so as to evaluate the remaining coefficients,

$$\text{E.g., } x_g(t) = b_1\epsilon^{-2t} + b_2\epsilon^{-3t} + [4/15]\epsilon^{-0.5t}$$

$$\dot{x}_g(t) = -2b_1\epsilon^{-2t} - 3b_2\epsilon^{-3t} - [2/15]\epsilon^{-0.5t}$$

Evaluating these equations at t= 0 and setting each result equal to the appropriate initial condition,

$$x_g(0) = b_1 + b_2 + [4/15] = 1$$

$$\dot{x}_g(0) = -2b_1 - 3b_2 - [2/15] = 2$$

Solving these equations simultaneously, we find $b_1 = 65/15$; $b_2 = -54/15$. So the complete solution is

$$x(t) = [65/15]\epsilon^{-2t} - [54/15]\epsilon^{-3t} + [4/15]\epsilon^{-0.5t}$$

9. Check the solution by making sure that it satisfies both the original differential equation and the initial conditions.

$$[D^2 + 5D + 6]\left\{\frac{65}{15}\,\varepsilon^{-2t} \quad \frac{54}{15}\,c^{-3t} + \frac{4}{15}\,\varepsilon^{-0.5t}\right\} =$$

$$[260/15]\varepsilon^{-2t} - [486/15]\varepsilon^{-3t} + [\ 1/15]\varepsilon^{-0.5t}$$

$$-[650/15]\varepsilon^{-2t} + [810/15]\varepsilon^{-3t} - [10/15]\varepsilon^{-0.5t}$$

$$+[390/15]\varepsilon^{-2t} - [324/15]\varepsilon^{-3t} + [24/15]\varepsilon^{-0.5t} = \varepsilon^{-0.5t}$$

So the differential equation is satisfied.

$$x(0) = [65/15] - [54/15] + [4/15] = 1$$

$$\dot{x}(0) = -[130/15] + [162/15] - [2/15] = 2$$

So the initial conditions are also satisfied.

These same steps will be applied, without discussion, to a second example, which will illustrate how to deal with multiple roots. D.E. = differential equation, I.C. = initial conditions.

$$\text{D.E.} \quad \frac{d^2x(t)}{dt^2} + 4\,x(t) = 3\sin 2t$$

$$\text{I.C.} \quad x(0) = 0; \ \dot{x}(0) = 1$$

1. $[D^2 + 4]\{x(t)\} = 3\sin 2t$

2. $\lambda^2 + 4 = 0$

3. $\lambda_1 = j2, \ \lambda_2 = -j2$

4. $x_c(t) = b_1\varepsilon^{j2t} + b_2\varepsilon^{-j2t} = A\sin 2t + B\cos 2t$

5. $r_1 = j2, \ r_2 = -j2$

6. $x_g(t) = A\sin 2t + B\cos 2t + \alpha t\sin 2t + \beta t\cos 2t$

7. $[D^2 + 4]\{A\sin 2t + B\cos 2t + \alpha t\sin 2t + \beta t\cos 2t\} = 3\sin 2t$

$$\dot{x}_g(t) = 2A\cos 2t - 2B\sin 2t + \alpha\sin 2t + 2\alpha t\cos 2t + \beta\cos 2t - 2\beta t\sin 2t$$

$$= [2A + \beta]\cos 2t + [\alpha - 2B]\sin 2t + 2\alpha t\cos 2t - 2\beta t\sin 2t$$

$$\ddot{x}_g(t) = -[4A + 2\beta]\sin 2t + [2\alpha - 4B]\cos 2t + 2\alpha\cos 2t - 4\alpha t\sin 2t$$

$$- 2\beta\sin 2t - 4\beta t\cos 2t$$

$\ddot{x}_g(t) = -[4A + 4\beta]\sin 2t + [4\alpha - 4B]\cos 2t - 4\alpha t \sin 2t - 4\beta t \cos 2t$

$4x_g(t) = 4A \sin 2t + 4B \cos 2t + 4\alpha t \sin 2t + 4\beta t \cos 2t$

$\ddot{x}_g(t) + 4x_g(t) = -4\beta \sin 2t + 4\alpha \cos 2t = 3 \sin 2t$

$-4\beta = 3 \qquad \beta = -3/4$

$4\alpha = 0 \qquad \alpha = 0$

$x_p(t) = -[3/4]t \cos 2t$

8. $x_g(t) = x_c(t) + x_p(t) = A\sin 2t + B\cos 2t - [3/4]t \cos 2t$

$\dot{x}_g(t) = 2A \cos 2t - 2B \sin 2t - [3/4]\cos 2t + [6/4]t \sin 2t$

$x_g(0) = B = 0$

$\dot{x}_g(0) = 2A - [3/4] = 1 \quad ; \quad A = [7/8]$

$x(t) = [7/8]\sin 2t - [6/8]t \cos 2t$

9. $\dot{x}(t) = [14/8]\cos 2t - [6/8]\cos 2t + [12/8]t \sin 2t$

$= \cos 2t + [12/8]t \sin 2t$

$\ddot{x}(t) = -2 \sin 2t + [12/8]\sin 2t + [24/8]t \cos 2t$

$4x(t) = [28/8]\sin 2t - [24/8]t \cos 2t$

$[D^2 + 4]\{[7/8]\sin 2t - [6/8]t \cos 2t\} = 3 \sin 2t$

$x(0) = [7/8]\sin 0 - [6/8][0]\cos 0 = 0$

$\dot{x}(0) = \cos 0 + [12/8][0]\sin 0 = 1$

Check

COMPLEX ROOTS, DAMPING RATIOS, AND WAYS OF EXPRESSING SINUSOIDAL SOLUTIONS

When the characteristic equation for a linear system has complex roots, the system will exhibit sinusoidal oscillations in response to initial conditions or to a step input applied at time zero. The following discussion is intended to explain the relationships among various ways of expressing the oscillatory solution.

Consider a system described by the equation

$$a \frac{d^2y(t)}{dt^2} + b \frac{dy(t)}{dt} + c = 0$$

where y(t) is the output and a, b, and c are constants.

The corresponding characteristic equation,

$$a\lambda^2 + b\lambda + c = 0$$

has the roots

$$\lambda_1 = \left\{ -b + \sqrt{b^2 - 4ac} \right\} / 2a$$

$$\lambda_2 = \left\{ -b - \sqrt{b^2 - 4ac} \right\} / 2a$$

The solution will therefore be of the general form

$$y(t) = A\epsilon^{\lambda_1 t} + B\epsilon^{\lambda_2 t}$$

where the coefficients A and B are determined by the initial conditions. We shall confine our discussion to stable systems in which the real parts of λ_1 and λ_2 are negative.

DAMPING RATIO. The damping ratio, ρ, for a second- or third-order system is defined as

$$\rho \equiv \sqrt{b^2/4ac}$$

(With systems of fourth or higher order, the characteristic polynomial will have two or more quadratic factors. Hence, a single damping ratio cannot be defined for such systems.)

If $b^2 > 4ac$, $\rho > 1$, and the system is overdamped. In response to a constant input imposed at time zero, y(t) will approach a constant value, $y(\infty)$, at some less-than-maximal rate.

If $b^2 = 4ac$, $\rho = 1$ and the system is critically damped. In response to a constant input imposed at time zero, y(t) will approach a constant value, $y(\infty)$, at the maximum rate possible without oscillations.

If $4ac > b^2 > 0$, $1 > \rho > 0$ and the system is underdamped. In response to a constant input imposed at time zero, y(t) will approach a constant value, $y(\infty)$, by undergoing a series of sinusoidal oscillations whose magnitude will decrease exponentially with time.

If $b^2 = 0$, $\rho = 0$, and the system is undamped. Undamped systems are called conservative systems because they conserve without loss all of the energy which they have stored, or which they receive from outside sources. Once an undamped system has been activated (e.g. in response to initial conditions) y(t) will continue to undergo sinusoidal oscillations of constant magnitude forever with no further input.

(For completeness, we may note that if b is negative and $\rho < 1$, the system would be unstable and y(t) would undergo sinusoidal oscillations whose magnitude would increase exponentially with time.)

COMPLEX ROOTS and SINUSOIDAL OSCILLATIONS. Complex roots of the characteristic equation will occur whenever $b^2 - 4ac$ is negative, i.e. whenever $b^2 < 4ac$ so that $\rho < 1$. Hence complex roots imply sinusoidal oscillations of the output. We shall illustrate this kind of oscillatory behavior with a specific example.

Consider a linear system described by

$$\frac{d^2y(t)}{dt^2} + 3\frac{dy(t)}{dt} + 27.25\, y(t) = 0$$

Subject to the initial conditions,

$$y(0) = 3,\ y'(0) = 20$$

The characteristic equation is

$$\lambda^2 + 3\lambda + 27.25 = 0$$

so the damping ratio, $\sqrt{9/109} \approx 0.287$, is substantially less than unity. The roots of the characteristic equation are the complex conjugate pair,

$$\lambda_1 = \frac{-3 + \sqrt{9-109}}{2} = -1.5 + j5$$

$$\lambda_2 = \frac{-3 - \sqrt{9-109}}{2} = -1.5 - j5$$

The solution will therefore be of the form

$$y(t) = A\epsilon^{[-1.5 + j5]t} + B\epsilon^{[-1.5 - j5]t}$$

$$= \epsilon^{-1.5t}\left\{A\epsilon^{j5t} + B\epsilon^{-j5t}\right\}$$

But by Euler's identity

$$\epsilon^{j5t} \equiv \cos 5t + j \sin 5t$$

$$\epsilon^{-j5t} \equiv \cos 5t - j \sin 5t$$

Hence,

$$y(t) = \epsilon^{-1.5t}\{[A[\cos 5t + j\sin 5t] + B[\cos 5t - j \sin 5t]\}$$

or, collecting terms,

$$y(t) = \epsilon^{-1.5t}\{[A + B]\cos 5t + j[A - B]\sin 5t\}$$

where A and B are still just unevaluated numbers which, in general, will be complex. For convenience, let us define two real numbers, a and b, such that

$$a \equiv A + B, \quad b \equiv j[A - B]$$

(PARENTHETICAL NOTE. Since a and b are real, these defining equations can be true if and only if A and B are complex. Let us accordingly define

$$A \equiv \alpha_A + j\beta_A; \quad B \equiv \alpha_B + j\beta_B$$

and substitute these values for A and B. We then have

$$a = \alpha_A + j\beta_A + \alpha_B + j\beta_B = [\alpha_A + \alpha_B] + j[\beta_A + \beta_B]$$

$$b = j[\alpha_A + j\beta_A - \alpha_B - j\beta_B] = j[\alpha_A - \alpha_B] + [-\beta_A + \beta_B]$$

By requiring the real parts to equal the real parts, and the imaginary parts to equal the imaginary parts in both of these equations, we obtain

$$\alpha_A + \alpha_B = a; \quad \alpha_A - \alpha_B = 0; \quad \text{so } \alpha_A = \alpha_B = a/2$$

and

$$\beta_A + \beta_B = 0; \quad -\beta_A + \beta_B = b; \quad \text{so } \beta_B = -\beta_A = b/2$$

Hence, $A = [a - jb]/2$ and $B = [a + jb]/2$)

We can now write our general solution, unencumbered by complex numbers, as

$$y(t) = \epsilon^{-1.5t}\{a \cos 5t + b \sin 5t\}$$

To evaluate a and b, we apply the initial conditions.

$$y(0) = a = 3$$

$$y'(t) = -1.5\epsilon^{-1.5t}\{a \cos 5t + b \sin 5t\}$$

$$+ \epsilon^{-1.5t}\{-5a \sin 5t + 5b \cos 5t\}$$

$$y'(0) = 1.5a + 5b = -4.5 + 5b = 20$$

$$b = 4.9$$

So our specific solution is

$$y(t) = \epsilon^{-1.5t}\{3 \cos 5t + 4.9 \sin 5t\}$$

The damped sinusoidal approach of y(t) to zero as $t \to \infty$ is shown in the following graph. (See also Figure 2.15.4).

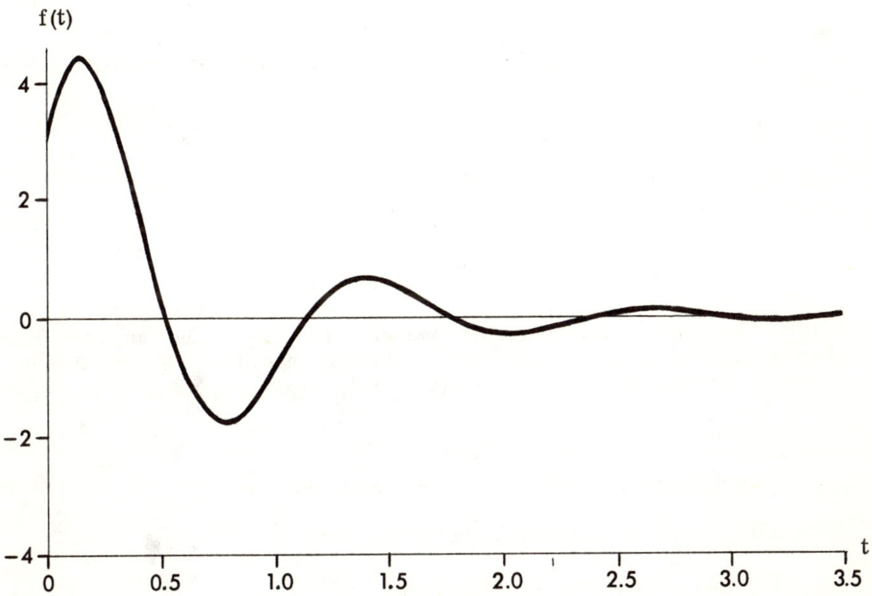

ALTERNATE WAYS OF EXPRESSING A SINUSOIDAL SOLUTION. In general, the sinusoidal portion of our solution will appear (as it did in the previous example) in the form

$$a \cos \omega t + b \sin \omega t, \tag{1.}$$

i.e. as the sum of a cosine wave of magnitude a, and a sine wave of magnitude b, both waves having the same frequency, namely ω radians per unit time. If we add any two such waves, we shall obtain a single sinusoidal wave of the same frequency which can be expressed either as a pure sine function or as a pure cosine function. To express the sum as a sine wave, we use the identity,

$$[\sin u][\cos v] + [\cos u][\sin v] \equiv \sin (u + v).$$

To put (1.) in this form, we define

$$a \equiv M \sin \theta ; \quad b \equiv M \cos \theta \tag{2.}$$

where M and θ are constants. Substituting in (1.)

$$M[\sin \theta \cos \omega t + \cos \theta \sin \omega t] = M \sin (\omega t + \theta)$$

But from (2.) we have

$$\sqrt{a^2 + b^2} = \left[M^2 \sin^2 \theta + M^2 \cos^2 \theta\right]^{1/2} = M\left[\sin^2 \theta + \cos^2 \theta\right]^{1/2} = M$$

so M, the magnitude of the sine wave, is $\sqrt{a^2 + b^2}$. Furthermore, from (2.) we have

$$a/b = \sin \theta / \cos \theta = \tan \theta$$

so θ, the phase angle of the sine wave, is $\tan^{-1}(a/b)$. Accordingly, we can rewrite (1.) as

$$a \cos \omega t + b \sin \omega t = \sqrt{a^2 + b^2} \sin (\omega t + \tan^{-1}(a/b)) \tag{3.}$$

NOTE. The positive phase angle means that $\sin(\omega t + \theta)$ reaches any particular phase of its cycle (e.g. its maximum value) θ/ω units of time earlier than $\sin \omega t$ would. In engineering jargon, $\sin(\omega t + \theta)$ leads $\sin \omega t$ by θ radians.

By using the identity,

$$[\cos u][\cos v] + [\sin u][\sin v] \equiv \cos (u + v),$$

in a precisely similar derivation, we can show that

$$a \cos \omega t + b \sin \omega t = M \cos (\omega t - \varphi)$$

where the magnitude, M, is $\sqrt{a^2 + b^2}$ as before, but the phase angle, φ, is $\tan^{-1}(b/a)$.

Accordingly we can rewrite (1.) as

$$a \cos \omega t + b \sin \omega t = \sqrt{a^2 + b^2} \cos(\omega t - \tan^{-1}(b/a)) \qquad (4.)$$

NOTE. The negative phase angle means that $\cos(\omega t - \varphi)$ reaches any particular phase of its cycle φ/ω units of time later than $\cos \omega t$ would. $\cos(\omega t - \varphi)$ is therefore said to lag $\cos \omega t$ by φ radians.

Finally, there is no reason why we could not express a pure sine function as a pure cosine function. Using (4.) we would have

$$b \sin \omega t = \sqrt{b^2 + 0^2} \cos(\omega t - \tan^{-1}(b/0)) = b \cos(\omega t - \pi/2)$$

which shows that a sine wave of given magnitude and frequency can be expressed as a cosine wave of the same magnitude and frequency but with a phase lag of $\pi/2$ radians. This is precisely what we would expect, for we know that $\sin \omega t$ reaches any point in its cycle exactly one-quarter of a cycle, i.e. $\pi/2$ radians later than $\cos \omega t$ does. For example, at $t = 0$, $\cos \omega t$ is at its maximum, while $\sin \omega t$ does not reach its maximum until $\omega t = \pi/2$ i.e. until $t = \pi/2\omega$.

The following figure should help to clarify the preceding discussion. It illustrates the solution of the differential equation

$$\frac{d^2y(t)}{dt^2} + 25\, y(t) = 0$$

Subject to the initial conditions, $y(0) = 3$, $y'(0) = 20$, with time measured in seconds. In exponential form, the solution is

$$y(t) = A\epsilon^{j\omega t} + B\epsilon^{-j\omega t}$$

where ω, the radian frequency, is 5 radian/sec. The corresponding period of the oscillation is

$$\frac{2\pi \text{ rad/cycle}}{\omega \text{ rad/sec.}} = \frac{2\pi}{5} \approx 1.2566 \text{ sec/cycle}$$

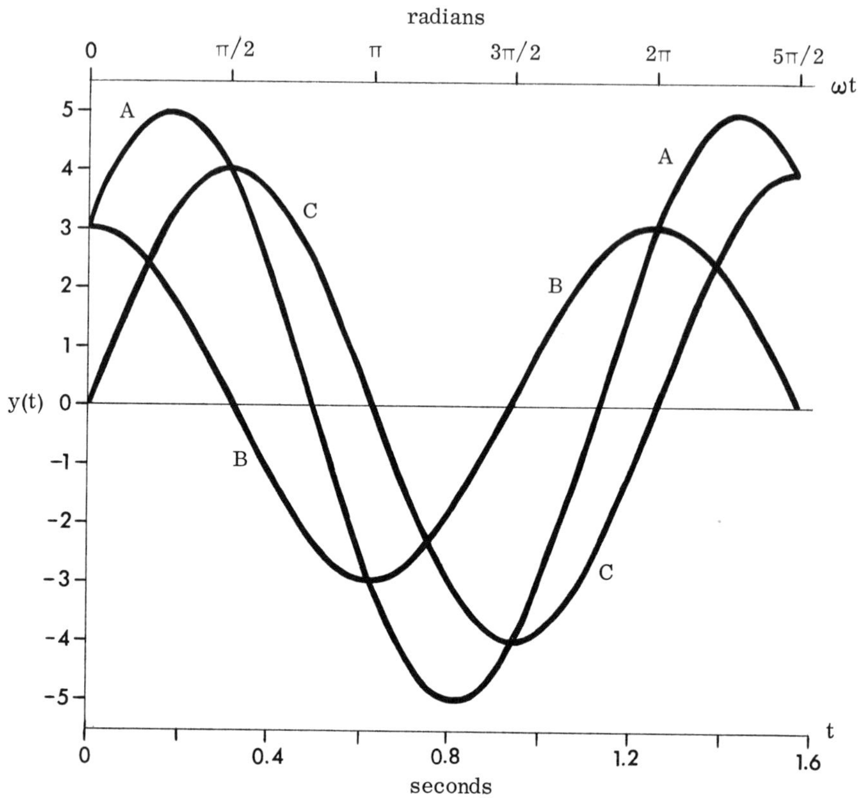

This solution (Curve A) can be expressed equally well as the sum of a cosine wave of magnitude 3 and a sine wave of magnitude 4,

$$y(t) = \underbrace{3 \cos 5t}_{\text{curve B}} + \underbrace{4 \sin 5t}_{\text{curve C}}$$

or as a cosine wave of magnitude $\sqrt{3^2 + 4^2} = 5$ with a phase lag of $\tan^{-1}(4/3) = 0.927$ radians,

$$y(t) = 5 \cos(5t - 0.927)$$

or as a sine wave of magnitude $\sqrt{3^2 + 4^2} = 5$ with a phase lead of $\tan^{-1}(3/4) = 0.644$ radians,

$$y(t) = 5 \cos(5t + 0.644)$$

NOTE that in accordance with these phase shifts, the maximum of curve A occurs 0.927 radians after the maximum of curve B, a cosine wave with no phase shift, and 0.644 radians before the maximum of curve C, a sine wave with no phase shift.

PARTIAL FRACTIONS

When solving differential equations by using Laplace transforms, we usually end up with a ratio of polynomials in s, the ratio being the Laplace transform of our unknown output function. To facilitate taking inverse Laplace transforms so as to express the output as a function of t, the ratio of polynomials must be broken into partial fractions corresponding to the factors of its denominator. It is then usually easy to find the inverse transform of each partial fraction. The output itself is simply the sum of the individual inverse transforms.

Consider the ratio N(s)/D(s) where N(s) and D(s) are both polynomials in s, and D(s) is of higher order than N(s). By factoring the denominator (see the section on factoring polynomials elsewhere in this Appendix) we can express it as a product of factors which may be 1.) linear (for example s, or s + 3), or 2.) quadratic with complex roots (for example $s^2 + 4$, or $s^2 + 3s + 6.25$). Furthermore, any given factor may appear more than once, so we must consider 3.) repeated factors. For clarity we shall deal with these three kinds of factors separately.

1.) Non-repeated linear factors. Let $D(s) = s[s + a_1][s + a_2][s + a_3] \cdots$ where $a_1, a_2, a_3, \cdots$ are all different real numbers. We set

$$\frac{N(s)}{D(s)} = \frac{A}{s} + \frac{B}{[s + a_1]} + \frac{C}{[s + a_2]} + \frac{D}{[s + a_3]} + \cdots$$

where A, B, C, D, $\cdots$ are to be evaluated by requiring the above equation to be an identity in s, i.e. by requiring it to be true for any arbitrary value of s. Clear of fractions by multiplying both sides of the equation by D(s).

$$N(s) \equiv A[s+a_1][s+a_2][s+a_3] \cdots + Bs[s+a_2][s+a_3] \cdots + Cs[s+a_1][s+a_3] \cdots + \text{etc.}$$

which is still an identity in s.

METHOD 1. Since the equation is an identity, we may equate the coefficients of like powers of s on both sides of the equation and solve the resulting set of simultaneous equations for the unknowns, A, B, C, D, $\cdots$

METHOD 2. Since the equation is an identity in s, we may substitute for s any value we please without destroying the validity of the equation. Accordingly, let us substitute for s the value which makes the first factor of D(s) zero. This will eliminate B, C, D, $\cdots$ from the equation and allow us to evaluate A. we repeat this procedure for each of the other factors. For linear factors, Method 2 is usually easier than Method 1.

Example 1. Break $1/[s^2 + 4s + 3]$ into partial factions.

$$\frac{N(s)}{D(s)} = \frac{1}{s^2 + 4s + 3} = \frac{1}{[s + 1][s + 3]}$$

Set

$$\frac{1}{[s+1][s+3]} \equiv \frac{A}{[s+1]} + \frac{B}{[s+3]}$$

Multiply both sides by $D(s) = [s+1][s+3]$

$$1 \equiv [s+3]A + [s+1]B$$

METHOD 1, Equate sums of coefficients of like powers of s on both sides of the equation

	right side		left side
1st power	$A + B$	=	0
0th power	$3A + B$	=	1

Solving simultaneously, $A = 1/2$; $B = -1/2$.

METHOD 2. Let $s = -1$. Then

$$[-1+3]A = 1;\ A = 1/2$$

Let $s = -3$. Then

$$[-3+1]B = 1;\ B = -1/2$$

CHECK by showing that $\dfrac{1}{s^2+4s+3} = \dfrac{1/2}{s+1} - \dfrac{1/2}{s+3}$.

Cross-multiplying, $\dfrac{1}{2}\left\{\dfrac{s+3-s-1}{s^2+4s+3}\right\} = \dfrac{1}{s^2+4s+3}$

Example 2. Break

$$\frac{N(s)}{D(s)} = \frac{s^2+11s+30}{s^4+2.5\,s^3+0.5s^2-s} = \frac{s^2+11s+30}{s[s+1][s+2][s-0.5]}$$

into partial fractions. We set

$$\frac{s^2+11s+30}{s[s+1][s+2][s-0.5]} = \frac{A}{s} + \frac{B}{s+1} + \frac{C}{s+2} + \frac{D}{s-0.5},$$

and clear of fractions by multiplying both sides by D(s).

$$s^2+11s+30 = [s+1][s+2][s-0.5]A + s[s+2][s-0.5]B + s[s+1][s-0.5]C$$
$$+ s[s+1][s+2]D$$

Using METHOD 2, we have,

when s = 0, -A = 30; A = -30 = -450/15

when s = -1, [-1][1][-1.5]B = 1-11 + 30; B = 20/1.5 = 200/15

when s = -2, [-2][-1][-2.5]C = 4-22 + 30; C = 12/-5 = -36/15

when s = 0.5, [0.5][1.5][2.5]D = 0.25+5.5+30; D = 35.75/[15/8]= 286/15

2.) Non-repeated quadratic factors with complex roots. It is much the wisest plan to keep these as quadratics, letting the numerator of each corresponding partial fraction be of the form, [as + b]. With such factors, METHOD 1 is the better.

Example 3. Break $[s + 1]/[s^2 + 2s + 2][s^2 + 4s + 13]$ into partial fractions.

$$\frac{N(s)}{D(s)} = \frac{s+1}{[s^2+2s+2][s^2+4s+13]} \equiv \frac{as+b}{s^2+2s+2} + \frac{cs+d}{s^2+4s+13}$$

Multiply both sides by D(s)

$$s + 1 = [s^3 + 4s^2 + 13s]a + [s^2 + 14s + 13]b + [s^3 + 2s^2 + 2s]c + [s^2+2s+2]d$$

Set coefficients of like powers of s on either side equal to each other.

3rd power $a + c = 0$

2nd power $4a + b + 2c + d = 0$

1st power $13a + 4b + 2c + 2d = 1$

0th power $13b + 2d = 1$

By putting the augmented matrix for these simultaneous equations,

$$\left[\begin{array}{cccc|c} 1 & 0 & 1 & 0 & 0 \\ 4 & 1 & 2 & 1 & 0 \\ 13 & 4 & 2 & 2 & 1 \\ 0 & 13 & 0 & 2 & 1 \end{array}\right]$$

into row-echelon form (see elsewhere in this Appendix) we find a = 18/170, b = 22/170, c = -18/170, d = -58/170.

CHECK. Show that

$$\frac{s+1}{[s^2+2s+2][s^2+4s+13]} \equiv \frac{1}{170}\left\{\frac{18s+22}{s^2+2s+2} - \frac{18s+58}{s^2+4s+13}\right\}.$$

$$170s + 170 \equiv [18s + 22][s^2 + 4s + 13] - [18s + 58][s^2 + 2s + 2]$$

$$= 18s^3 + 94s^2 + 322s + 286 - 18s^3 - 94s^2 - 152s - 116$$

$$= 170s + 170$$

<u>IMPORTANT NOTE</u> concerning inverse Laplace transforms corresponding to quadratic factors of D(s).

A partial fraction whose denominator is a quadratic with complex factors will be in the form

$$\frac{as + b}{s^2 + cs + d} \; ; \; c^2 < 4d$$

where some of the constants, a, b, c, d, may be zero.

Now $\mathcal{L}\{\alpha\epsilon^{-kt}\cos \omega t + \beta\epsilon^{-kt}\sin \omega t\} =$

$$\frac{\alpha[s + k] + \beta\omega}{[s + k]^2 + \omega^2} = \frac{[\alpha]s + [\alpha k + \beta\omega]}{s^2 + [2k]s + [k^2 + \omega^2]}$$

Hence

$$\frac{[\alpha]s + [\alpha k + \beta\omega]}{s^2 + [2k]s + [k^2 + \omega^2]} = \frac{as + b}{s^2 + cs + d}$$

Setting coefficients of like terms first in the numerators, then in the denominators equal to each other*

$$\alpha = a; \alpha k + \beta\omega = b; 2k = c; k^2 + \omega^2 = d$$

[*This is <u>NOT</u> in general a valid procedure. But here it yields the same results as would be obtained (more legitimately but much more laboriously) by cross-multiplying to get polynomials on both sides and solving the equations obtained by equating coefficients of like powers.]

Solving these equations simultaneously,

$$\alpha = a$$

$$k = c/2$$

$$\omega = [d - k^2]^{1/2} = \sqrt{d - [c^2/4]}$$

$$\beta = [b - \alpha k]/\omega = [2b - ac]/2\sqrt{d - [c^2/4]}$$

For a numerical example, see Example 3.6.1 in Chapter 3.

3.) Repeated Factors. Suppose that some factor of D(s), be it linear or quadratic, is repeated so that it is raised to some power, n, when D(s) is written as a product of factors. There are n partial fractions corresponding to any such repeated factor. The numerator of each of the n partial fractions is of the same form as for a single such factor, but the successive denominators are: $[\text{factor}]^1, [\text{factor}]^2, \cdots, [\text{factor}]^n$.

Example 4. Break $5/s^2[s+1]^3$ into partial fractions.

$$\frac{5}{s^2[s+1]^3} = \frac{A}{s} + \frac{B}{s^2} + \frac{C}{s+1} + \frac{D}{[s+1]^2} + \frac{E}{[s+1]^3}$$

Multiplying both sides by $s^2[s+1]^3$ as usual,

$$5 = s[s+1]^3A + [s+1]^3B + s^2[s+1]^2C + s^2[s+1]D + s^2E$$

First exploit METHOD 2.

when $s = 0$, $B = 5$
when $s = -1$, $E = 5$

which gives us

$$5 = s[s+1]^3A + 5[s+1]^3 + s^2[s+1]^2C + s^2[s+1]D + 5s^2$$

$$5 = [s^4+3s^3+3s^2+s]A + 5s^3 + 15s^2 + 15s + 5 + [s^4+2s^3+s^2]C + [s^3+s^2]D + 5s^2$$

Next, apply METHOD 1.

4th power $A + C = 0$	$C = 15$	
3rd power $3A + 2C + D = -5$	$-45 + 30 + D = -5;$	$D = 10$
2nd power $3A + C + D = -20$	$-45 + 15 + 10 = -20$	CHECK
1st power $A = -15$	$A = -15$	
0th power $5 = 5$ CHECK		

Therefore $A = -15$
$B = 5$
$C = 15$
$D = 10$
$E = 5$

CHECK. Show that

$$\frac{5}{s^2[s+1]^3} = -\frac{15}{s} + \frac{5}{s^2} + \frac{15}{s+1} + \frac{10}{[s+1]^2} + \frac{5}{[s+1]^3}$$

$$5 = -15s^4 - 45s^3 - 45s^2 - 15s + 5s^3 + 15s^2 + 15s + 5 + 15s^4 + 30s^3 + 15s^2 + 10s^3 + 10s^2$$

$$+ 5s^2 = 5$$

Example 5. A combination of all three types

$$\frac{N(s)}{D(s)} = \frac{1}{s[s+2]^2[s^2+4s+5]} \equiv \frac{A}{s} + \frac{B}{s+2} + \frac{C}{[s+2]^2} + \frac{ds+e}{[s^2+4s+5]}$$

Multiplying both sides by D(s)

$$1 \equiv [s+2]^2[s^2+4s+5]A + s[s+2][s^2+4s+5]B + s[s^2+4s+5]C + s[s+2]^2[ds+e]$$

Exploiting METHOD 1.

when $s = 0$, $20A = 1$; $A = 1/20$
when $s = -2$, $-2C = 1$; $C = -1/2 = -10/20$

We now have

$$1 \equiv \frac{1}{20}s^4 + \frac{2}{5}s^3 + \frac{5}{4}s^2 + \frac{9}{5}s + 1 + [s^4 + 6s^3 + 13s^2 + 10s]B - \frac{1}{2}s^3 - 2s^2 - \frac{5}{2}s +$$

$$+ ds^4 + [4d+e]s^3 + 4[d+e]s^2 + 4es$$

Using METHOD 2.

4th	$1/20 + B + d = 0$	$B + d = -1/20$	$B = -5/20$
3rd	$2/5 + 6B - 1/2 + 4d + e = 0$	$6B + 4d + e = 2/20$	$d = 4/20$
2nd	$5/4 + 13B - 2 + 4d + 4e = 0$	$13B + 4d + 4e = 15/20$	$e = 16/20$
1st	$9/5 + 10B - 5/2 + 4e = 0$	$10B + 4e = 14/20$	
0th	$1 = 1$ CHECK		

CHECK. Show that

$$\frac{1}{s[s+2]^2[s^2+4s+5]} \equiv \frac{1}{20}\left\{\frac{1}{s} - \frac{5}{s+2} - \frac{10}{[s+2]^2} + \frac{4s+16}{[s^2+4s+5]}\right\}$$

$$20 \equiv [s+2]^2[s^2+4s+5] - 5s[s+2][s^2+4s+5] - 10s[s^2+4s+5] + s[s+2]^2[4s+16]$$

$$= s^4 + 8s^3 + 25s^2 + 36s + 20 - 5s^4 - 30s^3 - 65s^2 - 50s - 10s^3 - 40s^2 - 50s$$

$$+ 4s^4 + 32s^3 + 80s^2 + 64s$$

$$= 20$$

INDEX

Most main entries are nouns. Attributive modifiers appear beneath the main entry. Thus:

Point	not	
operating		Operating point
reference		Reference point
etc.		etc.

Many terms embrace two or more nouns. If one is not indexed, look for another. Thus:

Origin of coordinates, shifting, (Indexed)
Coordinates, shifting origin of, (not indexed)
Shifting origin of coordinates, (not indexed)

Hyphenated expressions are indexed as if they were single words.

When several references are listed for a single entry, one or more may be underlined to indicate that they are the most important.

"Ex. 4.1" refers to Exercise 1 at the end of Chapter 4.

			GENERAL	TRANSLATIONAL	ROTATIONAL
VARIABLES	PRIMARY	**A** = Across variable	$v_E(t)$ = velocity, voltage (etc.) difference across element E	$v(t)$ = velocity	$\Omega(t)$ = angular veloc
		T = Through variable	$f_E(t)$ = force, flow, (etc.) through element E	$f(t)$ = force	$\tau(t)$ = torque
		AT = Product	$p_E(t)$ = power supplied to element E $p_E(t) = f_E(t)v_E(t)$	$p(t) = f(t)v(t)$	$p(t) = \tau(t)\,\Omega(t)$
	INTEGRATED	$\int$**A** $\cdot dt$	$x_E(t) \int v_E(t) \cdot dt$	$x(t)$ = displacement $x(t) = \int v(t) \cdot dt$	$\theta(t)$ = angular displace $\theta(t) = \int \Omega(t) \cdot dt$
		$\int$**T** $\cdot dt$	$h_E(t)$ = momentum or quantity of thing flowing $h_E(t) \int f_E(t) \cdot dt$	$h(t)$ = momentum $h_m(t) = \int f_m(t) \cdot dt = mv_m$	$h(t)$ = angular momen $h_J(t) \int \tau_J(t) \cdot dt = J\Omega$
		$\int$**AT** $\cdot dt$	$\mathcal{E}(t)$ = energy = work done on system $\mathcal{E}(t) = \int p(t) \cdot dt$	$\mathcal{E}_k$ = kinetic energy of moving mass $\mathcal{E}_p$ = potential energy of compressed spring	$\mathcal{E}_k$ = kinetic energy of rotating mass $\mathcal{E}_p$ = potential energy twisted spring
ELEMENTS	PASSIVE	RESISTORS (D-elements) Dissipate energy as heat	$f(t) = k_D v(t)$ $v(t) = \frac{1}{k_D} f(t)$	$k_D = D$ = damping constant $v_D = v_1 - v_2$ 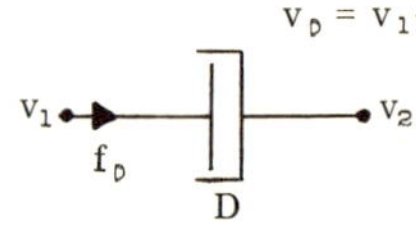	$k_D = D$ = damping con $\Omega_D = \Omega_1$ 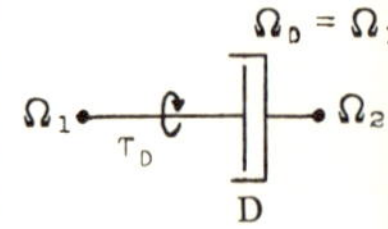
		CAPACITORS (A-elements) Store energy by virtue of their A-variable	$f(t) = k_A \frac{dv(t)}{dt}$ $v(t) = \frac{1}{k_A} \int f(t) \cdot dt$	$k_A = m$ = mass $v_m = v_1 - v_{Ref}$ 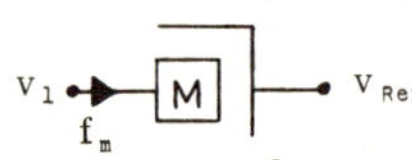 $\mathcal{E}_k = mv^2/2$	$k_A = J$ = Inertia $\Omega_J = \Omega_1 - \Omega_R$ 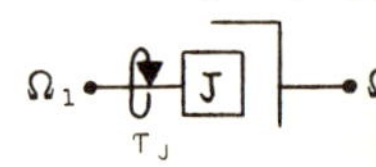 $\mathcal{E}_k = J\Omega^2/2$
		INDUCTORS (T-elements) Store energy by virtue of their T-variable	$f(t) = \frac{1}{k_T} \int v(t) \cdot dt$ $v(t) = k_T \frac{df(t)}{dt}$	$1/k_T = k$ = spring stiffness $v_k = v_1 - v_2$ 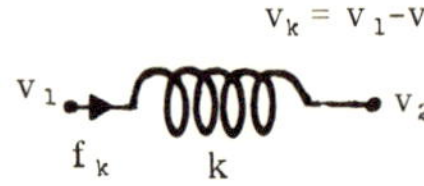 $\mathcal{E}_p = \frac{1}{2} kx^2 = \frac{1}{2} \cdot \frac{1}{k} f^2$	$1/k_T = k$ = torsional stiffne $\Omega_k = \Omega_1 -$ 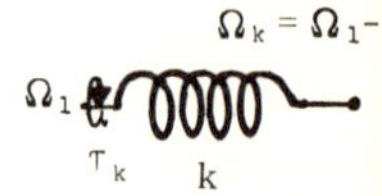 $\mathcal{E}_p = \frac{1}{2} k\theta^2 = \frac{1}{2} \cdot \frac{1}{k}$
	SOURCE	Constant input A-sources (examples)	$v_{in}(t)$ is determined by the source; $f_{in}(t)$ by the passive elements	Hydraulic piston moved by constant flow pump 	Large flywheel with constant angular velo
		Constant input T-sources (examples)	$f_{in}(t)$ is determined by the source; $v_{in}(t)$ by the passive elements	Weight of a given mass in constant gravitational field	Shaft turned by weight: 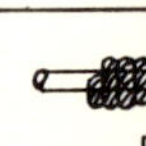

ELECTRICAL	HYDRODYNAMIC	THERMAL	CONCENTRATIONAL
e(t) = voltage	P(t) = pressure	T(t) = temperature	C(t) = concentration of solute
i(t) = current	$\dot{V}(t)$ = flow	$\dot{q}(t)$ = heat flow	$\dot{Q}(t)$ = solute flow
$p(t) = e(t)i(t)$	$p(t) = P(t)\dot{V}(t)$	N.B.! NOT "AT" $p(t) = \dot{q}(t)$	
) = magnetic flux $t) = \int e_L(t) \cdot dt$	$\Gamma(t)$ = pressure momentum $\Gamma_I(t) = \int P(t) \cdot dt$		
$q(t)$ = charge $q_c(t) = \int i_c(t) \cdot dt$	V(t) = volume $V_c(t) = \int \dot{V}_c \cdot dt$	H(t) = quantity of heat $H(t) = \int \dot{q}(t) \cdot dt$	Q(t) = quantity of solute $Q(t) = \int \dot{Q}(t) \cdot dt$
electrical energy of capacitor magnetic energy of inductor	$\mathcal{E}_k$ = kinetic energy of fluid flow $\mathcal{E}_p$ = potential energy of pressure head	$\mathcal{E}_p$ = thermal potential energy of stored heat	
, = R = resistance $e_R = e_1 - e_2$ 	$1/k_D$ = R = resistance to flow $P_R = P_1 - P_2$ 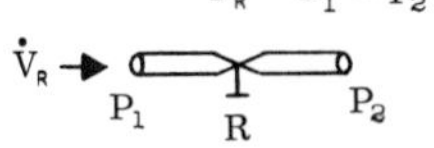	$1/k_d$ = R = resistance to flow of heat 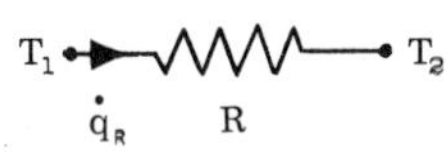	$k_d = [\dot{V}cl]$ = clearance (solute conductance) or $k_d = \dot{Q}$ per unit C diff. 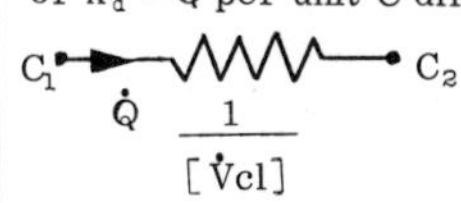
= C = capacitance $e_C = e_1 - e_2$ 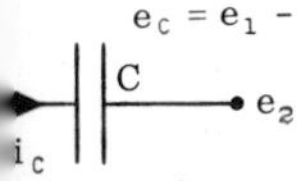 $\mathcal{E}_e = Ce^2/2$	k_A = C = fluid capacitance $P_C = P_1 - P_2$ 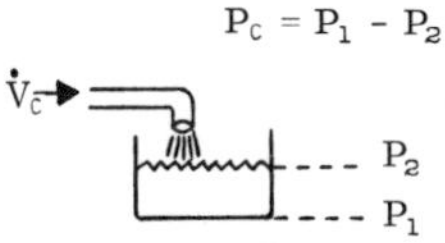 $\mathcal{E}_p = CP^2/2$	k_A = C = heat capacitance $T_C = T_1 - T_2$ 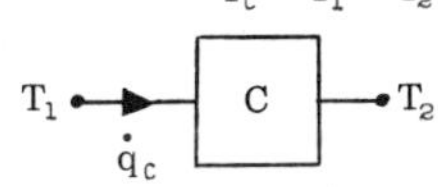 $\mathcal{E}_p = CT_C$	k_A = [Vdist] = solute capacitance (volume of distribution) $\Delta C = C_1 - C_2$ 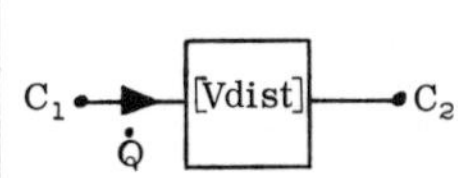
= L = inductance $e_L = e_1 - e_2$ 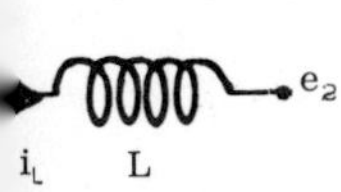 $\frac{1}{2} \cdot \frac{1}{L} \Psi^2 = \frac{1}{2} Li^2$	k_T = I = fluid inertance $P_I = P_1 - P_2$ 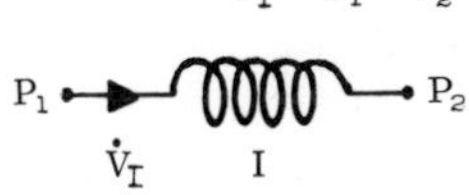 $\mathcal{E}_k = \frac{1}{2} I\dot{V}^2 = \frac{1}{2} \cdot \frac{1}{I} \Gamma^2$		
cell with "zero" rnal resistance	Large reservoir supplying constant pressure head	Constant-temperature bath of large volume	Large volume of dialyzing solution
oelectric cell with nite" internal stance.	Constant-volume pump (e.g., mammalian heart under certain conditions)	Heat production by metabolism	Intravenous infusion of solute